AF581897

Antennas and Propagation

Antennas and Propagation

Editors

Razvan D. Tamas
Alina Badescu
Tudor Palade
Florin Alexa
Ioan Nicolaescu

MDPI • Basel • Beijing • Wuhan • Barcelona • Belgrade • Manchester • Tokyo • Cluj • Tianjin

Editors

Razvan D. Tamas
Department of Electronics and Telecommunications
Constanta Maritime University
Constanta
Romania

Alina Badescu
Telecommunications
University Politehnica of Bucharest
Bucharest
Romania

Tudor Palade
Department of Communications
Technical University of Cluj-Napoca
Cluj-Napoca
Romania

Florin Alexa
Department of Communications
University Politehnica of Timisoara
Timisoara
Romania

Ioan Nicolaescu
Faculty of Military Electronic and Informatic Systems
Military Technical Academy - Ferdinand I
Bucharest
Romania

Editorial Office
MDPI
St. Alban-Anlage 66
4052 Basel, Switzerland

This is a reprint of articles from the Special Issue published online in the open access journal *Sensors* (ISSN 1424-8220) (available at: www.mdpi.com/journal/sensors/special_issues/iWAT2020).

For citation purposes, cite each article independently as indicated on the article page online and as indicated below:

LastName, A.A.; LastName, B.B.; LastName, C.C. Article Title. *Journal Name* **Year**, *Volume Number*, Page Range.

ISBN 978-3-0365-1828-2 (Hbk)
ISBN 978-3-0365-1827-5 (PDF)

Contents

Preface to "Antennas and Propagation"

Antennas are essentially transducers, as they convert electromagnetic fields into signals and vice versa. Moreover, remote sensing or sensor networks cannot be imagined without antennas, and radiowave propagation in complex environments is a crucial aspect for the operation of such systems. New technologies, such as 5G, generate further perspectives for sensor networks and raise additional challenges for antenna design.

This Special Issue gathers topics of utmost interest in the field of antennas and propagation, such as:

- New directions and challenges in antenna design and propagation;
- Innovative antenna technologies for space applications;
- Metamaterial, metasurface and other periodic structures;
- Antennas for 5G; and
- Electromagnetic field measurements and remote sensing applications.

We originally invited the authors who contributed to the 2020 International Workshop on Antenna Technology, held in Bucharest (Romania) on 25–28 March, to submit thoroughly extended versions of their work. Surprisingly, half of the submissions to the Special Issue through the deadline were not conference paper extensions, but completely new contributions.

Razvan D. Tamas, Alina Badescu, Tudor Palade, Florin Alexa, Ioan Nicolaescu
Editors

Editorial

Antennas and Propagation: A Sensor Approach

Razvan D. Tamas 

Department of Electronics and Telecommunications, Constanta Maritime University, Str. Mircea cel Batran nr. 104, 900663 Constanta, Romania; tamas@ieee.org

Antennas are essentially transducers, as they convert electromagnetic fields into signals and vice versa. Moreover, remote sensing or sensor networks cannot be imagined without antennas, and radiowave propagation in complex environments is a crucial aspect for the operation of such systems. New technologies such as 5G generate further perspectives for sensor networks and raise additional challenges to antenna design.

This Special Issue gathers topics of utmost interest in the field of antennas and propagation, such as:

- New directions and challenges in antenna design and propagation;
- Innovative antenna technologies for space applications;
- Metamaterial, metasurface and other periodic structures;
- Antennas for 5G;
- Electromagnetic field measurements and remote sensing applications.

We originally invited the authors who contributed to the 2020 International Workshop on Antenna Technology, held in Bucharest (Romania) on 25–28 March, to submit thoroughly extended versions of their work. Surprisingly, half of the submissions to the Special Issue through the deadline were not conference paper extensions but completely new contributions.

Citation: Tamas, R.D. Antennas and Propagation: A Sensor Approach. *Sensors* **2021**, *21*, 4920. https://doi.org/10.3390/s21144920

Received: 7 July 2021
Accepted: 15 July 2021
Published: 20 July 2021

1. Summary of the Special Issue

1.1. New Directions and Challenges in Antenna Design and Propagation

A slot-fed terahertz dielectric resonator antenna driven by an optimized photomixer is proposed in [1], and the interaction of the laser and photomixer is also studied. It is demonstrated that, in a continuous wave terahertz photomixing scheme, the generated THz power is proportional to the fourth power of the surface electric field on the photoconductive layer. The total efficiency was considerably improved due to enhancements in the laser-to-THz conversion as well as the radiation and matching efficiencies.

In Ref. [2], radio transmission and impedance matching in medical telemetry are investigated. Impedance matching inside a human body is studied both for electric and magnetic dipoles. The authors demonstrate that the implantation of a magnetic dipole is more beneficial than that of an electric dipole, as it provides impedance characteristics that are more appropriate to the human body as an antenna environment.

The simultaneous influence of the substrate anisotropy and substrate bending are numerically and experimentally investigated in [3] for planar resonators on flexible textile and polymer substrates. The effects are studied on various resonant structures with different types of slots and defected ground as well as on fractal resonators.

In Ref. [4], the authors present small, flexible, low-profile and light-weight all-textile antennas for wearable wireless sensor networks (W-WSNs) and investigate the impact of the textile materials on the antenna performance. A step-by-step procedure for design, fabrication and measurement of small wearable backed antennas for W-WSNs is also suggested.

The work in [5] proposes a random spreading of the scrambled timestamp sequence (STS) in ultra-wide-band (UWB) pulse communications. The timestamp estimation is improved by reducing the side lobes of the correlation. A transmission in a UWB channel with frequency selective fading shows low timestamp detection errors.

A modified, compact dipole antenna for energy harvesting applications is proposed in [6]. The design is based on a folded, short-circuited dipole radiator, and three feeding techniques are investigated. Such antenna configurations show good conversion efficiency when powering Bluetooth low-energy wireless sensors.

1.2. Innovative Antenna Technologies for Space Applications

In Ref. [7], multibeam antenna systems are proposed with the aim to provide multispot coverage for broadband satellite communications in the Ka-band. Two multibeam reflectarrays are used, making it possible to halve the number of required antennas onboard the satellite.

The work in [8] presents an antenna design that can be used as an array element for monitoring and detecting radio emissions resulting from cosmic particle interactions in the atmosphere. The proposed antenna provides a high gain over a large relative bandwidth, a narrow beamwidth, a small group delay variation and a very stable radiation pattern between 110 and 190 MHz.

1.3. Metamaterial, Metasurface and Other Periodic Structures

In [9], a new design method for a planar and compact dual-band dipole antenna is proposed. The antenna has a hybrid CRLH (composite right- and left-handed) structure with lumped elements for a dual-band operation. A design for 2.4 and 5.2 GHz mobile applications is presented.

A printed edge-fed counterpart of the Bruce wire antenna array for frequency scanning applications is presented in [10]. The unit cell of the proposed antenna consists of bowtie and semi-circular elements that cover a bandwidth between 22 GHz and 38 GHz.

The work presented in [11] focuses on design issues in the implementation of holey glide-symmetric periodic structures for waveguide-based components. An analysis of realistic hole structures is performed by using an effective hole depth method that can be used as a tool for designing electrically large waveguide-based components.

An integration of a microstrip slot antenna array for 5.8 GHz with dye-sensitized solar cells is proposed in [12]. It is shown that the antenna array has a slight influence on the solar cell performance, and the interference of the solar cells with the antenna feeding system is also negligible.

A new slot-based antenna system with horizontal polarization for unmanned aerial vehicle (UAV) ground control stations (GCS) is outlined in [13]. The proposed antenna system consists of coaxial cylinders and slots. The vertical slots are periodically placed around the outer cylinder and generate in-phase, horizontally polarized beams, resulting in an omnidirectional radiation pattern.

1.4. Antennas for 5G

In Ref. [14], a multiband antenna for microwave and millimeter-wave applications is presented. The proposed antenna consists of a slotted, conical patch connected to a small triangular patch. It can be used for wireless local area network (WLAN) applications and for fifth-generation (5G) communication devices.

In Ref. [15], the authors present a simple, compact and low-cost design method for low-profile, multi-band antennas. Such antennas can be employed in overcrowded, future generation networks in the K/Ka band. The proposed antenna structures consist of several monopoles, one for each operating frequency, along with a frequency selective feeding network. This concept leads to scalable structures suitable for 5G applications.

In Ref. [16], a small antenna for sub-6GHz 5G communications is proposed. The design consists of a wide-band antenna connected to a small multiplexer comprising three metamaterial channel filters. It can be used on channels within three frequency bands; channel selection is experimentally demonstrated so as to prove the validity of the presented design.

1.5. Electromagnetic Field Measurements and Remote Sensing Applications

In Ref. [17], the authors propose an ultra-wide band (UWB) antenna system and a direction-finding (DF) approach based on using energy-based descriptors instead of classical frequency domain parameters. The method can be applied for locating electric discharges in high-voltage power distribution systems through their electromagnetic signature in the radio frequency range.

A calibration method for high-resolution, hybrid MIMO turntable radars is presented in [18]. A line of small metal spheres is employed as a test pattern in the calibration process in order to measure the position shift caused by undesired antenna effects. The unwanted effects resulting from the near-field antenna response are analyzed, modelled and significantly mitigated, by exploiting the symmetry and response of the MIMO configuration.

In Ref. [19], the authors show that the distance averaging technique can be applied to reduce the effect of the common mode currents for measuring the field radiated by symmetrical antennas. Two measuring configurations are proposed, depending on the number of symmetry degrees of the antenna under test. A differential approach for extracting the field created by the common mode currents is also developed.

In Ref. [20], the authors elaborate on how a sensor using modified split ring resonators (SRRs) can be designed, simulated, fabricated and used for advanced investigation and accurate measurements of the complex permittivity of solid dielectrics.

The work in [21] proposes a near-field to far-field transformation algorithm based on spherical wave expansion for near-field RCS measurements. Each weight in this expansion is calculated by using an iterative least squares QR factorization method. The proposed NFFFT is verified for several types of scatterers.

Funding: This research received no external funding.

Institutional Review Board Statement: Not applicable.

Informed Consent Statement: Not applicable.

Data Availability Statement: Not applicable.

Conflicts of Interest: The author declares no conflict of interest.

References

1. Li, X.; Yin, W.; Khamas, S. An Efficient Photomixer Based Slot Fed Terahertz Dielectric Resonator Antenna. *Sensors* **2021**, *21*, 876. [CrossRef] [PubMed]
2. Kim, I.; Lee, S.-G.; Nam, Y.-H.; Lee, J.-H. Investigation on Wireless Link for Medical Telemetry Including Impedance Matching of Implanted Antennas. *Sensors* **2021**, *21*, 1431. [CrossRef] [PubMed]
3. Dankov, P.I.; Sharma, P.K.; Gupta, N. Numerical and Experimental Investigation of the Opposite Influence of Dielectric Anisotropy and Substrate Bending on Planar Radiators and Sensors. *Sensors* **2021**, *21*, 16. [CrossRef]
4. Atanasova, G.; Atanasov, N. Small Antennas for Wearable Sensor Networks: Impact of the Electromagnetic Properties of the Textiles on Antenna Performance. *Sensors* **2020**, *20*, 5157. [CrossRef]
5. Domuta, I.; Palade, T.P.; Puschita, E.; Pastrav, A. Timestamp Estimation in P802.15.4z Amendment. *Sensors* **2020**, *20*, 5422. [CrossRef]
6. Sidibe, A.; Takacs, A.; Loubet, G.; Dragomirescu, D. Compact Antenna in 3D Configuration for Rectenna Wireless Power Transmission Applications. *Sensors* **2021**, *21*, 3193. [CrossRef] [PubMed]
7. Martinez-de-Rioja, D.; Martinez-de-Rioja, E.; Rodriguez-Vaqueiro, Y.; Encinar, J.A.; Pino, A. Multibeam Reflectarrays in Ka-Band for Efficient Antenna Farms Onboard Broadband Communication Satellites. *Sensors* **2021**, *21*, 207. [CrossRef]
8. Tatomirescu, A.; Badescu, A. Wideband Dual-Polarized VHF Antenna for Space Observation Applications. *Sensors* **2020**, *20*, 4351. [CrossRef] [PubMed]
9. Khattak, M.K.; Lee, C.; Park, H.; Kahng, S. A Fully-Printed CRLH Dual-Band Dipole Antenna Fed by a Compact CRLH Dual-Band Balun. *Sensors* **2020**, *20*, 4991. [CrossRef] [PubMed]
10. Ahmed, Z.; McEvoy, P.; Ammann, M.J. A Wide Frequency Scanning Printed Bruce Array Antenna with Bowtie and Semi-Circular Elements. *Sensors* **2020**, *20*, 6796. [CrossRef]

11. Sipus, Z.; Cavar, K.; Bosiljevac, M.; Rajo-Iglesias, E. Glide-Symmetric Holey Structures Applied to Waveguide Technology: Design Considerations. *Sensors* **2020**, *20*, 6871. [CrossRef] [PubMed]
12. Bai, B.; Zhang, Z.; Li, X.; Sun, C.; Liu, Y. Integration of Microstrip Slot Array Antenna with Dye-Sensitized Solar Cells. *Sensors* **2020**, *20*, 6257. [CrossRef] [PubMed]
13. Sadiq, M.S.; Ruan, C.; Nawaz, H.; Ullah, S.; He, W. Horizontal Polarized DC Grounded Omnidirectional Antenna for UAV Ground Control Station. *Sensors* **2021**, *21*, 2763. [CrossRef] [PubMed]
14. Khan, Z.; Memon, M.H.; Rahman, S.U.; Sajjad, M.; Lin, F.; Sun, L. A Single-Fed Multiband Antenna for WLAN and 5G Applications. *Sensors* **2020**, *20*, 6332. [CrossRef] [PubMed]
15. Ramos, A.; Varum, T.; Matos, J.N. Compact N-Band Tree-Shaped Multiplexer-Based Antenna Structures for 5G/IoT Mobile Devices. *Sensors* **2020**, *20*, 6366. [CrossRef] [PubMed]
16. Khattak, M.K.; Lee, C.; Park, H.; Kahng, S. RF Channel-Selectivity Sensing by a Small Antenna of Metamaterial Channel Filters for 5G Sub-6-GHz Bands. *Sensors* **2020**, *20*, 1989. [CrossRef]
17. Tamas, R.D.; Bucuci, S. An Ultra-Wide Band Antenna System for Pulsed Sources Direction Finding. *Sensors* **2020**, *20*, 4659. [CrossRef]
18. Hoang, H.; John, M.; McEvoy, P.; Ammann, M.J. Calibration to Mitigate Near-Field Antennas Effects for a MIMO Radar Imaging System. *Sensors* **2021**, *21*, 514. [CrossRef]
19. Constantin, A.; Tamas, R.D. Evaluation and Impact Reduction of Common Mode Currents on Antenna Feeders in Radiation Measurements. *Sensors* **2020**, *20*, 3893. [CrossRef]
20. al-Behadili, A.A.; Mocanu, I.A.; Codreanu, N.; Pantazica, M. Modified Split Ring Resonators Sensor for Accurate Complex Permittivity Measurements of Solid Dielectrics. *Sensors* **2020**, *20*, 6855. [CrossRef]
21. Kim, W.; Im, H.-R.; Noh, Y.-H.; Hong, I.-P.; Tae, H.-S.; Kim, J.-K.; Yook, J.-G. Near-Field to Far-Field RCS Prediction on Arbitrary Scanning Surfaces Based on Spherical Wave Expansion. *Sensors* **2020**, *20*, 7199. [CrossRef] [PubMed]

Communication

Compact Antenna in 3D Configuration for Rectenna Wireless Power Transmission Applications

Alassane Sidibe [1,2,*], Alexandru Takacs [1], Gaël Loubet [1] and Daniela Dragomirescu [1]

1 Laboratoire d'Analyse et d'Architecture des Systèmes du Centre National de la Recherche Scientifique (LAAS-CNRS), Université de Toulouse, Centre National de la Recherche Scientifique (CNRS), Institut National des Sciences Appliqués de Toulouse (INSA), Université Paul Sabatier, Toulouse III (UPS), 31400 Toulouse, France; alexandru.takacs@laas.fr (A.T.); gael.loubet@laas.fr (G.L.); daniela.dragomirescu@laas.fr (D.D.)

2 Uwinloc, 9 Rue Humbert Tomatis, 31200 Toulouse, France

* Correspondence: alassane.sidibe@laas.fr

Abstract: This work presents methods for miniaturizing and characterizing a modified dipole antenna dedicated to the implementation of wireless power transmission systems. The antenna size should respect the planar dimensions of 60 mm × 30 mm to be integrated with small IoT devices such as a Bluetooth Lower Energy Sensing Node. The provided design is based on a folded short-circuited dipole antenna, also named a T-match antenna. Faced with the difficulty of reducing the physical dimensions of the antenna, we propose a 3D configuration by adding vertical metallic arms on the edges of the antenna. The adopted 3D design has an overall size of 56 mm × 32 mm × 10 mm at 868 MHz. Three antenna-feeding techniques were evaluated to characterize this antenna. They consist of soldering a U.FL connector on the input port; vertically connecting a tapered balun to the antenna; and integrating a microstrip transition to the layer of the antenna. The experimental results of the selected feeding techniques show good agreements and the antenna has a maximum gain of +1.54 dBi in the elevation plane (E-plane). In addition, a final modification was operated to the designed antenna to have a more compact structure with a size of 40 mm × 30 mm × 10 mm at 868 MHz. Such modification reduces the radiation surface of the antenna and so the antenna gain and bandwidth. This antenna can achieve a maximum gain of +1.1 dBi in the E-plane. The two antennas proposed in this paper were then associated with a rectifier to perform energy harvesting for powering Bluetooth Low Energy wireless sensors. The measured RF-DC (radiofrequency to direct current) conversion efficiency is 73.88% (first design) and 60.21% (second design) with an illuminating power density of 3.1 μW/cm^2 at 868 MHz with a 10 kΩ load resistor.

Keywords: compact antenna; wireless power transmission (WPT); energy harvesting; rectenna; wireless sensors

Citation: Sidibe, A.; Takacs, A.; Loubet, G.; Dragomirescu, D. Compact Antenna in 3D Configuration for Rectenna Wireless Power Transmission Applications. *Sensors* **2021**, *21*, 3193. https://doi.org/10.3390/s21093193

Academic Editor: Razvan D. Tamas

Received: 12 February 2021
Accepted: 28 April 2021
Published: 4 May 2021

1. Introduction

Over the last decades, we have been faced with the miniaturization of electronic devices, especially in the field of wireless systems. The aim is to have multiple functionalities on an ever-smaller surface area. Recent IoT applications (Internet of Things) tend to employ tiny and low power electronic components [1]. However, batteries are still widely used for powering the devices despite their significant size and the frequent need for replacement. An alternative is using a battery-free system powered by energy harvesting (EH) or wireless power transmission (WPT), for instance, based on a rectenna circuit.

A rectenna is a combination of a rectifier and an antenna used to scavenge ambient or specially generated far-field radiofrequency (RF) waves [2]. The implementation of a rectenna in a battery-free system can allow increasing its lifetime, reducing its manufacturing costs, while ensuring reliable performances for decades. Several kinds of application exist, such as, for instance, rectennas, which were developed for a biomedical device pasted on the human body [3] or used in the IoT domain [4–6].

The antenna is a ubiquitous element in IoT and other wireless applications. However, the required size stays important due to the important dependence of the geometrical elements with the targeted wavelength. Several miniaturization techniques are described in the literature. Structural modification on a Printed Circuit Board (PCB) antenna consists of acting on the geometry of the antenna or by adding another element on the antenna shape.

In this sense, a coupling element, such as a rectangular ring, can allow a reduction in size and an increase in bandwidth as presented in [7]. Traditionally, small size antennas are designed by using meander lines which reduce the resonant frequency [8]. Fractal geometry is also used to miniaturize antennas [9]. Another possibility is to add reactive loading elements. This technique is employed in [10] through an LC load (a combination of a lumped inductor and a distributed capacitor). Miniaturization can also be achieved by reducing the guided wavelength through the use of a higher permittivity material, such as a ceramic—polymer composite [11]. Recent research activities were focused on developing small antennas on metamaterial as presented in [12]. In this paper, we present a miniaturization technique which consists of shaping the antenna to form a three-dimensional (3D) structure tuned for the Industrial Scientific and Medical (ISM) 868 MHz frequency band. This configuration allows us to have an electrically small antenna as described in Section 2. A more compact antenna based on the previous one is designed to compare the trade-off between size and rectenna performances. Then, Section 3 presents the experimental results of the fabricated antennas with different feeding methods to validate the use of the U.FL connector for the next steps. Simulations were carried out on the Ansys HFSS software and verified with far-field measurements performed in an anechoic chamber. The final section of this paper describes an original concept of powering a Bluetooth Low Energy Sensing Node embedded in concrete element with a compact and efficient rectenna design.

2. Design of a Compact 3D Dipole Antenna

2.1. Antenna Miniaturization Methods and the First Design of the Compact 3D Dipole

In this study, we proposed a miniaturized antenna design for WPT applications. An antenna is a resonant structure with a proper frequency depending on its length. Therefore, there are size and performance limitations for small antennas [13,14]. The size reduction imposes a smaller radiation resistance, so a lower radiation efficiency and a bandwidth limitation. Wheeler defines a electrically small antenna as one defined by the formula given in Equation (1). It means that the antenna sphere is smaller than the radian sphere, also defined as Wheeler Cap. λ is the wavelength at the operating frequency, k represents the free-space wavenumber and a is the minimum antenna sphere radius. The choice of proposing an electrically small antenna should allow us to have an antenna design in the maximum planar size of 60 mm $\times$ 30 mm. Nevertheless, the antenna bandwidth reduction does not matter with the applications in the ISM 868 MHz frequency band. The targeted antenna gain is about +1 dBi less compared to a conventional dipole antenna gain (+2.15 dBi).

$$k{\cdot}a < 1 \; with \; k = \frac{2\pi}{\lambda} \tag{1}$$

The proposed design is based on a dipole antenna according to its multiple advantages, such as the symmetry of its shape and radiation pattern, its easy and low-cost fabrication, and the possibility of receiving balanced signals. The required dimensions of 60 mm $\times$ 30 mm allow us to have a conventional half-wavelength dipole antenna at 2 GHz, as presented in Figure 1.

Starting from this antenna at 2 GHz, we had investigated a technique for reducing the physical dimensions of the antenna without significantly degrading its performances (+1 dBi less of the gain). The planar structure of the final antenna design on an FR4 substrate is presented in Figure 2. Overall, a folded dipole antenna designed after size optimization

(Antenna in Figure 3 stamped "None") presents a resonant frequency close to 1.12 GHz regarding the return loss (S11) equal to −5.2 dB.

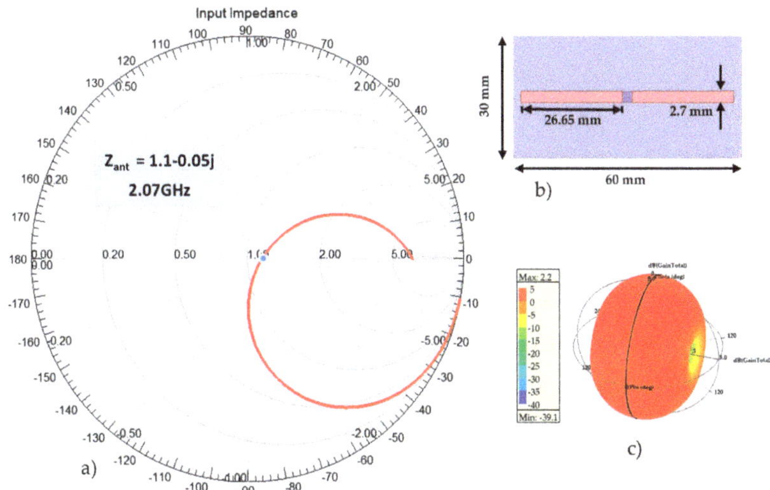

Figure 1. Conventional dipole antenna simulated on the required size dimensions. (**a**) The input impedance representation on Smith chart; (**b**) Half-wavelength antenna dimensions; (**c**) Simulated 3D radiation pattern at 2 GHz.

This antenna can be considered as a half-wavelength dipole antenna at 1.12 GHz. The total length of each folded arm ($L_1 + W_1 + L_2$ = 43.4 mm) is closed to the quarter wavelength (<$0.25 \cdot \lambda_g$ = 31.9 mm). Figure 3 presents the width tuning of the inductive shorting loop to make a T-match structure [15]. The resonance frequency is greatly influenced by the width and the length of the T-match structure. As seen in Figure 3, the absence of the T-match structure (None) shows a significant resonant frequency of the antenna higher than 1.12 GHz where the impedance is (17.35 + j·27.3) Ω. The shorting line (W = 0) connected to the two dipole arms downshifts the resonant frequency around 1.07 GHz. Thereafter, the geometric parameters (W and L) were adjusted to match the input impedance of the antenna (50 Ω) at lower frequency.

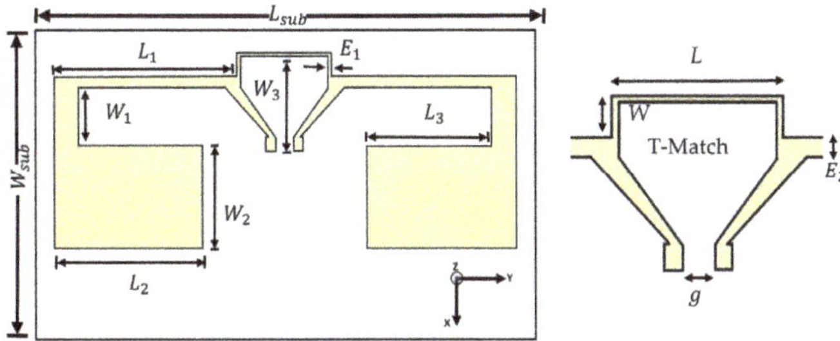

Figure 2. Detailed geometry of the compact, T-match dipole antenna on an FR4 substrate (L_{sub} = 56 mm, W_{sub} = 32 mm, L_1 = 20.65 mm, W_1 = 5.95 mm, L_2 = 16.8 mm, W_2 = 10.5 mm, L_3 = 14 mm, W_3 = 9.625 mm, E_1 = 0.35 mm, E_2 = 1.05 mm, g = 2.03 mm).

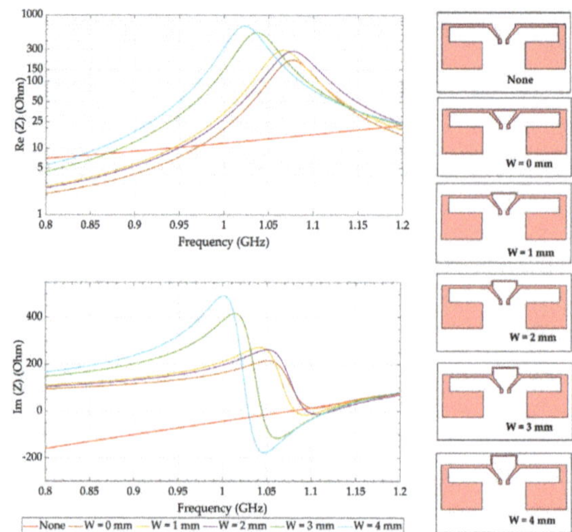

Figure 3. Evolution of the input impedance (Real and Imaginary parts) as function of the frequency with various shorting line width (W) values.

The simulated results of the optimization on HFSS software are displayed in Figures 3 and 4. The return loss (reflection coefficient S_{11}) indicates how much the power is reflected on the input port when the antenna is excited. A practical criterion for the antenna (impedance) input matching is generally specified at −10 dBm at least for a reference impedance of 50 Ω. As represented in Figure 4, the greater the length (L) of the short-circuited line, the more the antenna resonates at higher frequency.

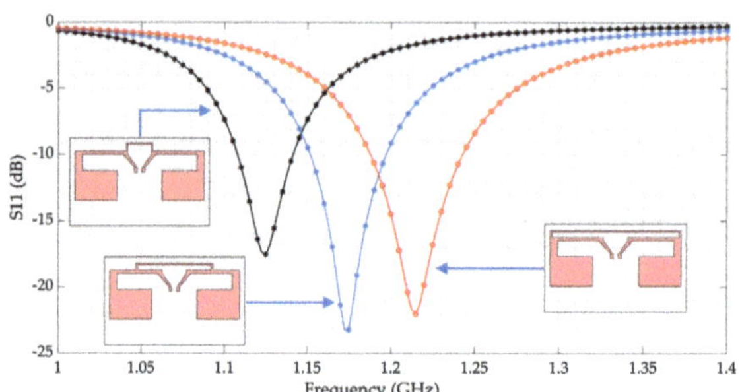

Figure 4. Simulated S11 (return loss) of the antenna for different configurations; L = 10 mm (black), L = 30 mm (blue) and L = 51.4 mm (red).

For the next miniaturization step, the T-match structure parameters were fixed to those which give the lower resonant frequency (W = 4 mm and L = 10 mm). To downshift the operating frequency obtained from the planar antenna from 1.12 GHz to 868 MHz, a 3D configuration was investigated. The operating/resonant frequency of the antenna can be downshifted by connecting two metal strands [16]. In our case, two capacitive metallic arms with a height of 10 mm are vertically connected to the planar antenna with a short transmission line (Figure 5a). This line's position at the edge of the planar antenna was tuned to obtain an operating frequency as close as possible to 868 MHz with a bandwidth of 30.7 MHz.

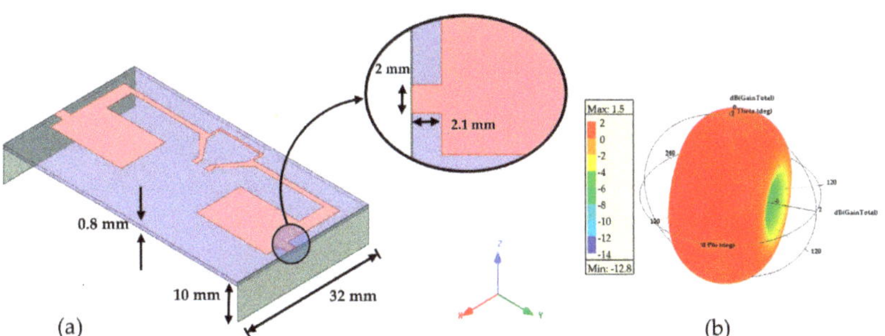

Figure 5. (**a**) Geometry of the 3D configuration antenna with the connected metallic arms; (**b**) Simulated 3D gain polar plot at 868 MHz.

In our application at the European ISM 868 MHz frequency band, a narrow band antenna is not critical. The radiation pattern looks like a doughnut shape and the maximum simulated gain is +1.5 dBi (Figure 5b).

2.2. Second Miniaturized Antenna Design

In terms of compactness, the previous antenna was modified and two different antennas (A1 and A2) were designed, respectively, the unconnected and connected antenna to the metallic arms. The horizontal arms of the planar dipole are folded in a spline shape to occupy the blue part display in Figure 6a and reduce the length (L_{sub}) of the planar antenna. By the way, the width of the arms was adjusted to the substrate width. This curving shape was obtained after several optimized simulations, and the size of A2 (represented in Figure 6b) is only 40 mm × 30 mm × 10 mm.

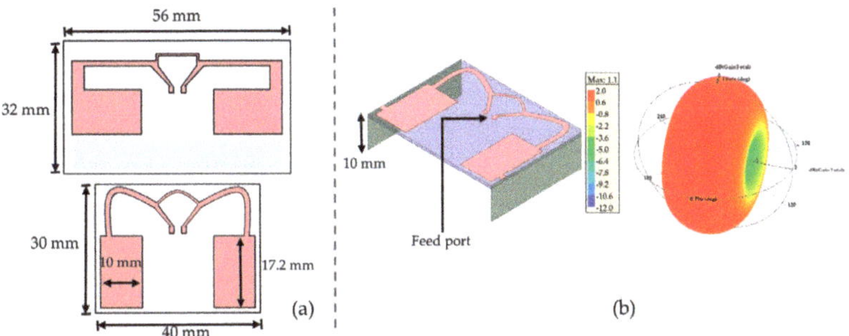

Figure 6. Designed antennas on HFSS: (**a**) The first miniaturized antenna and the second miniaturized antenna named A2: 3D FDA; (**b**) 3D polar plot of the radiation pattern of A2 antenna at the resonant frequency (HFSS results).

The simulations performed show an operating frequency at 1.15 GHz for A1 and 868 MHz for A2. However, there is a small gain reduction of 0.4 dBi compared to the previous designed antenna, so that we still respect the first condition of a maximum +1 dBi loss on the gain from a conventional dipole antenna. The radiation patterns in the E-plane and the H-plane are presented in Figure 7.

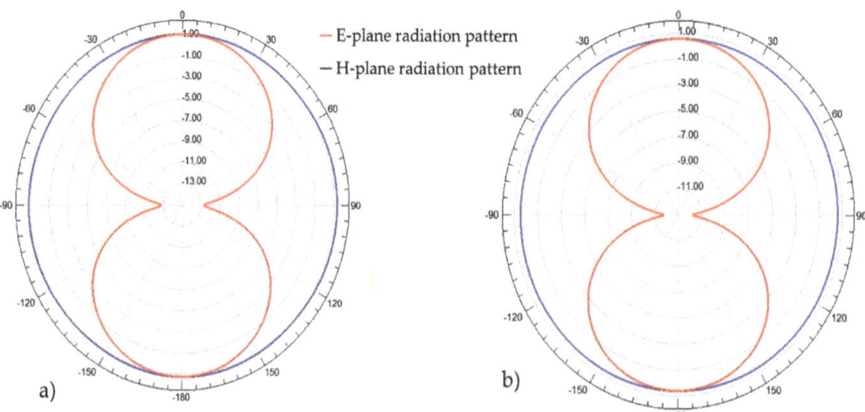

Figure 7. Radiation pattern on the E-plane and H-plane at 868 MHz. (**a**) The antenna first 3D dipole antenna in Section 2.1; (**b**) The modified 3D dipole antenna (A2).

3. Characterization of the Designed Antennas

The designed antenna was manufactured in an FR4 substrate with a thickness of 0.8 mm. Dipole antennas require a balanced-unbalanced (balun) circuit or a microstrip transition for a coaxial measurement. The balun allows canceling the flowing current on the outside surface of the outer coaxial conductor and then affects the measurement [17]. In the literature, several dipole antenna-feeding techniques were proposed. A microstrip tapered balun was used as a feeding line in [18] and a microstrip-to-coplanar (CPW) feed network balun for a flexible bowtie antenna in [19]. On the other hand, in many electronic devices especially in IoT, a surface mount coaxial U.FL connector is strongly used for characterization and RF connection for embedded antennas. Its advantages are its low cost, small size and light weight.

To well characterize our designed antenna, three feeding methods were selected. The first feeding technique carried in this paper is to use a U.FL connector [20]. The pads of the connector were soldered on the antenna feeding pads, as seen in Figure 8a. The second method consists of vertically connecting a designed tapered microstrip balun (Figure 8b). The last one is a microstrip transition: while one pad of the antenna is connected to a 50 Ω transmission line, the other balanced pad is connected to the ground plane (Figure 8c). For the return loss (S_{11}) and the gain measurements, a compatible coaxial cable was connected to the vector network analyzer (VNA). The measured return loss is plotted in Figure 9 (by using three different connection methods). They all fit well to each other but present a 15 MHz frequency shift compared to the simulation. The same frequency shift can be observed in Figure 10 between simulation and measurement for the curved 3D antenna A2. Without the two connected arms, the simulated and measured return loss fit perfectly, but once they are connected, a 50 MHz frequency shift appears. The difference is mainly due to the soldering effect and the vertical capacitive arms whose dimensions are not perfectly respected as compared with the simulated ones. The A2 antenna was measured with a U.FL connector and resonates at 878 MHz, as shown in Figure 10. The measured and simulated gain in the E-plane are shown in Figure 11.

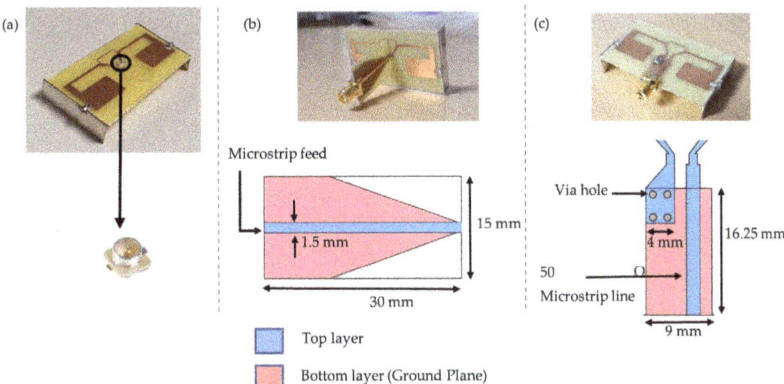

Figure 8. (**a**) Antenna with a U.FL connector named AC; (**b**) Antenna with connected tapered balun named ATB; (**c**) Antenna with integrated microstrip transition named AIT.

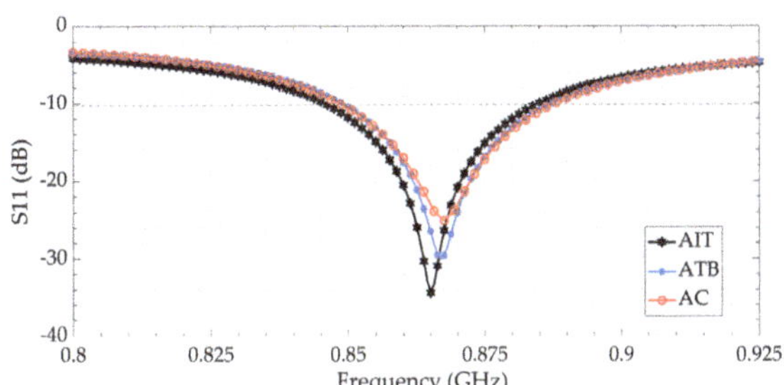

Figure 9. Comparison of the measured return loss (S11) of the D1 antenna with different feeding methods.

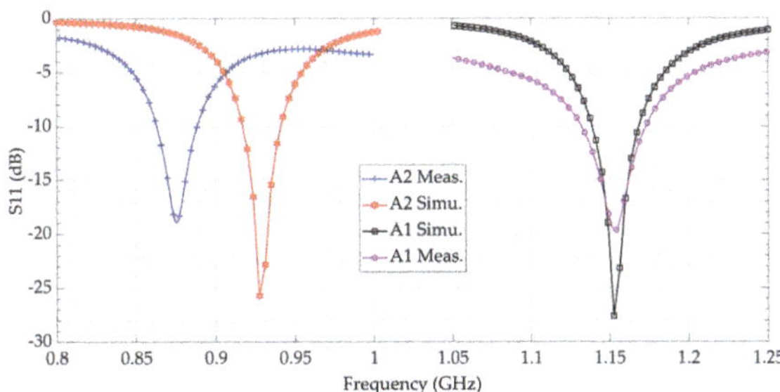

Figure 10. Measured and simulated reflection coefficient of the A1 and A2 antennas.

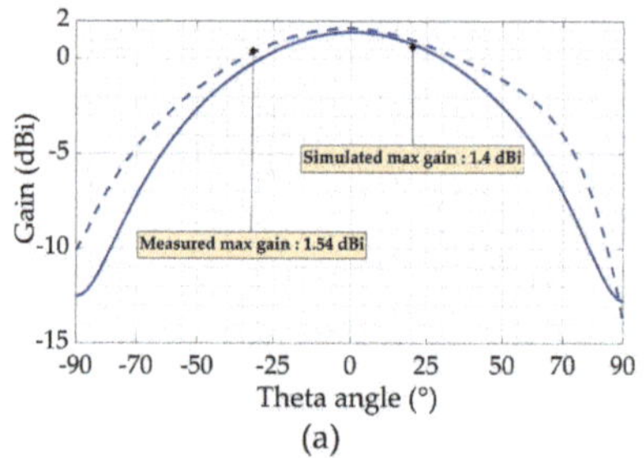

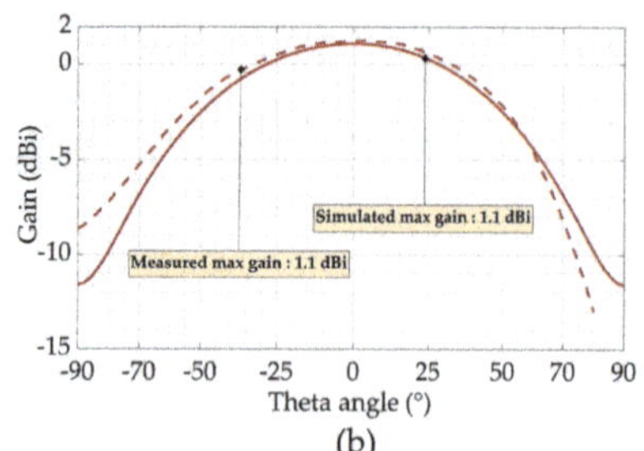

Figure 11. Simulated (dashed line) and measured (continuous line) radiation pattern (gain plot in the E-plane): (**a**) AC antenna; (**b**) A2 antenna at the resonant frequency (868 MHz).

Table 1 compares in terms of performances and size some state-of-the-art antennas and the presented solution operating in ISM 868 MHz or ISM 915 MHz frequency bands. The proposed compact 3D antenna was implemented on an FR4 substrate that is lossier as compared with the substrates used by other state-of-the-art designs.

Table 1. Comparison with the different compact antennas for rectenna of IoT devices in the state of the art.

Ref.	Freq. (MHz)	Type	Max Gain (dBi)	BW (MHz)	Substrate	Size (mm × mm × mm)
[21]	878	Dual Band PIFA	+1.8–+1.9	80 (855–937)	Duroid 5880	80 × 45 (0.03·λ^2)
[22]	868	UCA PIFA	+0.71	23 (857–880)	FR4	34 × 80 (0.02·λ^2)
[16]	915	Slot loaded DB folded dipole	+1.87	Not available	Arlon 25N	60 × 60 × 60 (0.006·λ^3)
[23]	868	3D single arm bowtie	+0.19	90	Ultralam	50 × 50 × 10 (0.0006·λ^3)
This work	868	3D modified T-match dipole	+1.54	32 (865–897)	FR4	56 × 32 × 10 (0.0004·λ^3)
			+1.1	26 (862–888)		40 × 30 × 10 (0.0003·λ^3)

BW: Bandwidth, PIFA: Planar Inverted F-Antenna, DB: Dual Band.

4. Rectenna and Wireless Power Transmission Experimentations

The antennas presented in this paper are more compact than the 3D antennas in [16,23], and exhibit the best tradeoff between size, gain and bandwidth and can be good candidates for WPT applications. They were integrated in rectennas used for battery-free wireless sensors embedded in concrete structures [24,25].

The rectifier part is designed using ADS/Momentum software from Keysight. The circuit is designed on a microstrip coupled transmission line allowing differential feeding by the dipole antenna (Figure 12). It is composed of a SMS7630-005LF Schottky diode, an LC matching network (L1-C1) ensuring 50 Ω input impedance at the input of the rectifier and a low pass filter formed by a shunt capacitor (C3) and a resistive load (R1). However, considering the nonlinear behavior of the Schottky diode, the rectifier was designed and optimized for a low input power of −15 dBm and a 10 kΩ resistor to emulate the load (that is the input impedance of the power management unit (PMU)). The simulation result was performed and described in [24]. Two rectennas, R1 and R2, in Figure 12 represent the rectifier with the antenna AC without any connector and A2, respectively. They were made by integrating the rectifier with the antennas on the same substrate. Their characterization was performed in an anechoic chamber at the distance of 1.5 m from the patch-transmitting antenna, and the harvested DC power was measured across a 10 kΩ load through a multimeter. The rectennas R1 and R2 allow a harvested DC voltage of 788 mV and 726 mV for 1 μW/cm^2.

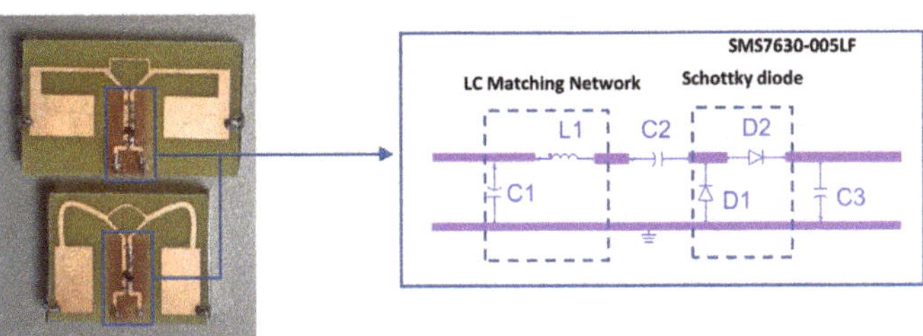

Figure 12. Manufactured rectenna with the rectifier schematic.

The conversion efficiency as a function of the power density is presented in Figure 13. It is obtained by computing the formula given in Equation (2), where:

- P_{DC} is the harvested DC power obtained across the load;
- Gt and Gr are the transmitting antenna gain and the receiver antenna gain;
- λ is the wavelength at the operating frequency;
- S is the electromagnetic (EM) power density and P_t is the RF power from the source;
- d is the distance between the transmitting and receiving antennas.

The variation by decade shows a difference between R1 and R2 of 13.67%, 6.01% and 1.48%. The difference can be explained by the non-linearity of the used Schottky diode. By reducing the planar antenna size of 30%, we reduce the antenna gain of 0.44 dBi and then the RF-DC conversion efficiency of 13% with a power density of 3.1 μW/cm^2 at 868 MHz. The second rectenna R2 will be preferred in cases where the power density is very low (e.g., 31 nW/cm^2). For our application, in the aim of supplying the power management unit, which requires a DC voltage of 600 mV at least, R1 was chosen.

$$\eta = 100 \cdot \frac{P_{DC}}{P_{RF}} = 100 \cdot \frac{4 \cdot \pi \cdot P_{dc}}{S \cdot G_r \cdot \lambda^2} \ with \ S = \frac{E^2}{120 \cdot \pi} = \frac{30 \cdot P_t \cdot G_t}{d^2 \cdot 120 \cdot \pi} \quad (2)$$

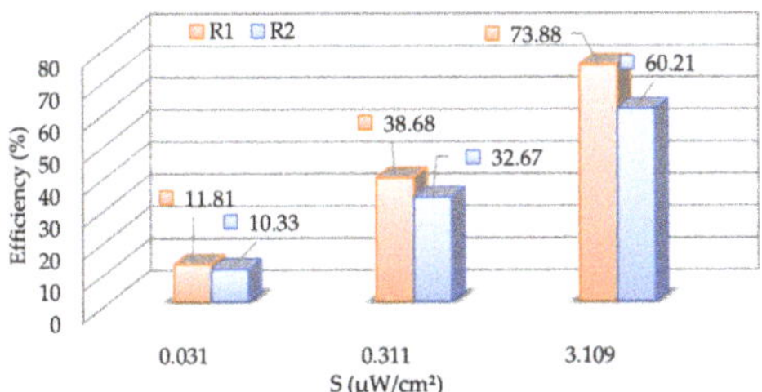

Figure 13. RF-DC conversion efficiency for the rectennas at 868 MHz for various decade power densities.

In the last set of tests, underrun in our laboratory, an innovative Bluetooth Low Energy (BLE) battery-free wireless sensor network (composed of battery-free sensing nodes and a communicating node [26]) was operated in a harsh environment (Figure 14). Once embedded, the main objectives were to periodically monitor the physical data (e.g., temperature and humidity) of the concrete and broadcast them (by BLE) to the communicating node.

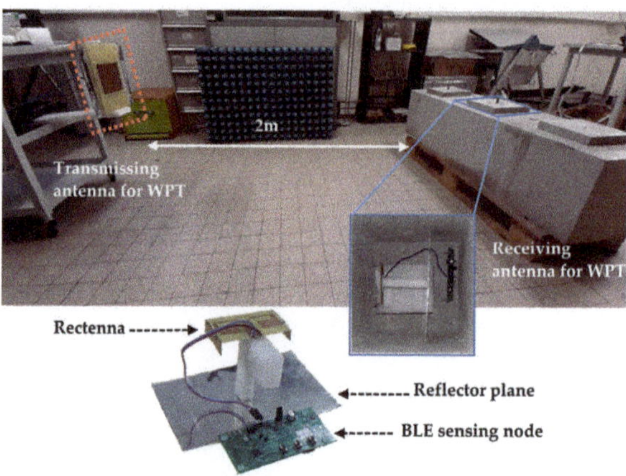

Figure 14. Photograph of the experimental setup for the sensing node embedded and powered by using WPT system in the concrete structure.

The sensing node in Figure 15 is composed of a rectenna (including the compact 3D antenna), an ultra-low-power BLE System on Chip (SoC) (NXP QN9080) [27], a power management unit microcontroller (Texas Instruments bq25570) [28], a capacitor of 100 μF acting as an energy storage element (Panasonic EEEFK0J101P) [29] and low-power temperature and humidity digital sensors (Texas Instruments HDC2080) [30]. The capacitor was chosen after evaluating the power needed by the PMU (870 μJ) to startup from deep-sleep mode. Through a far field WPT system, the illuminating power over a distance of two meters can be harvested by the rectenna and allow powering the sensing node. Initially empty, the capacitor is charging through the PMU on deep-sleep mode up to 1.5 V and switches on fast charging until the stored voltage reaches 5.3 V. Once it has enough energy stored by the capacitor, it discharges and allows powering the active component on the board (sensors and BLE chip) for a measurement and data transmission.

Figure 15. Developed sensing node for BLE communication.

Preliminary tests, with an EIRP power of +33 dBm, allow a periodicity of charging, measuring and wireless communication equal to at most 190 s, representing the first charge duration (cold start + fast charging + data measurement and transmission).

5. Conclusions

An electrically small antenna operating in the ISM 868 MHz frequency band and dedicated to wireless power transmission applications is proposed in this paper. Its design is based on a folded dipole antenna with a shorting line to form a T-match antenna. A 3D configuration allows the exploitation of the Z-plane in the final implementation. Thus, the connection of two metallic arms was operated by the planar antenna to downshift the operating frequency from 1.09 GHz to 868 MHz. The design was modified to apply a spline shape to the planar antenna arms in order to reduce the overall dimensions (in the horizontal plane) from 56 mm × 32 mm to 40 mm × 30 mm.

For the characterization, three feeding methods were used to provide accurate experimental results. In this application, the U.FL connector can be used for measurement but may have limitations at higher frequencies. The experimental results confirm that the proposed 3D antenna exhibits a gain of +1.54 dBi and +1.1 dBi at the operating frequency, respectively, for AC and A2. As shown in Table 1, a tradeoff between the compactness, the maximal gain and the bandwidth should be performed before the design. Our solution is also the most compact and has a maximum gain higher than +1 dBi at the price of a reduced bandwidth. The combination of these antennas and an optimized rectifier performs a high efficiency of 73.88% and 60.31%, respectively, for R1 and R2 with an illuminating power density of 3.1 $\mu W/cm^2$. Future works with a WPT system associated with a battery-free wireless sensor network using BLE are in progress.

Author Contributions: Conceptualization, A.S. and A.T.; methodology, A.S.; software, A.S.; validation, A.S. and G.L.; investigation, A.T.; writing—original draft preparation, A.S, A.T., G.L. and D.D.; writing—review and editing, A.S., A.T., G.L. and D.D.; supervision, D.D.; project administration, A.T. and D.D.; funding acquisition, A.T. and D.D. All authors have read and agreed to the published version of the manuscript.

Funding: Region OCCITANIE, France in the frame of the OPTENLOC project, supported this research. The system tests were performed in the frame of the ANR McBIM project.

Institutional Review Board Statement: Not applicable.

Informed Consent Statement: Not applicable.

Conflicts of Interest: The authors declare no conflict of interest.

References

1. Carminati, M.; Sinha, G.R.; Mohdiwale, S.; Ullo, S.L. Miniaturized Pervasive Sensors for Indoor Health Monitoring in Smart Cities. *Smart Cities* **2021**, *4*, 146–155. [CrossRef]
2. Donchev, E.; Pang, J.; Gammon, P.; Centeno, A.; Xie, F.; Petrov, P.; Alford, N. The rectenna device: From theory to practice (a review). *MRS Energy Sustain.* **2014**, *1*, E1. [CrossRef]
3. Lin, L.C.; Chiu, C.; Gong, J. A Wearable Rectenna to Harvest Low-Power RF Energy for Wireless Healthcare Applications. In Proceedings of the 11th International Congress on Image and Signal Processing, BioMedical Engineering and Informatics (CISP-BMEI), Beijing, China, 30 August–15 September 2018; pp. 1–5.
4. Takacs, A.; Okba, A.; Aubert, H. Compact Planar Integrated Rectenna for Batteryless IoT Applications. In Proceedings of the 48th European Microwave Conference (EuMC), Madrid, Spain, 23–27 September 2018; pp. 777–780.
5. Shafique, K.; Khawaja, B.A.; Khurram, M.D.; Sibtain, S.M.; Siddiqui, Y.; Mustaquin, M.; Chattha, T.H.; Yang, X. Energy Harvesting Using a Low-Cost Rectenna for Internet of Things (IoT) Applications. *IEEE Access* **2018**, *6*, 30932–30941. [CrossRef]
6. Pizzotti, M.; Perilli, L.; Del Prete, M.; Fabbri, D.; Canegallo, R.; Dini, M.; Masotti, D.; Costanzo, A.; Franchi Scarselli, E.; Romani, A. A Long-Distance RF-Powered Sensor Node with Adaptive Power Management for IoT Applications. *Sensors* **2017**, *17*, 1732.
7. Okba, A.; Takacs, A.; Aubert, H. Compact Flat Dipole Rectenna for IoT Applications. *Prog. Electromagn. Res. C* **2018**, *87*, 39–49. [CrossRef]
8. Rokunuzzaman, M.; Islam, M.T.; Rowe, W.S.T.; Kibria, S.; Singh, M.J.; Misran, N. Design of a Miniaturized Meandered Line Antenna for UHF RFID Tags. *PLoS ONE* **2016**, *11*, e0161293. [CrossRef] [PubMed]
9. Gianvittorio, J.P.; Rahmat-Samii, Y. Fractal antennas: A novel antenna miniaturization technique, and applications. *IEEE Antennas Propag. Mag.* **2002**, *44*, 20–36. [CrossRef]
10. Das, S.; Sawyer, D.J.; Diamanti, N.; Annan, A.P.; Iyer, A.K. A strongly miniaturized and inherently matched folded dipole antenna for narrowband applications. *IEEE Antennas Propag. Mag.* **2019**, *68*, 3377–3386. [CrossRef]

11. Babar, A.A.; Virtanen, J.; Bhagavati, V.A.; Ukkonen, L.; Elsherbeni, A.Z.; Kallio, P.; Sydänheimo, L. Inkjet-printable UHF RFID tag antenna on a flexible ceramic-polymer composite substrate. In Proceedings of the IEEE MTT-S International Microwave Symposium Digest (MTT '12), Montreal, QC, Canada, 17–22 June 2012; pp. 1–3.
12. Li, K.; Zhu, C.; Li, L.; Cai, Y.M.; Liang, C.H. Design of electrically small metamaterial antenna with ELC and EBG loading. *IEEE Antenn. Wirel. Pract. Lett.* **2013**, *12*, 678–681. [CrossRef]
13. Balanis, C.A. Frequency Independent Antennas, Antenna Miniaturization, and Fractal Antennas. In *Antenna Theory: Analysis and Design*, 3rd ed.; John Wiley: Hoboken, NJ, USA, 2005; pp. 637–640.
14. Wheeler, H.A. Fundamental Limitations of Small Antennas. *Proc. IRE* **1947**, *35*, 1479–1484. [CrossRef]
15. Marrocco, G. The art of UHF RFID antenna design: Impedance-matching and size-reduction techniques. *IEEE Antennas Propag. Mag.* **2008**, *50*, 66–79. [CrossRef]
16. Niotaki, K.; Kim, S.; Jeong, S.; Collado, A.; Georgiadis, A.; Tentzeris, M.M. A Compact Dual-Band Rectenna Using Slot-Loaded Dual Band Folded Dipole Antenna. *IEEE Antennas Wirel. Propag. Lett.* **2013**, *12*, 1634–1637. [CrossRef]
17. Balanis, C.A. Broadband Dipoles and Matching Techniques. In *Antenna Theory: Analysis and Design*, 3rd ed.; John Wiley: Hoboken, NJ, USA, 2005; pp. 538–541.
18. Ting, J.; Oloumi, D.; Rambabu, K. A miniaturized broadband bow-tie antenna with improved cross-polarization performance. *Int. J. Electron. Commun.* **2017**, *78*, 173–180. [CrossRef]
19. Durgun, A.C.; Balanis, C.A.; Birtcher, C.R.; Allee, D.R. Design, Simulation, Fabrication and Testing of Flexible Bow-Tie Antennas. *IEEE Trans. Antennas Propag.* **2011**, *59*, 4425–4435. [CrossRef]
20. U.FL Series Connector HIROSE Electric Group. Available online: https://www.hirose.com/product/series/U.FL?lang=en (accessed on 5 December 2019).
21. Zahid, S.; Quddious, A.; Tahir, F.A.; Vryonides, P.; Antoniades, M.; Nikolaou, S. Dual-Band Compact Antenna for UHF and ISM Systems. In Proceedings of the 13th European Conference on Antennas and Propagation (EuCAP), Krakow, Poland, 31 March–5 April 2019; pp. 1–5.
22. Trinh, L.H.; Nguyen, T.Q.K.; Phan, D.D.; Tran, V.Q.; Bui, V.X.; Truong, N.V.; Ferrero, F. Miniature antenna for IoT devices using LoRa technology. In Proceedings of the International Conference on Advanced Technologies for Communications (ATC), Quy Nhon, Vietnam, 18–20 October 2017; pp. 170–173.
23. Asadallah, F.A.; Eid, A.; Shehadeh, G.; Costantine, J.; Tawk, Y.; Tentzeris, M.M. Digital Reconfiguration of a Single Arm Three-Dimensional Bowtie Antenna. *IEEE Trans. Antennas Propag.* **2020**. [CrossRef]
24. Sidibe, A.; Loubet, G.; Takacs, A.; Dragomirescu, D. Energy Harvesting for Battery-Free Wireless Sensors Network Embedded in a Reinforced Concrete Beam. In Proceedings of the 50th European Microwave Conference (EuMC), Utrecht, The Netherlands, 12–14 January 2021; pp. 702–705.
25. Sidibe, A.; Takacs, A. Compact 3D Rectenna for Low-Power Wireless Transmission. In Proceedings of the XXXIV URSI General Assembly and Scientific Symposium (URSI GASS), Rome, Italy, 28 August–4 September 2021.
26. Loubet, G.; Takacs, A.; Gardner, E.; De Luca, A.; Udrea, F.; Dragomirescu, D. LoRaWAN Battery-Free Wireless Sensors Network Designed for Structural Health Monitoring in the Construction Domain. *Sensors* **2019**, *19*, 1510. [CrossRef] [PubMed]
27. NXP Semiconductors. QN908x Ultra-Low-Power BLE SoC. Available online: https://www.nxp.com/docs/en/nxp/data-sheets/QN908x.pdf (accessed on 12 January 2021).
28. Texas Instruments. bq25570 Ultra Low-Power Harvester power Management IC with boost charger and Nanopower Buck Converter. Available online: http://www.ti.com/lit/ds/symlink/bq25570.pdf (accessed on 12 January 2021).
29. Panasonic. EEEFK0J101P Aluminum Electrolytic Capacitors. Available online: https://industrial.panasonic.com/cdbs/www-data/pdf/RDE0000/ABA0000C1181.pdf (accessed on 12 January 2021).
30. Texas Instruments. Ultra-Low-Power, Digital Humidity Sensor with Temperature Sensor HDC2080. Available online: https://www.ti.com/lit/ds/symlink/hdc2080.pdf (accessed on 12 January 2021).

Communication

Horizontal Polarized DC Grounded Omnidirectional Antenna for UAV Ground Control Station

Muhammad Shahzad Sadiq [1], Cunjun Ruan [1,2,*], Hamza Nawaz [3], Shahid Ullah [1] and Wenlong He [4]

[1] School of Electronic and Information Engineering, Beihang University, Beijing 100191, China; shahzadsadiq@buaa.edu.cn (M.S.S.); shahidkhan@buaa.edu.cn (S.U.)
[2] Beijing Key Laboratory for Microwave Sensing and Security Applications, Beihang University, Beijing 100191, China
[3] School of Electrical, Information and Electronics Engineering, Shanghai Jiao Tong University, Shanghai 200240, China; hamza_nawaz@hotmail.com
[4] College of Electronics and Information Engineering, Shenzhen University, Shenzhen 518060, China; wenlong.he@szu.edu.cn
* Correspondence: ruancunjun@buaa.edu.cn; Tel.: +86-135-0120-5336

Abstract: A new slot-based antenna design capable of producing horizontal polarization for unmanned aerial vehicle (UAV) ground control station (GCS) applications is outlined in this paper. The proposed antenna consists of oversize coaxial cylinders, slots, and slot-feed assembly. Each of the four vertical slots, arranged periodically around the antenna's outer cylinder, emits a horizontally polarized broad beam of radiation, in phase, to produce an omnidirectional pattern. The antenna possesses a low-ripple ±0.5 dB in azimuth gain (yaw) due to its symmetric axis shape and an enclosed feed within itself, which does not radiate and interfere with the main azimuth pattern. This is crucial for a UAV GCS to symmetrically extend its coverage range in all directions against yaw planes. Simulation and measurement results reveal that the antenna maintains stable gain in the omnidirectional pattern (+0.5 dB) over the entire operational frequency band (2.55 GHz to 2.80 GHz), where S11 is lower than −10 dB. A further advantage of this configuration is its enhanced polarization purity of −40 dB over the full frequency band. The direct-current (DC) grounding approach used in this antenna is beneficial due to its electrostatic discharge (ESD) and lightning protection. Furthermore, its aerodynamic, self-supporting, and surface-mount structural shape makes this antenna a good and worthy choice for a UAV GCS.

Keywords: horizontal polarization; UAV ground station; Omni-directional

Citation: Sadiq, M.S.; Ruan, C.; Nawaz, H.; Ullah, S.; He, W. Horizontal Polarized DC Grounded Omnidirectional Antenna for UAV Ground Control Station. *Sensors* **2021**, *21*, 2763. https://doi.org/10.3390/s21082763

Academic Editors: Shuai Zhang and Razvan D. Tamas

Received: 16 January 2021
Accepted: 5 April 2021
Published: 14 April 2021

1. Introduction

Since the discovery of the interdependency between electrical parameters and electromagnetic radiation, antennas have been developed that actively exploit this phenomenon. Antennas convert electrical parameters (current and voltages) into electromagnetic parameters (electric and magnetic fields) and vice versa. Hence, an antenna can be regarded as a transducer or a sensor as it converts electrical energy to electromagnetic energy, or the opposite [1]. Antennas are always considered essential parts of communication systems, and their radiation and polarization characteristics play a vital role in defining such systems' performance and efficiency [2,3].

The use of unmanned aerial vehicles (UAVs) is rapidly expanding to commercial, scientific, agricultural, and military applications [4]. To overcome the difficulty of finding the exact location of mobile UAVs from ground control stations (GCS), omnidirectional antennas are utilized to resolve acquisition and pointing complications [5–10]. It has been proven that using horizontally polarized antennas can achieve a 10 dB improvement in terms of system gain as compared to vertically polarized antennas [11]. For GCS deployment, where the antenna is intended to cover a wide range of angles at variant distances, it is essential to utilize a low-gain ripple radiation pattern to ensure continuous coverage in the yaw plane (azimuth plane) [12–14] as gain ripple fluctuates and reduces the coverage range at variant horizontal plane angles [15].

The challenging aspect of designing horizontally polarized omnidirectional antennas is producing a uniform and in-phase current in the antenna's azimuth plane. That necessary condition can be fulfilled by utilizing a single loop or multi-element arrangements [16]. Three primary topological schemes are used with omnidirectional horizontally polarized antennas. In the first topology, a single radiating structure, such as a loop, is utilized to achieve horizontally polarized radiations [17–20], but it is inherently band-limited due to an open feed. The second group imitates the loop arrangement of first with dipole elements arranged in a ring or circular array format [21–26] at the expense of a complex open feeding arrangement. The third topology utilizes slots to complete the horizontally polarized antenna. There are a few slot-based omnidirectional antennas described in the literature. In [27], an omnidirectional antenna operating at X band used an array of slot doublets etched in the broadside wall of the rectangular waveguide. However, there was no mention of the azimuth gain, gain fluctuations, and operating band. A slot-based antenna capable of producing horizontal polarization was constructed by arranging alternate slots with opposite tilt angles along the axis with intervals of $\lambda g/2$. To improve the antenna's performance, alternate axial slot arrays were shifted by $\lambda g/4$ along the axis. Even then, it was not improved by more than -7 dB [28]. In [29], an omnidirectional antenna was proposed, but it was circularly polarized. Moreover, it was not direct-current (DC) grounded and had a built-in main beam frequency scanning problem. In [30], a slant polarized omnidirectional antenna was presented. All slot-based horizontal polarized topologies were arranged in a series of fed axial arrays to achieve the required polarization. The other two methodologies had open feeding networks that interfered with the radiating apertures and perturbed antenna radial symmetry causing an uneven azimuth gain pattern, which further reduced antenna coverage range.

This paper proposes an omnidirectional antenna capable of achieving low azimuthal gain variations of ± 0.5 dB. This work is the first single-element design based on slots capable of horizontal polarization and stable gain without making a complex axial array to achieve the required polarization. The flaunted antenna comprises four slot apertures evenly spaced around the antenna's outer circumference. It also encloses the feeding topology, so antenna symmetry is not disturbed. The device's compactness, ruggedness, and direct-current grounding are further important features of this antenna design. The proposed technique has improved polarization stability since the cross-pols are very weak relative to the co-pols. The antenna structure is exhibited and explained in Section 2. Section 3 elaborates on the slot-feed mechanism. In Section 4, simulation verification is performed. Section 6 describes the manufacturing and measurements of the antenna prototype. Section 6 presents a comparison of the proposed work with those published. Finally, Section 7 details our conclusions.

2. Antenna

The structure of the suggested horizontally polarized omnidirectional antenna is shown in Figure 1. It must be shaped like a pole due to vehicle-mounting requirements. It is based on the coaxial line and is composed of inner and outer conductors. There are etched slots around the coaxial cylinder, and the internal and external coaxial cylinders are separated by air. The primary radiation is emitted via the slots (each slot is matched to a dipole with the magnetic current source), which are periodically positioned along the antenna's outer cylinder as depicted in Figure 1a. According to Babinet's principle, the slots are complementary to the dipole antenna. The far field of the linear dipole [31], is found using:

$$E_\theta = \left(j60 I_m cos\ cos\ (klcos\theta)\ e^{-jkr}\right)/rsin\theta$$

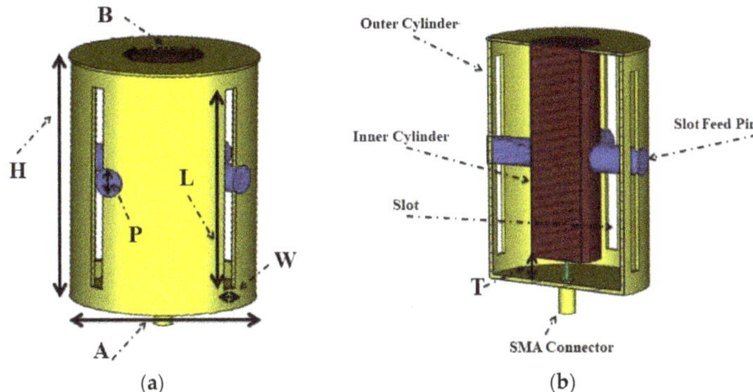

Figure 1. The geometry of the horizontally polarized omnidirectional antenna: (**a**) 3D view, (**b**) cross-sectional view.

In the equation, θ is the angle between the line direction and the dipole. This means the pattern function of the dipole is the same as the slot antenna.

$$F_{\theta} = (cos\ cos\ (klcos\theta)\ - coskl)/sin\theta$$

For idea half wavelength slot, 2l = λ/2, and

$$F_{\theta} = Cos(2\pi cos\theta)/sin\theta$$

The pattern of the slot antenna is the same as the dipole with the same length, but their elevation plane (E-plane) and omnidirectional plane (H-plane) are exchanged according to the duality principle. Each slot aperture produces horizontally polarized radiation. Four apertures around the circumference complete the antenna, as illustrated in Figure 1a,b, and radiate in an omnidirectional pattern. The SMA connector smoothly converts the TEM modes from SMA to a large antenna assembly with a matching structure that is an optimized inner pin height, as given in Figure 1b.

The diameter of the outer cylinder, the diameter of the feed pin that connects the inner cylinder to the outer part of the antenna, and the length of the slots are what primarily impact the performance of the antenna. The optimal specifications are listed in Table 1.

Table 1. Optimal parametric values of the antenna.

Parameter	Value (mm)
Outer Cylinder Diameter A	60
Inner Cylinder Diameter B	25
Slot feed Pin Diameter P	10
Antenna Height H	75
Slot Length L	65
Slot Width W	5
SMA Pin Height T	7.25

At UAV GCSs, there are relaxed limitations with regard to size and weight compared to aerial platforms [32]. For military operations, the UAV operator at the GCS is located in a harmless, secured place while the desired information or strategic data from the battlefield is gathered remotely. For such applications, antennas must be capable of withstanding all terrain operational area requirements and should be able to function correctly under extreme weather scenarios. So the required antenna should be mechanically robust and sturdy without external supports as these supports would increase the antenna's size [4] and result in more drag, which might weaken the antenna's structure due to rigorous terrain and weather conditions [4,33]. Thus, it is crucial to use a compact, aerodynamic design. There is often a chance that an instance of peak instantaneous power (PIP) happens inside the printed circuit board- (PCB) based feed network. Such an event would easily damage the PCB [34], so the feed must be capable of bearing sudden PIP. The antenna would also be the primary source to channel electrostatic discharge (ESD) and lightning into the electronic systems. An ESD incident would place the functionality and safety of these systems at risk, while a lightning bolt would annihilate them. Keeping the antenna DC grounded is the most feasible and efficient strategy used in combat [35]. This antenna design would circumvent all the problems described above. The axis-symmetrical, all-metal rugged antenna is primarily constructed of brass and is DC-grounded. The solid metal feed network is enclosed inside the antenna's conformal and compact shape.

3. Feed Mechanism

Horizontal slots induce vertical polarization as they can quickly interrupt the longitudinal surface current on the antenna's outer surface [36], as seen in Figure 2. Conversely, the longitudinal slots in the antenna's outer surface cannot be stimulated due to their orientations that are in line with the surface current, and even a short circuit would not modify the flow of the surface current [28]. So it is not easy to produce horizontal polarization using a slot configuration on a coaxial cylinder. In our design, feed pins are inserted to excite the vertical slots, which connect the outer conductor of the oversized coaxial cable with its inner conductor, as shown in Figure 1b. Thus, these slot apertures are energized sideways while the opposing sides are kept floating. The slot is regarded as a dipole having a magnetic current source [29], so the slot is $\lambda g/2$ long. Normally, the external feed has a built-in problem where it radiates along with the main radiating elements and causes a significant gain ripple in the omnidirectional pattern. Here, we have designed an internal feed that runs inside the radiating part and does not interfere or radiate. As for the actual feeding of the antenna, a standard SMA connector is used for feeding. The SMA connector is a coaxial structure and the antenna designed in this section is also based on coaxial structure, so the matching structure is designed and inserted between the radiation part and the feed part according to impedance transformation of coaxial transmission line [37],

$$Z_{oversize} = Z_{match}(Z_{sma} + jZ_{match}tan\beta T)/(Z_{match} + jZ_{sma}tan\beta T)$$

where the T is the length of the matching pin and Z_{sma}, Z_{match}, and $Z_{oversize}$ are the characteristic impedances of the SMA connector, matching pin, and oversize antenna assembly, respectively. The SMA connector's inner pin's optimized height ensures a seamless transition from normal Coaxial TEM mode to oversized TEM cable mode. The slot excitation of the proposed antenna is simple and easy without involving baluns or impedance transformers. Four pins join the inner cylinder to the vertical slots in the antenna's outer cylinder. The feed and antenna can be conveniently integrated by arranging the manufactured parts together around the central axis.

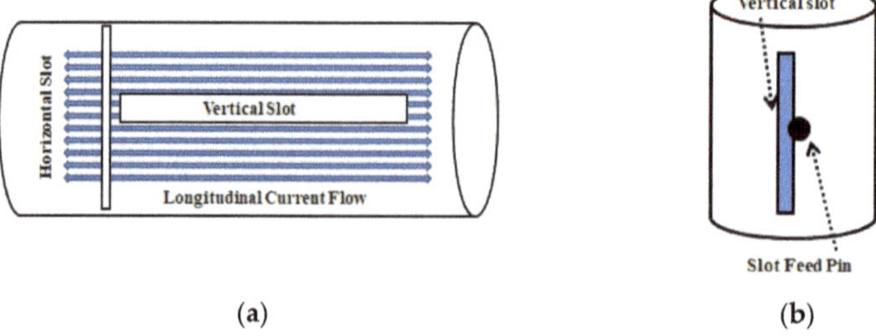

Figure 2. Slot configurations: (**a**) vertical and horizontal slot, (**b**) vertical slot feed.

4. Simulation Verification

CST Microwave Studio was been used to simulate and optimize the antenna design. Figure 3 demonstrates the mutual connection between the antenna azimuth gain and the total number of slots along the antenna's circumference. Each slot radiates a directed pattern. With each increment in the number of slots along the antenna's circumferential axis, these directional radiations widened, as shown in Figure 3. Four slots made the radiation patterns combine and generate a low-ripple horizontal polarized omnidirectional radiation pattern.

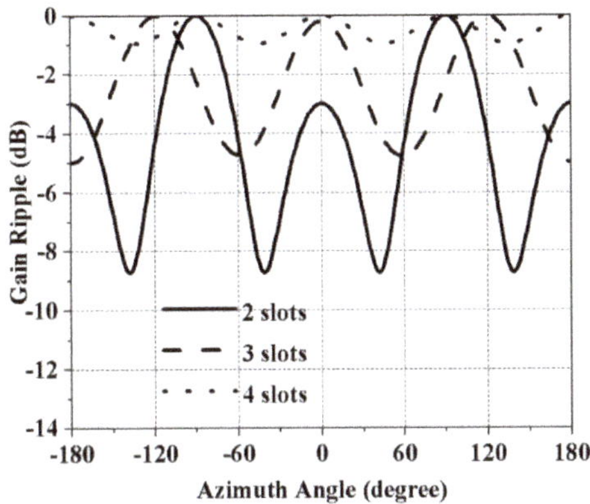

Figure 3. Azimuthal gain vs. the number of slots along the antenna's circumferential axis.

A. Determination of Pin Diameter and Slot Size Effect

Figure 4 helps us to see the impact of the slot-feed pin diameter on the antenna input reflection. The change of diameter changed the antenna's matching, as shown in Figure 4a. Figure 4b depicts how the slot-length variations shifted the antenna resonance region.

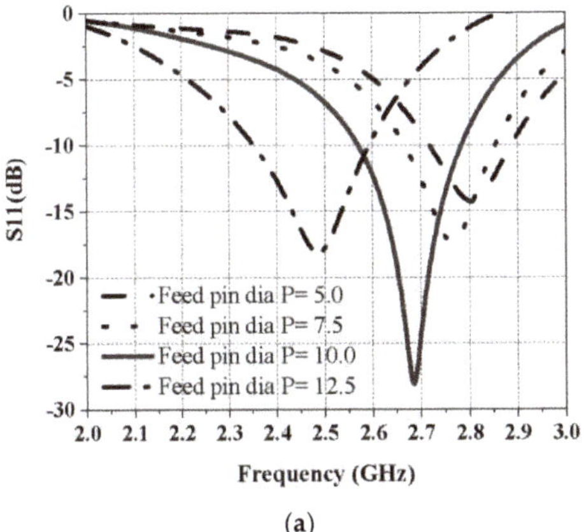

(a)

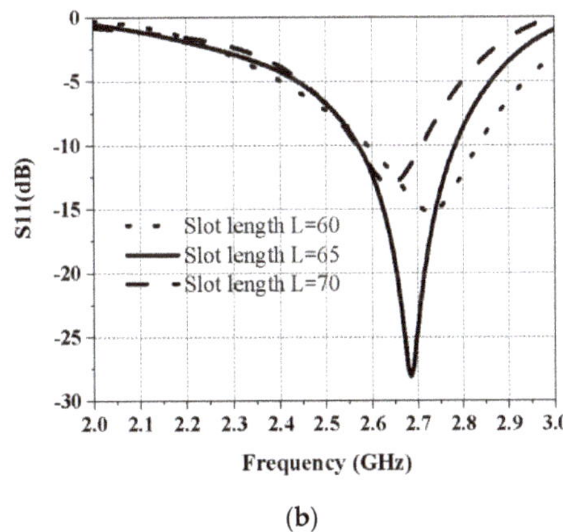

(b)

Figure 4. Effect on S_{11} (**a**) by changing the feed pin diameter and (**b**) by changing the slot length.

B. Field Verification

The field simulations were performed with the help of CST Microwave Studio software. The electric and magnetic fields' cross-sectional views through the SMA connector are shown Figure 5a,c. The cross-sectional views of the electric and magnetic fields through the pins that connect the inner coaxial cylinder to the slot apertures in the outer coaxial cylinder are shown in Figure 5b,d. At the input SMA connector of the antenna, the electric field is spread radially outward (TEM mode). At the edge of the oversized coaxial antenna assembly near the SMA connector, the electric field is again radially outward as that of the TEM mode. This demonstrates that the SMA connector's inner pin's adjusted length effectively converted the connector TEM mode into the oversized coaxial assembly TEM mode, as shown in Figure 5a. As this mode travels toward the pin connected to the slot, the field circulates the slot area. All four slots have the same circulation pattern, which indicates that they are all in phase, as seen in Figure 5b. The electric field steadily travels from the slot middle toward the slot end and thereby emits a horizontally polarized field, as seen in Figure 5a,b. The directional radiation slot patterns are large enough to converge to generate an omnidirectional, horizontally polarized outward wave. Correspondingly, the magnetic fields form closed loops (TEM mode) at the SMA feed and eventually transform to perpendicular loops corresponding to the E field outside the antenna, as clearly visible in the simulated field trajectories in Figure 5b,d.

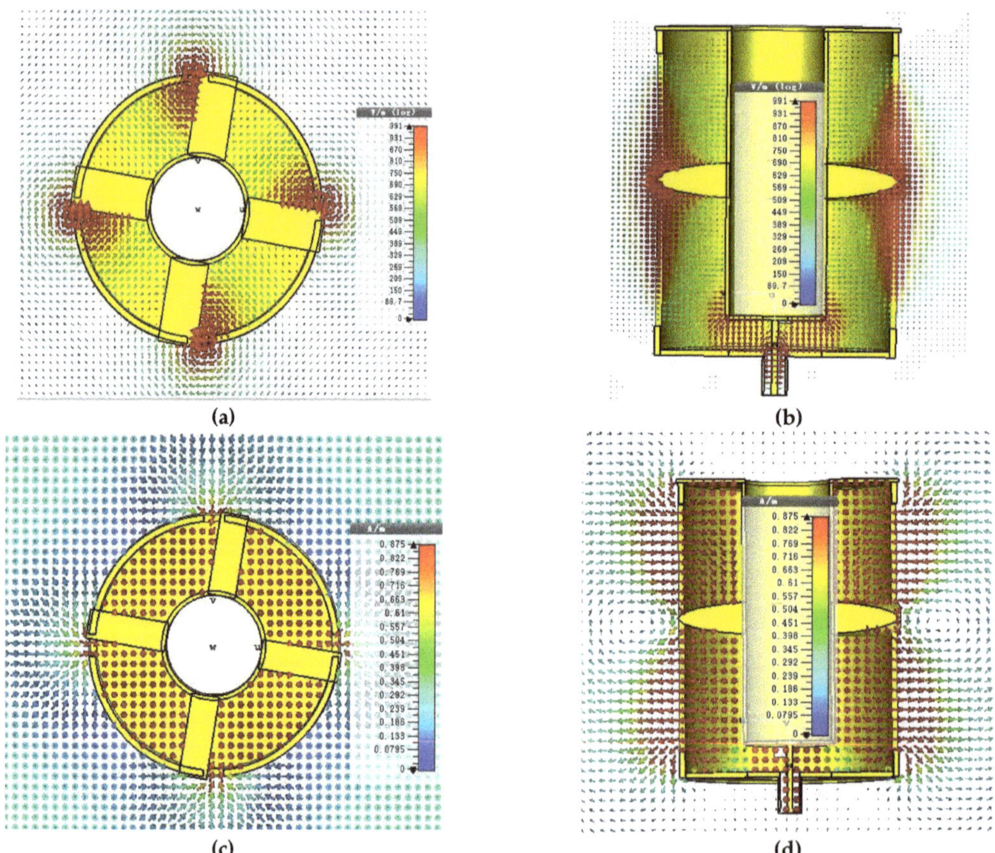

Figure 5. The cross-sectional views of the fields at the SMA connector: (**a**) the electric field, (**b**) the magnetic field cross-sectional views of the fields at the slot-feeding pins, (**c**) the electric field (**d**) the magnetic field.

5. Antenna Fabrication and Measurement Result

An antenna was manufactured using brass for the design validation. This antenna can be assembled using CNC-machined parts or expensive 3D printing. This antenna was built utilizing the first approach. The antenna had a reduced footprint and conformal shape to maintain low air resistance. The simulated and measured antenna test results are discussed in this section. Figure 6a displays an image of the prototype antenna. The input scattering parameter S_{11} of the manufactured antenna was measured with Agilent N5242A VNA's help. In Figure 6b, the simulated and measured S_{11} are plotted. Measurements were less than −10 dB from 2.5 GHz to 2.8 GHz, which were in good harmony with simulations. The antenna was a reasonably broadband structure (achieved bandwidth of 11.3%).

(a)

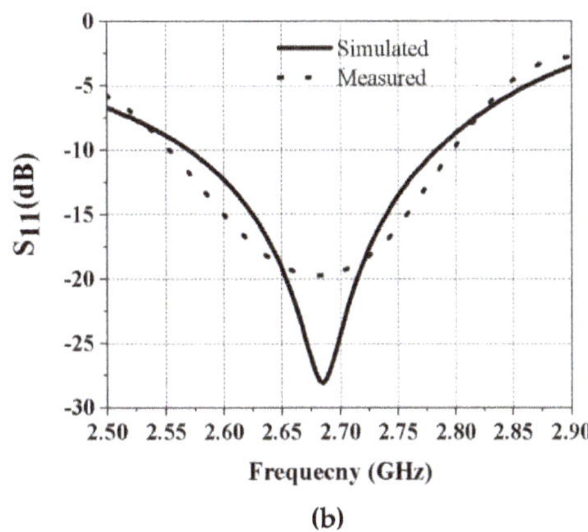

(b)

Figure 6. (**a**) The fabricated prototype antenna. (**b**) Simulated and measured S_{11} of the antenna.

Measured and simulated vertical elevation planes and horizontal azimuth planes of the antenna at 2.6 GHz and 2.7 GHz are plotted in Figure 7. Measurements were done in the the compact antenna test range (ATR) of March Microwave Systems B.V., which uses a source antenna that radiates a spherical wavefront and two secondary reflectors to collimate the radiated spherical wavefront into a planar wavefront within the desired test zone where the test antenna is placed and precalibrated standard gain antennas are used to determine the absolute gain of the AUT(antenna under test). The simulated and measured co-polarization (normalized) and cross-polarization (normalized) radiation patterns in the omnidirectional plane (H-plane) are shown in Figure 7a. The 360° radiation at the horizontal plane helps to maintain complete yaw plane operation. The measured cross-polarization levels in the azimuth plane are more than −40 dB down, which agrees with the simulation. Figure 7b depicts the simulated and measured co-polarization (normalized) radiation patterns in vertical elevation (E-plane). Figure 8a,b shows a measured azimuth gain ripple of ±0.5 dB, whereas the azimuth pattern phase ripple is only 10° peak-to-peak. These were measured at 2.6 GHz and 2.7 GHz, respectively. Both affirm the excellent stability of the antenna pattern.

In Figure 9a, the measured and simulated azimuth gain ripple are plotted for the entire frequency range for the clear visibility of the gain fluctuations. The maximum peak-to-peak value is 1 dB in the azimuth plane, confirming a good omnidirectionality. Figure 9b illustrates the simulated and measured gain of our antenna. This DC-grounded antenna demonstrated reliable gain within the entire band. These results indicate promising and effective radiation characteristics in the yaw plane, making this antenna an appealing option for GCS applications.

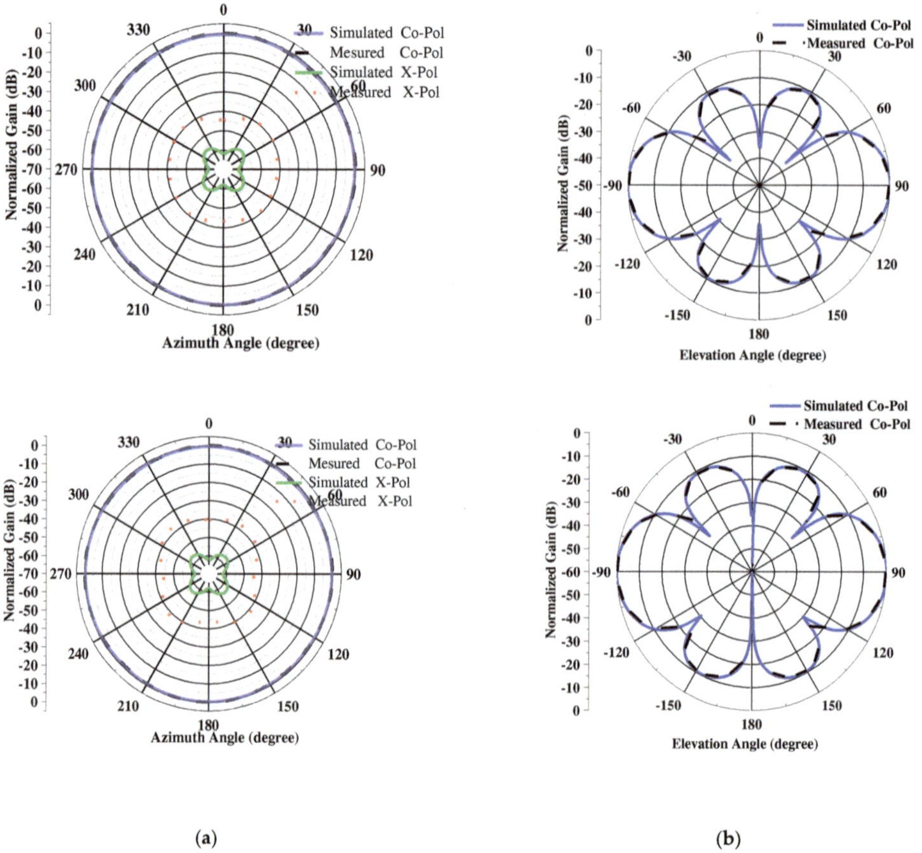

(a) (b)

Figure 7. (**a**) Measured and simulated normalized co-polarization and cross-polarization in the omnidirectional plane, or H-plane; top 2.6 GHz; bottom 2.7 GHz. (**b**) Measured and simulated normalized co-polarization in the elevation plane, or E-plane; top 2.6 GHz; bottom 2.7 GHz.

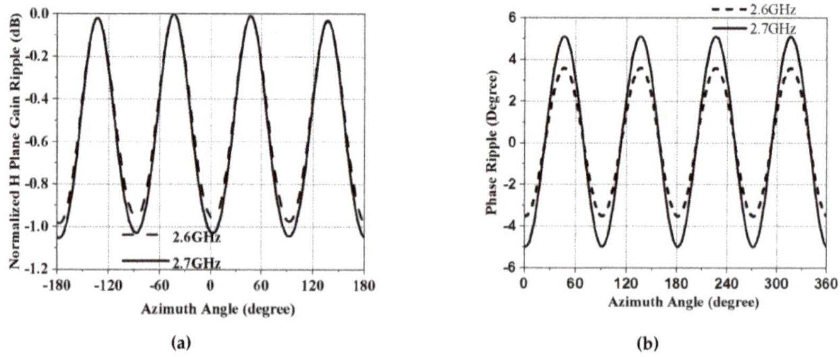

Figure 8. (**a**) Measured gain ripple vs. azimuth angle, (**b**) measured phase angle ripple vs. azimuth angle, at 2.6 GHz and 2.7 GHz, respectively.

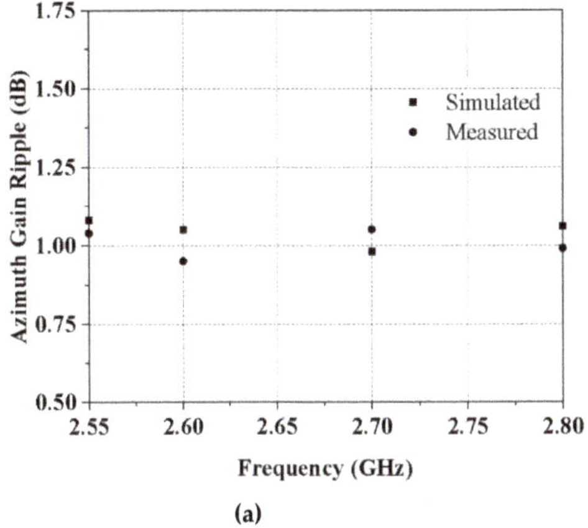

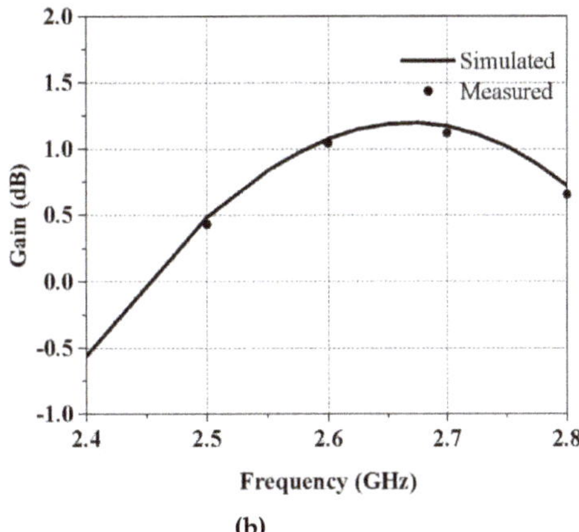

Figure 9. (**a**) Measured and simulated azimuth gain ripple vs. frequency. (**b**) Measured and simulated antenna gain vs. frequency.

6. Comparison

Table 2 compares this study to the prior works published in the literature that also featured horizontal polarization. All works tabulated were designed with external open feed and lack lighting protection capability. The polarization purity was also not so high. This work proposes for the first time an antenna that could produce horizontal polarization utilizing slot radiators and form a stable gain at the azimuth plane like traditional omnidirectional design topologies such as loop or printed dipole antenna arranged in the form of a circle. Due to vertical slot radiations, its cross-polarization levels are extremely low, as shown in the Table. The proposed antenna is novel because it has an internal axis-symmetric feeding system. Due to its enclosed nature, the feed does not radiate and interfere with the main radiating slots. So low-gain ripples in the azimuth plane are achieved as compared to the listed works. Moreover, it is DC-grounded, which is essential for any practical deployment.

Table 2. Performance comparison of the proposed antenna with the existing literature.

Reference	Polarization Purity	Gain Ripple (dB)	DC Ground	Feed Type
[17]	11	±5.5	No	Exposed
[18]	20	NA	No	Exposed
[19]	20	1.3	No	Exposed
[20]	25	1.0	No	Exposed
[21]	18	1.5	No	Exposed
[22]	20	1.5	No	Exposed
[23]	15	1.5	No	Exposed
[24]	15	NA	No	Exposed
[25]	20	2.0	No	Exposed
[26]	20	2.1	No	Exposed
Proposed work	40	±0.5	Yes	Enclosed

7. Conclusions

A novel horizontally polarized omnidirectional antenna that is built on the slot structures is introduced in this work. By positioning four vertical slots along the antenna circumference and energizing them with a robust central feeding mechanism, steady gain with improved polarization stability is realized in the antenna's azimuth axis. The internal axis-symmetric feed network itself radiates no power, hence it does not interfere with the radiating structure. Low ripples appear among the antenna azimuth gain. Steady yaw plane gain with low gain fluctuations enhances coverage area or increases the link efficiency. So it is desirable in numerous ground station-based applications, such as UAV communication and direction finding, to have a low azimuth gain ripple antenna. This antenna also possesses requisite mechanical features, which are crucial for its smooth operation. It has a sturdy, DC-grounded construction that does not need any exterior framework or support. Its conformal and aerodynamic shape minimizes air drag and any corresponding degradation attributed to military operations' environmental and terrain conditions. Altogether, this may make this antenna a favorable candidate for ground stations.

Author Contributions: M.S.S., in collaboration with H.N., proposed the antenna's configuration and simulated it in CST. M.S.S. executed the fabrication and tuning of the antenna prototype. H.N. did the measurements. M.S.S. and S.U. did the plotting and manuscript preparation. C.R. and W.H. provided step-by-step supervision, reviewed the work, and did manuscript refinement. All authors have read and agreed to the published version of the manuscript.

Funding: This work is supported by the National Natural Science Foundation of China (Grant No. 61831001) and the High-Level Talent Introduction Project of Beihang and the Youth-Top-Talent University (Grant No. ZG216S1878) and Support Project of Beihang University (Grant No. KG12060401).).

Institutional Review Board Statement: Not applicable.

Informed Consent Statement: Not applicable.

Data Availability Statement: Not applicable.

Conflicts of Interest: Authors declare no conflict of interest.

References

1. Sevgi, L. The Antenna as a Transducer: Simple Circuit and Electromagnetic Models. *IEEE Antennas Propag. Mag.* **2007**, *49*, 211–218. [CrossRef]
2. Sadiq, M.S.; Ruan, C.; Nawaz, H.; Abbasi, M.A.B.; Nikolaou, S. Mutual Coupling Reduction between Finite Spaced Planar Antenna Elements Using Modified Ground Structure. *Electronics* **2021**, *10*, 19. [CrossRef]
3. Niaz, M.W.; Sadiq, M.S.; Yin, Y.; Zheng, S.; Chen, J. Reduced aperture low sidelobe patch array. *Electron. Lett.* **2018**, *54*, 800–802. [CrossRef]
4. Cui, Y.; Luo, P.; Gong, Q.; Li, R. A Compact Tri-Band Horizontally Polarized Omnidirectional Antenna for UAV Applications. *IEEE Antennas Wirel. Propag. Lett.* **2019**, *18*, 601–605. [CrossRef]
5. Nawaz, H.; Liang, X.; Sadiq, M.S.; Abbasi, M.A.B. Ruggedized Surface-Mount Omnidirectional Antenna for Supersonic Aerial Platforms. *IEEE Antennas Wirel. Propag. Lett.* **2020**, *19*, 1439–1442. [CrossRef]
6. Nawaz, H.; Liang, X.; Sadiq, M.S.; Geng, J.; Jin, R. Circularly-Polarized Shaped Pattern Planar Antenna for Aerial Platforms. *IEEE Access* **2020**, *8*, 7466–7472. [CrossRef]
7. Nawaz, H.; Liang, X.; Sadiq, M.S.; Geng, J.; Zhu, W.; Jin, R. Ruggedized Planar Monopole Antenna With a Null-Filled Shaped Beam. *IEEE Antennas Wirel. Propag. Lett.* **2018**, *17*, 933–936. [CrossRef]
8. Choi, Y.-S.; Park, J.-S.; Lee, W.-S. Beam-Reconfigurable Multi-Antenna System with Beam-Combining Technology for UAV-to-Everything Communications. *Electronics* **2020**, *9*, 980. [CrossRef]
9. Lee, C.U.; Noh, G.; Ahn, B.; Yu, J.-W.; Lee, H.L. Tilted-Beam Switched Array Antenna for UAV Mounted Radar Applications with 360° Coverage. *Electronics* **2019**, *8*, 1240. [CrossRef]
10. Kim, K.; Yoo, J.; Kim, J.; Kim, S.; Yu, J.; Lee, H.L. All-Around Beam Switched Antenna With Dual Polarization for Drone Communications. *IEEE Trans. Antennas Propag.* **2020**, *68*, 4930–4934.
11. Sun, L.; Li, Y.; Zhang, Z.; Feng, Z. Compact Co-Horizontally Polarized Full-Duplex Antenna with Omnidirectional Patterns. *IEEE Antennas Wirel. Propag. Lett.* **2019**, *18*, 1154–1158. [CrossRef]
12. Liu, Y.; Li, X.; Yang, L.; Liu, Y. A Dual-polarized Dual-band Antenna with Omni-directional Radiation Patterns. *IEEE Trans. Antennas Propag.* **2017**, *65*, 4259–4262. [CrossRef]
13. SJun, Y.; Shastri, A.; Sanz-Izquierdo, B.; Bird, D.; McClelland, A. Investigation of Antennas Integrated into Disposable Unmanned Aerial Vehicles. *IEEE Trans. Veh. Tech.* **2019**, *68*, 604–612.
14. Echeveste, J.I.; de Aza, M.A.G.; Zapata, J. Shaped beam synthesis of real antenna arrays via finite-element method Floquet modal analysis and convex programming. *IEEE Trans. Antennas Propag.* **2016**, *64*, 1279–1286. [CrossRef]
15. Manoochehri, O.; Darvazehban, A.; Salari, M.A.; Khaledian, S.; Erricolo, D.; Smida, B. A dual-polarized biconical antenna for direction finding applications from 2 to 18 GHz. *Microw. Opt. Technol. Lett.* **2018**, *60*, 1552–1558. [CrossRef]
16. Chen, Z.N.; Liu, D.; Nakano, H.; Qing, X.; Zwick, T. (Eds.) *Handbook of Antenna Technologies*; Springer: Singapore, 2016.
17. Lin, C.; Kuo, L.C.; Chuang, H.R. A horizontally polarized omnidirectional printed antenna for WLAN applications. *IEEE Trans. Antennas Propag.* **2006**, *54*, 3551–3556. [CrossRef]

18. Wei, K.; Zhang, Z.; Feng, Z. Design of a wideband horizontally polarized omnidirectional printed loop antenna. *IEEE Antennas Wirel. Propag. Lett.* **2012**, *11*, 49–52.
19. Wei, K.; Zhang, Z.; Feng, Z.; Iskander, M.F. An MNG-TL loop antenna array with horizontally polarized omnidirectional patterns. *IEEE Trans. Antennas Propag.* **2012**, *60*, 2702–2710. [CrossRef]
20. Kajfez, D.; Elsherbeni, A.Z.; Demir, V.; Hasse, R. Omnidirectional square loop segmented antenna. *IEEE Antennas Wirel. Propag. Lett.* **2016**, *15*, 846–849. [CrossRef]
21. Quan, X.; Li, R. A broadband dual-polarized omnidirectional antenna for base stations. *IEEE Trans. Antennas Propag.* **2013**, *61*, 943–947. [CrossRef]
22. Fan, Y.; Liu, X.; Liu, B.; Li, R. A broadband dual-polarized omnidirectional antenna based on orthogonal dipoles. *IEEE Antennas Wirel. Propag. Lett.* **2016**, *15*, 1257–1260. [CrossRef]
23. Dai, X.-W.; Wang, Z.-Y.; Liang, C.-H.; Chen, X.; Wang, L.-T. Multiband and dual-polarized omnidirectional antenna for 2G/3G/LTE application. *IEEE Antennas Wirel. Propag. Lett.* **2013**, *12*, 1492–1495. [CrossRef]
24. Yu, Y.; Jolani, F.; Chen, Z. A wideband omnidirectional horizontally polarized antenna for 4G LTE applications. *IEEE Antennas Wirel. Propag. Lett.* **2013**, *12*, 686–689. [CrossRef]
25. Cai, X.; Sarabandi, K. A Compact Broadband Horizontally Polarized Omnidirectional Antenna Using Planar Folded Dipole Elements. *IEEE Trans. Antennas Propag.* **2016**, *64*, 414–422. [CrossRef]
26. LYe, H.; Zhang, Y.; Zhang, X.Y.; Xue, Q. Broadband Horizontally Polarized Omnidirectional Antenna Array for Base-Station Applications. *IEEE Trans. Antennas Propag.* **2019**, *67*, 2792–2797.
27. Sangster, A.J.; Wang, H. Moment method analysis of a horizontally polarised omnidirectional slot array antenna. *IEEE Proc. Microw. Antennas Propag.* **1995**, *142*, 1–6. [CrossRef]
28. Iigusa, K.; Tanaka, M. A horizontally polarized slot-array antenna on a coaxial cylinder. In Proceedings of the 2000 Asia-Pacific Microwave Conference. Proceedings (Cat. No.00TH8522), Sydney, Australia, 3–6 December 2000; pp. 1444–1447.
29. Zhou, B.; Geng, J.; Bai, X.; Duan, L.; Liang, X.; Jin, R. An Omnidirectional Circularly Polarized Slot Array Antenna with High Gain in a Wide Bandwidth. *IEEE Antennas Wirel. Propag. Lett.* **2015**, *14*, 666–669. [CrossRef]
30. Sadiq, M.S.; Ruan, C.; Nawaz, H.; Ullah, S.; He, W. Low Gain Ripple and DC-Grounded Slant-Polarized Formulation with 360° Broadbeam Coverage. *IEEE Access* **2020**, *8*, 224190–224199. [CrossRef]
31. Balanis, C.A. *Antenna Theory: Analysis and Design*; Wiley-Interscience: Hoboken, NJ, USA, 2005.
32. Costa, A.; Goncalves, R.; Pinho, P.; Carvalho, N.B. Design of UAV and ground station antennas for communications link budget improvement. In Proceedings of the 2017 IEEE International Symposium on Antennas and Propagation & USNC/URSI National Radio Science Meeting, San Diego, CA, USA, 9–14 July 2017; pp. 2627–2628.
33. Quan, X.L.; Li, R.-L.; Wang, J.Y.; Cui, Y.H. Development of a Broadband Horizontally Polarized Omnidirectional Planar Antenna and its Array for Base Stations. *Prog. Electromagn. Res.* **2012**, *128*, 441–456. [CrossRef]
34. Daly, R. Development of Omnidirectional Collinear Arrays with Beam Stability for Base Station and Mobile Applications. Master's Thesis, RMIT University, Melbourne, Australia, 2013.
35. Fujita, K. MNL-FDTD/SPICE method for fast analysis of short-gap ESD in complex systems. *IEEE Trans. Electromagn. Compat.* **2016**, *58*, 709–720. [CrossRef]
36. Ohmine, H.; Sunahara, Y.; Sato, S.; Katagi, T.; Wadaka, S. Omnidirectional Slot Antenna. U.S. Patent 5,717,410, 10 February 1998.
37. Pozar, D.M. *Microwave and RF Design of Wireless Systems*, 1st ed.; Wiley Publishing: Hoboken, NJ, USA, 2000.

Article

Investigation on Wireless Link for Medical Telemetry Including Impedance Matching of Implanted Antennas

Ilkyu Kim [1], Sun-Gyu Lee [2], Yong-Hyun Nam [2] and Jeong-Hae Lee [2,*]

[1] C4I Team, Defense Agency Technology and Quality, Jinju 52851, Korea; ilkyukim7@naver.com
[2] Department of Electronics and Electrical Engineering, Hongik University, Seoul 04066, Korea; gyul0206@gmail.com (S.-G.L.); namy129@naver.com (Y.-H.N.)
* Correspondence: jeonglee@hongik.ac.kr

Abstract: The development of biomedical devices benefits patients by offering real-time healthcare. In particular, pacemakers have gained a great deal of attention because they offer opportunities for monitoring the patient's vitals and biological statics in real time. One of the important factors in realizing real-time body-centric sensing is to establish a robust wireless communication link among the medical devices. In this paper, radio transmission and the optimal characteristics for impedance matching the medical telemetry of an implant are investigated. For radio transmission, an integral coupling formula based on 3D vector far-field patterns was firstly applied to compute the antenna coupling between two antennas placed inside and outside of the body. The formula provides the capability for computing the antenna coupling in the near-field and far-field region. In order to include the effects of human implantation, the far-field pattern was characterized taking into account a sphere enclosing an antenna made of human tissue. Furthermore, the characteristics of impedance matching inside the human body were studied by means of inherent wave impedances of electrical and magnetic dipoles. Here, we demonstrate that the implantation of a magnetic dipole is advantageous because it provides similar impedance characteristics to those of the human body.

Keywords: biomedical devices; wireless communication link; near-field region; impedance matching characteristics

Citation: Kim, I.; Lee, S.-G.; Nam, Y.-H.; Lee, J.-H. Investigation on Wireless Link for Medical Telemetry Including Impedance Matching of Implanted Antennas. *Sensors* **2021**, *21*, 1431. https://doi.org/10.3390/s21041431

Academic Editor: Razvan D. Tamas

Received: 31 January 2021
Accepted: 17 February 2021
Published: 18 February 2021

1. Introduction

Recent advancements have been made in fields related to the development of biomedical devices for the diagnosis and treatment of patients. In particular, pacemakers implanted in the chest offer real-time health care for patients who suffer from cardiac disease. Pacemakers provide the important capability of sensing intrinsic cardiac activity and transferring the pacemaker data using a wireless communication link. A class of antennas with different characteristics in terms of sizes and radiation patterns have been studied for viable in-body, on-body and off-body medical telemetries [1,2]. In addition to the antenna designs, studies on radio transmission have been performed with the aim to create reliable communication links between devices outside and inside the human body. Wireless power transfer (WPT) using electromagnetic waves can be classified into the following groups: (1) non-radiative, (2) radiative and non-radiative mid-field, and (3) radiative far-field [3]. In addition, WPT using an acoustic link can be classified as a non-EM (Electromagnetics) solution [4]. Active research has been taking place on the use of inductive coupling for non-radiative wireless links [5–7]. Inductive coupling has been studied with the aim to achieve a higher transfer efficiency by utilizing a hybrid microstrip and coil [5], in addition to resonators inside and outside body [6]. The design guidelines for inductive power transfer have been studied in terms of miniaturization, power consumption, and links of biomedical devices [7]. The antennas and coils used need to be interpreted as a circuit model, which requires an additional step to compute the wireless link. Radiative WPT has garnered a lot of attention due to its robustness against the misalignment of two antennas. Radio transmission of the different in-body, on-body and off-body telemetries has been investigated both by using numerical methods, such as the FDTD method, and via experimentation [8]. RFID (Radio Frequency Identification) technology for biomedical implants has been studied in terms of the wireless transfer of power and data [9]. The challenge is to calculate the wireless communication link with a reduced computational intricacy. In addition, although there have been a wide range of studies on radiative mid-field WPT [10,11], little attention has been directed towards the inclusion of accurate radiation characteristics in the near-field region. The antenna gain typically reduces as one antenna is located in the near-field region of the other antenna, which needs to be considered in the calculation of the wireless link. A simple quadratic correction term [12–14] to the Friis formula has been studied to increase the accuracy of the prediction in the Fresnel region. The transmission integral [15] with the plane–wave scattering matrix (PWSM) has facilitated the development of the near-field antenna measurement technique for various antennas. Based on the transmission integral, an integral form of coupling formula has been developed in order to provide the convenience of using an easily attainable far-field pattern [16]. The formula provides an enormous capacity to compute the near-field or far-field antenna coupling, which is not restricted to the type of antenna or the motion of the antenna [16–20]. Advancements in the formula have been made in terms of its widespread applicability to electromagnetic problems. It has been proven that the employment of a larger solid angle is beneficial for achieving converged results [19]. For microwave applications, it has been used to compute the coupling, including the dielectric structure between two antennas [19], and to provide near-field power densities of the array antenna [20]. The qualitative comparison among different techniques is summarized in Table 1. Another important aspect in realizing a reliable wireless link is the use of an optimal antenna to provide the best impedance matching in the environment of a human implant. In order to provide the best match, antennas must be electromagnetically characterized in the environment, for example, inside and outside the human body. In order to achieve improved matching characteristics, the design of a pyramidal horn antenna, including a composite material which is similar to the human skin, has been optimized [21]. The impedance characteristics of electrical and magnetic dipole antenna have been studied in detail [22,23]. The advantage of the magnetic dipole antenna is that it provides lower impedance characteristics, corresponding to those of human tissues. The use of a magnetic dipole is advantageous in terms of achieving characteristics that best match those inside the human body. In order to establish a stable wireless link, the wireless link needs to be estimated using an efficient method, including the selection of the optimal antenna to provide the best impedance characteristics inside the human body.

Table 1. Comparison among different methods for estimating the wireless link.

Methods	Effective Range	Accuracy	Computational Complexity
Inductive coupling [5–7]	Non-radiative reactive near-field	High	High
Radiative Mid-field [10,11]	Radiative near-field and far-field region	Low	High
Coupling formula [16–20]	Radiative near-field and far-field region	High	Low

In this paper, radio transmission between antennas that are placed inside and outside the human body is studied in the context of realizing a stable wireless communication link. The wireless link is used to transfer pacemaker data, which operates at 402–405 MHz for a medical implant communication system (MICS). Figure 1 depicts the evaluation scheme for radio transmission between antennas inside and outside the human body. The complete wireless communication link is evaluated in terms of the efficient estimation of the wireless link and the best matching characteristics inside the human body. The key milestones of this work are that: (1) The coupling formula is applied to estimate the wireless link for biomedical applications; (2) for use in biomedical applications, the formula is slightly modified in terms of including characterization of the far-field pattern inside the body and adding a reflection coefficient term between the implanted antenna and the human body; (3) impedance matching in the environment of the human body was studied through representative examples, such as electrical and magnetic dipole antennas. The wireless link was computed using the coupling formula and the result was compared with full-wave simulation, FEKO, and measurements. The wireless link was measured using phantom fluid that provides relative permittivity $\varepsilon_r = 46.4$ and conductivity $\sigma = 0.67$, which is similar to the characteristics of the human skin. The measured result agrees well with the computed result and FEKO simulations.

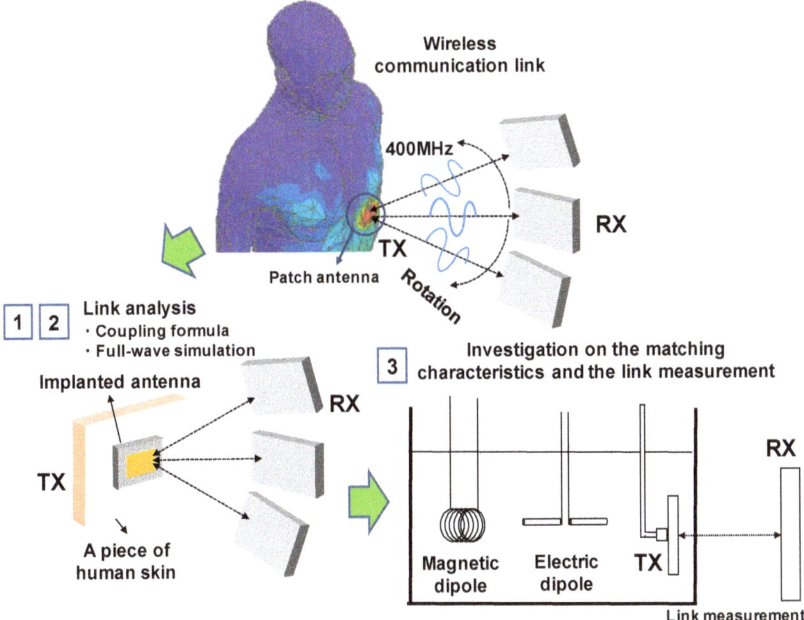

Figure 1. Overview of this research study for the evaluation of the communication link between two antennas.

2. Computation of the Antenna Coupling in the Near-Field and Far-Field Region

The power transmission is computed using the integral coupling formula based on the 3D vector far-field pattern. The formula computes the antenna coupling in the near-field and far-field region. The formula was modified slightly to provide an accurate link analysis for biomedical applications. The far-field pattern of the antenna placed inside human body was characterized using a piece of human skin, with the associated antenna. Furthermore, the impedance mismatch of the antenna in the human body environment was added in order to include the reflection coefficient between the antenna and the human body. The formula was used to evaluate the link between two biomedical antennas placed in various configurations.

2.1. Revisiting the Integral Coupling Formula

The transmission integral with the plane–wave scattering matrix (PWSM) theory has expanded its applicability to arbitrarily oriented antennas [15]. The coupling formula in [16,18], based on a normalized vector far-field pattern, was derived from the transmission integral. The formula utilizes the complex vector far-field pattern, which is easily attainable through numerical methods or measurements. In this section, the essentials of the coupling formula presented in [16,18] are revisited. The geometries of the two arbitrarily positioned antennas are shown in Figure 2. The coupling quotient can be determined based on the relationship between the amplitude of the input wave of the transmitting antenna, a_0, and the output wave of the receiving antenna, b'_0

$$\frac{b'_0}{a_0}(\vec{R}) = -\frac{C}{4\pi k} \iint_{\sqrt{(k_x^2+k_y^2)} < k} \frac{\vec{f}_{TX}(\vec{k}) \cdot \vec{g}_{RX}(-\vec{k})}{k_z} e^{i\vec{k}\cdot\vec{R}} \, dk_x \, dk_y \tag{1}$$

where $\vec{k} = k_x\hat{x} + k_y\hat{y} + k_z\hat{z}$ is the wave vector in free space and $\vec{f}_{TX}(\vec{k})$ and $\vec{g}_{RX}(-\vec{k})$ are the complex vector far-field pattern of the transmitting and receiving antenna, respectively. A double integral of the scalar product between the two vector far-field patterns was used to calculate the coupling quotient. The relationship between the coupling quotient and the S_{21} can be defined as $S_{21} = |b'_0/a_0|^2$.

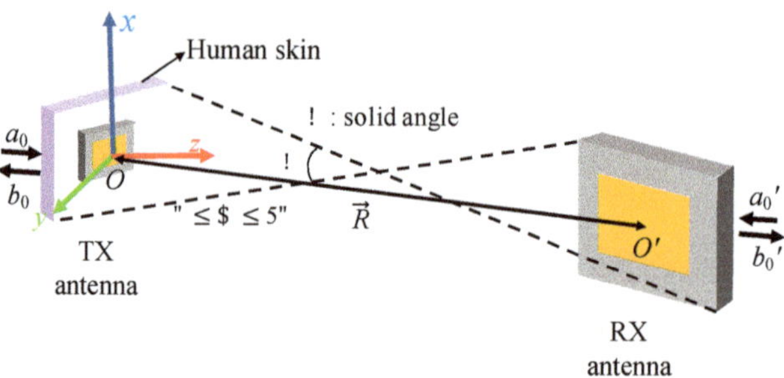

Figure 2. Geometries of the mutual coupling between the two antennas used for biomedical applications.

Reductions in the integration range, such as $\sqrt{(k_x^2 + k_y^2)} < k$, only provides propagating waves, while it neglects evanescent waves. It is worth noting that Formula (1) depends on the $e^{-i\omega t}$ time convention and neglects multiple reflections. The solid angle α can be defined as the angle subtended by two spheres enclosing the transmitting and receiving antennas. The first step is the acquisition of the 3D vector far-field patterns of the two antennas. The 3D vector far-field pattern can be acquired through numerical methods, full-wave simulations, and measurements. One must obtain the far-field pattern at the phase center of the antenna. The phase center of the antenna is placed at the point where the uniform phase response of the far-field pattern is observed. This procedure will be helpful in terms of achieving sufficient convergence of the calculation. In this study, the far-field pattern characterized in the human body environment was employed. It was demonstrated in [8], that the use of a small piece of human skin creates similar EM characteristics to those of the entire human body. Therefore, the antenna was implanted inside a small piece of human skin to characterize the far-field pattern. The next step is to perform the coordinate transformation in order to create the geometries of the two antennas. Eulerian angles were used to transform the coordinate system of each antenna into the global coordinate system. The relative orientation and separation distance between two antennas can be created using coordinate transformation. An interpolation process was used to evaluate the far-field pattern at the sampling points on the x–y plane.

The impedance matching of the antenna placed inside and outside human body represents an important consideration in realizing a stable communication link. Therefore, it is advantageous to contain the reflection coefficient between an antenna and the human body. Figure 3 shows the two-port network for the impedance mismatch term C. It was assumed that the transmitting antenna is placed inside the human skin, as discussed in Figure 1. The complete impedance matching term, constant C, can be expressed as

$$C = \frac{Z_{FL,RX}}{Z_0} \frac{1}{(1 - \Gamma_{0,\,TX}\Gamma_{0,\,Tissue})} \frac{1}{(1 - \Gamma_{0,\,RX})} \tag{2}$$

where $Z_{FL,RX}$ is the impedance of the receiving antenna feedline and Z_0 is the intrinsic impedance in a free space. For the transmitter, $\Gamma_{0,TX}$, and $\Gamma_{0,Tissue}$ indicate the reflection coefficient of the transmitting antenna and human tissue, respectively. For the receiver, $\Gamma_{0,RX}$, represents the reflection coefficient of the receiving antenna. It was assumed that the transmitting and the receiving antenna are perfectly matched to the source and the load, respectively. The detailed procedure to derive the impedance matching term C is provided in Appendix A.

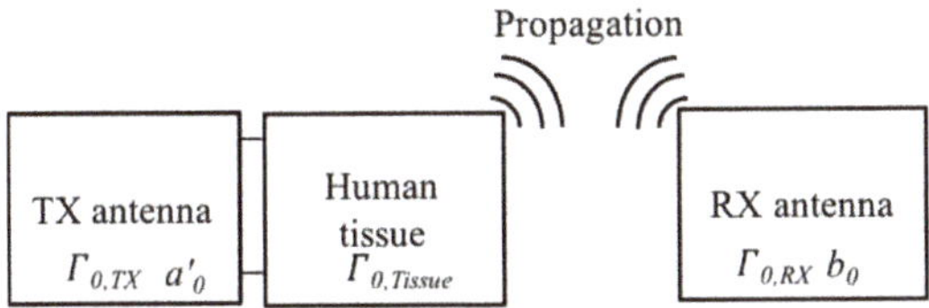

Figure 3. Description of the two-port network for the wireless communication link between the transmitting implanted antenna placed inside the human skin and the receiving antenna outside the human tissue.

The computer program in [18] was developed in order to compute the antenna coupling between two antennas, which allows for flexibility in terms of the relative axis, orientation, and arbitrary antenna movement. Furthermore, the sampling frequency has been introduced to adjust the size of the integration for sufficient convergence. The sampling frequency presented in [18] is

$$f_s = 2\kappa \times (D_{TX} + D_{RX}) \tag{3}$$

where D_{Tx} and D_{Rx} are the diameters of the transmitting and the receiving antennas, respectively, and κ is termed the oversampling ratio. One can use the constant κ in order to change the sampling frequency. Spectrum integration is confined to the solid angle α subtended by the diameter D_{Tx} and D_{Rx}. This confinement is an important requirement in reducing the necessary computational resources. The recent advances in computer capabilities have allowed for the testing of the converged results using different solid angles. It was demonstrated in [19] that the utilization of a larger solid angle α is beneficial for offering a degree of improvement in terms of the converged results. The upper and lower bounds for the effective separation distance can be defined as

$$\frac{D_{TX} + D_{RX}}{2} < R < \frac{(D_{TX} + D_{RX})^2}{\lambda} \tag{4}$$

The integral form of coupling formula was implemented using the double summation at the sampling points (k_x-k_y). The summation form of the formula makes it possible to implement through the computer program presented in [18]. The summation form of the formula is presented as

$$\frac{b'_0}{a_0}(\vec{R}) = -\frac{C}{k}(\Delta k)^2 \times \sum_m \sum_n \frac{\vec{f}_{TX}(k_x^{mn}, k_y^{mn}) \cdot \vec{g}_{RX}(k_x^{mn}, k_y^{mn})}{k_z^{mn}} \times e^{i\vec{k}^{mn} \cdot \vec{R}} \tag{5}$$

where $k = \hat{x}\, k_x^{mn} + \hat{y}\, k_y^{mn} + \hat{z}\, k_z^{mn}$ and $\Delta k = \frac{2\pi}{f_s} = \frac{\pi}{\kappa(D_{TX}+D_{RX})}$.

The index m, n for the double summation can be defined as

$$1 \leq m,\ n \leq \frac{2(D_{TX} + D_{RX})^2}{\lambda R} \tag{6}$$

Note that $\kappa = 4$ is used for the index m, n. The upper bound of the effective range is restricted to compute the antenna coupling in the far-field region. The complete form of the Friis formula, including the reflection coefficient between the implanted antenna and the human skin, can be defined as

$$\left|\frac{b'_0}{a_0}\right|^2 = \left(\frac{\lambda}{4\pi R}\right)^2 G_{TX}(\theta,\phi)G_{RX}(\theta,\phi)\times|\hat{\rho}_{TX}\cdot\hat{\rho}_{RX}|^2\left(1-|\Gamma_{0,\ TX}\Gamma_{0,Tissue}|^2\right)\left(1-|\Gamma_{0,\ RX}|^2\right) \tag{7}$$

where $G_{TX}(\theta,\phi)$ and $G_{RX}(\theta,\phi)$ represent the far-field gain of the transmitting antenna and the receiving antenna, respectively, and $\hat{\rho}_{TX}$ and $\hat{\rho}_{RX}$ are the polarization vectors of the transmitting antenna and the receiving antenna, respectively. It was assumed that both the transmitting and receiving antennas are connected to the identical feed line, and that there was no reflection from the source and the load.

2.2. *Electromagnetic Characterization of Implanted Antennas*

In order to establish a reliable communication link, implanted antennas need to be investigated in terms of impedance matching the characteristics present inside the human body. The characteristics of the wave impedance vary according to the different kinds of antennas. The human body typically has a large dielectric constant, causing low impedance characteristics. Therefore, the use of an antenna with lower wave impedance is beneficial in order to provide the best matching characteristics.

The wave impedance of the electric dipole $Z_E(r)$ and the one of the magnetic dipole $Z_H(r)$ can be derived from the ratio between the electric and magnetic field of each antenna [22,23], given by

$$Z_E(r) = \eta_0\left|\frac{\left[1+\frac{1}{j\beta r}+\frac{1}{(j\beta r)^2}\right]}{\left(1+\frac{1}{j\beta r}\right)}\right| \tag{8}$$

$$Z_H(r) = \eta_0\left|\frac{\left(1+\frac{1}{j\beta r}\right)}{\left[1+\frac{1}{j\beta r}+\frac{1}{(j\beta r)^2}\right]}\right| \tag{9}$$

where η_0 and β represent the characteristic impedance of the free space and the propagation constant of the free space, respectively. Figure 4 shows the comparison of the wave impedance of the electric dipole and the magnetic dipole. As previously discussed, the magnetic dipole provides low impedance characteristics, while the electric dipole shows an opposite trend in terms of impedance characteristics. The impedance inside the human body can be derived from taking into account the different kinds of electrical properties of the human body using the relationship

$$Z = \sqrt{\frac{\mu}{\varepsilon}} = \sqrt{\frac{1}{\varepsilon_r}}\sqrt{\frac{\mu_0}{\varepsilon_0}} = \sqrt{\frac{1}{\varepsilon_r}}\eta_0 \tag{10}$$

where μ_0 and ε_0 are the free space permeability and permittivity, respectively, and ε_r is the relative permittivity which can be determined by the electrical properties of the human tissue. The reflection coefficient between antenna and human tissue can be derived using

$$\Gamma_{0,\ Tissue} = \frac{Z_A - Z_{Tissue}}{Z_A + Z_{Tissue}} \tag{11}$$

where Z_A and Z_{Tissue} are the impedance of the antenna and the human tissue, respectively.

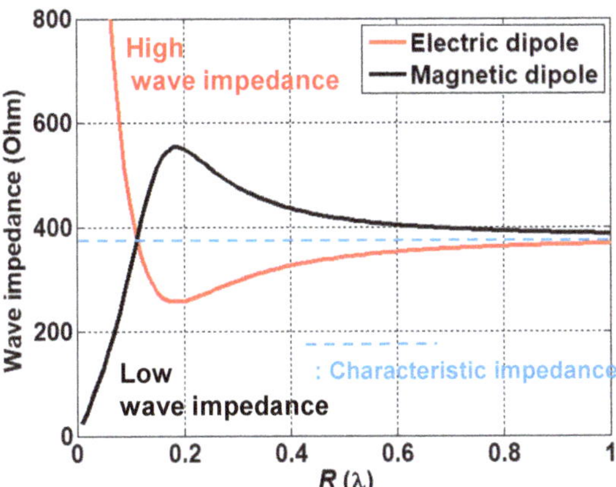

Figure 4. Wave impedance of the electric dipole and the magnetic dipole antenna.

3. Results

In this section, the complete wireless communication link is investigated including calculation of the wireless link and analysis of the matching characteristics inside and outside the human body. It was assumed that the TX antenna is located inside the skin, while the RX antenna is placed outside the human body to establish communication links for the pacemaker system. All evaluations were performed at the center frequency of 400 MHz.

3.1. Calculation of the Link between Two Antennas

The coupling formula was used to evaluate the link between the two antennas. This section contains the evaluations for: (1) the free space link, and (2) the link between the two antennas placed inside and outside the human body. The computed results are compared with the results obtained using a full-wave simulation FEKO. The calculated results show an excellent level of agreement with the results from the full-wave simulation FEKO.

3.1.1. Antenna Design Used for the Link Analysis

The microstrip patch antennas are designed to operate inside and outside the human body. The size of the patch antenna outside the human body was ($L_1 \times W_1$) = (40 cm × 40 cm), while the patch antenna inside the human body was smaller at ($L_2 \times W_2$) = (20 cm × 20 cm) in order to provide the best matching characteristics inside the human body. The patch antenna outside the human body was designed on a 3.2 mm thick substrate with ε_r = 2.2 tan δ (loss tangent) = 0.0009. For the patch antenna inside the human body, the thickness of the substrate was set as 6.4 mm and the microstrip antenna was printed in the middle of the substrate with ε_r = 10.2 and tan δ = 0.0025. The size of the skin tissue was set as ($L_3 \times W_3$) = (25 cm × 25 cm), with a thickness of 4 mm above the patch antenna. The characteristics of impedance matching for the patch antennas are shown in Figure 5a. The patch antennas placed inside and outside the human body provide reasonably good matching characteristics of less than −10 dB around 400 MHz, however, the characteristics of the implanted patch deteriorate when it operates in air. Figure 5b shows a comparison of the radiation patterns in diverse environments. The patch antenna placed outside the human body has a maximum gain of 6.8 dB, while the other one placed inside the human body has a maximum gain of −7.65 dB. The patch antennas were used to evaluate the free-space link and the link between the implanted antenna and the outside receiving antenna.

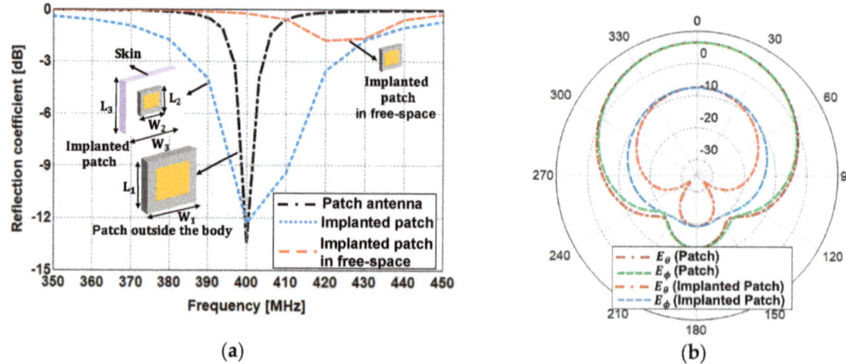

(a) (b)

Figure 5. (**a**) Impedance matching characteristics of the patch antennas used for the link analysis and (**b**) the radiation patterns of the patch antennas at 400 MHz.

3.1.2. Free-Space Link Analysis

Prior to estimating the link, including the implanted antenna, the link was evaluated using the two microstrip patch antennas placed in a free space. The coupling formula was used to compute the coupling versus the separation distance and the coupling versus the transverse displacement [16–18]. In this section, the coupling formula is applied to both of the antenna displacements.

For the coupling versus separation distance, the antenna coupling was evaluated by moving the RX patch antenna in terms of the R direction. For the coupling versus transverse displacement, the receiving antenna was located at different transverse displacements, and the antenna coupling was evaluated. The antenna coupling for the first scheme is shown in Figure 6a. It is shown that the coupling formula provides the coupling level which is roughly 1.5 dB lower than the one of the Friis formula in the closest distance. This is attributed to the gain reduction effect in the near-field region. The computed results show a good agreement with the full-wave simulation FEKO with a deviation less than 0.6 dB. The second scheme was evaluated as shown in Figure 6b. The coupling was evaluated for the transverse displacements of 0.5 and 0.75 λ. There is a significant difference in the coupling level for different transverse displacements in close proximity, while all curves converge to a coupling level in the far-field region. This is because the offset geometry in close proximity is more influential in determining mutual coupling. There was a 1 dB deviation between the computed and the simulated results.

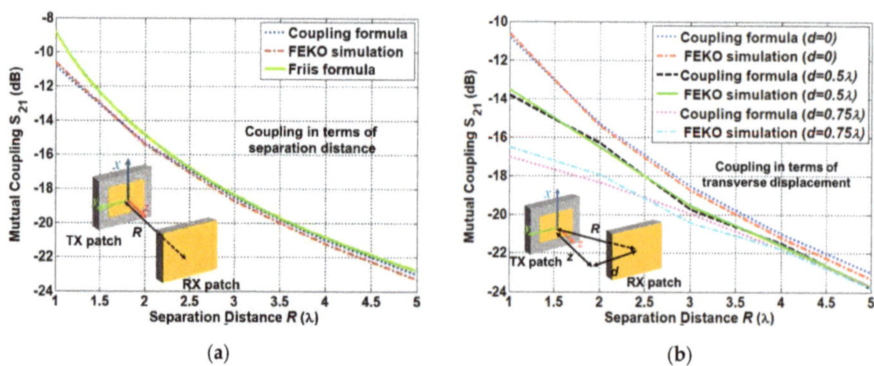

(a) (b)

Figure 6. Free-space link analysis for (**a**) the coupling in terms of separation distance normalized with the wavelength at 400 MHz and (**b**) the coupling in terms of transverse displacement normalized with the wavelength at 400 MHz.

3.1.3. Link between the Two Antennas Inside and Outside the Human Body

The link between two patch antennas placed inside and outside the human body was evaluated. The patch antenna embedded in a piece of human tissue was used to represent the implanted antenna inside the human skin. It was reported in [8], that the simplified model provides similar characteristics to those of the antenna implanted inside the large human body model. This will help reduce the required computational resources. The attenuation due to the propagation inside the human body is simply characterized through utilizing the difference between the maximum antenna gain of the implanted transmitting patch and that of the receiving patch. The antenna coupling was evaluated for the first scheme with the inclusion of a rotational one, as shown in Figure 7a. For an accurate estimation, the rotation angles are restricted to the angles which are smaller than $\theta_t = 30°$ [17]. Therefore, the coupling curves were obtained at the rotation angles $\theta_t = 0°$ and $\theta_t = 30°$. It can be observed that the deviation between the computed and simulated results is less than 0.8 dB for both scenarios. The difference between the two coupling curves agrees well with the one of the radiation pattern at the different rotation angles. Coupling curves for the second scheme are shown in Figure 7b. The deviation between the computed and simulated results is 1.2 dB, which is slightly higher than that of the free-space link.

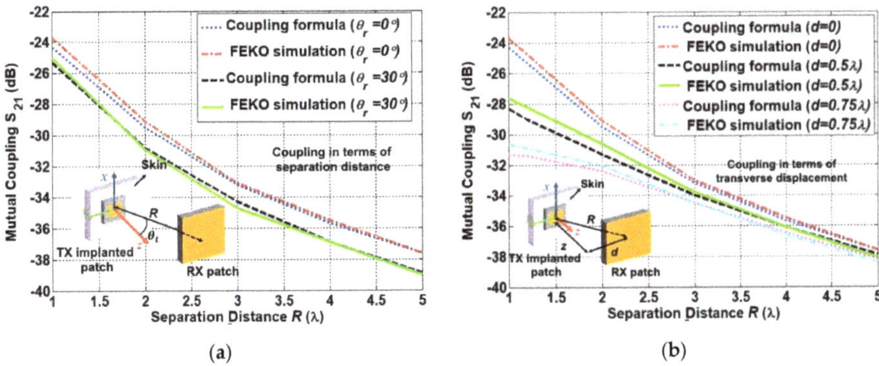

Figure 7. Link between the two antennas inside and outside the human body for (**a**) the coupling in terms of separation distance, normalized with the wavelength at 400 MHz and (**b**) the coupling in terms of transverse displacement, normalized with the wavelength at 400 MHz.

3.2. Matching Characteristics Inside and Outside the Human Body

The impedance matching inside the human body was investigated based on the characteristics of the implanted antennas. Owing to the electrical properties of the human body, the antenna design needs to compensate for the differences between free space and the human body. To investigate the characteristics of the antenna inside the body, we evaluated the design of a helix antenna on PEC (Perfect Electric Conductor) ground and a dipole antenna mounted on an EBG (Electromagnetic Band Gap) structure, which are representative examples of magnetic and electric dipole antennas, respectively. In particular, a small helix antenna is favorable—due to its miniaturized design—for implantation inside the human body. The wave impedance of the representative antennas was obtained using a full-wave simulation FEKO. The simulated results were compared with the theoretical results discussed in Section 2.2. As expected, the curve of the helix antenna resembles that of the magnetic dipole, while the curve of the dipole antenna is similar to that of the electric dipole. In particular, the helix antenna provides low impedance characteristics in close proximity. This corresponds to the characteristics of the magnetic dipole, and this antenna will possess optimal values that are analogous to the human body. Using Equation (10), the impedances of different parts of the body were calculated. The computed results are provided in Table 2. It can be seen that the impedance of the human body is lower than the characteristic impedance of the free space. Based on the impedance of the skin and the wave impedance of the antenna, the matching characteristics were acquired from Equation (11). Figure 8 shows the reflection coefficient in terms of the different wave impedance of the antenna. The best matching characteristics are obtained when the impedance of the antenna is similar to that of the tissue. The matching characteristics were evaluated through using representative antennas, such as magnetic dipole and electric dipole antennas. A small helical antenna on the PEC was selected as an example of the magnetic dipole antenna, while a dipole antenna on the EBG structure was chosen as an example of the electric dipole antenna. The configuration of the representative antennas is shown in Figure 9. For the helical antenna, the radius and the height were set as 2.1 and 9 mm, and the height from the PEC to the center of the helical antenna (H_1) was designed as 3.4 mm. The size of the PEC was set as ($L_4 \times W_4$) = (5.4 cm × 5.4 cm). The dipole-EBG was designed and scaled based on a previous study [24]. One difference is that the miniaturized 6 × 6 EBG structure was used in this study. The size of the EBG structure was set as ($L_5 \times W_5$) = (9.7 cm × 9.7 cm).

Table 2. Impedances of the different parts of the human body.

Tissue	Relative Permittivity (ε_r)	Impedance (Z_{Tissue})
Skin	46.7	55.2 Ω
Fat	11.6	110.7 Ω
Skull	17.8	89.4 Ω
Brain	49.7	53.5 Ω

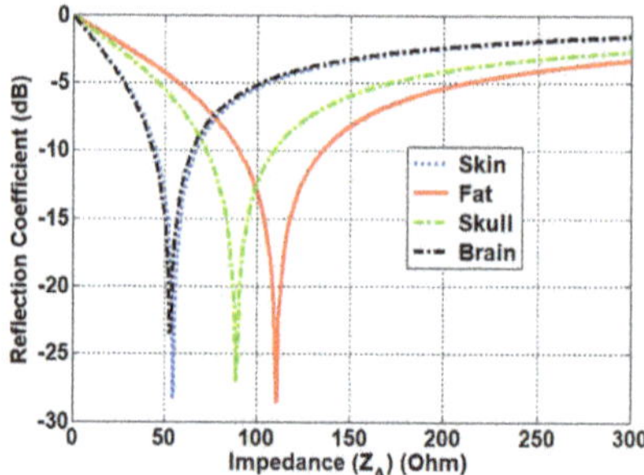

Figure 8. Matching characteristics of an antenna inside the different parts of the human body.

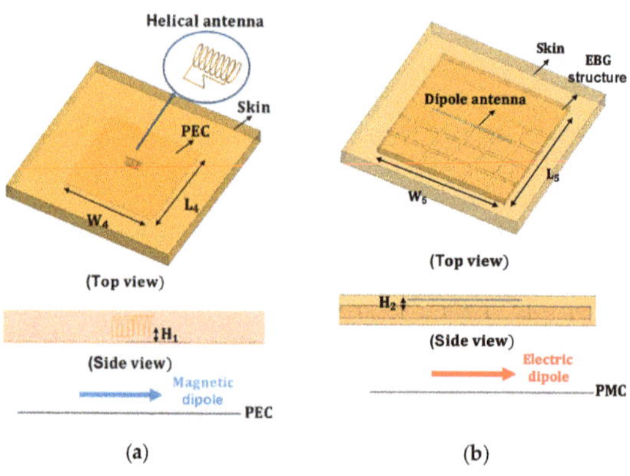

Figure 9. Configuration of the representative antennas: (**a**) helical antenna mounted on PEC and (**b**) dipole antenna mounted on EBG structure.

The height from the EBG structure to dipole antenna (H_2) was set as 2.2 mm. The helical antenna on the PEC corresponds to the magnetic dipole on the PEC, while the dipole antenna on the EBG structure represents the electric dipole on the PMC. The wave impedance of the two antennas was investigated in order to evaluate the matching characteristics inside the human body. Figure 10 shows the simulated results of the wave impedance and the calculated wave impedance of the electric and magnetic dipole antenna. It can be seen that the wave impedance of the helical antenna is similar to that of the magnetic dipole antenna, while the one of the dipole antenna resembles the trend of the electric dipole antenna. In particular, within close proximity, the magnetic dipole possesses low impedance, and the electric dipole has high impedance, when compared to the characteristics of impedance in air. The matching characteristics of the two kinds of antennas were investigated by placing the antennas inside the human body (the skin tissue). Figure 11 shows the simulated impedance characteristics inside and outside the skin tissue. As predicted, the helical antenna provides the best matching characteristics inside the skin tissue, while it shows the deteriorated one outside the skin tissue. In contrast, the electric dipole exhibits slightly degraded matching characteristics inside the skin tissue when compared to those outside the body. It was demonstrated that the magnetic dipole type antenna is advantageous for providing the best matching characteristics inside the human body.

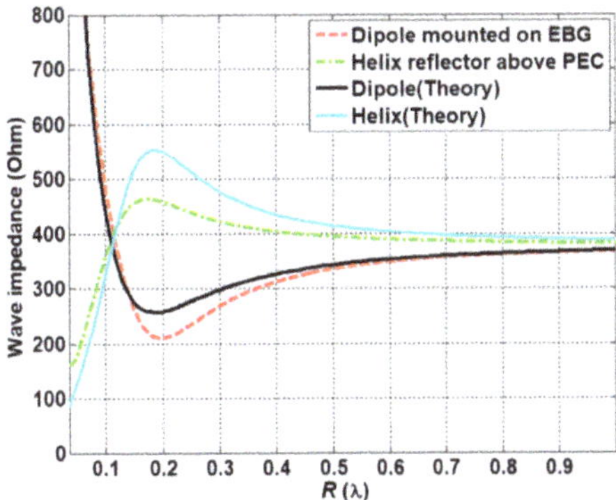

Figure 10. Wave impedance of the dipole on EBG and helix antenna on PEC, and a comparison to the calculated wave impedances of the dipole and the helix antenna.

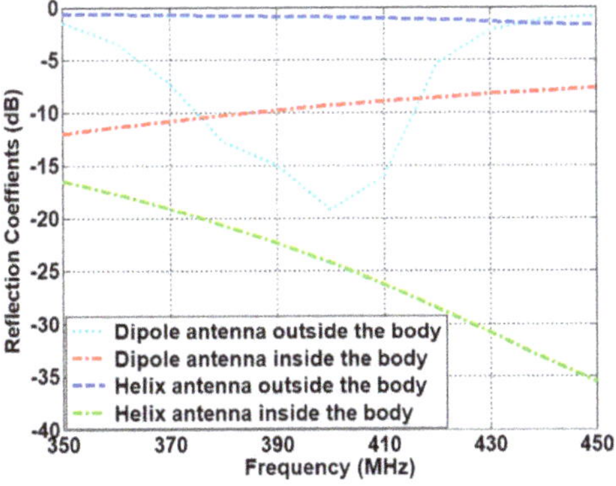

Figure 11. Simulated impedance matching characteristics inside and outside the body (the skin tissue).

4. Measurements

Experiments were conducted to verify the calculated and simulated results of the wireless link. After the experiment to verify the impedance matching performance of the patch antenna inside and outside the human body, the coupling between the antennas was measured by changing the separation distance. The MS46522B model vector network analyzer(VNA) from the company, Anritsu, was used for the experiment. Figure 12a shows the patch antenna outside the human body, fabricated for operating at a frequency of 400 MHz. The patch antenna was fabricated on a Taconic TLY-5 substrate (ε_r = 2.2, tanδ = 0.0009) with a thickness of 3.2 mm, and the size of the ground and the conducting patch were 40 cm × 40 cm and 24.82 cm × 24.82 cm, respectively. The reader patch was fed from the SubMiniature version A(SMA) connector from the company, WithWave, and the feed position was located 4 cm from the patch center to improve the impedance matching. The fabricated implant patch antenna operates with the performance of $Z_{11} = 55.7 - j7.07\ \Omega$ and $S_{11} = -22.3$ dB at 401 MHz, which is the result of a 1 MHz up shift compared to the center frequency of the simulated result. Nevertheless, the patch antenna outside the human body operates with excellent performance with $Z_{11} = 53.5 + j28.7\ \Omega$ and $S_{11} = -12$ dB at 400 MHz, and the impedance matching characteristic matches well with the simulated result, as shown in Figure 12b,c.

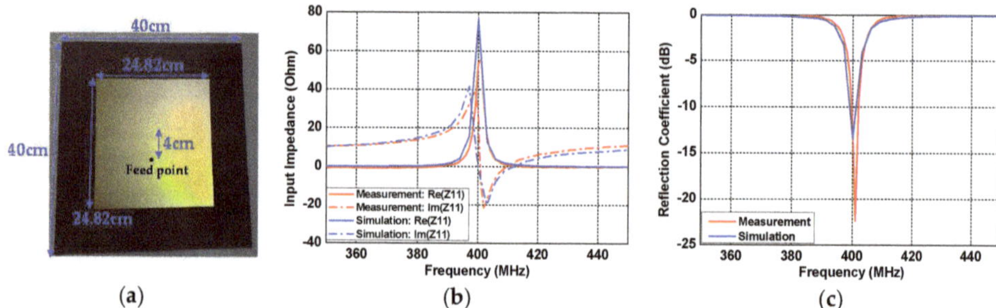

Figure 12. Patch antenna outside the human body: (**a**) fabricated structure, (**b**) input impedance Z_{11}, and (**c**) reflection coefficient S_{11}.

The measurement of the implant patch antenna was conducted both in skin tissue liquid (corresponding to inside the human body) and in free-space (corresponding to outside the human body). For the human body, 10 L of Skin Tissue Simulating Liquid (SKIN350-500V2) from Schmid & Partner Engineering AG company (Zürich, Switzerland) was used as shown in Figure 13a. The relative permittivity and conductivity values are ε_r = 46.4 and σ = 0.67 at room temperature (22 °C) and a frequency of 400 MHz, which are specified in the data sheet. Figure 13b shows the 10 L acrylic tank used to contain the skin tissue liquid. The size of the acrylic tank was made to be 31.6 cm × 31.6 cm × 10 cm considering the size of the implant patch antenna, and the wall thickness was set to 8 mm, considering the relative density of the skin tissue liquid (1.2–1.4 kg/L). It is worth noting that the 8mm of the wall thickness corresponds to about 0.01 λ, so its effect on the radiation performance of the implant patch antenna is negligible. The antenna was assumed to be located 4 mm from the surface of the acrylic tank, and M5 size polycarbonate (PC) screw hole structures were added to stably mount the implant patch.

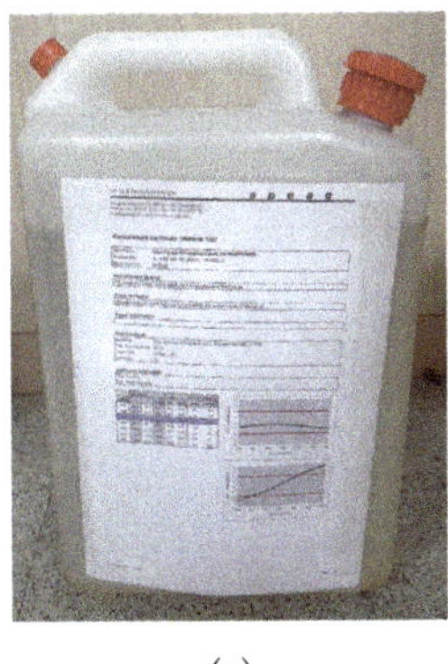

(a)

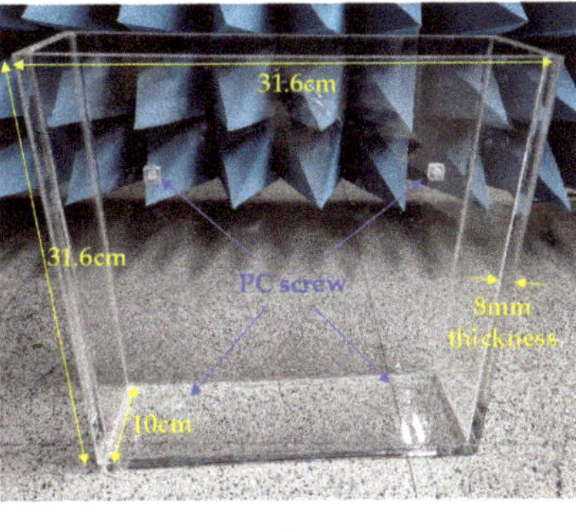

(b)

Figure 13. Implementation of the human body (**a**) skin tissue liquid and (**b**) acrylic tank.

The patch antenna inside the human body consisted of a dielectric substrate (metal free) on the upper layer and a conducting patch with a grounded dielectric slab on the lower layer. Figure 14a shows the lower layer of the patch antenna fabricated for an operating frequency of 400 MHz inside the skin tissue. The patch antenna was designed on Taconic RF-10 substrate (ε_r = 10.2, tanδ = 0.0025) with a thickness of 3.2 mm, and the sizes of the ground and the conducting patch were 20 cm × 20 cm and 10.8 cm × 10.8 cm, respectively. The feed position of the implant patch was located 5.2 cm from the patch center to improve the impedance matching. Figure 14b shows the structure in which the upper layer and the lower layer were assembled. Both layers were assembled with M5 size PC screws, and were mounted through screw holes in the acrylic tank. To prevent the occurrence of air gaps and leakage of skin tissue, waterproof tape was firmly attached to the four corners of the assembed structure. Finally, the assembled structure was fixed by screwing into an acrylic tank containing skin tissue liquid, as shown in Figure 14c. Figure 14d–g shows the measured results of the fabricated implant patch antenna. Similar to the simulation, the implant patch antenna operated at 400 MHz in the skin tissue, and operated near 420 MHz in the air, due to the decrease in relative permittivity. The measured input impedance and reflection coefficients were $Z_{11} = 44.4 + j11.2\ \Omega$ and $S_{11} = -17.5$ dB at a 400 MHz frequency in the skin tissue, as shown in Figure 14d and e. The frequency downshift occurred in the measured results due to a small air gap by soldering the SMA connector.

The measurements were conducted to verify the coupling formula using the fabricated patch antennas. Figure 15a shows the measurement setup of the coupling between one patch antenna inside the phantom fluid tank and the other one in air. The two antennas were set to face each other in the broadside direction, and the coupling was measured from S_{21} of the VNA as a separation distance R changes from 0.5–5λ (correspond to 37.5–375 cm). The measured coupling S_{21} shows a similar tendency to the results calculated from the coupling formula and full-wave simulation, as shown in Figure 15b. The ground effect caused by the floor generated at the far separation distance was minimized by the installed microwave absorber. However, at a near separation distance, a measurement error of 0.4 dB occurred due to the reflected wave from the table used in the measurement.

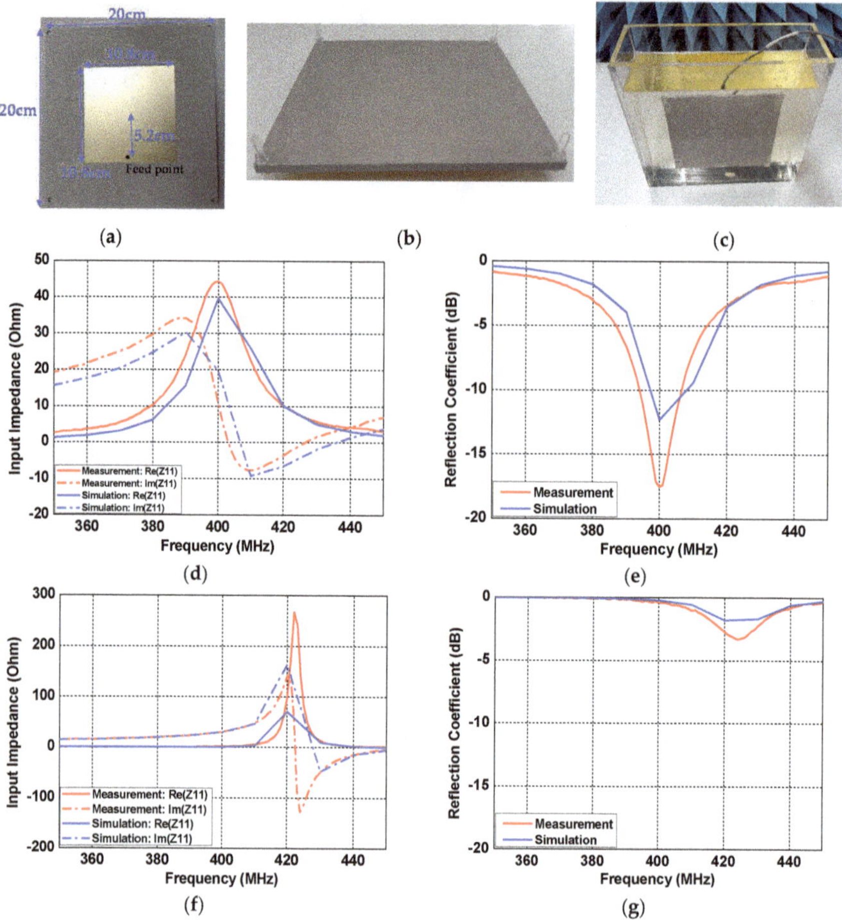

Figure 14. Patch antenna inside the human body: (**a**) lower layer of fabricated structure, (**b**) assembly of the fabricated structure, (**c**) antenna measurement setup in skin tissue liquid, (**d**) input impedance Z_{11} in skin tissue liquid, (**e**) reflection coefficient S_{11} in skin tissue liquid, (**f**) input impedance Z_{11} in free space, and (**g**) reflection coefficient S_{11} in free space.

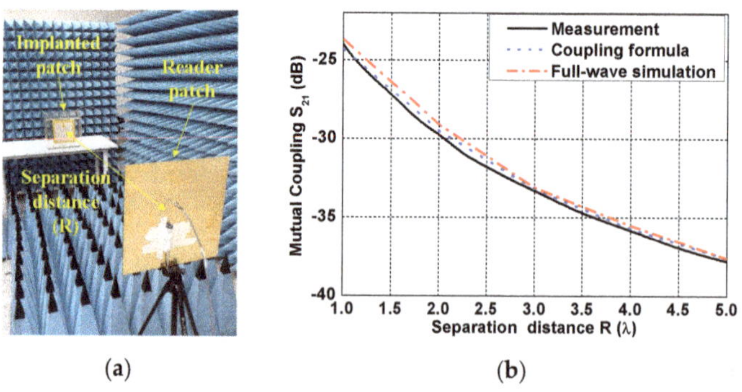

Figure 15. Measurement of coupling between the reader patch and the implant patch antennas (**a**) measurement setup and (**b**) measured result.

5. Discussion

The coupling formula is advantageous since it enables us to compute the near-field antenna coupling based on the far-field radiation pattern. The coupling formula was applied to compute the wireless link between two antennas inside and outside of the human body. The patch antenna inside the small part of the human tissue was used to characterize the far-field pattern of the implanted antenna. The computed results obtained from the coupling formula show good agreement with the full-wave simulation FEKO and measurements. The deviation between the computed results and full-wave simulation was less than 0.6–1.2 dB for both cases versus separation distance and transverse displacement. The matching characteristics inside the human body was investigated in terms of the electric and magnetic dipole antennas. The representative examples of the electric and magnetic dipole antennas were selected as the dipole antenna on the EBG structure and the helical antenna on the metal ground plane, respectively. It was found that the magnetic dipole provides low impedance characteristics which are similar to those of the human body, which is advantageous in terms of providing optimal impedance matching inside the human body. The indoor measurement was performed using one patch antenna inside the phantom fluid and the other one in air. The measured results show a good level of agreement with the simulated results in terms of matching characteristics and link performance. This study provides an important guideline for the creation of reliable wireless links based on an accurate numerical method and antenna design in terms of matching characteristics.

Author Contributions: Conceptualization, J.-H.L. and I.K.; methodology, I.K.; software, I.K.; validation, S.-G.L. and Y.-H.N.; formal analysis, I.K.; investigation, I.K.; resources, I.K.; data curation, I.K.; writing—original draft preparation, I.K. and S.-G.L.; writing—review and editing, S.-G.L. and Y.-H.N.; visualization, I.K.; supervision, J.-H.L.; project administration, J.-H.L.; funding acquisition, J.-H.L. All authors have read and agreed to the published version of the manuscript.

Funding: The research of three of the authors, S.-G.L., Y.-H.N. and J.-H.L., were supported by the Basic Science Research Program through the National Research Foundation of Korea (NRF), funded by the Ministry of Education (No. 2015R1A6A1A03031833).

Data Availability Statement: The data presented in this study are available on request from the corresponding author.

Acknowledgments: The authors would like to thank to Ahmed Akgiray for his initial support in the development of the coupling program.

Conflicts of Interest: The authors declare no conflict of interest.

Appendix A

The impedance mismatch of an antenna inside the human body was proposed in order to estimate an accurate coupling quotient. In this Appendix, the derivation of the antenna mismatch term is presented in detail. Figure A1 shows the two-port network, linked between the transmitting antenna and the receiving antenna. Except for the propagation part, the coupling quotient between the transmitting antenna and the receiving antenna is derived, which is identical to the impedance mismatching term. One can start with the mismatch term of transmitting antenna. The mismatch term between the TX antenna and the human tissue can be defined as

$$a_{TX} = \Gamma_{0,\ Tissue}\, b_{TX} \tag{A1}$$

$$a'_{TX} = \Gamma_{0,\ TX}\, b'_{TX} + a_0 \tag{A2}$$

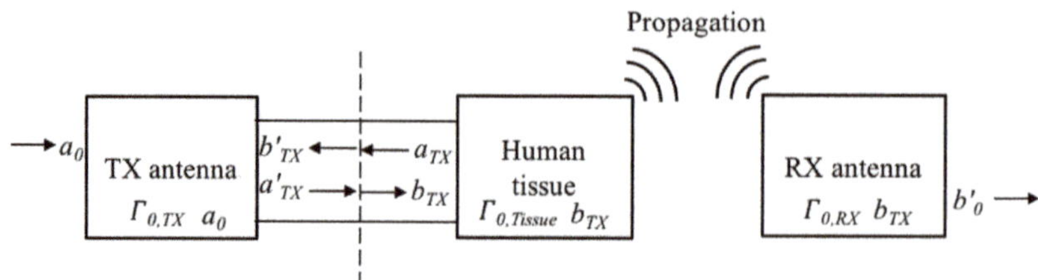

Figure A1. Description of the two-port network used for the derivation of the impedance mismatching term.

Substituting Equation (A1) into Equation (A2) based on the relationship $a'_{TX} = b_{TX}$ and $a_{TX} = b'_{TX}$ will produce

$$b_{TX} = \frac{a_0}{1 - \Gamma_{0,\ TX}\,\Gamma_{0,\ Tissue}} \tag{A3}$$

For the receiving antenna, the amplitude of the received wave can be derived using the following relationship

$$b_{TX} - b_{TX}\Gamma_{0,\ RX} = b'_0 \tag{A4}$$

Substituting Equation (A3) into Equation (A4), the coupling quotient can be derived as

$$\frac{b'_0}{a_0} = \left(\frac{1}{1 - \Gamma_{0,\ Tissue}}\right)\frac{1}{1 - \Gamma_{0,\ RX}} \tag{A5}$$

It was assumed that both the transmitting and receiving antennas are fed by an identical waveguide. Note that multiple reflection is ignored in the derivation. The Equation (A5) was applied to compute the coupling quotient presented in Equation (1).

References

1. Izdebski, P.; Rajagopalan, H.; Rahmat-Samii, Y. Conformal Ingestible Capsule Antenna: A Novel Chandelier Meandered Design. *IEEE Trans. Antennas Propag.* **2009**, *57*, 900–909. [CrossRef]
2. Rao, S.; Llombart, N.; Moradi, E.; Koski, K.; Bjorninen, T.; Sydanheimo, L.; Rabaey, J.; Carmena, J.; Rahmat-Samii, Y.; Ukkonen, L. Miniature implantable and wearable on-body antennas: Towards the new era of wireless body-centric. *IEEE Antennas Propag. Mag.* **2014**, *56*, 271–291. [CrossRef]
3. Khan, S.R.; Pavuluri, S.K.; Cummins, G.; Desmulliez, M.P.Y. Wireless Power Transfer Techniques for Implantable Medical Devices: A Review. *Sensors* **2020**, *20*, 3487. [CrossRef]
4. Basaeri, H.; Christensen, D.B.; Roundy, S. A Review of Acoustic Power Transfer for Bio-medical Implants. *Smart Mater. Struct.* **2016**, *25*, 123001. [CrossRef]
5. Shadid, R.; Haerinia, M.; Roy, S.; Noghanian, S. Hybrid Inductive Power Transfer and Wireless Antenna System for Biomedical Implanted Devices. *Prog. Electromagn. Res. C* **2018**, *88*, 77–88.
6. Monti, G.; Tarricone, L.; Trane, C. Experimental Characterization of a 434 MHz Wireless Energy Link for Medical Applications. *Prog. Electromagn. Res. C* **2012**, *30*, 53–64. [CrossRef]
7. Yakovlev, A.; Kim, S.; Poon, A. Implantable Biomedical Devices: Wireless Powering and Communication. *IEEE Commun. Mag.* **2012**, *50*, 152–159. [CrossRef]
8. Kim, J.; Rahmat-Samii, Y. Implanted Antennas inside a Human Body: Simulations, Designs and Characterizations. *IEEE Trans. Microw. Theory Tech.* **2004**, *52*, 1934–1943. [CrossRef]
9. Song, L.; Rahmat-Samii, Y. An End-to-End Implanted Brain–Machine Interface Antenna System Performance Characterizations and Development. *IEEE Trans. Antennas Propag.* **2017**, *65*, 3399–3408. [CrossRef]
10. Abid, A. Wireless Power Transfer to Millimeter-Sized Gastrointestinal Electronics Validated in a Swine Model. *Sci. Rep.* **2017**, *7*, 46745. [CrossRef] [PubMed]
11. Shaw, T.; Mitra, D. Metasurface-based Radiative Near-field Wireless Power Transfer System for Implantable Medical Devices. *IET Microw. Antennas Propag.* **2019**, *13*, 1974–1982. [CrossRef]
12. Pace, J.R. Asymptotic formulas for coupling between two antennas in the Fresnel region. *IEEE Trans. Antennas Propag.* **1969**, *AP-17*, 285–291. [CrossRef]
13. Kim, I.; Xu, S.; Rahmat-Samii, Y. Generalised Correction to the Friis Formula: Quick Determination of the Coupling in the Fresnel Region. *IET Microw. Antenna Propag.* **2013**, *7*, 1092–1101. [CrossRef]

14. Kim, I.; Lee, S.-G.; Lee, J.-H. Derivation of Far-field Gain Using the Gain Reduction Effect in a Fresnel Regions. *Appl. Comput. Electromagn. Soc. J.* **2020**, *35*, 3399–3408.
15. Kerns, D.M. *Plane-Wave Scattering Matrix Theory for Antenna-Antenna Interaction*; Technical Report; National Bureau of Standards: Boulder, CO, USA, 1981.
16. Yaghjian, A.D. Efficient Computation of Antenna Coupling and Fields within the Near-Field Region. *IEEE Trans. Antennas Propag.* **1982**, *AP-30*, 113–128. [CrossRef]
17. Francis, M.H.; Stubenrauch, C.F. Comparison of Measured and Calculated Antenna Sidelobe Coupling Loss in the Near Field Using Approximate Far-field Data. *IEEE Trans. Antennas Propag.* **1988**, *AP-36*, 438–441. [CrossRef]
18. Akgiray, A.H.; Rahmat-Samii, Y. Mutual Coupling between Two Arbitrarily Oriented and Positioned Antennas in Near- and Far-field Regions. In Proceedings of the 2010 URSI International Symposium on Electromagnetic Theory, Berlin, Germany, 16–19 August 2010.
19. Kim, I.; Lee, C.-H.; Lee, J.-H. On Computing the Mutual Coupling between Two Antennas. *IEEE Trans. Antennas Propag.* **2020**, *68*, 6557–6565. [CrossRef]
20. Kim, I.; Lee, J.-H. Theoretical Calculation of Power Densities in the Near-field Region of Phased Array Antenna. *Microw. Opt. Technol. Lett.* **2021**, *63*, 367–371. [CrossRef]
21. Sarjoghian, S.; Rahimian, A.; Alfadhl, Y.; Saunders, T.G.; Liu, J.; Parini, C.G. Hybrid Development of a Compact Antenna Based on a Novel Skin-Matched Ceramic Composite for Body Fat Measurement. *Electronics* **2020**, *9*, 2139. [CrossRef]
22. Yamaguchi, T.; Okumura, Y.; Amemiya, Y. Wave Impedance in the Near Field around the Fundamental Electromagnetic Radiating Elements. *Electron. Commun. Jpn.* **1991**, *74*, 86–95. [CrossRef]
23. Paul, C.R. *Introduction to Electromagnetic Compatibility*; John Wiley: New York, NY, USA, 1992.
24. Yang, F.; Rahmat-Samii, Y. Reflection Phase Characterizations of the EBG Ground Plane for Low Profile Wire Antenna Applications. *IEEE Trans. Antennas Propag.* **2003**, *51*, 2691–2703. [CrossRef]

Letter

An Efficient Photomixer Based Slot Fed Terahertz Dielectric Resonator Antenna †

Xiaohang Li [1,*], Wenfei Yin [2] and Salam Khamas [1]

1 Department of Electronic and Electrical Engineering, University of Sheffield, Sheffield S10 2TN, UK; s.khamas@sheffield.ac.uk
2 School of Computer Science and Information Engineering, Hefei University of Technology, Hefei 230009, China; wenfeiyin@hfut.edu.cn
* Correspondence: xli54@sheffield.ac.uk
† This paper is an extended version of our paper: Li, X.; Yin, W.; Khamas, S. An Efficient Photomixer Based Slot Fed Terahertz Dielectric Resonator Antenna. In Proceedings of the 2020 International Workshop on Antenna Technology (iWAT), Bucharest, Romania, 25–28 February 2020; pp. 1–4, doi:10.1109/iWAT48004.2020.1570618183.

Abstract: A slot fed terahertz dielectric resonator antenna driven by an optimized photomixer is proposed, and the interaction of the laser and photomixer is studied. It is demonstrated that in a continuous wave terahertz photomixing scheme, the generated THz power is proportional to the 4th power of the surface electric field of photocondutive layer. Consequently, the optical to THz conversion efficiency of the proposed photomixer has an enhancement factor of 487. This is due to the fact that the surface electric field of the proposed photomixer with a 2D-Photonic Crystal (PhC) superstrate has been improved from 2.1 to 9.9 V/m, which represents a substantial improvement. Moreover, the electrically thick Gallium-Arsenide (GaAs) supporting substrate of the device has been truncated to create a dielectric resonator antenna (DRA) that offers a typical radiation efficiency of more than 90%. By employing a traditional coplanar strip (CPS) biasing network, the matching efficiency has been improved to 24.4%. Therefore, the total efficiency has been considerably improved due to the enhancements in the laser-to-THz conversion, as well as radiation and matching efficiencies. Further, the antenna gain has been improved to 9dBi at the presence of GaAs superstrate. Numerical comparisons show that the proposed antenna can achieve a high gain with relatively smaller dimensions compared with traditional THz antenna with Si lens.

Keywords: photomixer; terahertz source; two dimensional photonic crystal; frequency selective surface superstrate; terahertz antenna; dielectric resonator antenna

Citation: Li, X.; Yin, W.; Khamas, S. An Efficient Photomixer Based Slot Fed Terahertz Dielectric Resonator Antenna . *Sensors* **2021**, *21*, 876. https://doi.org/10.3390/s21030876

Academic Editors: Razvan D. Tamas and Shuai Zhang
Received: 24 November 2020
Accepted: 25 January 2021
Published: 28 January 2021

1. Introduction

Terahertz (THz) spectrum extends from 300 GHz to 10 THz, which covers the frequency range between mm-wave and infra-red bands. In addition, the corresponding wavelengths represent the transition between photonics and electronics. Higher attention has been paid to the development of THz technologies, owing to the variety of THz spectrum potential applications including monitoring and spectroscopy in pharmaceutical industry [1,2], imaging [3,4], material spectroscopy [5], security [6,7], biology and medicine [8,9], and high-speed communication [10].

However, the main limit to the development of THz technologies is the lack of available THz emitters and detectors [11]. To date, most of the THz systems that utilize time domain techniques employs bulky and expensive femtosecond lasers. In this case, the optical excitation from the lasers can generate and detect sub-picosecond electrical pulses. On the other hand, frequency domain techniques can achieve higher resolutions and high scanning speed in a low cost and portable devices [12]. So far, it has been demonstrated by numerous that continuous wave THz sources can either be generated directly or converted up and down from microwave and optical frequencies, respectively [11]. Nevertheless, there is a lack of efficient room temperature THz sources without the need of cryogenic cooling system and external magnetic field [13].

One of the most promising continuous wave THz sources that utilizes Optical Heterodyne Generation (OHG) [14] is known as photomixer that is capable of generating tunable and coherent THz signals with low-cost and low power consumption in a compact devices [15–18]. A photomixer consists of two set of metal electrodes, a photoconductive layer and a bulky supporting dielectric substrate. In the photomixer, as two interfering laser beams incident on the non-linear photoconductive medium, electrons are excited from valence band to conduction band, hence, spatiotemporal electrons and holes are generated. Owing to the applied DC biasing voltage, induced photocurrent is driven at the beating frequency of two incident laser beams [19]. Typically, a conventional photomixer can only achieve 0.1% optical to THz power conversion efficiency [20,21]. It has been demonstrated earlier that an enhancement factor of 4 in the optical to THz conversion efficiency can be achieved by using plasmonic material as interdigital electrodes [22]. Additionally, more than 4 times of THz power has been achieved by using optical antenna array of ZnO nanorods [23]. Further, an enhancement factor of 25 in terahertz radiation has been demonstrated by utilizing transparent-conducting oxides nanocylinders between photomixer electrodes [24]. Additionally, double the effective electric energy can be generated by utilizing embedded electrodes [25]. In addition, to date, a highest reported laser to THz conversion efficiency is 7.5%, where replaced the conventional photomixer electrodes have been replaced by three-dimensional plasmonic contact electrodes [26,27]. However, though many researchers have attempted to optimize the optical to THz conversion efficiency, it is still significantly less than tenth of the theoretical maximum of 100% [28].

As mentioned previously, a photomixer is implemented on a photoconductive layer that is supported by a bulky dielectric substrate. The THz antenna's input resistance is expected to be reduced by a factor of $\sqrt{((\varepsilon_r + 1)/2)}$ due to the presence of the bulky dielectric supporting layer that has a dielectric constant of ε_r. On the other hand, the output resistance of a photomixer is in the order of ~10 kΩ [29]. Therefore, the reduced input resistance of the THz antenna leads to a poorer matching efficiency. A full wavelength dipole has been used to drive a Yagi-Uda array to achieve an input resistance of 2.6 kΩ [30]. Moreover, a 3.3 kΩ input resistance has been achieved by implementing an isolating metallic ground plane with a dipole placed on a thin dielectric slab [31].

In addition, it has been reported that THz communications are more likely to be influenced by the atmosphere, especially, the humidity [32]. Therefore, the radiation power enhancement becomes another challenge. In this case, multiple types of Si lenses have been utilized to achieve a higher gain [33,34]. Beyond that, with the presence of a thick supporting dielectric substrate, Si lens can be used to collect and collimate the generated THz power to minimize the power dissipation in the substrate. However, the usage of Si lens makes the entire antenna configurations even larger on the top of the usage of a thick supporting dielectric substrate.

In this study, the optical to THz conversion efficiency of the photomixer has been optimized based on a numerical study and the utilization of a two-dimensional photonic crystal optical frequency selective surface (FSS) superstrate. Then, a dielectric resonator antenna (DRA) with the appealing features of low cost, small size, and high radiation efficiency, as well as gain [35], has been truncated from the bulky dielectric supporting GaAs substrate, in this case, the low temperature grown GaAs, LT-GaAs, to reduce the size of the antenna configuration and enhances the radiation efficiency. Coplanar stripline and THz dielectric superstrate have been implemented for the further optimization of matching efficiency and antenna gain, respectively. The simulations have been conducted using computer simulation technology (CST) microwave studio.

2. Photomixer Design

In this section, the generated THz power from a photomixer, that is biased using a DC voltage and is illuminated by two laser beams with a difference in their central frequencies in the THz range has been calculated analytically. According to the numerical analysis, the generated THz power can be enhanced substantially by optimizing the photomixer configuration.

2.1. Derivation of the Generated THz Power from the Photomixer

Figure 1a illustrates two linearly polarized continuous wave laser beams with beating frequency falling in the THz spectrum that are incident on the DC biased photoconductive layer. On the other hand, the structure of a typical photomixer electrodes is presented in Figure 1b. Since the applied biased voltage cannot change the absorption coefficient, mobility and recombination time of the low temperature grown GaAs, LT-GaAs photoconductive layer, the electrodes of the photomixer can be considered as an ohmic conductances while the time varying source conductance is electrically modeled as a photoconductance. Consequently, the photomixer based THz antenna can be modeled using the equivalent circuit illustrated in Figure 2 from which it can be noted that the photomixer consists of a photoconductance, $G_s^{-1}(\Omega,t)$, and a paralleled capacitance, $C_{electrodes}$. The capacitance depends on the structure of the photomixer and the dielectric constant of the photoconductive layer. The generated photocurrent is driven by the biased voltage excites and the THz radiating antenna, which is represented in Figure 2 by a resistance of $R_{antenna}$ [36].

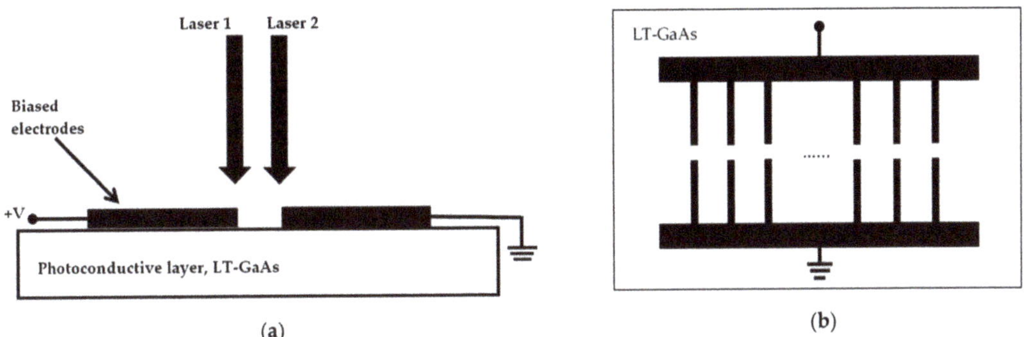

Figure 1. (**a**) Typical photoconductive photomxing scheme; (**b**) Top view of photomixer electrodes.

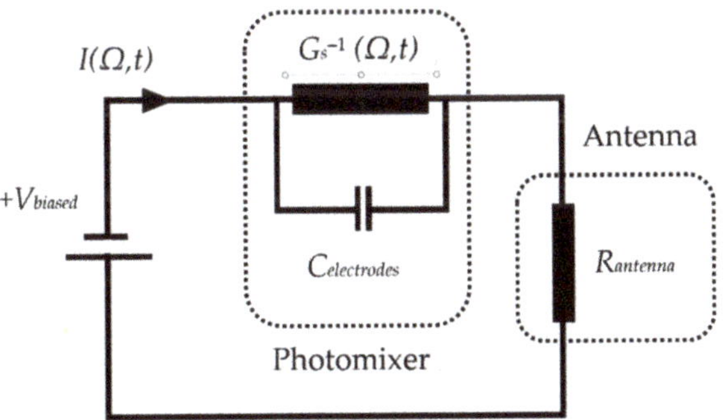

Figure 2. Equivalent circuit of photomixer based THz antenna.

The electric fields of the two incident laser beams on the LT-GaAs' surface can be expressed as:

$$E = |E_i|e^{j\omega_i t} \tag{1}$$

where ω is the lasers' angular frequency and I = 1,2, represents laser 1 and laser 2, respectively. The laser intensity been absorbed by the LT-GaAs is proportional to the square of the total incident electric field on the LT-GaAs' surface:

$$I(\Omega,t) = (1-\Gamma)\sum_i |E|^2 = I_0(1-\Gamma)[1+2\frac{\sqrt{mI_1I_2}}{I_0}\cos(\Omega t)] \tag{2}$$

where I_0 is the maximum optical intensity on the LT-GaAs' surface, Γ is the reflection coefficient at the LT-GaAs-air interface, m describes the overlap of the laser beams, which is known as the mixing efficiency and Ω is angular beat frequency, $(\omega_1 - \omega_2)$.

The induced photo-carriers generated from the incident laser beams as a function of time is:

$$\frac{dn(t)}{dt} = -\frac{n(t)}{\tau_c} + \frac{\alpha(T)}{hf_l}I(\Omega,t) \tag{3}$$

in which h is the Plank's constant, f_l is the mean frequency of the laser beams and τ_c is the carrier lifetime. In addition, $\alpha(T)$ is the temperature-dependent absorption coefficient and T is the temperature in kelvin. For a GaAs layer with a direct band gap, $\alpha(T)$ can be expressed as [37]:

$$\alpha(T) \approx K_{abs}\sqrt{\frac{hf_l - E_g(T)}{q}} \tag{4}$$

where K_{abs} is a certain frequency-independent constant which is approximately 9.7×10^{15} for GaAs [37], and $E_g(T)$ is the LT-GaAs' temperature dependent band gap energy defined as:

$$E_g(T) = E_g(0) - \frac{\alpha_E T^2}{T+\beta_E} \tag{5}$$

in which $E_g(0)$ is the GaAs' gap energy at 0 °K which is about 1.519 eV, α_E and β_E are material constants of GaAs which are approximately 5.41×10^{-4} eV/K and 204 K, respectively [38].

By assuming that $I_1 = I_2 = I_0$ and $t/\tau_c >> 1$, then substituting (2) into (3), the generated carrier density can be obtained as:

$$n(\Omega,t) = \frac{\alpha(T)}{hf_1}I_0(1-\Gamma)\tau_c(1+\sqrt{m}\frac{\cos(\Omega t)+\Omega\tau_c\sin(\Omega t)}{1+(\Omega\tau_c)^2}) \tag{6}$$

The conductance of the photomixer can be expressed as:

$$G_s(t) = \int dG_s(t) = \int_0^{T_{sub}} \sigma(t)e^{-\alpha(T)z}\frac{W}{L}dz = \frac{W}{\alpha(T)L}\sigma(t)(1-e^{-\alpha(T)\mathrm{T_{sub}}}) \tag{7}$$

in which T_{sub} is the depth of photoconductive region, W is the width of the electrode, L is the length of the electrode and $\sigma(t)$ is the conductivity. The electrical conductivity is defined as:

$$\sigma(t) = e\mu_e n(t) = \frac{\alpha(T)e\mu_e}{hf_l}I_0(1-\Gamma)\tau_c(1+\sqrt{m}\frac{\cos(\Omega t)+\Omega\tau_c\sin(\Omega t)}{1+(\Omega\tau_c)^2}) \tag{8}$$

where e is the electron charge and μ_e is the electron mobility. Therefore, the photomixer's conductance can be derived by substituting (8) into (7):

$$G_s(\Omega,t) = \frac{We\mu_e I_0\tau_c}{hLf_l}(1-\Gamma)(1-\mathrm{e}^{-\alpha(T)\mathrm{T_{sub}}})(1+\sqrt{m}\frac{\cos(\Omega t)+\Omega\tau_c\sin(\Omega t)}{1+(\Omega\tau_c)^2}) \tag{9}$$

The impedance of the system can be given by analyzing the equivalent circuit shown in the Figure 3:

$$Z_t(\Omega,t) = \frac{1}{j\Omega C_{electrodes} + G_s(\Omega,t)} + R_{antenna} \tag{10}$$

and the radiation power can be defined as:

$$P_{THz}(\Omega,t) = R_{antenna}\left(\frac{V_{biased}}{Z_t(\Omega,t)}\right)^2 \tag{11}$$

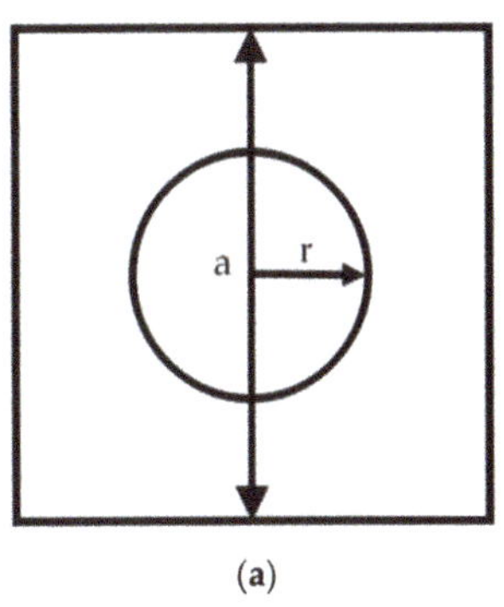

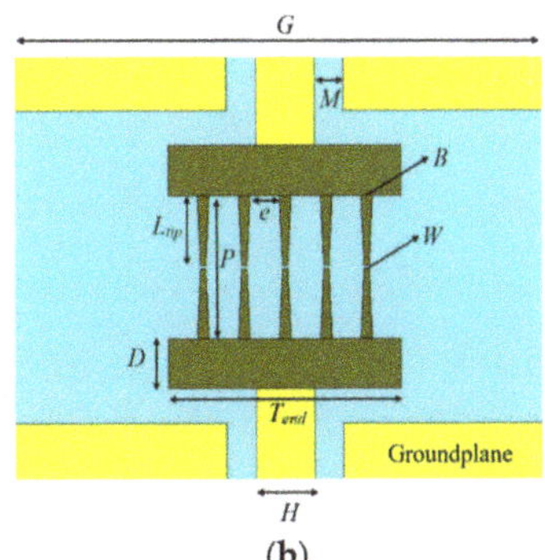

(a) (b)

Figure 3. (**a**) Configurations of the 2D-PhC unit cell; (**b**) Top view of the photomixer based slot.

Therefore, as $R_{antenna}G_s$ is much smaller than 1 and by replacing system impedance by (8), as well as neglecting the imaginary part, the radiation power can be expressed as:

$$P_{THz}(\Omega,t) \approx R_{antenna}\frac{V_{biased}^2 G_s^2(\Omega,t)}{1+(\Omega R_{antenna} C_{electrodes})^2} \tag{12}$$

In addition, by employing (9) and averaging the power, the mean generated THz power can be expressed as:

$$P_{THz} \approx \left[\frac{We\mu_e\tau_c}{hLf_l}(1-\Gamma)(1-e^{-\alpha(T)T_{sub}})\right]^2 \left(\frac{mR_{antenna}V_{biased}^2}{[1+(\Omega R_{antenna}C_{electrodes})^2][1+(\Omega\tau_c)^2]}\right) I_0^2 \tag{13}$$

It can be noted from (13) that the generated THz power depends on three main factors, (Ω $R_{antenna}$ $C_{electrode}$), ($\Omega\tau_c$) and I_0^2. As mentioned previously, $R_{antenna}$ and $C_{electrodes}$ depend on the photomixer's configuration and dielectric constant of the photoconductive layer. In addition the carrier lifetime is a function of the applied bias voltage [38–40]. Hence, it can be demonstrated that for the same photoconductive material and photomixer configuration that are used with the same biased voltage, the generated THz power is proportional to the square of the incident laser intensity on the surface of the LT-GaAs layer. Since the intensity is proportional to the square of electric field, the generated THz power is proportional to the 4th power of the electric field at the surface of LT-GaAs. Consequently, a design that optimizes the laser intensity is proposed instead of manipulating the photoconductive material and photomixer electrodes' configuration.

2.2. Photomixer Modeling

In order to optimize the electric field on the surface of LT-GaAs, two dimensional photonic crystal (2D-PhC) has been introduced by utilizing a periodic plane and a non-periodic third dimension to provide a pass, or stop, band frequency response [41]. The unit cell of the 2D-PhC is illustrated in Figure 3a. The 2D-PhC has been used as an optical frequency selective surface (FSS) superstrate that is placed at an optimum height above the photomixer. The electromagnetic wave bounces between the FSS and ground plane surrounding the photomixer, therefore, the cavity created by FSS superstrate and ground plane can enhance the optical intensity on the surface of the LT-GaAs.

Since the photomixer is used as a source to excite the truncated GaAs THz DRA, the metallic ground plane has been deployed on top of the LT-GaAs photoconductive layer. However, in order to illuminate the photomixer by the laser beams, a central slot feed is used to accommodates the photomixer as shown in Figure 3b. Moreover, the electrodes have been defined as optical gold (Palik) for CST simulation purposes. The dimensions of the parameters shown in Figure 3b have been defined as: M = 0.5 μm, B = 0.2 μm, W = 0.1 μm, e = 0.5 μm, P = 2.3 μm, L = 1.13 μm, D = 0.8 μm, T_{end} = 4 μm, H = 1 μm, G = 13.68 μm, and thickness of 0.1 μm. It should be noted that the thickness of the LT-GaAs photoconductive layer is 0.44 μm with a relative dielectric constant of 12.9. The material of the 2D-PhC FSS has been assumed as GaAs while the periodicity, central air hole radius, and thickness have been chosen as a = 0.76 μm, r = 0.3 a and h = 0.2 a, respectively. Figure 4 illustrates the reflectivity of the 2D-PhC FSS, where it can be noted that there is a stopband at wavelength range of 750 to 780 nm.

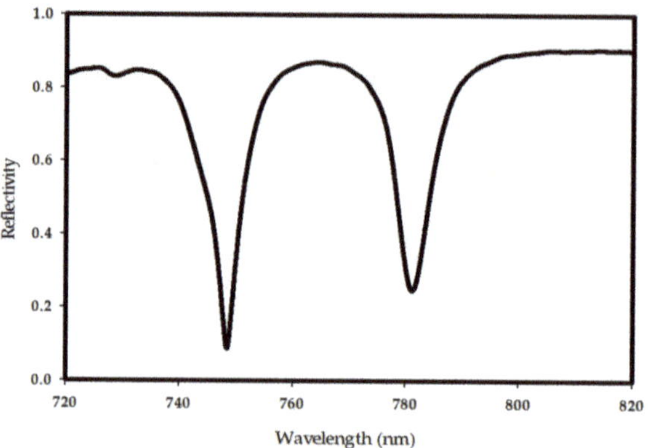

Figure 4. Reflectivity of the 2D-PhC FSS.

A 2D-PhC layer with 19 × 19 unit cells has been suspended at a height of 0.3 μm above the photomixer to act as an FSS superstrate. The incident laser beams have been modeled as a linearly polarized plane wave with a 1 V/m electric field component along the direction of photomixer electrodes. The electric field magnitudes between the central electrode pair on the LT-GaAs' surface is illustrated in Figure 5 with the comparison to that at the absence of 2D-PhC FSS superstrate. In addition, the cross-section of the surface electric field distribution at the central pair of photomixer electrodes are presented in Figure 6. From these results, it can be observed that the utilization of the 2D-PhC superstrate has improved the electric field on the electrodes from 2.1 to 9.9 V/m, which represents an enhancement factor of 4.7. As explained earlier, the generated THz power is proportional to the 4th power of the electric field, therefore the corresponding enhancement factor of the generated THz power is 487. Besides, the same methodology has been applied to an identical photomixer albeit with a InGaAs photoconductive layer, where the electric field on the InGaAs's surface has increased form 2.42 to 11.5 V/m by utilizing an FSS superstrate with unit cell's dimension of a = 0.72 μm r = 0.27 a and h = 0.19 a at a height of 0.23 μm above the photomixer. The results are presented in Figure 5, where it can be noted that approximately same enhancement factor has been achieved compared to the LT-Gaas photoconductive layer. Compared with the cases of increasing the electrodes E-field from 2.1 to 3.4 V/m and 4.365 V/m using 2D-PhC, with central hole [42] and plasmonic rod [43], respectively, as reflectors underneath the photoconductive layer, employing a

2D-PhC as FSS superstrate has substantially enhanced the optical to THz power conversion efficiency. However, the overall efficiency of the system depends on the laser to THz power conversion efficiency as well as the antenna's radiation and matching efficiencies that will be investigated next. In the following section, the optimized photomixer will be used to excite a THz DRA that is truncated from the supporting bulky GaAs substrate. A DRA has been chosen due to the high radiation efficiency of more than 90% at the frequency range of interest.

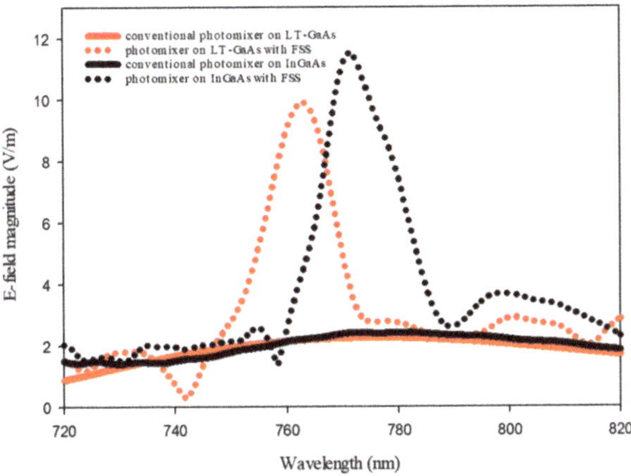

Figure 5. Optical E-field magnitude between the central electrodes of a photomixer on the surfaces of LT-GaAs and I GaAs photoconductive layers.

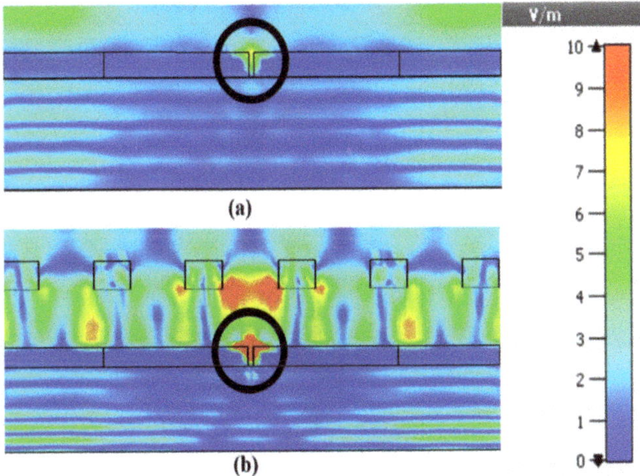

Figure 6. The optical E-field distribution (**a**) without FSS (**b**) with FSS.

3. THz Dielectric Resonator Antenna Design

3.1. Antenna Configuration

The presence of the bulky GaAs supporting substrate reduces the input impedance and absorbs most of the generated THz power, which impairs the matching and radiation efficiencies. A typical radiation efficiency for a dipole above a thick dielectric substrate is 40% or less depending on the thickness and dielectric constant of the substrate [44]. On the other hand, a rectangular dielectric resonator antenna offers a considerable enhancement in the radiation efficiency owing to the absence of surface waves and ohmic losses. Therefore, the bulky GaAs substrate can be truncated to act as a dielectric resonator antenna that operates at the higher order resonance mode. However, the truncated DRA should be large enough to maintain the required physical support to the photomixer device. For fabrication purposes, the width to height aspect ratio of the truncated DRA should be greater than 3. Otherwise, a fragile configuration will be achieved that is difficult to fabricate. Furthermore, the utilization of a DRA as a substrate results in a much smaller configuration compared to traditional structures that are based on utilizing a hemispherical Si lens to extract the THz power. As the configurations of the photomixer and corresponding 2D-PhC FSS superstrate are relatively small enough at the THz spectrum, they will have a negligible impact on the performance at the THz frequency range.

As a result, the GaAs substrate has been employed as the THz antenna that also provides the mechanical support to the device at the same time. The GaAs DRA is illustrated in Figure 7 with dimensions of W_{DRA} = 250 µm, H_{DRA} = 60 µm, as well as a relative dielectric constant of 12.9, and has been placed on a gold ground plane with a size of W_{ground} = W_{sup} = 400 µm. For further gain enhancement, an additional GaAs dielectric superstrate has been employed with dimensions of W_{sup} = 400 µm and T_{sup} = 60 µm. As illustrated in Figure 7, the original optical superstrate has been placed on the feed side of the DRA to capture the illuminating laser beams, while this THz superstrate is placed above the opposite side of the DRA to enhance the radiated THz power. Therefore, the two superstrates will not impact each other as they separated by the DRA and the gold ground plane that accommodates the photomixer. The distance between the DRA and the new THz superstrate can be determined as $H_{sup} = (0.25^*((\varphi_1 + \varphi_2)/\pi) + 0.5)\,\lambda$ [45], where φ_1, φ_2 represents the reflection coefficient phases of the superstrate and ground plane. Therefore, the distance between the GaAs DRA and the THz superstrate has been calculated as H_{sup} = 30 µm.

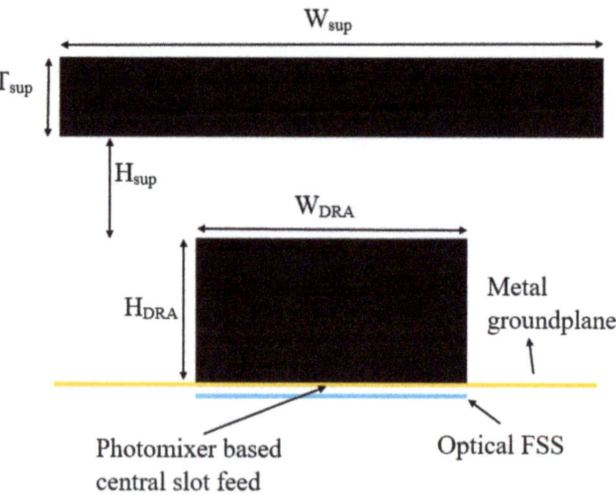

Figure 7. THz DRA and superstrate configuration.

DC bias is required to generate the THz power, therefore, the ground plane has been divided into two halves by a narrow slot with a width of $W_{seperate}$ = 0.5 µm to work as two large DC biasing pads as illustrated in Figure 8. Since the generated THz power can leak through the DC biased pads and transmission line, a coplanar stripline (CPS) network has been employed to work as a choke filter to minimize the THz current leakage, as well as improving the matching. The configuration of the CPS and feeding slot, which accommodates the photomixer and excites the DRA, has been included in Figure 8.

The feeding slot has a length of L_{slot} = 65 μm and width of W_{slot} = 5 μm. The dimensions of the CPS network have been chosen as L_{Tx} = 120 μm, W_{Tx} = 1 μm, L_{stub} = 91 μm, W_{stub} = 0.5 μm, g_{stub} = 50 μm, W_{gap} = 0.5 μm, and g_{Tx} = 3 μm, respectively.

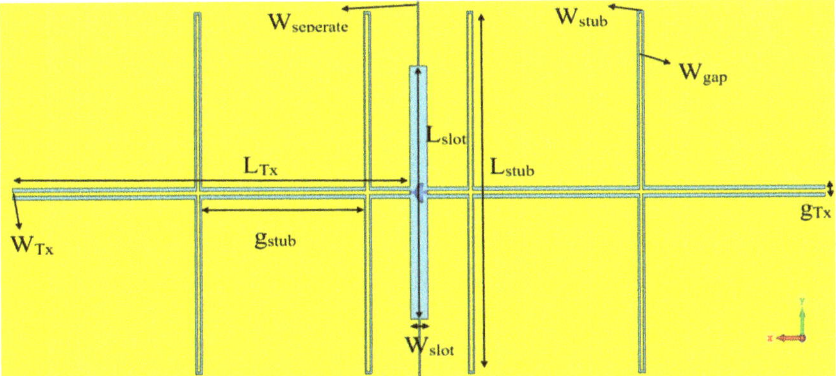

Figure 8. Top view of feeding slot and CPS.

Finally, the feeding photomixer has been modeled as a discrete port with a 10 kΩ input resistance that is in parallel with a 3fF lumped capacitance. Both of the discrete port and lumped capacitance have been deployed at the center of feeding slot in order to be connected with the CPS and DC bias pads.

3.2. Results and Discussion

The input impedance of the DRA with and without CPS network has been studied as shown in Figure 9, where it can be noted that the input resistance has been improved from 430 to 700 Ω by utilizing the CPS, which corresponding to an enhancement of matching efficiency from 15.8 to 24.5%. Furthermore, the resonance mode of the DRA has been investigated as illustrated in Figure 10, where it can be noted that the TE_{711} mode has been excited. The radiation patterns of the DRA are presented in Figure 11, where the broadside gain has been improved from 6.5 to 9 dBi by incorporating the THz GaAs superstrate. As a result, the radiated THz power has been enhanced by a factor of 2. Therefore, the performance of the THz photomixer based antenna has been improved considerably by combining several factors such as the improving the optical to THz power conversion efficiency as well as enhancing the radiation efficiency by utilizing a DRA and employing a CPS that improved the matching efficiency.

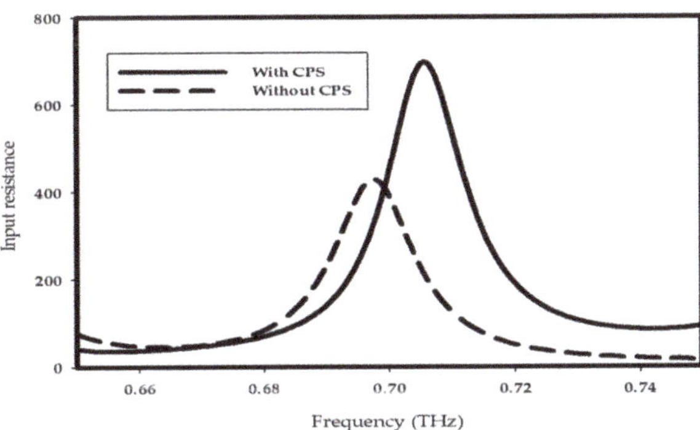

Figure 9. Input resistance of photomixer based slot fed THz DRA with or without CPS.

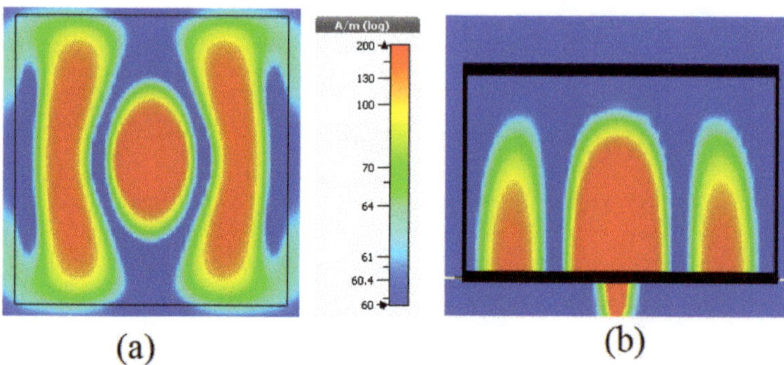

Figure 10. The H-field component inside the DRA with TE_{711}mode at (**a**) XY plane; (**b**) XZ plane.

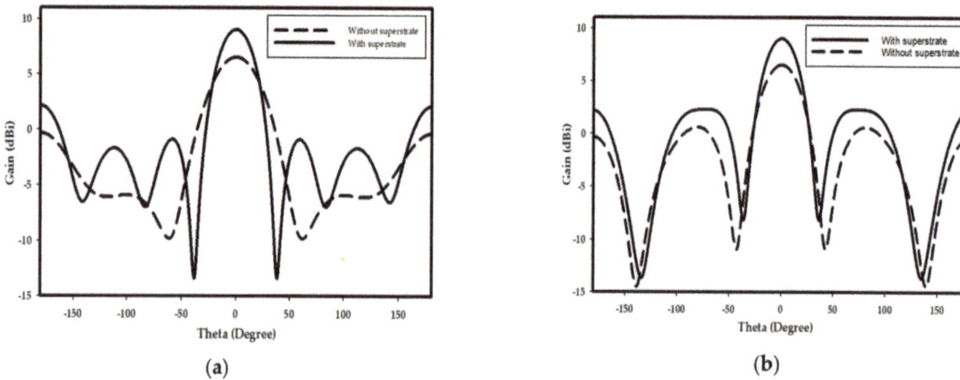

Figure 11. Radiation pattern of DRA with and without superstrate at (**a**) φ = 0°; (**b**) φ = 90°.

Table 1 compares the performance of the proposed DRA with published THz antennas and THz DRAs, where it can be observed that presented DRA offers a higher antenna gain as compared with other DRAs though, it is slightly larger than the reported DRAs. Compared with other antenna types, the proposed THz DRA achieves a similar gain with much smaller dimensions. Such a miniaturized high gain antenna can be used to optimize the performance of any THz application with a limited system space.

Table 1. Performance summary and comparison with prior works.

Reference	[34]	[46]	[47]	[48]	This Work
Antenna Type	Small lens with Leaky-wave Slit Dipole Antenna	Dipole Antenna with Horn Cavity	Slot Fed stacked DRA	Patch Fed Higher Order Mode DRA	Slot Fed GaAs Substrate Truncated DRA
Frequency (THz)	0.2	1	0.13	0.34	0.7
Antenna Gain	10.3	9.07	4.7	7.9	9
DR Material/ε_r	-	-	Alumina/10	Silicon/11.9	GaAs/12.9
DR type	-	-	Rectangular	Rectangular	Rectangular
Antenna Aperture (λ^2)	1.44	4.55	0.72	0.2	0.87
Antenna Height (λ)	1.2	1.16	1.28	0.5	0.35

4. Conclusions

The presented work introduces a photomixer based slot fed terahertz dielectric resonator antenna with enhanced optical to THz power conversion, as well as improved matching and radiation efficiencies. The interaction of two incident continuous wave laser beams and photomixer has been studied. A general analytical expression for the generated THz power has been derived, which demonstrates that the generated THz power is proportional to the 4th power of the electric field on the surface of photoconductive layer. Therefore, by utilizing a 2D-PhC as FSS, the optical to THz conversion efficiency has been improved by a factor of 487. Consequently, the optimized photomixer has been accommodated in a central slot, which has been used to excite the THz DRA that has been truncated from the thick supporting GaAs substrate. A coplanar stripline has been implemented to minimize the leakage of THz power through DC bias pads and transmission line, as well as improving the input resistance of the antenna. As a result, the input resistance of the DRA has been improved from 430 to 700 Ω, which corresponds to 15.8 and 24.4% matching efficiency, respectively. Finally, a THz GaAs superstrate has been employed with the THz DRA, which leads to an enhancement of the antenna gain from 6.5 to 9 dBi. The presented results demonstrate that the proposed design outperforms other counterparts reported in the literature. As demonstrated by (13), further enhancement of the optical to THz conversion efficiency can be achieved by reducing the carrier lifetime of the photoconductive layer such as changing the configuration of the photomixer electrodes to minimize the traveling distance of carriers. The performance can be improved further by altering the shape and dimensions of the slot to increase the order of the excited mode as this has the potential of providing higher gain. In addition the packaging of the proposed configuration needs to be considered so that a physical support is provided to the Thz superstrate in conjunction with improving the handling and stability of the device.

Author Contributions: Conceptualization, X.L., W.Y., and S.K.; methodology, X.L. and W.Y.; software, X.L.; validation, X.L.; formal analysis, X.L.; investigation, X.L.; writing—original draft preparation, X.L.; writing—review and editing, S.K.; supervision, S.K. All authors have read and agreed to the published version of the manuscript.

Funding: This research received no external funding.

Institutional Review Board Statement: Not applicable.

Informed Consent Statement: Not applicable.

Data Availability Statement: Data is contained within the article or supplementary material.

Acknowledgments: The authors would like to thank Professor Richard Hogg, James Watt School of Engineering, University of Glasgow for the valuable discussions and advice with respect to the practical dimensions of the truncated THz DRA.

Conflicts of Interest: The authors declare no conflict of interest.

References

1. Taday, P.F. Applications of terahertz spectroscopy to pharmaceutial sciences. *Phil. Trans. R. Soc.* **2004**, *362*, 351–364. [CrossRef] [PubMed]
2. Wallace, V.P.; Taday, P.F.; Fitzgerald, A.J.; Woodward, R.M.; Cluff, J.; Pye, R.J.; Arnone, D.D. Terahertz pulsed imaging and spectroscopy for biomedical and pharmaceutical applications. *Faraday Discuss.* **2004**, *126*, 255–263. [CrossRef]
3. Wang, S.; Ferguson, B.; Abbott, D.; Zhang, X.-C. T-ray imaging and tomography. *J. Biol. Phys.* **2003**, *29*, 247–256. [CrossRef] [PubMed]
4. Hu, B.B.; Nuss, M.C. Imaging with terahertz waves. *Opt. Lett.* **1995**, *20*, 1716–1718. [CrossRef]
5. Naftaly, M.; Foulds, A.P.; Miles, R.E.; Davies, A.G. Terahertz transmission spectroscopy of nonpolar materials and relationship with composition and properties. *Int. J. Infrared Millim. Waves* **2005**, *26*, 55–64. [CrossRef]
6. Karpowicz, N.; Zhong, H.; Zhang, C.; Lin, K.-I.; Hwang, J.S.; Xu, J.; Zhang, X.-C. Compact continuous-wave subterahertz system for inspection applications. *Appl. Phys. Lett.* **2005**, *86*, 054105. [CrossRef]
7. Choi, M.K.; Bettermann, A.; Van Der Weide, D.W. Potential for detection of explosive and biological hazards with electronic terahertz systems. *Phil. Trans. R. Soc.* **2004**, *362*, 337–349. [CrossRef]
8. Siegel, P.H. Terahertz technology in biology and medicine. *IEEE Trans. Microw. Theory Tech.* **2004**, *52*, 2438–2447. [CrossRef]
9. Crowe, T.W.; Globus, T.; Woolard, D.L.; Hestrer, J.L. Terahertz sources and detectors and their application to biological sensing. *Phil. Trans. R. Soc.* **2004**, *362*, 365–377. [CrossRef]
10. Akyildiz, I.F.; Jornet, J.M.; Han, C. Terahertz band: Next frontier for wireless communications. *Phys. Commun. AMST* **2014**, *12*, 16–32. [CrossRef]

11. Siegel, P.H. Terahertz technology. *IEEE Trans. Microw. Theory Technol.* **2002**, *50*, 910–928. [CrossRef]
12. Karpowicz, N.; Zhong, H.; Zhang, C.; Xu, J.; Lin, K.-I.; Hwang, J.S.; Zhang, X.-C. Comparison between pulsed terahertz time-domain imaging and continuous wave terahertz imaging. *Semicond. Sci. Technol.* **2005**, *20*, S293–S299. [CrossRef]
13. Belkin, M.A.; Wang, Q.J.; Pflügl, C.; Belyanin, A.; Khanna, S.P.; Davies, A.G.; Linfield, E.H.; Capasso, F. High-temperature operation of terahertz quantum cascade laser sources. *IEEE J. Sel. Topics Quantum Electron.* **2009**, *15*, 952–967. [CrossRef]
14. Seeds, A.J.; Williams, K.J. Microwaves photonics. *J. Lightw. Technol.* **2006**, *24*, 4628–4641. [CrossRef]
15. Teich, M.C. Field-theoretical treatment of photomixing. *Appl. Phys. Lett.* **1969**, *14*, 201–203. [CrossRef]
16. Brown, E.R.; McIntosh, K.A.; Nichols, K.B.; Dennis, C.L. Photomixing up to 3.8 THz in low-temperature-grown GaAs. *Appl. Phys. Lett.* **1995**, *66*, 285–287. [CrossRef]
17. Gu, P.; Chang, F.; Tani, M.; Sakai, K.; Pan, C.-L. Generation of coherent cw-Terahertz radiation using a tunable dual-wavelength external cavity laser diode. *Jpn. J. Appl. Phys.* **1999**, *38*, L1246–L1248. [CrossRef]
18. Brown, E.R. THz generation by photomixing in ultrafast photoconductors. *Int. J. High Speed Electron. Syst.* **2003**, *13*, 497–545. [CrossRef]
19. Plinski, E.F. Terahertz photomixer. *Bull. Pol. Acad. Sci. Technol.* **2010**, *58*, 463–470. [CrossRef]
20. Pačebutas, V.; Biciunas, A.; Balakauskas, S.; Krotkus, A.; Andriukaitis, G.; Lorenc, D.; Pugžlys, A.; Baltuška, A. THz time-domain-spectroscopy system based on femtosecond Yb: Fiber laser and GaBiAs photoconducting components. *Appl. Phys. Lett.* **2010**, *97*, 031111. [CrossRef]
21. Smith, P.R.; Auston, D.H.; Nuss, M.C. Sub-picosecond photoconducting dipole antenns. *IEEE J. Quantum Electron.* **1988**, *24*, 255–260. [CrossRef]
22. Eshagi, A.; Shahabadi, M. Plasmonic nanostructures for increasing the efficiency of terahertz large-area photomixing. In Proceedings of the Second Conferencce on Millimeter-Wave and Terahertz Technologies (MMWaTT), Tehran, Iran, 24–26 December 2012.
23. Bashirpour, M.; Forouzmehr, M.; Hosseininejad, S.E.; Kolahdouz, M.; Neshat, M. Improvement of terahertz photoconductive antenna using optical antenna array of ZnO nanorods. *Sci Rep.* **2019**, *9*, 1414. [CrossRef] [PubMed]
24. Gric, T.; Gorodetsky, A.; Trofimov, A. Tunable plasmonic properties and absorption enhancement in terahertz photoconductive antenna based on optimized plasmonic nanostructures. *J. Infrared Milli. Terahz. Waves* **2018**, *39*, 1028–1038. [CrossRef]
25. Sajak, A.A.B.; Shen, Y.C.; Huang, Y. Analysis of a photoconductive antenna using COMSOL. In Proceedings of the 10th UK-Europe-China Workshop on Millimetre Wave and Terahertz Technilogies (UCMMT), Liverpool, UK, 11–13 September 2017.
26. Yang, S.-H.; Hashemi, M.; Berry, C.; Jarrahi, M. 7.5% optical-to-terahertz conversion efficiency offered by photoconductive emitters with three dimensional plasmonic contact electrodes. *IEEE T. Terahertz Sci. Technol.* **2014**, *4*, 575–581. [CrossRef]
27. Yachmenev, A.; Lavrukhin, D.; Glinskiy, I.; Zenchenko, N.; Goncharov, Y.; Spektor, I.; Khabibullin, R.; Otsuji, T.; Ponomarev, D. Metallic and dielectric metasurfaces in photoconductive terahertz devices: A review. *Opt. Eng.* **2019**, *59*, 061608. [CrossRef]
28. Berry, C.W.; Jarrahi, M. Terahertz generation using plasmonic photoconductive grating. *New J. Phys.* **2012**, *14*, 105029. [CrossRef]
29. Ryu, H.C.; Kim, S.I.; Kwak, M.H.; Kang, K.Y.; Park, S.O. A folded dipole antenna having extremely high input impedance for continuous-wave terahertz power enhancement. In Proceedings of the 33rd International Conference on Infrared, Millimeter and Terahertz Waves, Pasadena, CA, USA, 15–19 September 2008.
30. Han, K.; Park, Y.; Kim, S.; Han, H.; Park, I.; Lim, H. A terahertz Yagi-Uda antenna for high input impedance. In Proceedings of the 33rd International Conference on Infrared, Millimeter and Terahertz Waves, Pasadena, CA, USA, 15–19 September 2008.
31. Yin, W.; Kennedy, K.; Sarma, J.; Hogg, R.A.; Khamas, S. A photomixer driven terahertz dipole antenna with high input resistance and gain. *Prog. Electromagn. Res.* **2015**, *44*, 13–20. [CrossRef]
32. Yang, Y.; Mandehgar, M.; Grischkowsky, D. THz-TDS characterization of the digital communication channels of the atmosphere and the enabled applications. *J. Infrared Millim. Technol.* **2014**, *36*, 97–129. [CrossRef]
33. Hussain, N.; Park, I. Optimization of a small lens for a leaky-wave slit dipole antenna at the terahertz band. In Proceedings of the International Symposium on Antennas and Propagation (ISAP), Okinawa, Japan, 24–28 October 2016.
34. Jha, K.R.; Singh, G. Ring resonator-intergrated hemi-elliptical lens antenna at terahertz frequency. In Proceedings of the International Conference on Communication Systems and Network Technologies, Katra, India, 3–5 June 2011.
35. Petosa, A.; Thirakoune, S. Rectangular dielectric resonator antennas with enhanced gain. *IEEE Trans. Antennas Propag.* **2011**, *59*, 1385–1389. [CrossRef]
36. Brown, E.R.; Smith, F.W.; McIntosh, K.A. Coherent milimeter-wave generation by heterodyne conversion in low-temperature-grown GaAs photoconductors. *J. Appl. Phys.* **1993**, *73*, 1480–1484. [CrossRef]
37. Zangeneh-Nejad, F.; Barani, N.; Safian, R. Temperature dependance of electromagnetic radiation from terahertz photoconductive antennas. *Microw. Opt. Technol. Lett.* **2015**, *57*, 2475–2479. [CrossRef]
38. Thim, H.W. Computer study of bulky GaAs Devices with random one-dimensional doping fluctuations. *J. Appl. Phys.* **1968**, *39*, 3897–3904. [CrossRef]
39. De Murcia, M.; Gasquet, D.; Elamri, A.; Nougier, J.P.; Vanbremeersch, J. Diffusion and noise in GaAs material and devices. *IEEE Trans. Electron Devices* **1991**, *38*, 2534–2539. [CrossRef]
40. Zamdmer, N.; Hu, Q.; McIntosh, K.A.; Verghese, S. Increase in response time of low-temperature-grown GaAs photoconductive switches at high voltage bias. *Appl. Phys. Lett.* **1999**, *7*, 2313–2315. [CrossRef]
41. McCall, S.L.; Platzman, P.M.; Dalichaouch, R.; Smith, D.; Schultz, S. Microwave propagation in two-dimensional dielectric lattices. *Phys. Rev. Lett.* **1991**, *67*, 2017–2020. [CrossRef] [PubMed]

42. Li, X.; Khamas, S.K. Enhance the optical intensity of a photomixer based THz antenna using a two dimensional photonic crystal. In Proceedings of the International Symposium on Antenna and Propagation (ISAP), Xi'an, China, 27–30 October 2019.
43. Li, X.; Khamas, S.K. Enhance the optical intensity of a THz photomixer using a plasmonic material filled two dimensional photonic crystal. In Proceedings of the IET's Antennas and Propagation Conference, Birmingham, UK, 11–12 November 2019.
44. Khamas, S.K.; Starke, P.L.; Cook, G.G. Design of a printed spiral antenna with a dielectric superstrate using an efficient curved segment moment method with optimisation using marginal distribution. *IEE Proc. Microw. Antenn. Propagat.* **2004**, *151*, 315–320. [CrossRef]
45. Frydrych, J.; Petrzela, J.; Pitra, K. Dual slot-superstrate terahertz antenna without Si-lens. In Proceedings of the Conference on Microwave Techniques (COMITE), Brno, Czech Republic, 20–21 April 2017.
46. Sun, M.; Chen, N.; Teng, J.H.; Tanoto, H. A membrane supported photomixer driven antenna with increased continuous-wave Terahertz output power. In Proceedings of the 2011 International Topical Meeting on Microwave Photonics Jointly held wiht the 2011 Asia-Pacific Microwave Photonics Conference, Singapore, 18–21 October 2011.
47. Hou, D.; Xiong, Y.-Z.; Goh, W.-L.; Hu, S.; Hong, W.; Madihian, M. 130-GHz on-chip meander slot antennas with stacked dielectric resonators in standard CMOS technology. *IEEE Tans. Antenn. Propag.* **2012**, *60*, 4102–4109. [CrossRef]
48. Li, C.; Chiu, T. 340-GHz low-cost and high-gain on-chip higher order mode dielectric resonator antenna for THz applications. *IEEE Trans. THz Sci. Technol.* **2017**, *7*, 284–294. [CrossRef]

Article

Calibration to Mitigate Near-Field Antennas Effects for a MIMO Radar Imaging System †

Ha Hoang [1,2,*], Matthias John [1], Patrick McEvoy [1] and Max J. Ammann [1]

1 Antenna & High Frequency Research Centre, Technological University Dublin, D08 NF82 Dublin, Ireland; matthias.john@tudublin.ie (M.J.); patrick.mcevoy@tudublin.ie (P.M.); max.ammann@tudublin.ie (M.J.A.)
2 Department of Telecommunications Engineering, Ho Chi Minh City University of Technology, Ho Chi Minh City, Vietnam
* Correspondence: hmanhha@gmail.com

† This paper is an extended version of our paper published in 2020 International Workshop on Antenna Technology (iWAT), Bucharest, Romania, 25–28 February 2020.

Abstract: A calibration method for a high-resolution hybrid MIMO turntable radar imaging system is presented. A line of small metal spheres is employed as a test pattern in the calibration process to measure the position shift caused by undesired antenna effects. The unwanted effects in the antenna near-field responses are analysed, modelled and significantly mitigated based on the symmetry and differences in the responses of the MIMO configuration.

Keywords: near-field antenna effect; radar calibration; MIMO radar; turntable radar; UWB radar; radar system; scattering imaging; inverse scattering problem; radar resolution

Citation: Hoang, H.; John, M.; McEvoy, P.; Ammann, M.J. Calibration to Mitigate Near-Field Antennas Effects for a MIMO Radar Imaging System. *Sensors* **2021**, *21*, 514. https://doi.org/10.3390/s21020514

Received: 7 December 2020
Accepted: 10 January 2021
Published: 13 January 2021

1. Introduction

Turntable radar imaging systems for high-resolution imaging of complex objects can observe objects in arbitrary orientations and with a minimum of equipment [1]. No relative motion takes place between sensors and other objects in the environment, which can be utilised to eliminate environment effects in order to increase system accuracy. The long duration due to mechanical spatial-scanning is a disadvantage of a single-sensor or single-input single-output (SISO) turntable system. In a multiple-sensor or multiple-input multiple-output (MIMO) system [2–7], the sensors can be spread spatially to excite/observe the object in different spatial positions including position, direction and polarisation. In a high-speed capturing system, measurement data about the object can be captured with only one snapshot [8–11] with multiple spatial positions in the excitation/observation. Complexity and cost of the system are proportional to the number of spatial sensor positions.

A hybrid combination of a turntable and a minimalistic MIMO system is a suitable trade-off between increasing capture speed and reducing complexity and system cost. Moreover, differential features in the MIMO channel responses are very important and cannot be identified in a SISO system. These can be utilised for system calibrations to mitigate the system errors and/or system imperfections.

The calibration for a practical system plays a vital role in the improvement of reconstructed image accuracy [12]. The accuracy of nonideal electromagnetic acquisition systems is affected by many factors, in which near-field characteristics of antennas are a significant factor. The imperfection of S_{11} characteristic of a bidirectional mode antenna and mutual near-field coupling between transmitting-receiving pairs of monodirectional mode antennas in the MIMO system are factors degrading the system accuracy. To eliminate these, the switching methods to turn off the receivers in the transmitting periods can be used, however this method is not suitable for a near-field radar system. Both the imperfection in S_{11} and the mutual near-field coupling could be considered as an unknown mutual coupling of the system antennas. In [13,14], this effect in the MIMO systems was mitigated by a subtraction between the two measurements with and without the object. While, blind calibration processes were implemented in [15,16] assuming known antennas positions and direction from the objects. An experimental study of antenna array calibration [17] revealed a mismatch if only the coupling effect of antennas was considered. This can be explained by another undesirable effect in antenna responses that is magnitude, latency/phase and polarisation in transmitting/receiving responses of the antennas dependent on direction. In [13], an adaptive weighting technique was proposed to calibrate this directional dependence error based on the measurement data at the exact positions. To tackle both the unknown mutual coupling and the antenna directional-dependence, the reflected signals from the metal plane in different positions were measured and processed in the calibration in [18,19]. Besides using the passive/static objects in calibration, the active/reconfigurable objects were also used in calibration processes as a beacon in [20] or a rotatable double-antenna polarimetric active radar calibrator in [21]. However, the measurement data at the exact spatial positions was also a requirement of these methods.

In this article, a hybrid MIMO turntable radar imaging system [12] and a calibration method to reduce the undesirable effects of the antennas on the system performance are reported. The undesired antennas effects in the MIMO system configuration are analysed in a perspective of near-field propagation [22]. Additionally, the effects of the object rotation using the system turntable are analysed and the impacts on the estimation of object position is modelled, investigated and measured. The calibration scheme is proposed to mitigate these effects in order to improve system accuracy with a minimisation of the complexity in measurement arrangement.

2. Radar Imaging System

Figure 1 shows the configuration of the MIMO turntable radar imaging system [12]. It includes two vertical fixed-mounted Vivaldi antennas and a turntable facilitating the rotation of measured objects around the system axis with an angular step size $\Delta\beta$. The two antennas Ant.1 and Ant.2 are parallel, with a distance of s_0 between them and have the same distance to the system axis. The two antennas axes are in the system plane. The system axis is perpendicular to the system plane and cuts this plane at the system origin O. The distance from the system axis to the plane containing the two antennas reference planes is r_0. The two antennas are connected with two bidirectional ports of a Rohde & Schwarz ZVA 40 vector network analyser (VNA) playing the role of a frequency-sweeping transceiver of the radar system. The system parameters are shown in Table 1.

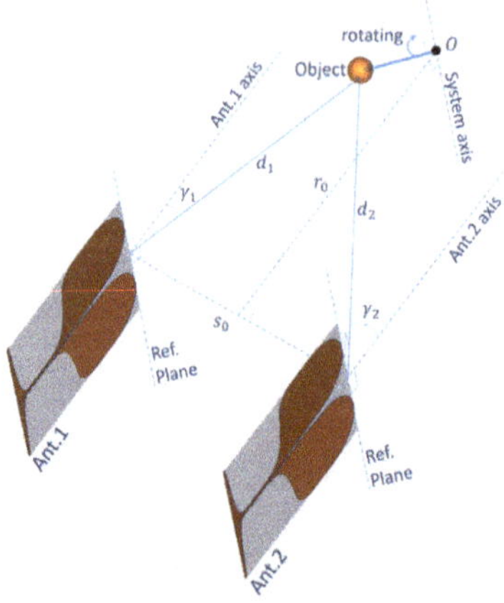

Figure 1. System configuration.

Table 1. System parameters.

Parameters	Values
Antennas space (s_0)	180 mm
System distance (r_0)	690 mm
Angular step size ($\Delta\beta$)	1.5°
Rotating step number	240
Turntable tolerance	$\pm 0.1^{\circ}$
Antenna size (L × W)	130 × 120 mm^2
System bandwidth	10 MHz–39.9 GHz
Frequency step	10 MHz
Transmitting power	10 dBm

Spatial tolerances in system configuration and object arrangements impact the accuracy of the imaging system. Reducing the spatial tolerances is challenging and increases the cost in setting up the system as well as arrangement of the objects. However, the calibration scheme proposed in the Section 4 can reduce the effect of some of these spatial tolerances.

In this MIMO system, each antenna plays the role of a transmitting antenna, a receiving antenna or both and time division multiplexing (TDM) of the VNA is used to divide each measurement period into the two time slots. In each time slot, concurrently, one antenna is in transmitting (Tx) and receiving (Rx) mode while the other is in Rx mode. The modes of the antennas alternate in the next time slot.

With two antennas and the turntable, there are four combinations for spatial observing channels (corresponding to four active radar virtual observing angles to the objects space) in the measurement of object scattering characteristics at each position of the mechanical rotation. These channels are presented in Table 2. The mechanical rotation of the turntable includes 240 steps with 1.5°/step. Thus, the total number of (virtual) observing angles to the objects space can reach to 4 × 240.

Table 2. Spatial observing channels vs. antenna modes.

Channel	Antenna Mode
C11	Tx: Ant.1, Rx: Ant.1
C12	Tx: Ant.1, Rx: Ant.2
C21	Tx: Ant.2, Rx: Ant.1
C22	Tx: Ant.2, Rx: Ant.2

In practice, the speed of electrical mode switching for the antennas is faster than mechanical state change for the turntable. Thus, this hybrid MIMO configuration is able to increase the density of (virtual) observing directions to the objects space and/or increase data acquirement speed when compared to a turntable single-channel configuration. The time-domain inverse scattering algorithm [23–25] is applied to reconstruct the object scattering image from radar measurement data. The object scattering images can be produced based on measured data of all of mechanical rotation angles and corresponding to each mechanical rotation angle, data of one or all of four channels are used for this reconstruction.

The inverse scattering algorithm is based on the propagating waves in the system model to identify scattering sources. The accurate modelling for the real propagation process on/between the antennas and the objects is necessary [22]. However, in this algorithm, each antenna is considered as working in the far-field relative to the objects, isolated from other antennas and with ideal S_{11}, so the effects of the real operational antenna conditions significantly degrade the system accuracy. These antenna effects are addressed, modelled and mitigated in the next sections. Additionally, the rotation of the objects in the measurement process and the reconstruction algorithm is also concerned and investigated in terms of its effect on estimated object-position in the next section.

3. Antennas Effects and Shift Modelling

In the case of an antenna operating concurrently in both transmitting or receiving mode (C11 or C22), the proportion of electromagnetic (EM) energy reflecting/scattering back to the antenna port at discontinuities in the structure (e.g., at the end edge, lateral edge or at the connector) is the main factor causing imperfection in S_{11} antenna characteristic [22]. This effect can occur many times between parts of the structure, forming higher-order reflection components visible in the received signal in time domain. Both first-order and higher-order reflection components of the dominant transmitted signal can mask the small amount of received EM energy scattered from the target object and is received by the antenna. The imperfection in S_{11} is illustrated in Figure 2 by the fact that $|S_{11}|$ parameter is always greater than zero in practice.

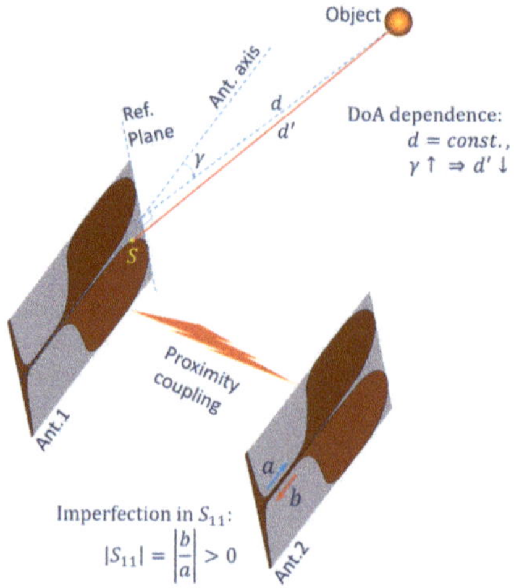

Figure 2. Undesired antennas effects.

Another undesired effect in the system configuration is the proximity coupling between the two antennas. In the case of the observing channel being C12 or C21, this can be considered as a mutual coupling channel between the transmitting and receiving antennas of distance s_0. The first-order scattering components from the transmitting antenna can be received by the receiving antenna. In the case of the observing channel being C11 or C22, mutual coupling also occurs, the inactive antenna appears as a distributed target in close proximity to the active antenna. Only second- and higher-order scattering components from the inactive antenna can be received by the active antenna. Thus, in the case of C11 or C22, the received signal is affected by the both S_{11} imperfection and mutual coupling effects. In all cases C11, C22, C12 or C21, the coupling EM energy proportion received at the receiving antenna can mask the desired signal scattered from the object. The proximity coupling between the two antennas of the system is also illustrated in Figure 2.

Figure 3 shows the impulse signal and the measured received signals at the antenna ports. It is observed that the intrinsic antenna structure causes significant reflections, even from the regions of the connection port as the reflecting signal is formed from the beginning of the impulse. The amplitude of this reflection signal (Received Sig. with C22) is significantly greater than the amplitude of received signal caused by the proximity coupling (Received Sig. with C12). Due to the distance s_0 between the two antennas, there is a corresponding latency in the proximity coupling signal (Received Sig. with C12). Another observation is that the higher-order scattering components of the two antennas lead to the elongation of both received signals in the time domain.

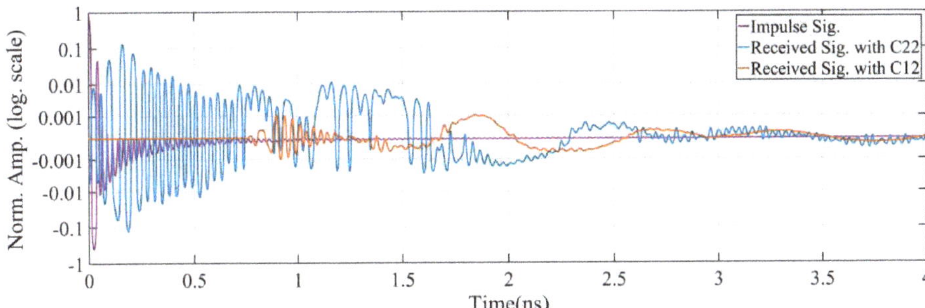

Figure 3. Received signals measured at the antenna port with the effects of imperfection in S_{11} and proximity coupling.

Another undesired effect of the Vivaldi antennas is dependence on direction of arrival (DoA). An example of a ray-model propagation path for part of the scattered EM energy propagating from the object to the antenna port is shown in Figure 2. When the antenna 1 (Ant.1) acts as a receiver, this path starts from the object, propagates over the subpath in the air d' to the scattering point S on the Vivaldi edge and propagates over the rest of the Vivaldi edge toward the antenna port. The direction of the path is reversed when the antenna is in transmitting mode. Assuming that the distance d from the reference plane of the antenna to the object is not changed, when the arrival angle γ increases, while the far-field model shows that d is constant versus γ, the subpath in the air d' of the ray-model decreases. This decrease phenomenon also happens to any arbitrary ray from the object to any point on the antenna element behind the referent plane. Thus, the practical equivalent length of the subpath in the air by multipath superposition of all rays scattering from the object to the antenna patches also has a corresponding decline versus γ. This DoA dependence effect leads to a significant error when the inverse scattering algorithm is applied to reconstruct the scattering image of the object space if only the far-field model is used.

The effect of DoA dependence of the Vivaldi antennas to a shift in object position in the measurement result is explained in Figure 4. When Ant.1 measures an object at P, the position in the measurement result is shifted to M by Δd, which is a function of the angle γ. For objects at points in the segment $[-s_0/2 \;\; s_0/2]$ of the x axis around the system origin O, the dependence of the shift Δd on γ can be approximated by a linear function versus $\tan(\gamma)$ or the distance BP, this is shown by the blue line (BM) in Figure 4. Symmetrically, the function of the shift when measured by Ant.2 is represented by the orange line and the angle between the two lines is α. This angle α can be considered as a differential characteristic parameter of the DoA dependence effect of the two antennas in the MIMO system.

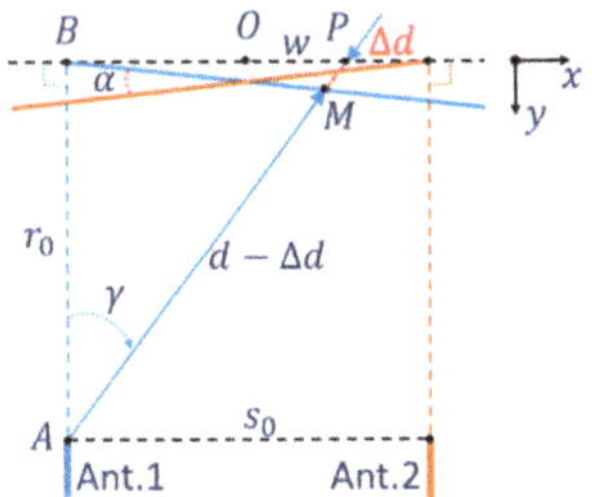

Figure 4. Shift model of antenna DoA dependence effect.

The relationship between the shift Δd and the system parameters, the differential parameter α, and the observing angle γ from Ant.1 to the object can be formulated based on the trigonometric relation of the edge lengths and the vertex angle in the triangle ABP:

$$\tan(\gamma) = \frac{\left(\frac{s_0}{2} + w\right)}{r_0}. \tag{1}$$

When combined with the relationship in the triangle BPM and the projections of Δd on the x axis—$\Delta d\ \sin(\gamma)$ and on the y axis—$\Delta d \cos(\gamma)$:

$$\tan\left(\frac{\alpha}{2}\right) = \frac{\Delta d \cos(\gamma)}{r_0 \tan(\gamma) - \Delta d \sin(\gamma)}. \tag{2}$$

Thus, the equation for the shift Δd versus γ and the system parameters can be written:

$$\Delta d = \frac{r_0\ \tan(\gamma) \tan\left(\frac{\alpha}{2}\right)}{\cos(\gamma) + \sin(\gamma) \tan\left(\frac{\alpha}{2}\right)}. \tag{3}$$

In this work, the estimation of the position of the object is based on measuring the distance to the object rotating around the system axis and applying the inverse scattering algorithm to the measurement data set in order to reconstruct the object image. Thereby, the shift caused by the antenna DoA dependence effect can be evaluated. However, effects of the rotation in the measurement can affect significantly the estimation results. Figure 5 shows the shift model with the rotation effect. In each step of the measurement process by Ant.1, when the turntable rotates by an angle β from the initial position, the object at P is moved to P$'$ and the measurement result for the position of P$'$ is moved to M$'$ with the shift Δd_β caused by the antenna DoA dependence effect. This shift can be considered as a function of the variable β.

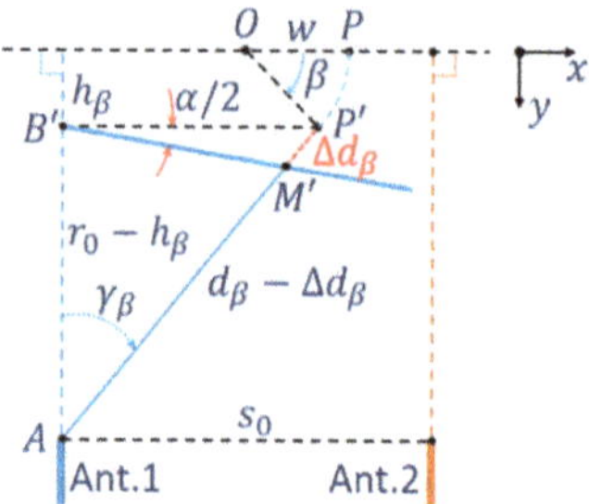

Figure 5. Rotation effect on the shift model of antenna DoA dependence effect.

Assuming that the differential parameter α does not depend on the distance from the antenna to the measured position or the distance AB$'$. The relationship between the shift Δd_β and the parameters and rotating variable β can be formulated as follows.

The projections of the segment OP$'$ on the x and y directions are:

$$w_\beta = w\ \cos(\beta),\ \ h_\beta = w\ \sin(\beta). \tag{4}$$

In the triangle AB$'$P$'$,

$$\frac{s_0}{2} + w_\beta = \frac{s_0}{2} + w\ \cos(\beta), \tag{5}$$

$$d_\beta = \frac{r_0 - h_\beta}{\cos(\gamma_\beta)} = \frac{r_0 - w\ \sin(\beta)}{\cos(\gamma_\beta)} \tag{6}$$

and

$$\tan(\gamma_\beta) = \frac{\frac{s_0}{2} + w\ \cos(\beta)}{r_0 - w\ \sin(\beta)} \tag{7}$$

or

$$\gamma_\beta = \text{artan}\left\{\frac{\frac{s_0}{2} + w\ \cos(\beta)}{r_0 - w\ \sin(\beta)}\right\}. \tag{8}$$

Considering the trigonometric relationship in the triangle B′P′M′, the projections of Δd_β on the x and y directions and the Equation (7) then

$$\tan\left(\frac{\alpha}{2}\right) = \frac{\Delta d_\beta \cos(\gamma_\beta)}{\{r_0 - w\ \sin(\beta)\}\tan(\gamma_\beta) - \Delta d_\beta \sin(\gamma_\beta)}. \tag{9}$$

Thus, the shift Δd_β can be formulated as a function of the variable β, the object position and the system parameters:

$$\Delta d_\beta = \frac{\left\{\frac{s_0}{2} + w\cos(\beta)\right\}\tan\left(\frac{\alpha}{2}\right)}{\cos\left(\text{artan}\left\{\frac{\frac{s_0}{2}+w\cos(\beta)}{r_0 - w\sin(\beta)}\right\}\right) + \sin\left(\text{artan}\left\{\frac{\frac{s_0}{2}+w\cos(\beta)}{r_0 - w\sin(\beta)}\right\}\right)\tan\left(\frac{\alpha}{2}\right)}. \tag{10}$$

An investigation of the shift Δd_β versus angle β with different initial x-axis object positions is implemented. With $w = 0$, the objects initial position is at the origin O, with $w < 0$ on the left and with $w > 0$ on the right of the origin. This investigation is implemented with the differential parameter α = 1.35°. The plots in Figure 6 show that the shift Δd_β of each point varies with the rotation angle β depending on distance w from the origin to the initial position. Another feature is that at $\beta = 0$ the average of Δd_β is greater than the initial value (Δd without the rotation effect) for points on the left of the origin and less for points on the right. Thus, if the estimation of Δd is based on averaging over β, then the estimated value of Δd tends to increase on the left of the origin and to decrease on the right. This demonstrates that the estimated function of Δd is nonlinear versus $\tan(\gamma)$ or distance BP.

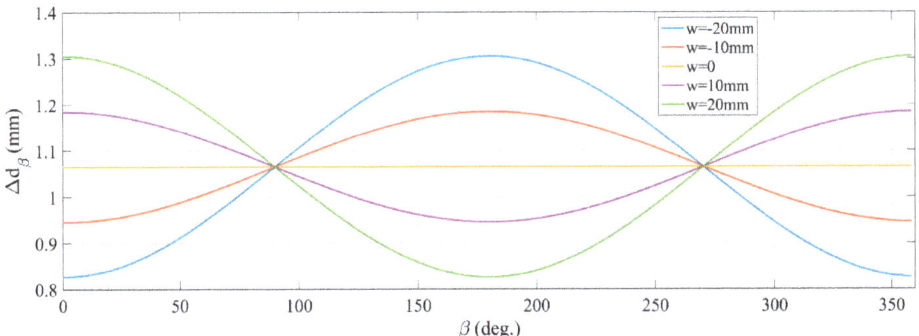

Figure 6. An investigation of rotation effect to the shift.

Considering the superposition in the reverse scattering algorithm to the response signals scattered from the objects with rotation, local peaks in the reconstructed image tend to spread out and shift with measured spatial errors Δd_β. However, because of nonlinear or non-sawtooth shape around peaks of the time-domain response signals corresponding to the measuring frequency band, this superposition leads to shrinking of the spatial errors in mapping to the reconstructed image. Thus, the value of the shift Δd estimated from the reconstructed image tends to be smaller than its value in the model. The above analyses show that there are differences between the shift model in Equation (3) and the measured and estimated result of the shift with the effects of the rotation and the feature of the inverse scattering reconstruction algorithm.

An adjusted model for the shift with effects of the rotation and the feature of the reconstruction algorithm is proposed. Considering the feasibility for the measurement and estimation of the shift caused by the DoA dependence of the antennas in this work, the adjusted model for the shift Δd is still a proportional function versus $\tan(\gamma)$ as Equation (3), but the differential parameter α is replaced by α_a with a k factor, $0 < k < 1$ as per Equation (11). The model is used in a calibration scheme for the system in next section.

$$\alpha_a = k\alpha \tag{11}$$

4. Calibration Scheme and Results

To mitigate the effects of the imperfection of S_{11} characteristic and mutual proximity coupling between the two antennas, the background subtraction method [23,25] is applied. However, the slow ripple over environment temperature in the response of the system transceiver can reduce the effectiveness of this method. The distance r_0 from the antenna reference planes to the system x axis is chosen large enough that undesired scattering components described above arrive earlier than the scattered signals from the object. Thereby, the error of higher-order components at the object-scattering period, caused by the slow ripple in the transceiver, is small enough compared to the intensity of the scattered signals from the object. Limiting the range of the angle γ considering the width of the target object is also a factor in the choice of the lower bound of r_0. The upper bound of r_0 depends on the intensity of scattering signals from the objects compared to the system noise level. In this work, r_0 is chosen as 690 mm.

Spatial tolerance in the alignment and positioning of the system components and calibration objects significantly affects the system accuracy. In the system, the fixed connected components such as the two antennas can be aligned accurately together with little additional effort. However, high-accuracy alignment and positioning for the separate parts of the system can require a lot of effort and high-cost measurement equipment.

To simplify alignment and positioning for the system origin, a calibration for the nominal distance r_0 of the system is implemented based on an equivalent distance calculated from the measured propagation time (from the antennas to the system origin). This measurement is implemented with a planar reflector placed at the system origin. To reduce the rigour in alignment of the reflecting plane in the measurement, the highly directional characteristic in radar cross-section of the reflecting plane and the high-resolution in rotation of the system turntable are utilised. The plane is rotated to find the balance in azimuth angle of the plane to the two antennas corresponding to the peak of the received signal with the observing channel C12 or C21, in which the propagation path starts from one antenna, propagates to the plane and reflects to the other antenna. The propagation time of the path corresponding to the case of the balance in azimuth angle of the plane is used to estimate the equivalent distance from the antennas to the system origin and correct the r_0 parameter. Figure 7 shows the 230×230 mm^2 reflecting plane aligned on the rotating axis of the turntable used in this calibration.

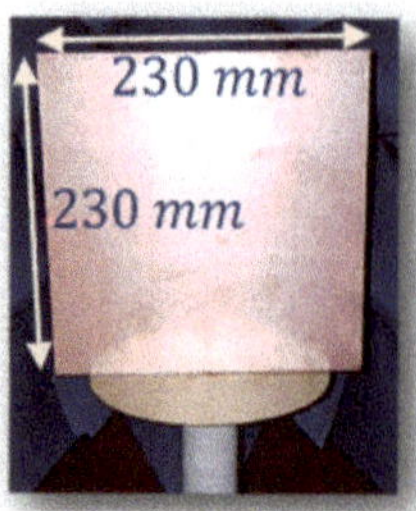

Figure 7. Reflector aligned at rotating axis of the turntable for the system origin calibration.

As mentioned in Section 3, the antenna DoA dependence effect causes a shift in the estimated object position when the estimation is based on reconstructed images. This shift in the MIMO system was modelled and characterised by the differential parameter α_a as presented in the Equations (3) and (11). The measurement for the differential parameter α_a uses a calibration pattern of seven steel spheres of 11 mm diameter evenly spaced 40 mm apart and aligned close to the horizontal line at the system origin. Figure 8 shows the pattern and its location in the system. Radar measured data is acquired based on the received signals of both antennas operating in monostatic mode (observing channel C11 and C22) at each rotating step of the turntable.

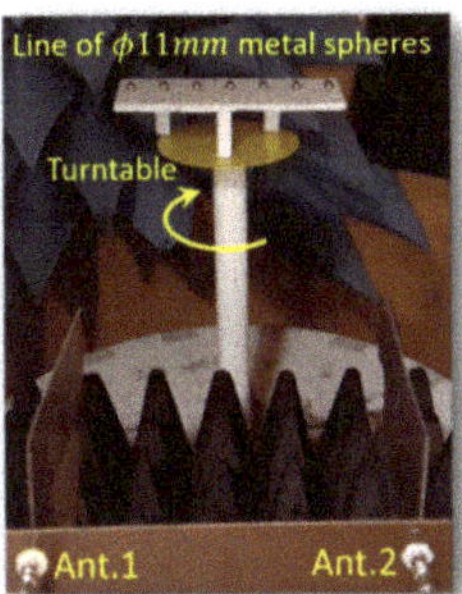

Figure 8. Measurement of the differential parameter by a line of metal spheres.

Figure 9a,b shows the reconstructed images of the calibration pattern based on two sets of radar data measured by channels C11 and C22 with the DoA dependence effect. The first observation from the images is that the peak values corresponding to object positions far from the system origin (image centre) are smaller than the values close to the origin. This can be explained by the rotation effect on the superposition of the reconstruction algorithm, there is a proportional increase in the fluctuation in the shift of object positions farther from the system origin, this leads to an increase in the spread of image energy around these peak positions and degrades these peak values.

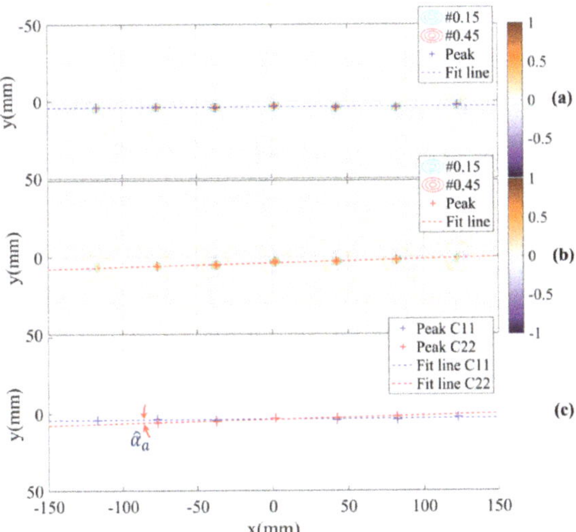

Figure 9. Reconstructed images with the DoA dependence effect based on measurement channels (**a**) C11, (**b**) C22 and (**c**) the estimation of differential parameter of the shift.

The shift phenomenon caused by the antenna DoA dependence effect can be evaluated based on these two images. Firstly, the positions of the objects in the two images are estimated using a local peak finding algorithm. Next, in each image, a line across the objects positions is estimated by fitting a linear function with the peak positions set. The angle of the two lines from the two images represents the differential parameter α_a of the shift in the adjusted model. In this experiment, the estimated angle of the parameter α_a is $\hat{\alpha}_a = 1.16^{\circ}$. The peak positions, fitted lines and the angle between the two lines are shown in Figure 9c.

Due to the symmetry of the MIMO system and the use of differential angle of the two lines for the estimation, the evaluation of the shift parameter is not sensitive to tolerances in direction and space between the line of the calibration pattern and the system horizontal line. These tolerances can be caused by inaccurate alignment of the calibration pattern on the turntable. Additionally, by fitting a linear function on the peaks corresponding to the sphere positions, this function is a characteristic line of the peaks set with relative distance errors to the peaks. This suggests that in the alignment of the calibration pattern, straightness of the line of spheres is not a rigorous requirement.

For a comparison, the images of the calibration pattern reconstructed from radar data collected by channels C12, C21 and all of four channels are also presented in Figure 10.

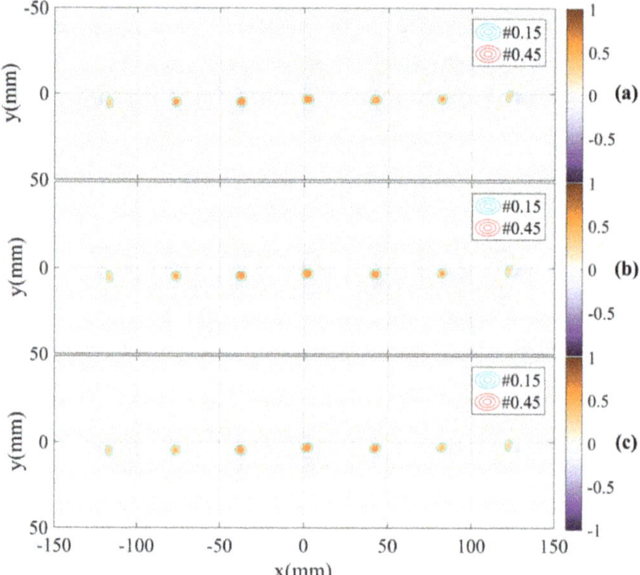

Figure 10. Reconstructed images with DoA dependence effect based on measurement channels (**a**) C12, (**b**) C21 and (**c**) all four channels.

Considering Equation (3) and the adjusted model of the shift with the effects of the rotation and the inverse scattering algorithm of the system, the estimated shift $\Delta\hat{d}$ for Δd in the model can be calculated from the estimated differential parameter $\hat{\alpha}_a$ by the equation:

$$\Delta\hat{d} = \frac{r_0 \tan(\gamma) \tan\left(\frac{\hat{\alpha}_a}{2}\right)}{\cos(\gamma) + \sin(\gamma) \tan\left(\frac{\hat{\alpha}_a}{2}\right)}. \tag{12}$$

To mitigate the antennas DoA dependence effect, the shifts at each position in the object space corresponding to the distance of the paths from the object to the two antennas are compensated by the estimated shifts calculated by Equation (12) in the inverse scattering algorithm. The results of applying the compensation to the calibration pattern are shown in Figure 11. The images show the improvement in focus specifically the significant increase and regularity of the peaks. The antennas DoA dependence effect is significantly mitigated. This is demonstrated by the overlap of the peak positions of the two images and the reduction of the estimated angle between the two lines down to approximately 0° in the reconstructed images measured by channels C11 and C22. Additionally, the images of the calibration pattern reconstructed from radar data collected by channels C12, C21 and all of the four channels with DoA dependence calibration are also presented in Figure 12. These figures also show the effectiveness of the calibration.

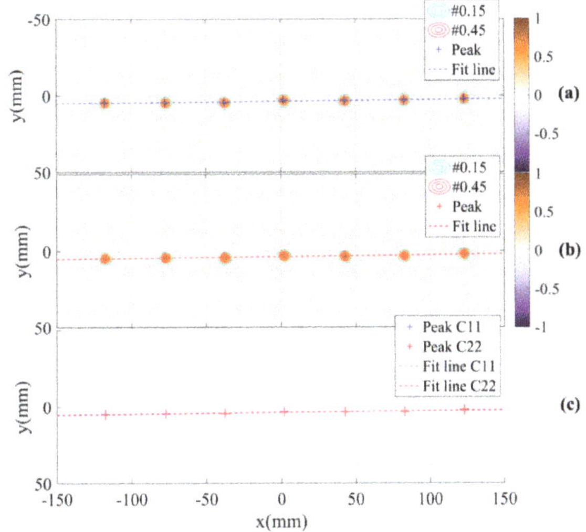

Figure 11. Reconstructed images with DoA dependence calibration based on measurement channels (**a**) C11, (**b**) C22 and (**c**) estimation of angle between the two lines.

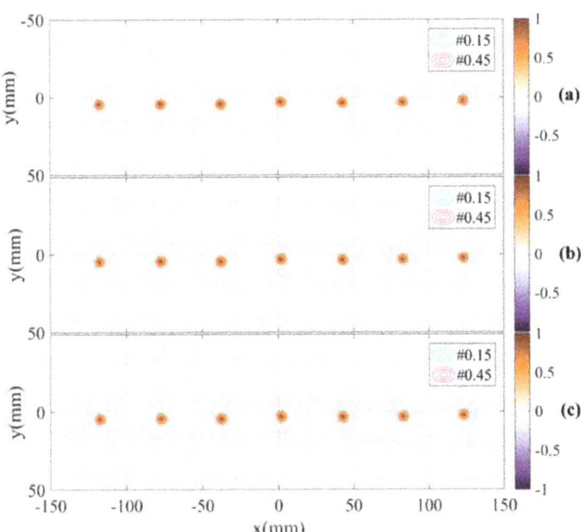

Figure 12. Reconstructed images with DoA dependence calibration based on measurement channels (**a**) C12, (**b**) C21 and (**c**) all four channels.

An investigation of relationship between the angle $-\alpha_a$ used to compensate for the differential parameter α_a in the adjusted model and the angle between the two lines practically estimated from the reconstructed images is implemented based on the calibration pattern measured radar data. This investigates how the compensation affects the practical estimation of the angle between the two lines and whether a multisolution in the algorithm to eliminate the DoA dependence effect exists. The result in Figure 13 shows that if there is no compensation ($-\alpha_a = 0$) then the estimated angle between the two lines is 1.16°. The angle between the two lines is suppressed when the compensation is implemented by a value of $-\alpha_a = 1.11^{\circ}$. This also shows that the error between the differential parameter α_a of the model and the measured and estimated differential parameter $\hat{\alpha}_a$ is approximately 0.05°. In this investigation with the range of 6° (-2° to 4°, step of 0.1°) of the parameter $-\alpha_a$, there is only one solution for elimination of the antenna DoA dependence effect corresponding to $-\alpha_a = 1.11^{\circ}$. Therefore, the calibration algorithm has a univalent convergence in the range of the parameter $-\alpha_a$.

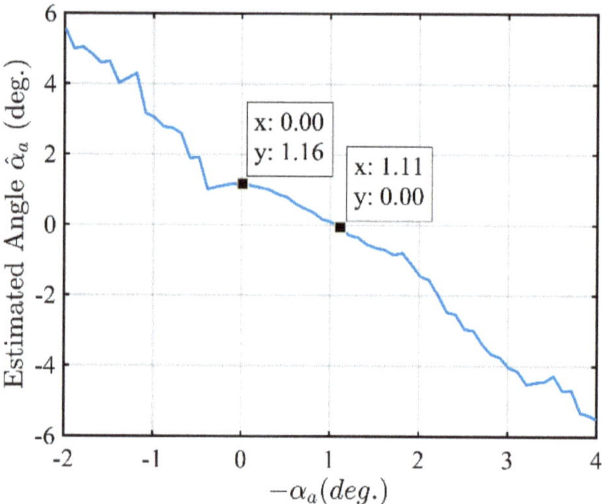

Figure 13. Angle between the two lines estimated from the reconstructed images vs. the angle $-\alpha_a$ used to compensate for the differential parameter α_a of the model.

When the calibration parameter has been determined, the system can be used to measure other objects such as a pattern with 31 steel spheres of 11 mm diameter arranged into a shape of "TUD" characters. The distance between two adjacent centres of spheres is 20 mm. The pattern and reconstructed images with and without DoA dependence calibration are shown in Figure 14. The results show that with DoA dependence calibration, intensities of the peaks are high and moderately regular. The positions of the spheres in the "TUD" pattern can be identified accurately based on these peaks as seen in Figure 14b. While without DoA dependence calibration, the peaks or the positions of the objects cannot be identified as shown in Figure 14c. This comparison demonstrates a significant effectiveness of the calibration method.

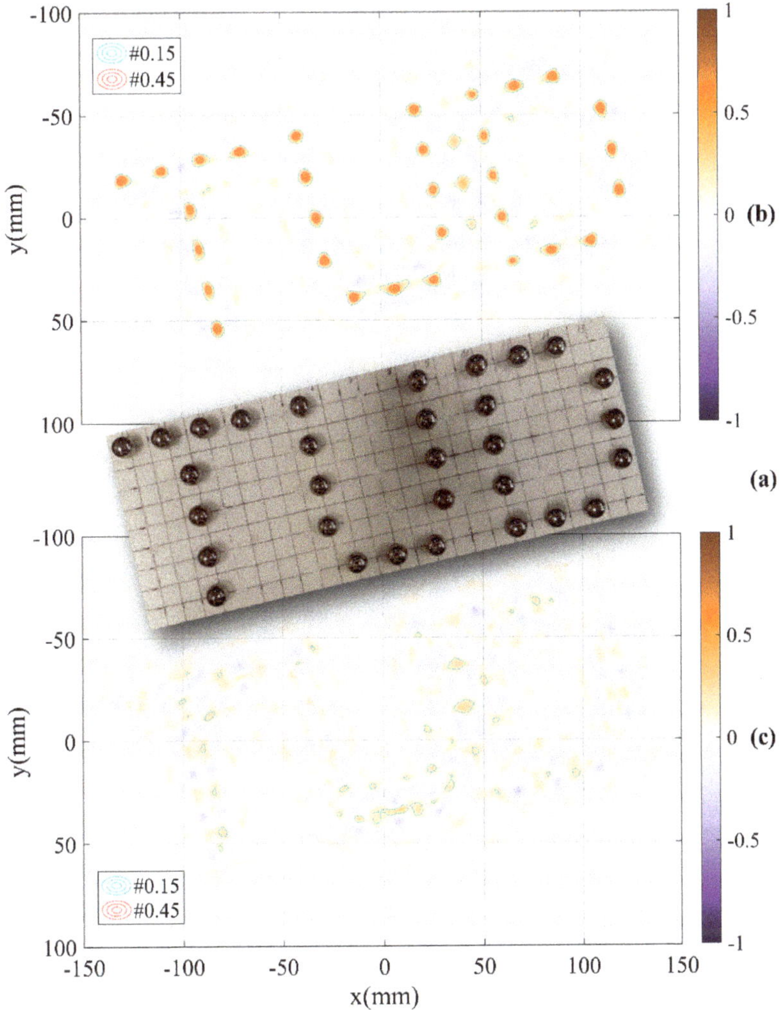

Figure 14. Testing the system with (**a**) "TUD" pattern and results (**b**) with DoA dependence calibration and (**c**) without DoA dependence calibration based on measurement data of all of four channels.

However, the quality of the reconstructed image in Figure 14b is lower than that of Figure 12c. We observe the appearance of phantom peaks (e.g., between the "U" and "D" characters) and a reduced and irregular intensity of object peaks. This is caused by the increase in number of spheres (from 7 to 31), the decrease in distance between the objects (40 mm down to 20 mm) and the distribution of the objects in two dimensions of the system plane in the "TUD" pattern, when compared to the calibration pattern. These differences lead to more complexity in the propagation progress [22] between objects and antennas at each observed angle. The increased probability of an object being occluded by others is the main factor in the quality reduction of the reconstructed image.

5. Conclusions

This article presented a hybrid MIMO radar imaging system associated with the undesired near-field antennas effects and demonstrated the effectiveness of a calibration method to mitigate these effects. The calibration scheme was based on the analysis and modelling of the propagation process and differential features of the MIMO system configuration as well as tolerated the errors in the measurement arrangement. The advantage of the method was demonstrated in improved focus of image energy at object peaks in the reconstructed scattering images. This facilitates highly accurate near-field detection of small objects using antennas which are large compared to the object size.

Author Contributions: Conceptualisation, H.H.; methodology, H.H.; software, H.H.; validation, H.H.; formal analysis, H.H.; investigation, H.H.; resources, H.H. and M.J.; data curation, H.H.; writing—original draft preparation, H.H.; writing—review and editing, H.H., M.J., P.M. and M.J.A.; visualisation, H.H.; supervision, M.J., P.M. and M.J.A. All authors have read and agreed to the published version of the manuscript.

Funding: This research was funded by Science Foundation Ireland (SFI) and European Regional Development Fund under grant number 13/RC/2077.

Acknowledgments: The authors are with the CONNECT Centre. This publication has emanated from research conducted with the financial support of Science Foundation Ireland (SFI) and is co-funded under the European Regional Development Fund under Grant Number 13/RC/2077.

Conflicts of Interest: The authors declare no conflict of interest.

References

1. Vaupel, T.; Eibert, T.F. Comparison and Application of Near-Field ISAR Imaging Techniques for Far-Field Radar Cross Section Determination. *IEEE Trans. Antennas Propag.* **2006**, *54*, 144–151. [CrossRef]
2. Bliss, D.W.; Forsythe, K.; Hero, A.; Yegulalp, A. Environmental issues for MIMO capacity. *IEEE Trans. Signal Process.* **2002**, *50*, 2128–2142. [CrossRef]
3. Marzetta, T.L. Noncooperative Cellular Wireless with Unlimited Numbers of Base Station Antennas. *IEEE Trans. Wirel. Commun.* **2010**, *9*, 3590–3600. [CrossRef]
4. Rusek, F.; Persson, D.; Lau, B.K.; Larsson, E.G.; Marzetta, T.L.; Tufvesson, F. Scaling Up MIMO: Opportunities and Challenges with Very Large Arrays. *IEEE Signal Process. Mag.* **2013**, *30*, 40–60. [CrossRef]
5. Ngo, H.Q.; Larsson, E.G.; Marzetta, T.L. Energy and Spectral Efficiency of Very Large Multiuser MIMO Systems. *IEEE Trans. Commun.* **2013**, *61*, 1436–1449. [CrossRef]
6. Larsson, E.G.; Edfors, O.; Tufvesson, F.; Marzetta, T.L. Massive MIMO for next generation wireless systems. *IEEE Commun. Mag.* **2014**, *52*, 186–195. [CrossRef]
7. Zeng, T.; Mao, C.; Hu, C.; Yang, X.; Tian, W. Multi-Static MIMO-SAR Three Dimensional Deformation Measurement System. In Proceedings of the 2015 IEEE 5th Asia-Pacific Conference on Synthetic Aperture Radar (APSAR), Singapore, 1–4 September 2015.
8. Li, J.; Stoica, P. MIMO Radar with Colocated Antennas. *IEEE Signal Process. Mag.* **2007**, *24*, 106–114. [CrossRef]
9. Haimovich, A.M.; Blum, R.; Cimini, L. MIMO Radar with Widely Separated Antennas. *IEEE Signal Process. Mag.* **2008**, *25*, 116–129. [CrossRef]
10. Frankford, M.T.; Stewart, K.B.; Majurec, N.; Johnson, J.T. Numerical and experimental studies of target detection with MIMO radar. *IEEE Trans. Aerosp. Electron. Syst.* **2014**, *50*, 1569–1577. [CrossRef]
11. Ma, C.; Yeo, T.; Liu, Z.; Zhang, Q.; Guo, Q. Target imaging based on l_1 l_0 norms homotopy sparse signal recovery and distributed MIMO antennas. *IEEE Trans. Aerosp. Electron. Syst.* **2015**, *51*, 3399–3414. [CrossRef]
12. Hoang, H.; Ahmed, Z.; John, M.; McEvoy, P.; Ammann, M. Calibration for a Hybrid MIMO Near-field Imaging System to Mitigate Antennas Effects. In Proceedings of the 2020 International Workshop on Antenna Technology (iWAT), Bucharest, Romania, 25–28 February 2020.
13. Liu, Y.; Xu, X.; Xu, G. MIMO Radar Calibration and Imagery for Near-Field Scattering Diagnosis. *IEEE Trans. Aerosp. Electron. Syst.* **2018**, *54*, 442–452. [CrossRef]
14. Korner, G.; Oppelt, D.; Adametz, J.; Vossiek, M. Novel Passive Calibration Method for Fully Polarimetric Near Field MIMO Imaging Radars. In Proceedings of the 2019 12th German Microwave Conference (GeMiC), Stuttgart, Germany, 25–27 March 2019.
15. Lin, M.; Yang, L. Blind Calibration and DOA Estimation with Uniform Circular Arrays in the Presence of Mutual Coupling. *IEEE Antennas Wirel. Propag. Lett.* **2006**, *5*, 315–318. [CrossRef]
16. Wei, H.; Wang, D.; Zhu, H.; Wang, J.; Sun, S.; You, X. Mutual Coupling Calibration for Multiuser Massive MIMO Systems. *IEEE Trans. Wirel. Commun.* **2016**, *15*, 606–619. [CrossRef]
17. Gupta, I.; Baxter, J.; Ellingson, S.; Park, H.-G.; Oh, H.S.; Kyeong, M.G. An experimental study of antenna array calibration. *IEEE Trans. Antennas Propag.* **2003**, *51*, 664–667. [CrossRef]
18. Nakhkash, M.; Huang, Y.; Al-Nuaimy, W.; Fang, M.T.C. An improved calibration technique for free-space measurement of complex permittivity. *IEEE Trans. Geosci. Remote. Sens.* **2001**, *39*, 453–455. [CrossRef]

19. Mikhnev, V.A.; Vainikainen, P. Single-reference near-field calibration procedure for step-frequency ground penetrating radar. *IEEE Trans. Geosci. Remote. Sens.* **2003**, *41*, 75–80. [CrossRef]
20. Sippel, E.; Lipka, M.; Geib, J.; Hehn, M.; Vossiek, M. In-Situ Calibration of Antenna Arrays Within Wireless Locating Systems. *IEEE Trans. Antennas Propag.* **2020**, *68*, 2832–2841. [CrossRef]
21. Kong, L.; Xu, X. Calibration of a Polarimetric MIMO Array with Horn Elements for Near-Field Measurement. *IEEE Trans. Antennas Propag.* **2020**, *68*, 4489–4501. [CrossRef]
22. Hoang, M.; John, M.; McEvoy, P.; Ammann, M.J. Near-Field Propagation Analysis for Vivaldi Antenna Design: Insight into the Propagation Process for Optimizing the Directivity, Integrity of Signal Transmission, and Efficiency. *IEEE Antennas Propag. Mag.* **2020**. [CrossRef]
23. Hoang, H.; Phan, P.; Dien, H.; Nguyen, L. Design and Experimental Study of an Ultra-Wideband Radar System. In Proceedings of the International Conference on Advanced Technologies for Communications (ATC 2014), Hanoi, Vietnam, 15–17 October 2014.
24. Broquetas, A.; Palau, J.; Jofre-Roca, L.; Cardama, A. Spherical wave near-field imaging and radar cross-section measurement. *IEEE Trans. Antennas Propag.* **1998**, *46*, 730–735. [CrossRef]
25. Tran, T.D.; Do, T.H.; Hoang, M. Development of system for detecting hidden objects based on UWB pulse radar. *Sci. Technol. Dev. J.* **2015**, *18*, 111–121. [CrossRef]

 sensors

Article

Multibeam Reflectarrays in Ka-Band for Efficient Antenna Farms Onboard Broadband Communication Satellites †

Daniel Martinez-de-Rioja [1,*], Eduardo Martinez-de-Rioja [2], Yolanda Rodriguez-Vaqueiro [3], Jose A. Encinar [1] and Antonio Pino [3]

1 Information Processing and Telecommunications Center, Universidad Politécnica de Madrid, 28040 Madrid, Spain; jose.encinar@upm.es
2 Area of Signal Theory and Communications, Universidad Rey Juan Carlos, 28942 Madrid, Spain; eduardo.martinez@urjc.es
3 AtlanTTic Research Center, Universidade de Vigo, 36310 Vigo, Spain; yrvaqueiro@com.uvigo.es (Y.R.-V.); agpino@com.uvigo.es (A.P.)
* Correspondence: jd.martinezderioja@upm.es

† This paper is an extended version of the paper entitled "Advanced Reflectarray Antennas for Multispot Coverages in Ka Band" presented at the 2020 International Workshop on Antenna Technology (iWAT 2020).

Abstract: Broadband communication satellites in Ka-band commonly use four reflector antennas to generate a multispot coverage. In this paper, four different multibeam antenna farms are proposed to generate the complete multispot coverage using only two multibeam reflectarrays, making it possible to halve the number of required antennas onboard the satellite. The proposed solutions include flat and curved reflectarrays with single or dual band operation, the operating principles of which have been experimentally validated. The designed multibeam reflectarrays for each antenna farm have been analyzed to evaluate their agreement with the antenna requirements for real satellite scenarios in Ka-band. The results show that the proposed configurations have the potential to reduce the number of antennas and feed-chains onboard the satellite, from four reflectors to two reflectarrays, enabling a significant reduction in cost, mass, and volume of the payload, which provides a considerable benefit for satellite operators.

Keywords: reflectarray antennas; multibeam antennas; dual band reflectarrays; communication satellites; Ka-band

Citation: Martinez-de-Rioja, D.; Martinez-de-Rioja, E.; Rodriguez-Vaqueiro, Y.; Encinar, J.A.; Pino, A. Multibeam Reflectarrays in Ka-Band for Efficient Antenna Farms Onboard Broadband Communication Satellites . *Sensors* **2021**, *21*, 207. https://doi.org/10.3390/s21010207

Received: 25 November 2020
Accepted: 28 December 2020
Published: 31 December 2020

1. Introduction

In the past years, the continental contoured beams traditionally used for broadcast satellite applications in Ku-band are being replaced by cellular coverages to provide broadband services typically in Ka-band [1]. The cellular coverages are formed by between 50 and 100 slightly overlapping spot beams, generated with a frequency and polarization reuse scheme of four colors, where each color is associated to a unique combination of frequency and polarization [2]. Thus, the four-color scheme requires splitting the available user spectrum into two different frequency sub-bands and two orthogonal circular polarizations (CP) [2,3]. The generation of four-color coverages makes it possible to spatially isolate the spots generated with same frequency and polarization, so the interference between spots is reduced and the throughput of the users can be increased without modifying the operational bandwidth of the system. The high-gain beams generated by the satellite antennas produce circular spots on the Earth's surface. The directions of the beams are adjusted to produce a triangular lattice of circular spots. Therefore, the service area can be split into hexagonal cells, each of which is covered by a circular beam [2]. Figure 1 shows an example of a four-color coverage based on hexagonal cells. The diameter of the spots is defined depending on the capacity need of the system. Typically, the spots cover a circular area of around 250–300 km in diameter, which corresponds to a beamwidth of about 0.5–0.65° for the beams produced by the satellite antennas [3].

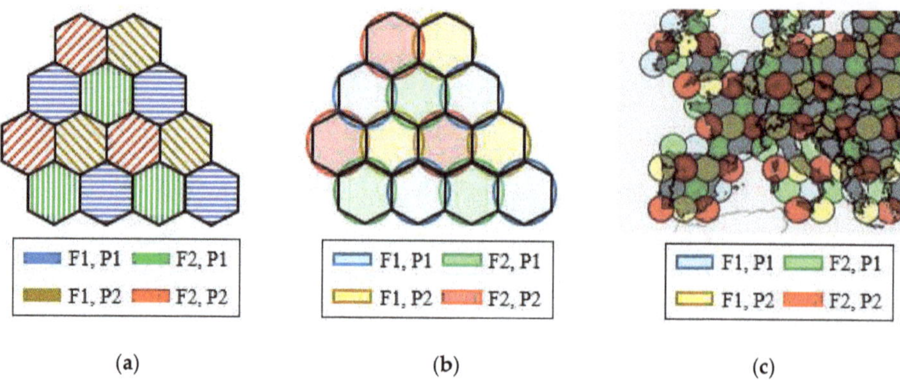

Figure 1. Example of a four-color multispot coverage. (**a**) Hexagonal service cells working in four different colors. (**b**) Hexagonal cells covered by circular beams generated in two orthogonal polarizations (P1, P2) and two frequencies (F1, F2). (**c**) Example of a European four-color coverage that would be generated by a geostationary satellite.

The generation of the multispot coverage from the satellite is typically accomplished by using four multi-fed single offset reflectors operating by a single feed per beam (SFPB) architecture [3,4]. Reflectors offer a reliable and simple operation; however, they cannot generate the beams with an angular separation between adjacent beams as small as required for this application, since overlapping feeds would be required. Thus, four multi-fed reflectors are usually employed onboard the satellite, where each reflector produces all the beams in a specific frequency and polarization, which are spatially isolated from each other. In this way, each reflector generates the beams associated to one color in a lattice of non-contiguous spots (the spots are separated by gaps that will be covered by the beams generated by the other reflectors). Therefore, the interlaced beams produced by the four reflectors form the final four-color cellular coverage of contiguous spots.

Due to the severe constraints in weight and volume of satellites, the use of four reflectors can be seen as a suboptimal antenna farm. Different multibeam antenna solutions have been proposed lately to reduce the number of antennas onboard the satellite [5]. The use of lenses [6] or array antennas [7] reduces the radiation efficiency and increases the complexity of the antenna architecture, while the highly oversized reflector proposed in [8] comprises a prohibitive stowage volume. In this paper, four different multibeam antenna solutions are proposed based on reflectarray antennas [9], following the study introduced in [10]. Each multibeam reflectarray is intended to generate half the required multispot coverage, making it possible to reduce the number of antennas onboard the geostationary satellite from four reflectors to two reflectarrays. The antenna specifications commonly required in real scenarios (described in Section 2) will be used to analyze the performance of the proposed reflectarrays. In Section 3, a 1.8 m flat reflectarray and a 1.8 m parabolic reflectarray will be proposed to generate a four-color coverage only for the transmission link with a single antenna aperture. Then, a dual-reflectarray system and a 1.8 m parabolic reflectarray will be shown in Sections 4 and 5, respectively, to generate half the required spots (two colors) simultaneously in transmission (Tx) and reception (Rx). The main characteristics of the four reflectarrays will be compared in Section 6, proving the great potential of reflectarrays for multibeam satellite applications and their capability to halve the number of antennas and feed-chains required onboard the satellite to generate a four-color coverage.

2. Mission Scenario and Requirements of the Antenna System

The main specifications of the cellular coverage, shown in Table 1, have been established from those of current multibeam satellites in Ka-band [2,3,11,12] and the Rec. ITU-R S.672-4 [13]. The coverage must be formed by around 100 spots in a triangular lattice, generated with a four-color reuse scheme based on two different frequencies and two orthogonal CP. The four-color coverage must be simultaneously generated at Tx and Rx user frequencies in Ka-band, which comprise from 19.2 to 20.2 GHz for Tx and from 29.0 to 30.0 GHz for Rx. Thus, both Tx and Rx frequency bands must be divided into two sub-bands to provide the four-color coverage with two orthogonal CP simultaneously in Tx and Rx. Figure 2 shows a schematic representation of a multispot coverage with a four-color reuse scheme of two frequencies (F1, F2) and two polarizations (P1, P2) together with the operating scheme of the current multi-fed reflectors used onboard the satellite, where each reflector generates a lattice of non-contiguous spots in a single color, simultaneously in Tx and Rx (the interlaced beams produced by the four reflectors form the final four-color coverage of contiguous spots). The diameter of the spots is set to 0.65° to cover a circular area of around 300 km on the Earth. The angular separation between adjacent spots from center to center to form the appropriate triangular lattice of spots must be 0.56° (computed as sin(60°)·0.65°). The minimum end of coverage (EOC) gain of the beams is set to 45 dBi, the minimum single-entry carrier over interference ratio (C/I) at 20 dB, and the minimum co-polar over cross-polar discrimination (XPD) is also fixed to 20 dB.

Table 1. Antenna system requirements.

Parameter	Requirement
Number of spots	100
Reuse scheme	4 colors
Spot lattice	Triangular
Spot diameter	0.65°
Spot separation	0.56°
EOC gain	45 dBi
Single-entry C/I	20 dB
XPD	20 dB
Tx frequency band	19.2–20.2 GHz
Rx frequency band	29.0–30.0 GHz

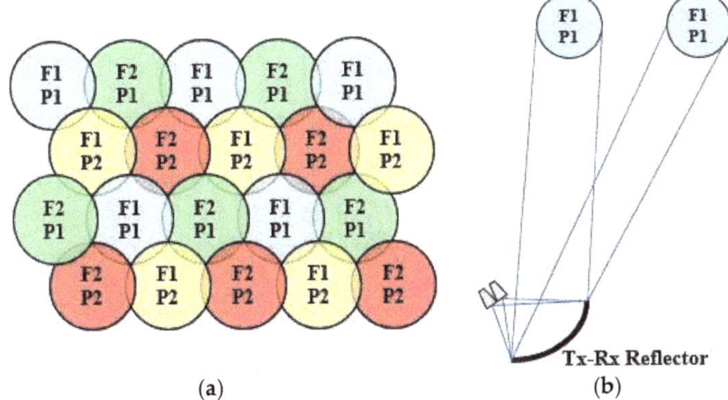

Figure 2. (**a**) Four-color multispot coverage required simultaneously in Tx and Rx. (**b**) Operating principle of current reflectors used onboard the satellite, operating simultaneously in Tx and Rx.

The proposed antenna configurations operate by an improved SFPB configuration, taking advantage of the ability of reflectarrays to generate independent beams at different frequencies [14] or polarizations [15] with a single feed. The proposed reflectarrays will be illuminated by a cluster of 27 feeds defined with a triangular lattice, in order to provide the required multibeam coverage with a triangular grid of spots. The feeds have been modelled considering the 54 mm Ka-band feed-chain reported in [4], thus the separation between adjacent feeds has been set to 55 mm. The radiation pattern of the feeds has been modelled in the simulations by an ideal $\cos^q(\theta)$ distribution. Figure 3 shows the proposed cluster of 27 feeds using the local coordinate system (x_f, y_f), the origin of which matches the center of the aperture of the central feed (C3 in Figure 3). The symmetry of each antenna configuration will be used to reduce the number of feeds considered in simulations (the lines of feeds A, B, C, plotted with thicker lines, or the array of 3x5 feeds plotted with solid lines in Figure 3).

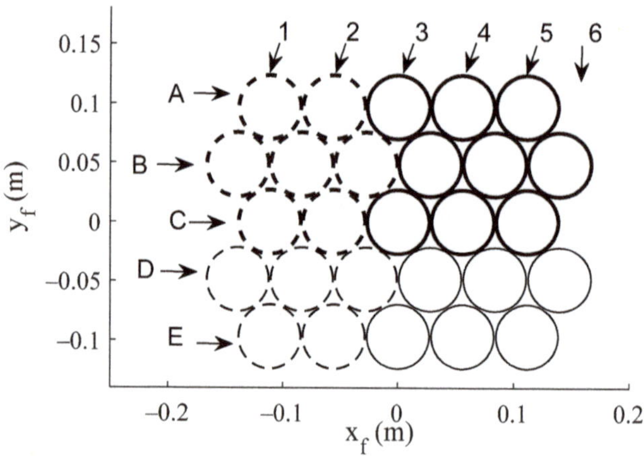

Figure 3. Cluster of 27 feeds defined to illuminate the different antenna farms.

3. Antenna Farm Based on Two Single-Band Reflectarrays

The first strategy introduced in [10] to generate a complete four-color multispot coverage simultaneously in Tx and Rx is based on the design of a reflectarray antenna to generate four spaced beams per feed in two different operating frequencies and two orthogonal polarizations (i.e., four spaced beams in four different colors). In this way, the limitation of conventional reflectors to provide such closely spaced beams is overcome by the generation of four adjacent beams per feed. Thus, a reflectarray illuminated by the proposed cluster of 27 feeds would generate a complete four-color coverage of 108 spots only for a single band (Tx or Rx). To provide multispot coverage both in Tx and Rx, two single-band reflectarrays designed with the same technique to produce four adjacent beams per feed would be required onboard the satellite (one for Tx and the other for Rx). The operating scheme of the proposed solution is shown in Figure 4 for a flat reflectarray to operate in the Tx band in Ka-band.

This solution requires a reflectarray with independent operation at close frequencies (within the same band for Tx or Rx in Ka-band) and also independent operation in orthogonal polarizations, which increases the complexity of the reflectarray cells and the design technique. The controlled application of the beam squint effect with frequency has been used to reduce the design complexity of the reflectarray antenna, as shown in [16]. The method to generate four spaced beams in four different colors per feed has been experimentally validated in [17] for a 43 cm reflectarray antenna. The prototype operates in linear polarization (LP), but the same design technique can be applied to produce the beams in CP by an appropriate selection of the reflectarray cells. Figure 5 shows the 43 cm prototype in the anechoic chamber and the measured radiation patterns of the four spaced beams in four different colors, according to the normalized angular coordinates $u = \sin\theta\cdot\cos\varphi$, $v = \sin\theta\cdot\sin\varphi$.

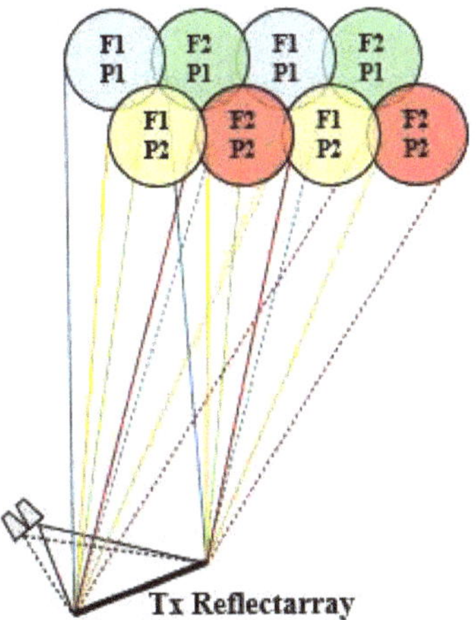

Figure 4. Operating principle of the proposed single-band reflectarray.

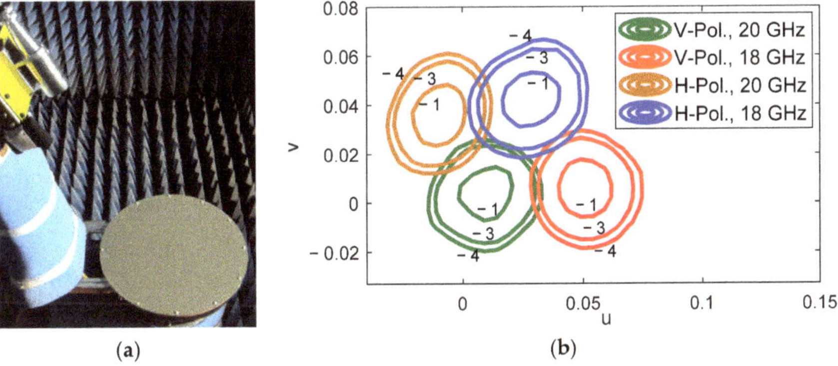

Figure 5. Reflectarray prototype to generate four spaced beams per feed [17]. (**a**) Picture of the prototype, (**b**) measured pattern contours at −1, −3, and −4 dB of the four beams generated by the same feed.

Two different approaches have been evaluated to produce a multispot coverage following the design technique developed in [17]: first, using a flat reflectarray as proposed in [10], and second, using a reflectarray with a parabolic surface. The two reflectarrays are expected to generate a four-color coverage for Tx in Ka-band using the feed cluster shown in Figure 3. Due to the symmetry of the antenna system, the simulations have considered the lines of feeds A, B, and C in Figure 3 (plotted with thicker lines in Figure 3), since there will be minimal differences between the beams generated by the lines of feeds B, C and D, E. The two reflectarrays have a diameter of 1.8 m and operate at 19.45 and 19.95 GHz in orthogonal polarizations. The simulations have considered ideal reflectarray elements to provide the required phase-shift in each color without phase errors or ohmic losses. The preliminary results will be used to evaluate the strengths and weaknesses of the design method. The conclusions reached in this study can also be applied to the associated reflectarrays used for Rx, since the design method would be identical.

3.1. Flat Transmit Reflectarray to Generate Four Spaced Beams per Feed in Ka-Band

A flat reflectarray has been proposed to generate four spaced beams per feed according to the design technique shown in [17]. The reflectarray has a diameter of 1.8 m and it is formed by 44125 reflectarray cells arranged in a 239 × 235 lattice, the period of which has been set to 7.5 mm to avoid grating lobes. As a result of the design method, four different phase distributions are implemented in the reflectarray surface (one for each combination of frequency and polarization). Figure 6 shows the required phase distributions at 19.45 and 19.95 GHz for one CP (there are no appreciable differences with the phase distributions for the orthogonal CP). The large size of the reflectarray and its flat configuration are the reason why the phase distributions depicted in Figure 6 show a large number of 360° cycles. The abrupt variations in the phase distributions limit the antenna performance and reduce the operational bandwidth.

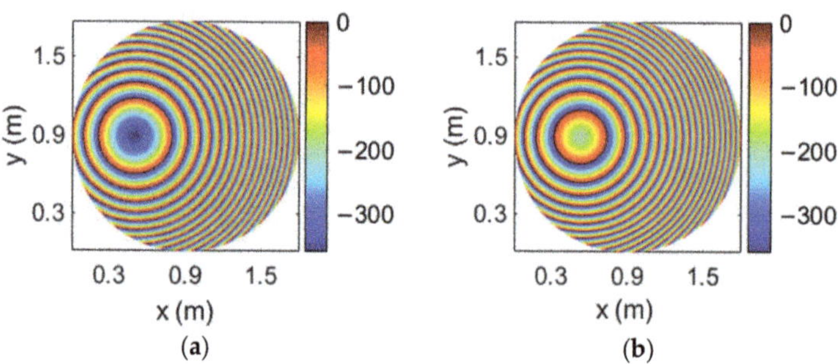

Figure 6. Required phase distributions of the 1.8 m flat reflectarray at the (**a**) lower and (**b**) upper operating frequencies for the same CP.

The results of the conducted simulations when the 1.8 m reflectarray is illuminated by the 16 feeds placed in the lines A, B, and C in Figure 3 are shown in Figure 7, which presents the pattern contours of the beams for a 46 dBi gain. The simulated pattern contours show a coverage of 64 beams generated in a four-color reuse scheme. The distribution of beams meets the required triangular lattice of spots, as well as the separation and diameter of the circular spots (0.56° and 0.65°, respectively).

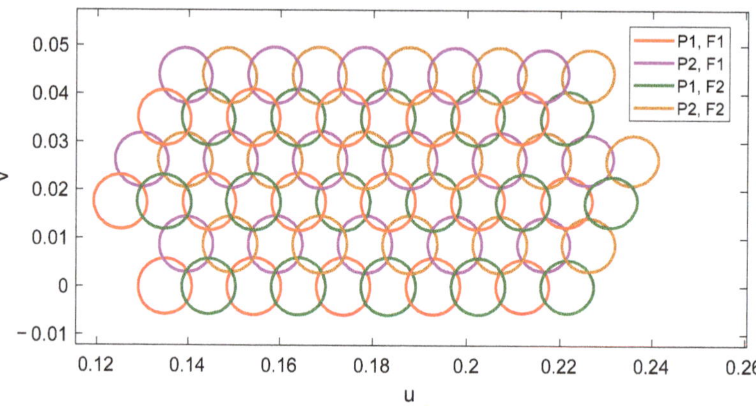

Figure 7. Simulated 46 dBi pattern contours of the 64 beams generated by 16 feeds in four different colors with a 1.8 m flat reflectarray.

The main cuts of the radiation patterns have been evaluated to estimate the interferences between beams generated in the same color. Figure 8 shows the horizontal cut of the radiation patterns in the plane v = 0 for the beams generated at both frequencies. Due to the monofocal antenna design, the extreme beams of the coverage show some aberrations that increase the C/I. Since the side-lobe levels of the beams can be reduced by slightly increasing the antenna aperture and reducing the edge illumination, this drawback can be easily overcome (space reflectors in Ka-band usually have a diameter of 2.3 m). Thus, the diameter of 1.8 m has been maintained for the rest of the reflectarray antennas proposed in this paper, expecting similar C/I levels. The use of ideal reflectarray cells prevents an accurate characterization of the cross-polar radiation, which is mainly produced by the reflectarray elements and the offset antenna configuration. Previous reflectarray prototypes with offset configurations have provided XPD above 25 dB [17,18].

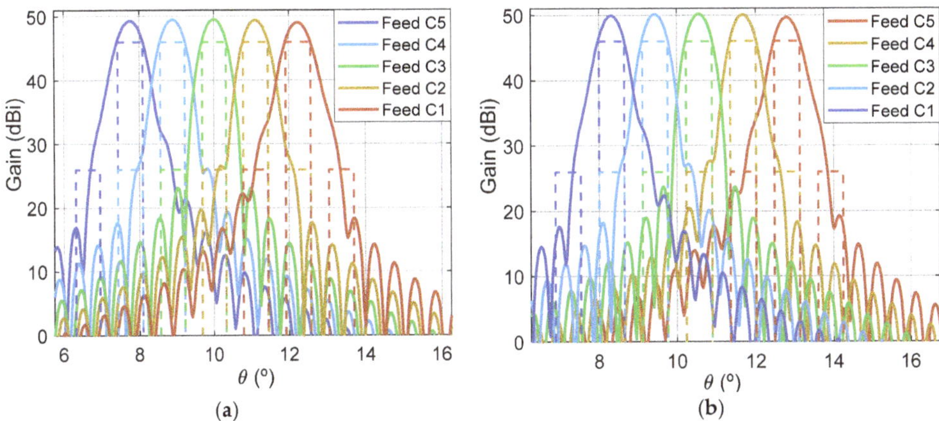

Figure 8. Cut of the simulated radiation patterns in v = 0 for the beams generated at the (**a**) lower and (**b**) upper frequency in the same circular polarization (CP).

The main constraint of this strategy is the operational bandwidth of the reflectarray, limited by the abrupt phase variations shown in Figure 6 and the independent operation at two close frequencies within the same link (Tx or Rx) in Ka-band.

3.2. Parabolic Transmit Reflectarray to Generate Four Spaced Beams per Feed in Ka-Band

The design of large and flat reflectarrays results in inappropriate phase distributions (with fast phase variations) that limit the antenna performance. The previous design of a multibeam reflectarray to generate four spaced beams per feed has been improved to fit a 1.8 m parabolic reflectarray. In this configuration, the parabolic surface shapes the beams as a conventional reflector, while the reflectarray elements only introduce a small phase correction to slightly deviate the beams produced in orthogonal polarizations and/or at different operating frequencies.

The four required phase distributions that must be implemented in the parabolic reflectarray have been computed and Figure 9 shows the required phase distributions at 19.45 and 19.95 GHz for one CP (the phase distributions for the orthogonal CP have a similar behavior). The phase distributions in Figure 9 show a great improvement with respect to the equivalent distributions for a flat reflectarray (see Figure 6), presenting smooth phase variations and a reduced number of 360° cycles. The 1.8 m parabolic reflectarray antenna has been simulated considering the same cluster of feeds as in the previous design, producing a four-color coverage of 64 beams similar to that shown in Figure 7. There are no appreciable differences between the simulated radiation patterns of both flat and parabolic reflectarrays (a comparison between the antenna performances of flat and parabolic reflectarrays to produce four beams per feed can be seen in detail in [19]).

The results obtained for this (flat or parabolic) antenna farm show that a single reflectarray with 27 feeds would produce a four-color coverage of 108 beams only in the Tx or Rx link of the Ka-band, so two reflectarrays would be needed to generate a complete multispot coverage both in Tx and Rx. As a result, the number of antennas would be halved with respect to the conventional four-reflector configuration, and the same would happen with the number of feed-chains (since each feed produces four beams in Tx or Rx).

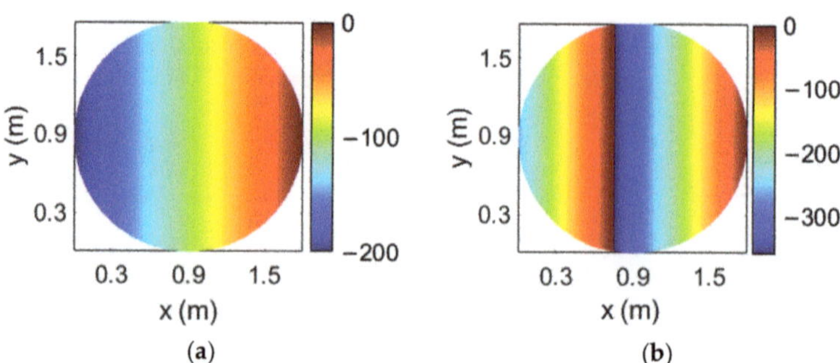

Figure 9. Required phase distributions of the 1.8 m parabolic reflectarray at the (**a**) lower and (**b**) upper operating frequencies for the same CP.

Moreover, this solution makes it possible to reduce the complexity of the feed-chains, since each reflectarray would be illuminated by single-band feeds, instead of the current dual-band feed-chains [4]. The use of independent reflectarrays to operate in Tx and Rx can be exploited to reduce the size of the Rx antenna, due to its higher operating frequency. However, the main limitation is related to the independent operation at two close frequencies. Although the parabolic surface has overcome the drawback associated with the required phase distributions, further research is needed on the reflectarray cell that should be used in a real scenario to achieve dual-band operation at near frequencies. The cell proposed in [20] could be scaled to operate at two close frequencies within the Tx band in Ka-band in dual CP. Once the reflectarray cell is defined, intense optimizations should be applied to provide an independent operation in both frequencies with a stable performance in band.

4. Antenna Farm Based on Two Dual-Band Dual Reflectarray Configurations

The constraints arising from the design of reflectarrays with independent operation at two close frequencies (within the same band for Tx or Rx) can be overcome by implementing the dual-frequency operation in two separate frequency bands. Then, the reflectarray must generate two spaced beams per feed in orthogonal CP simultaneously at Tx and Rx frequencies in Ka-band. In this way, the reflectarray is expected to generate all the beams in two different colors simultaneously in Tx and Rx. A second dual-band reflectarray with slightly different operating frequencies would generate the second half of the coverage, and the interlaced coverages of the two reflectarrays would form the complete four-color coverage in Tx and Rx. However, the design of a reflectarray with independent operation in dual CP simultaneously at two frequency bands is a complex task.

In this strategy, the reflectarray has been defined with a dual-antenna configuration based on a flat subreflector reflectarray and a main parabolic reflectarray. The use of a dual configuration makes it possible to simplify the implementation of the dual-band operation in dual CP thanks to the larger number of degrees of freedom provided by the use of two reflectarrays. In this way, the feeds will operate in dual-LP, which implies a lesser complexity of the feed-chain than in the case of dual-CP operation (there is no need for a polarizer in the feed-chain). Then, the subreflectarray will be designed to deviate (without focusing) the dual-LP beams radiated by the feeds by $\pm 0.28^{\circ}$ (in opposite directions for each LP) simultaneously in Tx and Rx. The main reflectarray will be designed to convert the incident dual-LP field into dual-CP, at the same time as its parabolic surface focuses high-gain beams. Figure 10a shows the operating principle of the dual reflectarray antenna to generate two spaced beams in left-handed and right-handed CP (LHCP and RHCP, respectively) from a single feed operating in horizontal and vertical LP (H and V, respectively). Figure 10b shows the operating scheme of the proposed antenna to generate half the required multispot coverage (two colors) simultaneously in Tx and Rx.

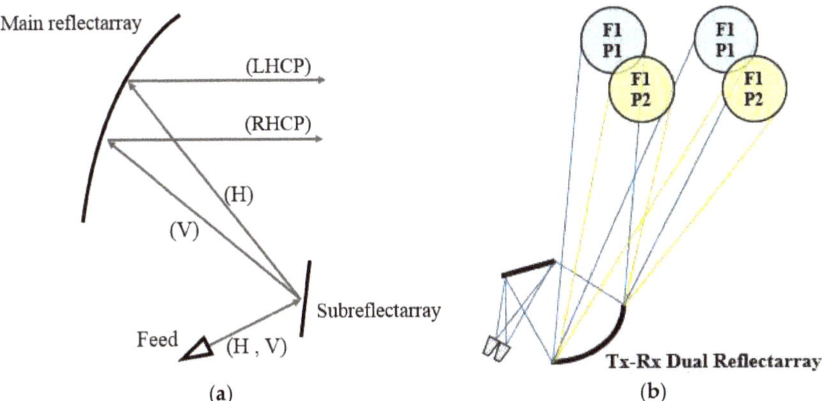

Figure 10. Dual reflectarray system proposed to operate simultaneously in Tx and Rx. (**a**) Operating principle of the dual-configuration, (**b**) functional scheme of the proposed antenna on the satellite.

The operating principles behind this solution have been experimentally validated separately. The design of a dual-band reflectarray with independent operation in dual LP at two separated frequencies can be reached in a direct way, as proposed in [21], by means of reflectarray cells based on orthogonal sets of parallel dipoles for each band. A polarizing reflectarray to convert dual LP into dual CP with broadband operation from 20 to 30 GHz (covering both Tx and Rx bands) has been demonstrated in [22].

The dual reflectarray designed to produce multiple spot beams in Ka-band has been defined on the basis of a Cassegrain system. The flat subreflectarray has a diameter of 0.65 m and it is formed by 11,497 cells disposed in a 117 × 125 grid, considering the same type of reflectarray cells used in [23] (based on two orthogonal set of parallel dipoles) with a period of 5.3 mm. The main parabolic reflectarray has a diameter of 1.8 m, formed by 62,654 cells disposed in a 286 × 279 grid. It has been designed using the polarizing cells shown in [22] (based on three coplanar parallel dipoles) with a period of 5 mm. In contrast to the complex phase distributions computed for the previous flat 1.8 m reflectarray (see Figure 6), the flat subreflectarray only deviates the orthogonal LP beams radiated by the feeds in opposite directions, so the required phase distributions present a smooth phase variation at both frequencies. Figure 11 shows the phase distributions implemented in the subreflectarray at 19.7 GHz and 29.5 GHz in one LP; note that the phase distributions in the orthogonal LP show the opposite phase variation at each frequency to deviate the orthogonal LP beam in the opposite direction. Then, the main parabolic reflectarray only converts the dual-LP incident field into dual-CP, so their cells have been designed to introduce a phase difference of 90° between the two orthogonal components of each incident LP [22].

The design of the subreflectarray and the main parabolic reflectarray has been carried out separately. The radiation patterns of the designed subreflectarray (without the main reflectarray) have been simulated and the results have been compared with those obtained considering ideal reflectarray cells (providing the required phase distributions without any phase errors and zero dielectric losses). As an example, Figure 12a shows the comparison between the ideal and realized patterns in the azimuth plane (φ = 90°) for the subreflectarray in H polarization at 19.7 and 29.5 GHz. Note that the patterns in Figure 12a show a wide main lobe since the subreflectarray does not focus the beam radiated by the feed (the parabolic surface of the main reflectarray will focus the beam). In the same way, the simulated radiation patterns of the designed polarizing main reflectarray (without the subreflectarray, so that the main reflectarray is illuminated from the virtual focus of the dual system) have been computed and compared with the simulations performed with ideal cells. Figure 12b shows the comparison between the ideal and realized patterns in the azimuth plane at 19.7 and 29.5 GHz for the polarizing main reflectarray in RHCP (obtained from the H polarization radiated by the feed).

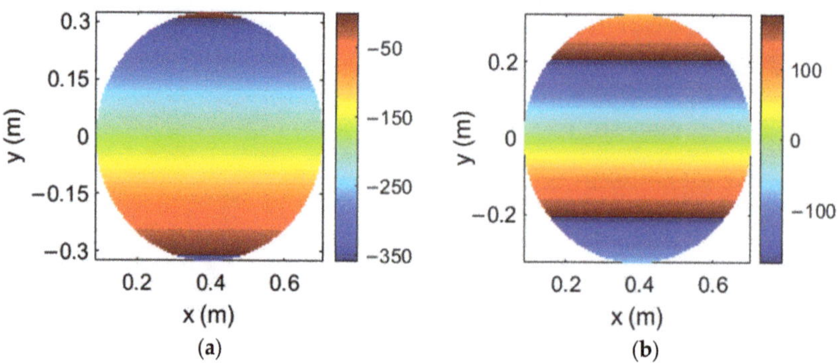

Figure 11. Required phase distributions of the 0.65 m flat subreflectarray at the (**a**) lower and (**b**) upper operating frequencies for the same linear polarization (LP).

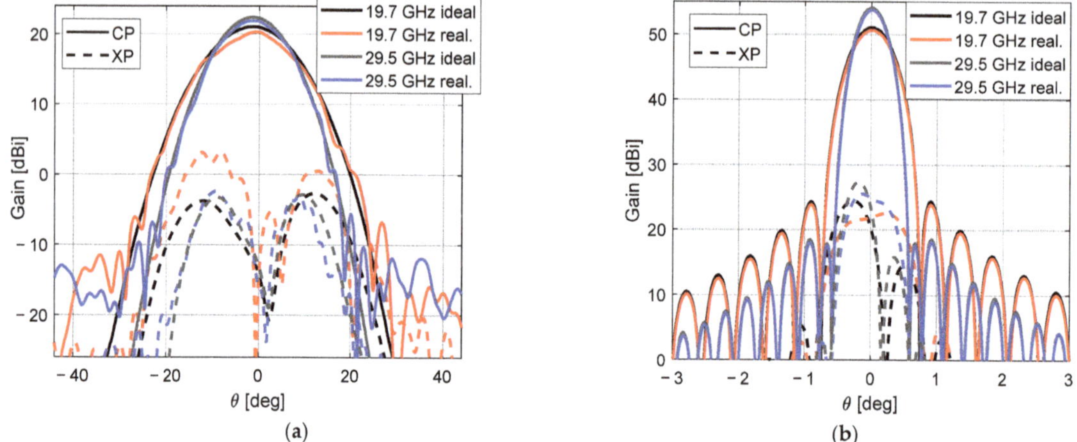

Figure 12. Simulated radiation patterns at 19.7 and 29.5 GHz in the azimuth plane of the ideal and realized antenna components. (**a**) Subreflectarray for H polarization, (**b**) main reflectarray for right-handed circular polarizations (RHCP).

The patterns in Figure 12 show a satisfactory agreement between the designed and ideal performances of both reflectarrays. The dual system configuration results in relatively large incidence angles from the feed on the subreflectarray, which complicates the design of the elements of the subreflectarray. As a result, the simulated patterns of the designed subreflectarray show a slightly lower gain and higher cross-polar levels than in the ideal case (see Figure 12a). On the other hand, the simulated results of the designed main reflectarray show an excellent performance, which is practically identical to the ideal behavior (see Figure 12b).

Finally, the dual antenna configuration is illuminated by the cluster of 27 feeds shown in Figure 3 rotated 90° in the $x_f y_f$-plane. The simulations of the complete dual reflectarray have considered 15 of the 27 feeds (an array of 3 × 5 feeds, plotted with solid lines in

Figure 3), since the beams produced by the remaining feeds are expected to have similar behavior. The simulated pattern contours of the 30 beams generated by the designed dual reflectarray simultaneously at 19.7 GHz and 29.5 GHz are depicted in Figure 13 at the EOC levels of the beams (between 45 and 46 dBi). The simulated beams match with the required spot distribution, where there is room for the beams that would be generated by a second dual reflectarray operating at slightly different frequencies to form the complete four-color multispot coverage in Tx and Rx.

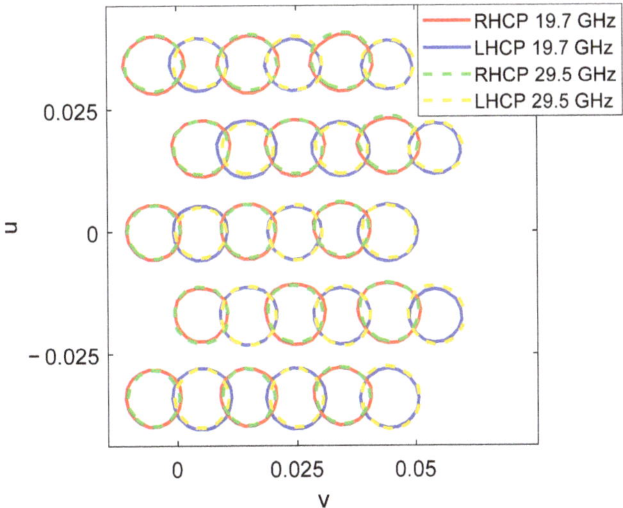

Figure 13. Simulated pattern contours for the 30 beams generated by 15 feeds illuminating the proposed dual reflectarray at 19.7 GHz and 29.5 GHz.

The main cuts of the radiation patterns have revealed similar C/I levels as those of the previous 1.8 m reflectarray. Figure 14 shows the cut of the radiation patterns in the plane u = 0 for the beams produced by the central row of feeds at both frequencies. The simulated cross-polar radiation provides an XPD lower than 20 dB for some beams. The analysis of the results has shown that the increase of the cross-polar component is mainly produced in the subreflectarray, so an optimization procedure should be applied to the dual-band dual-LP cells to reduce the cross-polarization. Moreover, the use of a dual configuration could be used to implement a bifocal technique to improve the shaping of the beams generated in the edge of the coverage [24].

In conclusion, the simulated results of the dual reflectarray configuration prove the capability of the proposed multibeam antenna to generate half the required four-color coverage simultaneously in Tx and Rx. The dual-band operation involving separate frequencies simplifies the design of the reflectarray elements, while the dual configuration simplifies the process of generating two spaced beams in orthogonal CP per feed, splitting it into two straightforward stages (deviation of the LP beams in opposite directions and dual-LP to dual-CP conversion), the principles of which have been experimentally validated previously. The simulated results show that a further optimization is required for the subreflectarray to improve the XPD, although the flexibility provided by the dual configuration could be also used to improve the overall performance of the antenna.

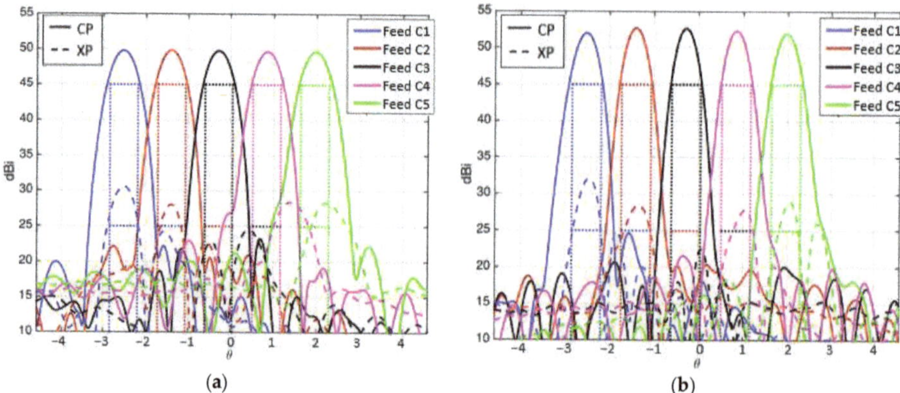

Figure 14. Cut of the simulated radiation patterns in u = 0 for the beams generated at (**a**) 19.7 GHz and (**b**) 29.5 GHz in same CP.

5. Antenna Farm Based on Two Dual-Band Offset Parabolic Reflectarrays

Bearing in mind the better performance of the previous dual-band dual reflectarray with respect to the single-band reflectarray, the final proposed antenna farm is also based on the use of a dual-band reflectarray to generate all the beams associated to two colors simultaneously in Tx and Rx. In this case, a single offset parabolic reflectarray is proposed instead of the previous dual configuration, reducing the number of components in the antenna configuration.

The generation of two spaced beams in orthogonal CP per feed is now completely managed by the reflectarray elements placed on the parabolic reflectarray surface. The operating principle of the antenna is based on the concept proposed in [25], where a single-band parabolic reflectarray deviates the orthogonal CP beams focused by the parabolic surface in opposite directions by means of the variable rotation technique (VRT) applied to the reflectarray elements. In our solution, a dual-band offset parabolic reflectarray is proposed to focus the beams by its parabolic surface and deviate the orthogonal CP by the application of VRT simultaneously at Tx and Rx frequencies. The dual-band operation by VRT is achieved using the reflectarray cells proposed in [20]. As a result, a single offset parabolic reflectarray illuminated by dual-CP feeds generates two spaced beams in orthogonal CP per feed at Tx and Rx frequencies in Ka-band. The operating scheme of the proposed reflectarray to generate half the required four-color coverage is shown in Figure 15.

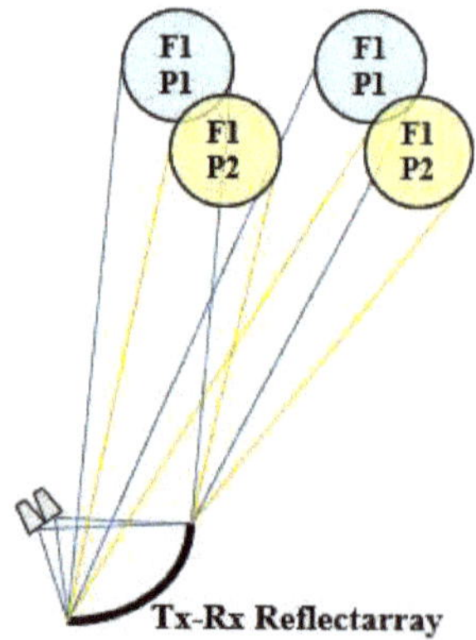

Figure 15. Operating principle of the proposed single-band parabolic reflectarray.

Recently, the proposed operating principle has been experimentally validated in [26] by a 0.9 m parabolic reflectarray prototype that generates two spaced beams per feed at Tx and Rx frequencies using the reflectarray cell described in [20]. Figure 16 shows the prototype in the anechoic chamber and the 40.6 dBi pattern contours of the two measured beams generated simultaneously at 19.7 GHz and 30 GHz with a single dual-CP feed.

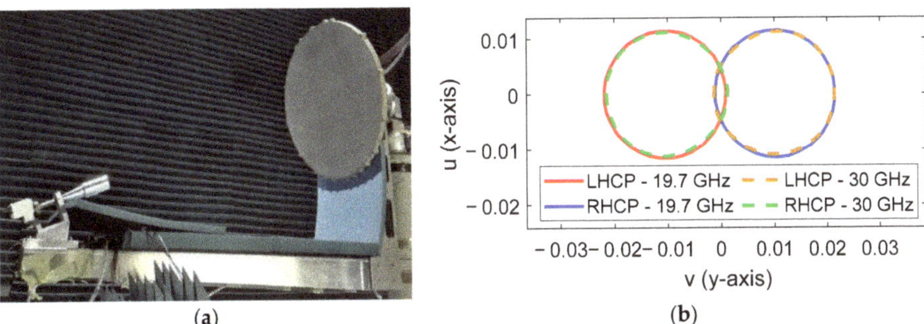

Figure 16. Dual band parabolic reflectarray prototype to generate two spaced beams per feed [26]. (**a**) Picture of the prototype, (**b**) measured 40.6 dBi pattern contours of the two beams generated at 19.7 and 30 GHz by the same feed.

A 1.8 m parabolic reflectarray formed by 62,654 reflectarray cells with a period of 6.5 mm has been designed to deviate the beams in orthogonal CP $\pm$ 0.28° simultaneously at 19.7 and 29.5 GHz. Since the parabolic surface of the antenna focuses the beams radiated by the feeds, the required phase distributions on the reflectarray to split the orthogonal CP beams present a similar aspect as those of the subreflectarray shown in Section 4. Figure 17 shows the required phase distributions on the parabolic surface at 19.7 GHz and 29.5 GHz in one CP (the phase distributions in the orthogonal CP show the opposite phase variation to deviate the orthogonal CP beams in the opposite direction).

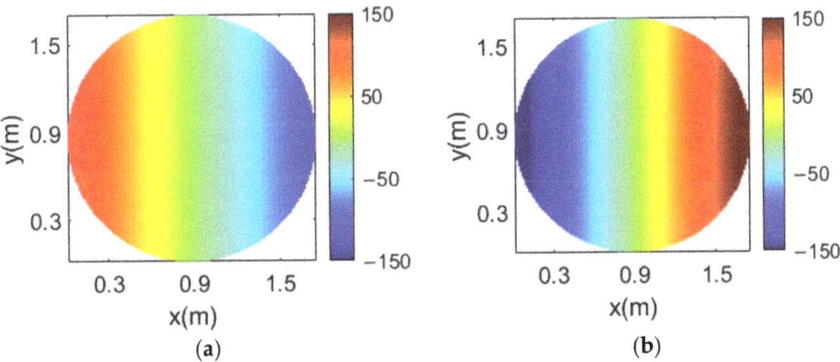

Figure 17. Required phase distributions of the 1.8 m parabolic reflectarray at the (**a**) lower and (**b**) upper operating frequencies for the same CP.

As in the first antenna solution shown in Section 3, the proposed dual-band parabolic reflectarray has been illuminated by the cluster of 27 feeds shown in Figure 3, and the simulations have considered three lines of feeds (16 feeds). Thus, the designed parabolic reflectarray is expected to generate 32 beams in orthogonal CP simultaneously in Tx and Rx. The simulated pattern contours of the 32 beams generated by the 1.8 m parabolic reflectarray at 19.7 and 29.5 GHz are depicted in Figure 18, where the contours of the beams at 19.7 and 29.5 GHz correspond to a gain level of 46 and 45 dBi, respectively. In a similar way to the previous dual reflectarray configuration, a second parabolic reflectarray operating at slightly different frequencies would produce the second half of the required four-color coverage.

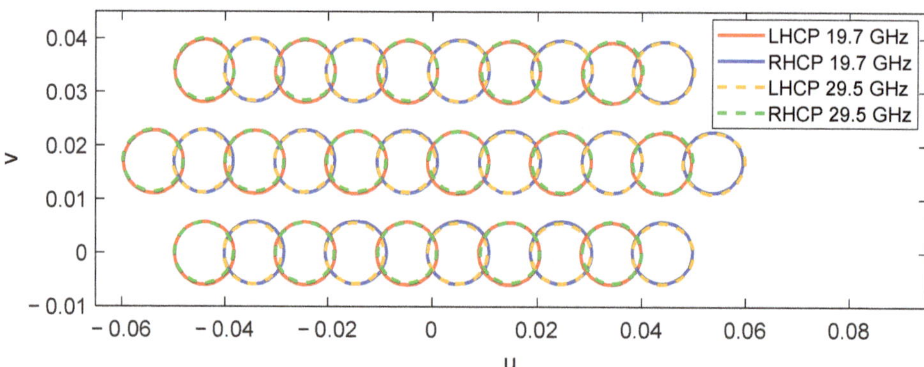

Figure 18. Contours of 32 beams generated simultaneously at 19.7 GHz and 29.5 GHz by the 1.8-m dual-band parabolic reflectarray illuminated by 16 feeds.

Figure 19 shows the horizontal cut in the plane v = 0 of the radiation patterns for the beams generated at 19.7 and 29.5 GHz. The main cuts of the radiation pattern have shown similar values of C/I to those of the previous configurations, with a XPD larger than 20 dB. Due to the large electrical size of the antenna at 29.5 GHz, the beams in Rx show a higher maximum gain than the beams in Tx; however, the shaping of the beams in Tx and Rx can be matched by the optimization procedure applied in [26].

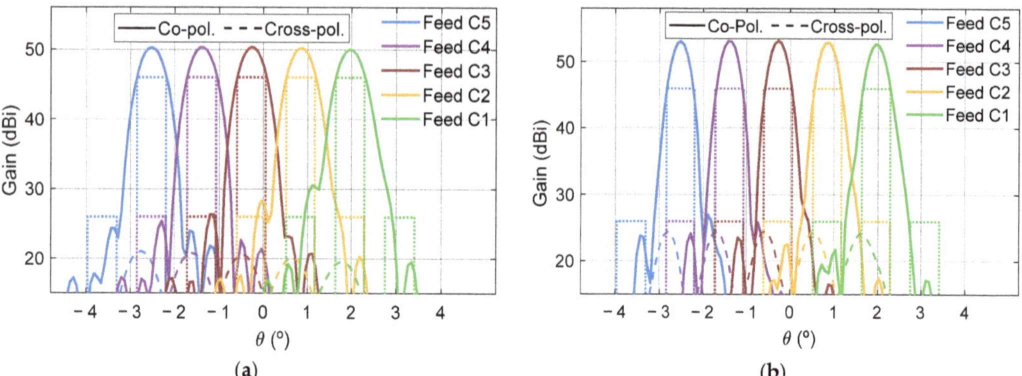

Figure 19. Cut of the simulated radiation patterns in v = 0 for the beams generated at (**a**) 19.7 GHz and (**b**) 29.5 GHz in same CP.

In conclusion, the simulated results of the dual-band parabolic reflectarray have demonstrated a very satisfactory performance. The simulations show a suitable distribution of the beams with appropriate values of gain and XPD. As in the previous configurations, the C/I can be improved by a small increase of the antenna size to reduce the edge illumination levels on the reflectarray. Moreover, the reflectarray cell considered in the design makes it possible to implement additional optimization to shape the beams in Rx and to maintain the cross-polar levels in band, as in [26], by a proper adjustment of the dimensions and rotation angle of the elements [27].

6. Conclusions

In this paper, four novel antenna farms based on reflectarrays for broadband communication satellites in Ka-band have been proposed and evaluated. The exclusive characteristics of reflectarrays, which are able to operate independently at different frequencies or polarizations, have been applied in different ways to improve the SFPB operation compared to conventional reflectors. The impact of using flat or curved surfaces, single or dual-band operation, and single or dual antenna configurations for large reflectarrays in Ka-band has been assessed to achieve the best antenna performance. Table 2 summarizes the main characteristics of the proposed antenna solutions.

Table 2. Main characteristics of the proposed antenna solutions.

Parameter	Flat Single-Band Reflectarray	Parabolic Single-Band Reflectarray	Dual-Band Dual Reflectarray	Parabolic Dual-Band Reflectarray
Antenna configuration	Single aperture	Single aperture	Dual configuration	Single aperture
Aperture type	Flat	Parabolic	Parabolic (main) + flat (sub)	Parabolic
Beams per feed	4	4	2	2
Frequency bands	1 (Tx or Rx)	1 (Tx or Rx)	2 (Tx and Rx)	2 (Tx and Rx)
Required antennas	2	2	2	2
Complexity of the cells	Very high	Very high	Low	Low
Operation in frequency	Very hard to implement	Hard to implement	Readily achievable	Readily achievable
Operation in CP	Hard to implement	Hard to implement	Readily achievable	Readily achievable

The preliminary simulations of single band reflectarrays to generate four beams per feed permits operation with a single antenna for Tx and an independent antenna for Rx, which simplifies the simultaneous operation in both bands as well as the design of the feed-chains. However, this solution involves a major difficulty in providing a stable performance in band due to the independent operation at close frequencies, requiring complex reflectarray cells to which intense optimizations must be applied. On the other hand, the two proposed dual-band reflectarrays show promising results. The simulations of the dual reflectarray system would require a further optimization of the subreflectarray to slightly reduce the cross-polar radiation, although the use of a dual configuration with two reflectarrays provides greater flexibility of operation than single aperture systems, which can be used to correct the shaping of the beams in the edge of the coverage. Finally, the dual-band parabolic reflectarray achieves an excellent antenna performance with a simple antenna configuration similar to that of the currently used reflectors. Moreover, the reflectarray elements can be optimized to properly shape the beams in Tx and Rx without the need to shape the antenna surface.

The core technologies of the proposed reflectarrays have been experimentally validated, which confirms the capabilities of reflectarrays to provide novel efficient multibeam antenna farms for broadband communications satellites in Ka-band. All the proposed antenna solutions make it possible to halve the number of antennas and feed-chains required onboard the satellite, from four reflectors to two reflectarray antennas, and from a total of N feed-chains (to produce a coverage of N spots) to N/2 feeds, allowing an important reduction in cost, as well as in mass and volume of the payload, which are severely limited in the satellite.

Author Contributions: Conceptualization, J.A.E. and A.P.; methodology, J.A.E. and A.P.; software, D.M.-d.-R., E.M.-d.-R. and Y.R.-V.; validation, D.M.-d.-R., E.M.-d.-R. and J.A.E.; formal analysis, D.M.-d.-R., E.M.-d.-R. and Y.R.-V.; investigation, D.M.-d.-R., E.M.-d.-R., Y.R.-V., J.A.E. and A.P.; data curation, Y.R.-V. and A.P.; writing—original draft preparation, D.M.-d.-R.; writing—review and editing, E.M.-d.-R., Y.R.-V., J.A.E. and A.P.; visualization, D.M.-d.-R., E.M.-d.-R. and Y.R.-V.;

supervision, J.A.E. and A.P.; project administration, J.A.E.; funding acquisition, J.A.E. All authors have read and agreed to the published version of the manuscript.

Funding: This research has been supported in part by the Spanish Ministry of Economy and Competitiveness under the project TEC2016-75103-C2-1-R, and under the project FJCI-2016-29943; by the European Regional Development Fund (ERDF), and by the European Space Agency (ESA) under contract 4000117113/16/NL/AF.

Conflicts of Interest: The authors declare no conflict of interest.

References

1. Minoli, D. High Throughput Satellites (HTS) and KA/KU Spot Beam Technologies. In *Innovations in Satellite Communication and Satellite Technology;* Wiley: Hoboken, NJ, USA, 2015; pp. 95–159. [CrossRef]
2. Fenech, H.; Amos, S.; Tomatis, A.; Soumpholphakdy, V. High throughput satellite systems: An analytical approach. *IEEE Trans. Aerosp. Electron. Syst.* **2015**, *51*, 192–202. [CrossRef]
3. Fenech, H.; Tomatis, A.; Amos, S.; Soumpholphakdy, V.; Serrano Merino, J.L. Eutelsat HTS systems. *Int. J. Satell. Commun. Netw.* **2016**, *34*, 503–521. [CrossRef]
4. Schneider, M.; Hartwanger, C.; Wolf, H. Antennas for multiple spot beams satellites. *CEAS Space J.* **2011**, *2*, 59–66. [CrossRef]
5. Toso, G.; Mangenot, C.; Angeletti, P. Recent advances on space multibeam antennas based on a single aperture. In Proceedings of the European Conference on Antennas and Propagation, Gothenburg, Sweden, 8–12 April 2013; pp. 454–458.
6. Reiche, E.; Gehring, R.; Schneider, M.; Hartwanger, C.; Hong, U.; Ratkorn, N.; Wolf, H. Space fed arrays for overlapping feed apertures. In Proceedings of the German Microwave Conference, Hamburg-Harburg, Germany, 10–12 March 2008; pp. 1–4.
7. Kaifas, T.N.; Babas, D.G.; Toso, G.; Sahalos, J.N. Multibeam antennas for global satellite coverage: Theory and design. *IET Microw. Antennas Propag.* **2016**, *10*, 1475–1484. [CrossRef]
8. Balling, P.; Mangenot, C.; Roederer, A.G. Shaped single-feed-per-beam multibeam reflector antenna. In Proceedings of the European Conference on Antennas and Propagation, Nice, France, 6–10 November 2006; pp. 1–6. [CrossRef]
9. Huang, J.; Encinar, J.A. *Reflectarray Antennas;* IEEE Press: Piscataway, NJ, USA; Wiley: Hoboken, NJ, USA, 2008.
10. Martinez-de-Rioja, D.; Encinar, J.A.; Martinez-de-Rioja, E.; Rodriguez-Vaqueiro, Y.; Pino, A. Advanced Reflectarray Antennas for Multispot Coverages in Ka Band. In Proceedings of the 2020 International Workshop on Antenna Technology, Bucharest, Romania, 25–28 February 2020; pp. 1–4. [CrossRef]
11. Fenech, H.; Amos, S.; Tomatis, A.; Soumpholphakdy, V. KA-SAT and Future HTS Systems. In Proceedings of the IEEE International Vacuum Electronics Conference, Paris, France, 21–23 May 2013; pp. 1–2. [CrossRef]
12. Chandler, C.; Hoey, L.; Peebles, A.; Em, M. Advanced Antenna Technology for a Broadband Ka-Band Communication Satellite. *Technol. Rev.* **2002**, *10*, 37–55.
13. Rec. ITU-R S.672-4 (1997). Satellite Antenna Radiation Pattern for Use as a Design Objective in the Fixed-Satellite Service Employing Geostationary Satellites (ITU-R). Available online: https://www.itu.int/rec/R-REC-S.672-4-199709-I/en (accessed on 30 December 2020).
14. Smith, T.; Gothelf, U.V.; Kim, O.S.; Breinbjerg, O. Design, manufacturing, and testing of a 20/30 GHz dual-band circularly polarized reflectarray antenna. *IEEE Antennas Wireless Propag. Lett.* **2013**, *12*, 1480–1483. [CrossRef]
15. Encinar, J.A.; Datashvili, L.; Agustín Zornoza, J.; Arrebola, M.; Sierra–Castañer, M.; Besada, J.L.; Baier, H.; Legay, H. Dual-Polarization Dual-Coverage Reflectarray for Space Applications. *IEEE Trans. Antennas Propag.* **2006**, *54*, 2827–2837. [CrossRef]
16. Martinez-de-Rioja, D.; Martinez-de-Rioja, E.; Encinar, J.A. Multibeam reflectarray for transmit satellite antennas in Ka band using beam-squint. In Proceedings of the IEEE International Symposium on Antennas and Propagation, Fajardo, Puerto Rico, 26 June–1 July 2016; pp. 1421–1422. [CrossRef]
17. Martinez-De-Rioja, D.; Martinez-De-Rioja, E.; Encinar, J.A.; Florencio, R.; Toso, G. Reflectarray to Generate Four Adjacent Beams per Feed for Multispot Satellite Antennas. *IEEE Trans. Antennas Propag.* **2019**, *67*, 1265–1269. [CrossRef]
18. Florencio, R.; Encinar, J.A.; Boix, R.R.; Losada, V.; Toso, G. Reflectarray Antennas for Dual Polarization and Broadband Telecom Satellite Applications. *IEEE Trans. Antennas Propag.* **2015**, *63*, 1234–1246. [CrossRef]
19. Martinez-de-Rioja, D.; Martinez-de-Rioja, E.; Encinar, J.A.; Rodriguez-Vaqueiro, Y.; Pino, A. Preliminary simulations of flat and parabolic reflectarray antennas to generate a multi-spot coverage from a geostationary satellite. *IET Microw. Antennas Propag.* **2020**, *14*, 1742–1748. [CrossRef]
20. Martinez-de-Rioja, D.; Florencio, R.; Encinar, J.A.; Carrasco, E.; Boix, R.R. Dual-Frequency Reflectarray Cell to Provide Opposite Phase Shift in Dual Circular Polarization With Application in Multibeam Satellite Antennas. *IEEE Antennas Wireless Propag. Lett.* **2019**, *18*, 1591–1595. [CrossRef]
21. Martinez-de-Rioja, E.; Encinar, J.A.; Barba, M.; Florencio, R.; Boix, R.R.; Losada, V. Dual Polarized Reflectarray Transmit Antenna for Operation in Ku- and Ka-Bands With Independent Feeds. *IEEE Trans. Antennas Propag.* **2017**, *65*, 3241–3246. [CrossRef]
22. Martinez-de-Rioja, E.; Encinar, J.A.; Pino, A.; Rodriguez-Vaqueiro, Y. Broadband Linear-to-Circular Polarizing Reflector for Space Applications in Ka-Band. *IEEE Trans. Antennas Propag.* **2020**, *68*, 6826–6831. [CrossRef]

23. Martinez-de-Rioja, E.; Encinar, J.A.; Florencio, R.; Boix, R.R. Reflectarray in K and Ka bands with independent beams in each polarization. In Proceedings of the IEEE International Symposium on Antennas and Propagation, Fajardo, Puerto Rico, 26 June–1 July 2016; pp. 1199–1200. [CrossRef]
24. Martinez-de-Rioja, E.; Encinar, J.A.; Florencio, R.; Tienda, C. 3-D Bifocal Design Method for Dual-Reflectarray Configurations With Application to Multibeam Satellite Antennas in Ka-Band. *IEEE Trans Antennas Propag.* **2019**, *67*, 450–460. [CrossRef]
25. Zhou, M.; Sørensen, S.B. Multi-spot beam reflectarrays for satellite telecommunication applications in Ka-band. In Proceedings of the European Conference on Antennas and Propagation, Davos, Switzerland; 2016; pp. 1–5. [CrossRef]
26. Martinez-De-Rioja, D.; Martinez-De-Rioja, E.; Rodriguez-Vaqueiro, Y.; Encinar, J.A.; Pino, A.; Arias, M.; Toso, G. Transmit-Receive Parabolic Reflectarray to Generate Two Beams per Feed for Multi-Spot Satellite Antennas in Ka-Band. *IEEE Trans. Antennas Propag.* **2020**. [CrossRef]
27. Florencio, R.; Martinez-de-Rioja, D.; Martinez-de-Rioja, E.; Encinar, J.A.; Boix, R.R.; Losada, V. Design of Ku- and Ka-Band Flat Dual Circular Polarized Reflectarrays by Combining Variable Rotation Technique and Element Size Variation. *Electronics* **2020**, *9*, 985. [CrossRef]

Article

Numerical and Experimental Investigation of the Opposite Influence of Dielectric Anisotropy and Substrate Bending on Planar Radiators and Sensors

Plamen I. Dankov [1,*], Praveen K. Sharma [2] and Navneet Gupta [2]

1 Faculty of Physics, Sofia University "St. Kliment Ohridski", 1164 Sofia, Bulgaria
2 Department of EEE, Birla Institute of Technology and Science (BITS), Pilani 333031, India; p2016502@pilani.bits-pilani.ac.in (P.K.S.); ngupta@pilani.bits-pilani.ac.in (N.G.)
* Correspondence: dankov@phys.uni-sofia.bg; Tel.: +359-899-052-097

Abstract: The simultaneous influences of the substrate anisotropy and substrate bending are numerically and experimentally investigated in this paper for planar resonators on flexible textile and polymer substrates. The pure bending effect has been examined by the help of well-selected flexible isotropic substrates. The origin of the anisotropy (direction-depended dielectric constant) of the woven textile fabrics has been numerically and then experimentally verified by two authorship methods described in the paper. The effect of the anisotropy has been numerically divided from the effect of bending and for the first time it was shown that both effects have almost comparable but opposite influences on the resonance characteristics of planar resonators. After the selection of several anisotropic textile fabrics, polymers, and flexible reinforced substrates with measured anisotropy, the opposite influence of both effects, anisotropy and bending, has been experimentally demonstrated for rectangular resonators. The separated impacts of the considered effects are numerically investigated for more sophisticated resonance structures—with different types of slots, with defected grounds and in fractal resonators for the first three fractal iterations. The bending effect is stronger for the slotted structures, while the effect of anisotropy predominates in the fractal structures. Finally, useful conclusions are formulated and the needs for future research are discussed considering effects in metamaterial wearable patches and antennas.

Keywords: anisotropy; dielectric constant; material characterization; planar resonators; substrate bending; textile fabrics; wearable radiators

Citation: Dankov, P.I.; Sharma, P.K.; Gupta, N. Numerical and Experimental Investigation of the Opposite Influence of Dielectric Anisotropy and Substrate Bending on Planar Radiators and Sensors. *Sensors* **2021**, *21*, 16. https://dx.doi.org/10.3390/s21010016

Received: 25 November 2020
Accepted: 18 December 2020
Published: 22 December 2020

1. Introduction

Recently, many artificial materials known with their traditional applications in the human life can be considered as electrodynamic media due to the propagation of waves through them. Textile fabrics are typical examples. Most of these materials and some of their flexible polymer substitutes have been transformed into a new type of electronic components—antenna/sensor substrates due to their new applications in the wearable communication systems (antennas, sensors, radio-frequency identification or RFID, millimetre-wave identification or mmID, etc.) [1–5]. In this role, they look like the commercial reinforced substrates with PCB (printed circuit board) applications (a comparison has been given in [6], chapter IV). From a long time, the PCB designers have required manufacturers of the traditional reinforced substrates to provide up-to-date and reliable information on their dielectric parameters. Nowadays, the situation with the antenna designers of the wearable textile devices is almost the same—they must know the right information for the actual dielectric parameters of these specific materials as substrates. Our observations show that a lot of research papers appeared in the last several years concerning the characterization of the dielectric parameters (dielectric constant ε_r and dielectric loss tangent tan δ_ε) of the most popular textile fabrics [7–12]. The used methods are quite different—resonance and non-resonance—and most of them are implemented in the traditional ISM bands (typically around 2.45 GHz). However, the textile substrates differ from the reinforced substrates consist of natural and/or synthetic fibres (threads, yarns, filaments, etc.) in air and form fibrous structures with a considerably bigger variety of different cross-section views [13,14] in comparison with the simple woven or non-woven reinforced substrates. Thus, depending on the used fibre materials, their density, applied fabrication technology, and selected stitch, they act as porous materials with relatively low permittivity (ε_r ~1.2–2.0), which is quite comfortable for antenna applications (the minimal dielectric constant for the reinforced substrate is typically ε_r ~3.0). The other differences are that the textile fabrics are more flexible and compressible materials, the thickness and density of which can be easily changed by low mechanical pressure. One property seems common—the existence of an intrinsic planar anisotropy due to the predominant orientation of the fibres. However, the anisotropy the woven/knitted fabrics is mainly related to their mechanical properties (e.g., tensile coefficients) [1,13–15] and very rarely to their dielectric parameters.

Like typical artificial materials, the textile fabrics can be considered as mixtures between two or more dielectrics (reinforced fibre nets with an appropriate filling/air). In such cases, effective-media models have been developed [16], which can predict numerically the resultant isotropic dielectric constant and the dielectric loss tangent of these materials. However, the variety of technologies used for the manufacturing of these fibrous materials and the complex cross sections [5] can provoke a measurable dielectric uni- or biaxial anisotropy—direction-dependent dielectric parameters ($\varepsilon_{xx} \neq \varepsilon_{yy} \neq \varepsilon_{zz}$; tan $\delta_{\varepsilon\_xx} \neq$ tan $\delta_{\varepsilon\_yy} \neq$ tan δ_{ε_zz}). Our previous research [17] showed that most of the textile fabrics have typical uniaxial anisotropy: different dielectric parameters in parallel and perpendicular directions regarding to the sample surface ($\varepsilon_{par} = \varepsilon_{xx}$ or $\varepsilon_{yy} \neq \varepsilon_{perp} = \varepsilon_{zz}$; and tan $\delta_{\varepsilon\_par} \neq$ tan $\delta_{\varepsilon\_perp}$), which is also typical for the wide-spread microwave reinforced substrates [6,18]. Actually, the anisotropy of both types of woven materials is an undesired property, but it should be taken into account in the RF design of different microwave (incl. antenna/sensor) components especially in the mm-wavelength range. Exactly here is the difference. Nowadays, the major manufacturers of reinforced substrate started to share information about the possible anisotropy of some of their commercial products, while the designers of wearable antennas usually completely ignore this property for the textile substrate. We found only a few papers, which comment on the dielectric anisotropy of textile fabrics. The authors of the review paper [1] considered this problem for the textile fabrics without presenting concrete data. A recent paper ([19], Table 6) investigated the influence of the percentage of the normal and in-plane (parallel) components (fibres) in the woven fabrics (at microstructural level) on the resultant dielectric constant, but without to give separate values of ε_{par} and ε_{perp}.

At the same time, we found another interesting fact—many papers, devoted to the dielectric characterization of textile materials, give different results for similar materials. A typical example is the measured dielectric constant of denim textile substrates, ε_{r_Denim}. Our survey shows that the used values vary from 1.4 to 2.0 (~35% scatter) [4]. One of the reasons is the possible different types of applied weaving stitch in different cases. However, we additionally encountered a relationship between the measured dielectric constant ε_{r_Denim} of denim fabrics on the applied measurement method. Researchers, who derive the dielectric constant from the resonance parameters of standards rectangular flat patches give values ε_{r_Denim} ~1.59–1.67 [12,20–22]. In this case, the extracted dielectric constant should be close to the perpendicular one, ε_{perp_Denim}. When the applied method is the popular coaxial dielectric probe (DAK, Dielectric Assessment Kit), the obtained parameters are typically ε_{r_Denim} ~1.78–1.8 and beyond [23,24]. The free-space method confirms these values 1.75–2 in the frequency range 14–40 GHz [25]. Both considered methods give values close to the parallel one, ε_{par_Denim}. Finally, the extracted dielectric constants from the microstrip ring resonator or other planar methods are typically ε_{r_Denim} ~1.69–1.73 [23,24]. In this case, the planar methods extract the equivalent dielectric constant (see the concept developed in [26]). The equivalent dielectric constant ε_{eq} appears for characterization of the whole substrate when the real anisotropic structure has been replaced with an isotropic equivalent. That's why, we observe the inequality $\varepsilon_{par\_Denim} > \varepsilon_{eq\_Denim} > \varepsilon_{perp\_Denim}$, which is a typical situation for the woven materials, e.g., for the reinforced substrates [18,27]. Very interesting are the obtained results in [23]; they confirm the assumption above because the measured values for the equivalent dielectric constants ε_{eq} for three textile fabrics by a ring-resonator method always are smaller than the corresponding values measured by the DAK method (ε_{par}). Therefore, we can conclude that the anisotropy of the textile fabrics is a natural property, and its existence can explain the behaviour of their permittivity. Our investigations show that the anisotropy of the materials in the antenna project directly influences mainly the matching conditions of the patches and radome transparency [28], while then at the working frequency it indirectly slightly changes the gain, radiation patterns, efficiency and even polarization thought the anisotropic radome. The most common circumstance in the research papers considering wearable radiating components is the observation of small, moderate, and sometimes big differences between the simulated and measured resonance characteristics, explained by the authors with different experimental and simulation conditions. Our opinion is that in the most cases this effect depends on the selected by the authors values of the dielectric constant—close to ε_{perp} (small changes for the patches resonances are observed), ε_{eq} (suitable for the microstrip feeding lines, transformers, steps, filters, etc.) or ε_{par} (applicable for the coplanar and slotted wearable structures) [26]. If the actual anisotropy of the used substrates is smaller than 2–3% (see below for this parameter), its influence is usually negligible.

The other important issue for the wearable antennas is the bending effect (for conformal patches) [29]. A part of the research papers dealing with the wearable antennas on textile fabrics and polymers usually include additional information for the effect of bending at typical radii, compliant with the human body [30–33]. A measure for the degree of bending is the curvature radius R_b (R_b is the radius of an imaginary cylinder to which the antenna is bent) or bending angle $\theta_b = L$ (or $W)/R_b$, where L and W are the length and width of the rectangular patch antenna [34]. Most of the papers simply registered the bending effect on the working frequency and/or frequency bandwidth (usually a decrease of the resonant frequency) and rarely on the gain and radiation pattern. Sometimes, unexpected discrepancies are detected between the simulated and measured results from the bending [32] explained by imperfect measurements. Only a few researchers provide discussions for the nature of the bending effect. When the measurements are well performed, the obtained results are useful for understanding the bending effect. For example, the results obtained in the paper [30] give the information that the thickness of the flexible substrate is important for the degree of the bending influence. For the substrate as a flexible felt ($\varepsilon_r = 1.3$) with thickness $h_S = 0.5$–12 mm, the optimal thickness for minimizing the effect of bending over the frequency shift is about 6 mm. Very helpful results for the bending effect on rectangular patch antenna on denim substrate are presented in [35]. The parameters of this material with thickness $h_S = 2$ mm are chosen to be $\varepsilon_{r\_Denim} = 1.6$ and tan $\delta_{\varepsilon\_Denim} = 0.01$ at 2.4 GHz. For the first time, the authors definitely show by simulations that the resonance frequency of the lowest-order TM_{10} mode in the rectangular patch antenna should continuously increase with increasing of the bending radius R_b—with a relatively low degree for the width-bent patches and with a higher degree—for the length-bent patches. However, the measurement results slightly differ from the simulations, as relatively big ripples appear in the experimental frequency shifts: ±2.5 MHz for width-bent and ±85 MHz for length-bent patches (compared to the resonance frequency ~2.4 GHz for the flat patches). Nevertheless, the tendency for increasing of the resonance frequency is visible. The authors commented that this behaviour was not expected from simulations. They attribute this discrepancy to other physical properties that the conductive textile was subjected to upon bending that were not correctly replicated in simulations.

In our paper [36], we supposed for the first time that both effects (bending of the flexible substrate and its anisotropy) can simultaneously affect the resonance behaviour of the resonance patches. There we presented some preliminary experimental results for a rectangular resonator with isotropic substrates, but the influence of the substrate anisotropy was not separately investigated. We cannot find other research papers, where the anisotropy and bending are considered in parallel, excepting some calculations of the input impedance [37] and additionally return losses S_{11} and mutual coupling [38] in cylindrically conformal patch antennas on anisotropic substrates.

In this paper, we continue to investigate more deeply the opposite impacts of the dielectric anisotropy and bending of the substrate on the resonance characteristics of planar radiators. We follow the same strategy in this paper—not to consider fed patches and antennas, but to examine pure resonant structures and to avoid any parasitic influence of the feeding lines. This paper includes new experimental and simulation results for the frequency shift of the modes in planar rectangular resonators and their modifications, which makes possible the separation between the effects of anisotropy and bending and the independent characterization of the degree of these effects. In the Materials and Methods section, two experimental and numerical methods have been used for determination of the uniaxial anisotropy of the textile fabrics. A methodology for accurate measurements of the bending effects on the resonance characteristics of the planar resonator has been described. Then, an efficient procedure is introduced for creating suitable 3D models of planar resonators for separate numerical investigations of the bending and anisotropy and both effects together. Data for the measured anisotropy of several selected for the research flexible anisotropic and isotropic materials are presented. In the Results and Discussions section, very interesting results are obtained and discussed for the separate and simultaneous influence of the anisotropy and bending for materials with different anisotropy and for conformal resonance structures bent at different radii. The results for the influence of the anisotropy and bending on several planar resonators with sophisticated shapes are added—for slotted rectangular patches, fractal structures, and resonators with defected grounds. Finally, the origins of the considered competitive effects on the resonance planar structures are discussed and explained and useful conclusions are offered. A possible future work has been formulated.

2. Numerical and Experimental Methods and Materials

The aim of this research is to investigate numerically and experimentally the possible competitive influences of the uniaxial anisotropy and bending of textile substrates on the resonance performances of wearable planar radiators. Therefore, in this section, we describe all applied experimental and numerical methods for the determination of the substrate anisotropy and reliable characterization of the bending effect in these radiating structures. The selected materials and their important characteristics for the research have been obtained.

2.1. Two-Resonator Method for Measurement of the Uniaxial Anisotropy of Textile Fabrics

The considered below method has been proposed in [27] and applied for anisotropy characterization of a variety of materials [19]. In this paper, it has been applied for the determination of the pairs of parameters, ε_{par}; tan δ_{ε_par} and ε_{perp}; tan $\delta_{\varepsilon\_perp}$, of textile fabrics. Figure 1a schematically presents the idea of the used method: a textile disk sample is placed sequentially in two resonators, which are designed to support either symmetrical TE_{0mn} modes (m = 1, 2, 3, … ; n = 1, 2, 3, …) in the cylinder marked as R1 or symmetrical TM_{0m0} modes (m = 1, 2, 3, …) in the cylinder marked as R2 with mutually perpendicular E fields—parallel to the sample surface in R1 or perpendicular to this surface in R2. The sample is placed in the middle of R1 and on the bottom of R2 ensuring the best conditions for the excited TE or TM modes to be influenced by the sample and these modes to be maximally separated (e.g., the resonators heights to be $H_1 \sim D_1$ and $H_2 < D_2$ and the coupling probes to be orientated to excite only TE modes in R1 or TM modes in R2). The sample diameter d_S is chosen to coincide with the resonator diameters $d_S \sim D_{1,2}$. In this case, the extraction of the dielectric parameters can be accurately performed by the analytical model described in [27,39]. In short, the measurement procedure is as follows. First, the resonance characteristics are measured (resonance frequency f_0 and unloaded quality factor Q_0) of each TE or TM mode under interest in the empty R1 or R2 resonator. This step makes possible a fine determining the equivalent resonator diameters $D_{1,2eq}$ and equivalent wall conductivity $\sigma_{1,2eq}$ of both resonators, which considerably increases the accuracy of the next measurements. The second step includes measurements of the resonance characteristics (f_ε and Q_ε) of the same TE or TM modes (well-identified) in the R1 or R2 resonators with a sample. Finally, the set of obtained data ensures the determination of the parallel dielectric constant ε_{par} and dielectric loss tangent tan $\delta_{\varepsilon par}$ in resonator R1 and determination of the perpendicular dielectric constant ε_{perp} and dielectric loss tangent tan $\delta_{\varepsilon perp}$ in resonator R2. The measurement uncertainty has been evaluated as relatively small [27]: 1–1.5% for ε_{par}, 3–5% for ε_{perp}, 5–7% for tan $\delta_{\varepsilon par}$ and 10–15% for tan $\delta_{\varepsilon perp}$ in the case of 0.5–1.5 mm thick substrates with dielectric constants ~1.3–5 in the Ku band. The main source of the pointed inaccuracy is the uncertainty for the determination of the sample thickness. Another circumstance is the selectivity of the considered method; due to the E-fields orientation the cylinder resonators measure the corresponding "pure" parameters (parallel ones in R1 and perpendicular ones in R2) with selectivity uncertainty less than ±0.3–0.4% for the dielectric constant and less than ±0.5–1.0% for the dielectric loss tangent in a wide range of substrate anisotropy and thickness [39].

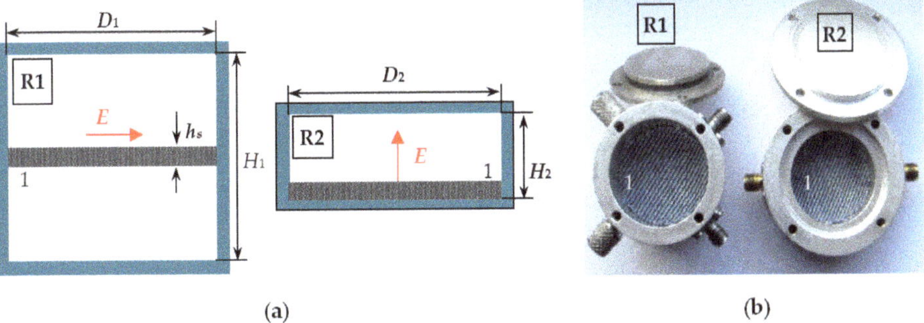

Figure 1. Two-resonator method: (**a**) Pair of resonators for measurement of parallel (R1) and perpendicular (R2) dielectric parameters of disk samples in cylindrical TE (R1) and TM-mode (R2) resonators; (**b**) Photography of resonators R1 and R2 with denim textile sample 1; E—electric field.

2.2. *Numerical Models for Determination of the Dielectric Constant and Anisotropy of Textile Fabrics as Dielectric Mixtures*

2.2.1. Limits for the Dielectric Parameters of Mixed Textile Threads

There exists a big variety of technologies to mix or blend two or more types of textile threads from different materials and with different mechanical properties, which makes possible to obtain new fabrics with desired specific elasticity moduli and stiffness and to control the stability of these properties. Due to these purposes, textile engineers have developed different models and effective- medium theories for reliable characterization of these structures [13–16,40–42]. Our survey shows that these models with some modifications could be successfully applied also to the electromagnetic properties of the textile fabrics dielectric mixtures, as it has been done in [16].

The simplest models (as the first stage of approximation, if the details of geometry are ignored) give the so-called upper and lower bounds of the resultant dielectric constant and loss tangent on the base of the modified Reuss (iso-strain) and Voigt (iso-stress) bound models (for series or parallel layered mixtures). The Bruggman formula [14] presents relatively accurate approximation for near-to-isotropic materials:

$$\varepsilon_{eq} = \frac{(\varepsilon_1 + u)(\varepsilon_2 + u)}{V_1(\varepsilon_2 + u) + V_2(\varepsilon_1 + u)} - u; \quad 0 \le u \le \infty, \quad V_1 + V_2 = 1 \tag{1}$$

where ε_{eq} is the scalar isotropic equivalent dielectric constant of the mixture, ε_1 and ε_2 are the dielectric constants of the mixed threats, V_1 and V_2 are the corresponding normalized volumes, and $u \subset (0; \infty)$ is a parameter which depends on the method of mixing. Three cases could be derived from this expression depending on the type of mixing: for series mixing $u = 0$ (Reuss bound); for parallel mixing $u = \infty$ (Voigt bound) and for random mixing, $u = (\varepsilon_1\varepsilon_2)^{1/2}$ (Bruggman curve). All these curves for the normalized dielectric constant are plotted in Figure 2a.

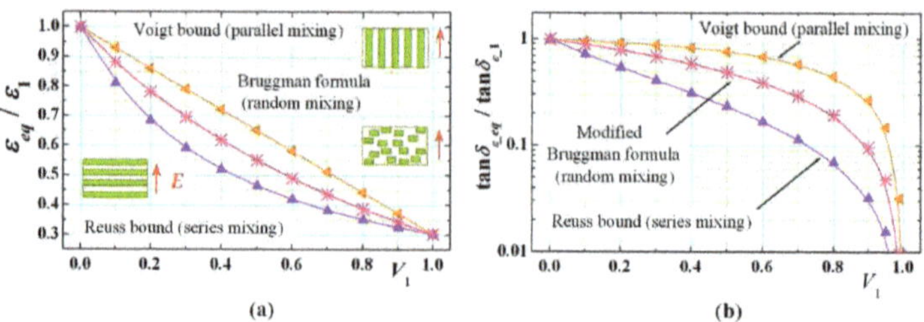

Figure 2. Minimal and maximal bounds for the normalized resultant equivalent dielectric constant $\varepsilon_{eq}/\varepsilon_1$ (**a**) and normalized equivalent dielectric loss tangent tan $\delta_{\varepsilon\_eq}/\tan\delta_{\varepsilon\_1}$ (**b**) of two mixed dielectrics with isotropic parameters $\varepsilon_1/\tan\delta_{\varepsilon 1}$ and $\varepsilon_2/\tan\delta_{\varepsilon 2}$ and normalized volumes V_1 and V_2 ($V_1 + V_2 = 1$) (insets: of parallel, series and random mixing regarding the direction of the applied E field).

The Equation (1) is a complex one; it can be rewritten also for a direct calculation of the corresponding dielectric loss tangents bounds (modified Bruggman formula); see the dependencies in Figure 2b:

$$\tan\delta_{\varepsilon\_eq} = \frac{(\tan\delta_{\varepsilon 1} + v)(\tan\delta_{\varepsilon 2} + v)}{V_1(\tan\delta_{\varepsilon 2} + v) + V_2(\tan\delta_{\varepsilon 1} + v)} - v; \quad 0 \le v \le \infty, \tag{2}$$

where tan δ_{ε_eq} is the isotropic equivalent dielectric loss tangent of the resultant fabrics, tan $\delta_{\varepsilon 1}$ and tan $\delta_{\varepsilon 2}$ are the dielectric loss tangent of the mixed/blended threats, and $v \subset (0; \infty)$ is a new parameters; now we have again $v = 0$ for series mixing, $v = \infty$ for parallel mixing, however, $v = [\varepsilon_1\varepsilon_2 (\tan\delta_{\varepsilon 1} + \tan\delta_{\varepsilon 2})]^{-1/2}$ for random mixing. We have selected a concrete synthetic material for the presented examples in Figure 2a,b—Polyester threads ($\varepsilon_1 \cong 3.4$; tan $\delta_{\varepsilon 1} \cong 0.005$) mixed with air ($\varepsilon_2 = 1.0$; tan $\delta_{\varepsilon 2} = 0$). However, the predicted anisotropy by Equations (1) and (2) is too large, does not take into account the concrete sizes and shapes of the threads and therefore, the results do not correspond to the realistic textile fabric. The survey of other effective-media analytical expressions for the resultant permittivity in different mixtures, presented in [43], show that they also cannot give the actual anisotropy.

Therefore, in this paper, we accepted another more realistic approach. Most of the textile fabrics can be considered as complex fabrics of cylindrical single or multi-fibre threads (a short survey on the Internet of the free microscopic images of the popular fabrics illustrates well the predominant existence of the cylindrical cross-section shape of the threads). Such an approach is very popular in the mechanical models of the textile fabrics [13–15], but also applicable for characterization and modelling of their dielectric properties [10,19,44]. In this research, a similar approach has been accepted. In the next subsection, we present an effective numerical model for accurate prediction of the real anisotropy of textile fabrics on the base of cylindrical unit cells.

2.2.2. Numerical Models for Evaluation of the Dielectric Anisotropy of the Textile Fabrics

The degree of anisotropy can be predicted for artificial textile fabrics by the numerical method introduced in [17]. The idea of this method is to build a unit cell by two or more isotropic cylindrical fibres (threads), to reproduce it in a hosting isotropic substrate (e.g., air) and to put the whole sample in a rectangular resonator, which supports TE and TM modes with exited E fields in three mutually perpendicular directions. The simulations are performed by electromagnetic simulators (HFSS® in this case). Figure 3 illustrates the selected unit cells with three mutually perpendicular cylinders of equal diameter d. The concrete unit cell is a prism with sides $a = b = 1.0$; $c = 1.5$ mm. They form a rectangular sample with dimensions 9.5 × 8 × 1.5 mm and this sample is placed in the middle of a rectangular box with dimensions 9.5 × 8 × 10 mm. Actually, this box is one-quarter part of a rectangular resonator with dimensions 19 × 16 × 10 mm, which support TE and TM mode in the Ku and K bands depending on the diameters, filling and dielectric constant of the threads. The resonator with a sample is solved in "eigenmode" option of the used HFSS simulator (calculating the resonance frequency f_r and the unloaded quality factor Q), where appropriate symmetrical boundary conditions are accepted at side A and B of the box: "symmetrical E-field" for TE modes and "symmetrical H-field" for TM modes. Thus, the considered resonator with 1.5-mm thick artificial textile sample supports the following mode of interest, illustrated in Figure 4 for a 3D-woven textile sample: (a) TE_{011} mode with resonance frequencies in the interval 19.3–21 GHz (E field along $0x$); (b) TE_{101} mode; 21.9–23.5 GHz (E field along $0y$); (c) TM_{010} mode; 11.6–12.2 GHz (E field along 0z); (d) TE_{111} mode; 17.6–18.7 GHz (E field in plane $0xy$). The considered set of modes makes it possible the extraction of the dielectric constants and dielectric loss tangent of the investigated textile samples in different directions, considered as samples with bi- or uniaxial symmetry as it is shown in Figure 4. The concrete resonator dimensions are chosen relatively small (to facilitate simulations). However, larger dimensions can be selected for lower-frequency ISM bands. The only rule is the size of the unit cell to be smaller than the free-space wavelength to ensure homogenization of the artificial structure at a given frequency. As quantitative measures are used, the parameters ΔA_ε, $\Delta A_{\tan\delta\varepsilon}$ for the degree of the dielectric anisotropy of the resulting (equivalent) dielectric constant/loss tangent for bi-/uni-axial anisotropy are calculated by the following expressions:

$$\Delta A_{\varepsilon\_xx,yy} = 2(\varepsilon_{xx,yy} - \varepsilon_{zz})/(\varepsilon_{xx,yy} + \varepsilon_{zz}), \quad (3)$$

$$\Delta A_{\tan\delta\varepsilon\_xx,yy} = 2(\tan\delta_{\varepsilon\_xx,yy} - \tan\delta_{\varepsilon\_zz})/(\tan\delta_{\varepsilon\_xx,yy} + \tan\delta_{\varepsilon\_zz}). \quad (4)$$

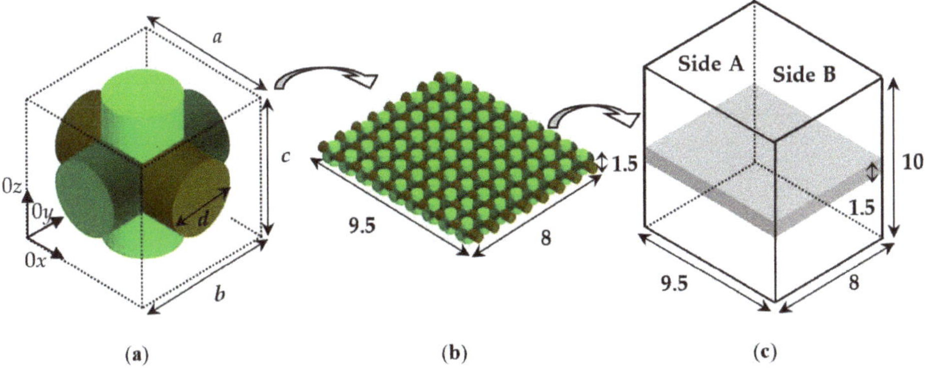

Figure 3. (**a**) Unit cell $a \times b \times c$ with cylindrical threads of diameter d; (**b**) constructed artificial sample with repeated unit cells in a hosting isotropic substrate (air); (**c**) equivalent sample in a rectangular box, which is a quarter part of the whole resonator with symmetrical boundary conditions on Side A and B.

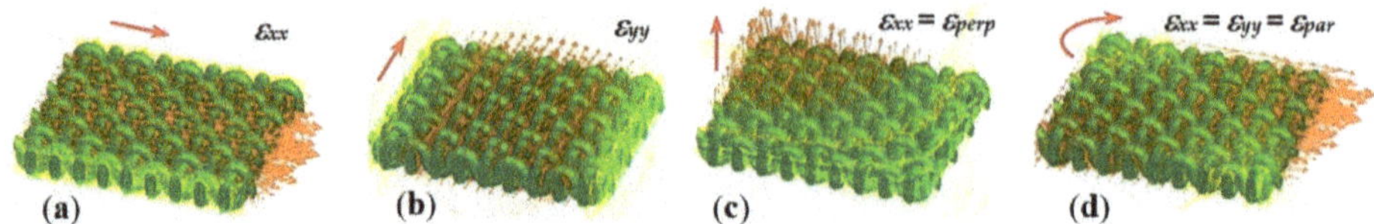

Figure 4. Artificial sample (3D-woven fabrics) in a rectangular resonator (not shown), which supports different modes with mutually perpendicular E fields (red arrows) along the axes: $0x$, $0y$, 0z and in the plane $0xy$ in the Ku-band (modes: TE_{011} (**a**); TE_{101} (**b**); TM_{010} (**c**); TE_{111} (**d**).

Independent extraction of the resultant dielectric constant along all three axes is possible after the replacing of the anisotropic sample under test with an equivalent isotropic sample (as in Figure 3c). The procedure is as follows. After the selection of each unit cell and the construction of the whole artificial 3D sample (see below), placed in the middle of the selected resonator, the resonance frequency f_r and Q factor of the corresponding mode can be obtained by simulations in "eigenmode option". Then, the anisotropic structure is replaced with an equivalent prism of the same dimensions and by tuning the corresponding isotropic values ε_{eq} and tan $\delta_{\varepsilon eq}$, a coincidence should be reached (typically <1%) between both simulated pairs f_r and Q for the anisotropic sample and its isotropic equivalent. Thus, the corresponding dielectric parameters of the biaxial anisotropic textile samples can be obtained by using the excited modes in Figure 4a–c, while the parameters for the uniaxial anisotropic textile samples (most of the cases) can be obtained by using the modes in Figure 4c,d.

The described procedure is effective and enough accurate for preliminary prediction of the anisotropy of different artificial materials. In the paper [17], some preliminary results have been obtained for several artificial woven and knitted tactile fabrics, but the method has been successfully applied for many other materials (incl. 3D printed) [6]. In this paper, we present new results to show how the anisotropy depends on the structure and threads' orientation of the textile fabrics and to establish the origin of this property. Some attempts to find such relations have been done in [19] but without to present quantitative results. In the beginning, three simple structures have been constructed as in paper [17] based on ordered single cylinders built from the unit cell in Figure 3 of diameter 0.5 mm and distance between their axes 1.0 mm. The cylinders are consistently orientated along $0x$, $0y$ and 0z axes. Applying the modes from Figure 4a,c with electric fields orientated along $0x$ or 0z axis, the described model makes it possible to determine the dependence of the dielectric constant anisotropy $\Delta A_{\varepsilon\_xx}$ versus the ratio $\varepsilon_{thread}/\varepsilon_{air}$ presented in Figure 5 (a). The parameter $\Delta A_{\varepsilon\_xx}$ increases with the ratio $\varepsilon_{thread}/\varepsilon_{air}$ increasing, but in different ways. When the cylinders are orientated along $0x$ (as the electric field E_{TE} of the exited mode), $\Delta A_{\varepsilon\_xx}$ has large positive values (~8% for ε_{thread} = 3.4). Contrariwise, when the cylinders are orientated along 0z (E_{TM}), $\Delta A_{\varepsilon\_xx}$ has negative values (~−4.5%). These values strictly correspond to the relative volume portions of the treads orientated along $0x$ (E_{TE}) and 0z (E_{TM}) in these simple cases (detailed geometrical calculations are not performed at this stage of the research). Only when the cylinders are orientated along $0y$ (perpendicularly to E_{TE} and E_{TM}, the parameter $\Delta A_{\varepsilon\_xx}$ is close to 0 (i.e., the sample behaves as almost isotropic one).

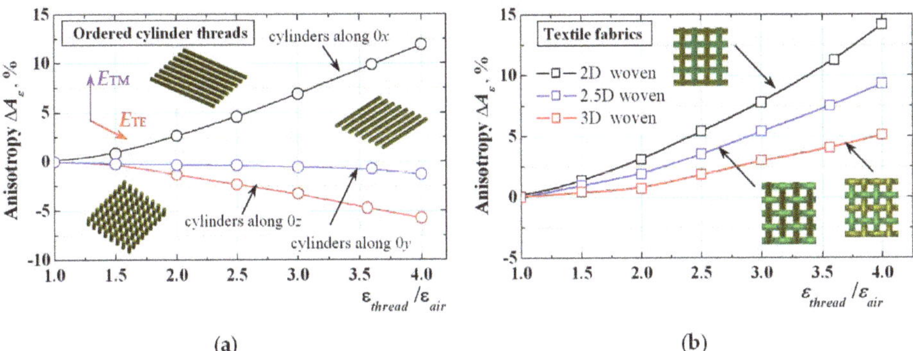

Figure 5. Dielectric constant anisotropy of artificial textile samples versus the ratio between the dielectric constant of the threads and air $\varepsilon_{thread}/\varepsilon_{air}$: (**a**) for ordered straight cylinders, orientated along axes $0x$, $0y$ or $0z$; (**b**) for 2D, $2\frac{1}{2}$D and 3D woven fabrics in Ku-band (see Figure 6) (Arrows: E fields of the used TE and TM modes).

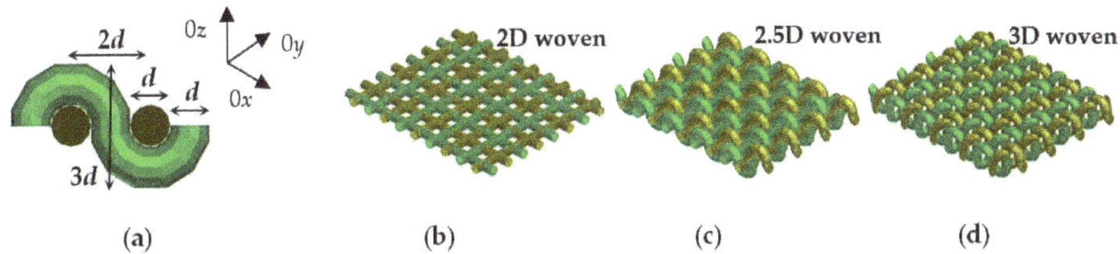

Figure 6. (**a**) Woven threads of corresponding dimensions; three types of woven fabrics: (**b**) 2D woven; straight threads along $0x$, $0y$; (**c**) 2.5D woven; straight threads along $0y$, wavy threads along $0x$; (**d**) 3D woven; wavy threads along $0x$, $0y$ (all threads are with equal dielectric constant $\varepsilon_{thread} = 3.4$).

Very interesting are the results for the uni-axial anisotropy (obtained by the modes from Figure 4c,d) of constructed three artificial woven fabrics (shown in Figure 6). They are conditionally named 2D, 2.5D and 3D woven samples due to the applied straight and/or wavy threads (see the figure captions of Figure 6). The behaviour of the uni-axial parameter ΔA_ε of the 2D-woven sample is close to this one of the pure cylinders along $0x$—Figure 5b. However, when the portion of the threads with orientation along 0z axis increases (for 2.5D and especially for 3D-woven samples) applying wavy threads, the anisotropy becomes smaller, from 10% (for 2D woven samples) to 7% (2.5D) and 3.5% (3D) for $\varepsilon_{thread} = 3.4$. This result shows that the dense woven fabrics have relatively small anisotropy, close to the realistically measured values of 4–6% for most of the textile fabrics. However, their anisotropy exists and can be taken into account in the design of different wearable devices, when the final design accuracy is important and for the higher 5G frequency bands.

2.3. Procedure for Accurate Measurements of Bent Planar Resonators on Textile Fabrics

The accurate measurement of bent wearable structures is not an easy task. There appear strong mechanical changes during the bending—deformations in the substrates; deformations in the metal layout (it should always tightly cover the substrate); the feeding lines can affect the resonance behaviour. Following the strategy in this paper to investigate only pure resonance structures, we apply coaxial probes to excite the lowest-order resonances in the planar structures. Figure 7 represents the simulated E-field pattern of the first two planar modes in flat and bent microstrip resonators. Coaxial probes from electric type (short coaxial pin orientated along the E field) should be put close to the E-maximums. However, in this research, we apply more stable coaxial magnetic loops placed close to the H-maximums of the magnetic field of the corresponding mode (Figure 8a for TM_{10}

mode; Figure 8b for TM_{01} mode and Figure 8c for both modes). The measurements are performed by a vector network analyzer in the L and S bands in transmission regime. The place and the orientation of the loops are tuned during the measurements until the transmission losses S_{21} increase more than −40 dB. At these conditions, the resonance frequency practically does not depend on the loop proximity and the measured resonance frequencies are enough accurate.

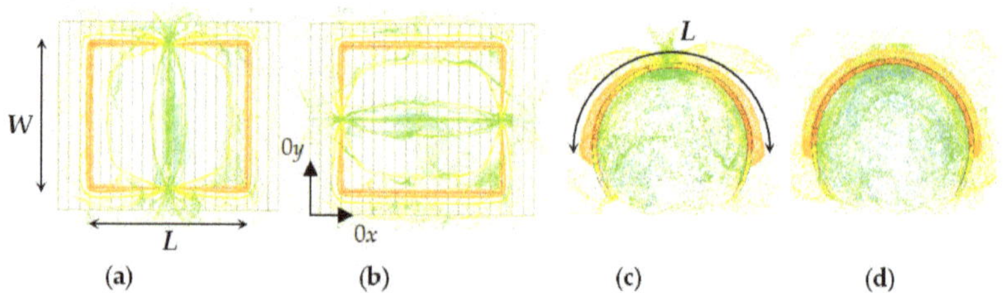

Figure 7. Simulated E-field pattern in microstrip resonators: (**a**,**b**) TM_{10} and TM_{01} in a flat resonator; (**c**,**d**) TM_{10} and TM_{01} in a bent resonator. Legend: *L*—length; *W*—width.

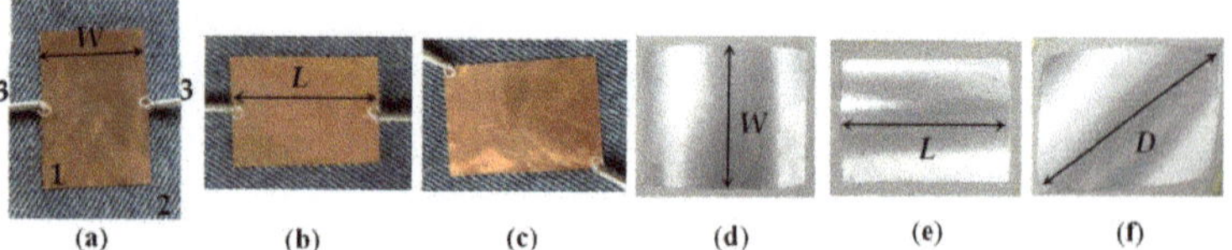

Figure 8. (**a**–**c**) Pair of magnetic coaxial loops placed on the length, width, and diagonal of the planar resonator; (**d**–**f**) length (L)-bent, width (W)-bent and diagonal (D)-bent microstrip resonators. Legend: 1—resonator; 2—substrate; 3—pair of magnetic coaxial probes.

In this research, we apply self-adhesive 0.05-mm thick metal (Al or Cu) folio to form the resonator layout. We start measurements in a flat position of the resonator and then measure the bent resonator with continuously decreasing bending radius. The resonator substrates are bending over a set of smooth metallic cylinders with radii R_b from 80 to 12.5 mm. Three types of bending are applied—length-(L), width-(W) and diagonal-bent (D) resonators—see the illustrations in Figure 8. When we bend, special care is taken to ensure that the metallization remains well adhered to the substrate and that it does not detach itself. Therefore, measurements are performed only for decreasing bending radius and not in reverse order. Each of the pointed types of bending is realized with a new fresh resonator folio. In this research, the results are presented for the ratio between the resonance frequencies for the bent and flat resonators.

2.4. *Numerical Models for Investigations of Bent Planar Resonators on Anisotropic Substrates*

Most of the modern electromagnetic simulators have options for the introduction of anisotropic materials. However, in the case of conformal planar structures, this is not easy to perform directly, when the substrate has been introduced as a single object and to be sure that the anisotropy is accurately described. Therefore, we chose a geometrical approach. The anisotropic substrate is divided into several equal slices with a form of prisms (with rectangular cross-section view for the flat resonators and with trapezoidal cross-section view for the flat resonators). The slices have equal anisotropic properties as the whole substrate, but the parallel and perpendicular directions used to determine the uniaxial anisotropic dielectric parameters can be controlled now for each slice with the change of the bending radius—as it is sown in Figure 9 for the half of structures. In this research, the concrete width w_s of the slices is chosen to be w_s = 2 mm but can be decreased for thicker substrates or smaller bending radii for better fitting of the cross-section of the bent substrate. The other sizes are height h_S and length $l_S = W_S$. The 3D views of flat and bent microstrip resonators on sliced anisotropic substrates are presented in Figure 10. During the bending, we satisfy the rule to keep the resonator dimensions L and W. However, the ground and the slices may undergo some deformations.

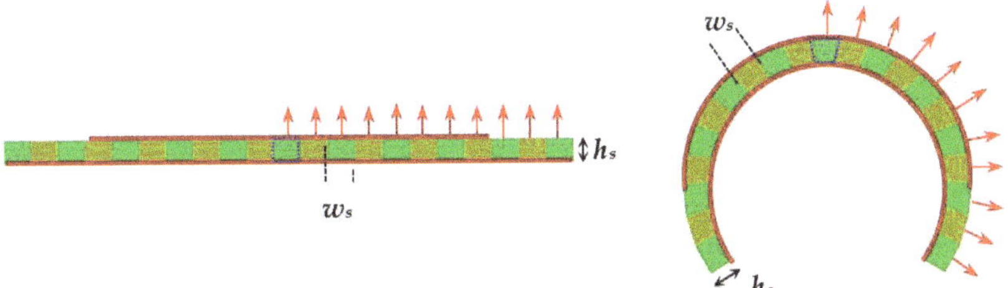

Figure 9. Flat and bent microstrip resonators on substrate constructed by sliced prisms, each with own anisotropic properties. Arrows represent the normal direction in each slice in flat and curved substrates.

Figure 10. 3D view of flat and bent microstrip resonators of length L = 30 and width W = 26 mm on sliced substrates with length L_S = 42 and width W_S = 34 mm (last two cases with bending radius R_b = 14.3 and 9.6 mm).

2.5. Materials Used in the Research

Based on the purposes of the paper, several types of materials have been selected. One of the groups consists of several textile and polymer samples with different measured degrees of anisotropy (ΔA_ε from 4.3 to 10.3) by the two-resonator method. The measured results for the pairs of parameters $\varepsilon_{par}/\tan\delta_{\varepsilon par}$ and $\varepsilon_{perp}/\tan\delta_{\varepsilon perp}$, as well as for the uniaxial anisotropy $\Delta A_\varepsilon/\Delta A_{\tan\delta\varepsilon}$ are presented in the upper part of Table 1. The other group includes several flexible isotropic substrates, selected for measurement of the pure bending effect. The measured anisotropy of these materials is very small, $\Delta A_\varepsilon < 1\%$. The last two groups have representatives of relatively flexible reinforced substrates and soft artificial ceramics. Their anisotropy ΔA_ε varies in a big interval—8.2–24.5%.

Table 1. Measured dielectric parameters and anisotropy of selected materials for this research (averaged values for the frequency interval 6–13 GHz).

Material	h_s, mm	ε_{par}/tan δ_{ε_par}	$\varepsilon_{\varepsilon perp}$/tan $\delta_{\varepsilon\_perp}$	$\Delta A_{\varepsilon}/\Delta A_{\tan\delta\varepsilon}$, %
Textile and polymer samples				
Denim	0.90	1.74/0.048	1.61/0.030	7.8/38
Linen	0.65	1.65/0.043	1.58/0.044	4.3/−2.3
Waterproof fabric with breathability GORE-TEX®	0.20	1.53/0.0057	1.38/0.0043	10.3/28
Polydimethylsiloxane (PDMS)	0.70	2.73/0.022	2.57/0.019	6.00/15
Flexible isotropic and near-to-isotropic samples				
Polytetrafluoroethylene (PTFE)	0.45	2.05/0.00027	2.04/0.00026	0.49/3.8
Polycarbonate (PC)	0.50	2.77/0.0056	2.76/0.0055	0.36/1.8
Silicone elastomer	0.90	2.21/0.0010	2.19/0.0008	0.91/22
Ro3003	0.51	3.00/0.0012	2.97/0.0013	1.0/−8
Relatively flexible anisotropic reinforces substrates				
Ro4003	0.21	3.67/0.0037	3.38/0.0028	8.2/28
NT9338	0.52	4.02/0.005	3.14/0.0025	24.6/67
Relatively flexible anisotropic soft ceramics				
Ro3010	0.645	11.74/0.0025	10.13/0.0038	14.7/−41

3. Results and Discussion

Three types of results and corresponding discussions are presented in this section. First, numerical and experimental results are presented for the pure bending effect in planar resonators on flexible isotropic and near-to-isotropic substrates (Section 3.1). The next step is to verify with results the assumption that the bending effect and substrate anisotropy have opposite impacts on the wearable radiators and sensors (Section 3.2). Finally, the simultaneous bending and anisotropy influence is investigated for several sophisticated planar resonators with magnetic slots, defected grounds and for Koch fractal resonators (Section 3.3).

3.1. Pure Bending Effect

As we mentioned in the Introduction, the investigated bending effect in wearable planar patches and devices usually has been masked by other phenomena, not considered in the simulations [32,33]. We try to solve these problems applying experimentally-proven pure flexible isotropic substrates, using pure resonance structures (to minimize the effects of the feeding lines) and follow an accurate measurement procedure described in Section 2.3.

First, Figure 11a presents the dependencies of the ratio f_{bent}/f_{flat} between the resonance frequencies for the lowest-order TM_{10} mode for bent and flat rectangular resonators on pure isotropic substrate versus the curvature angle α_C between the neighbour slices used to construct the substrate. This is a new measure for the bending degree, which is more comfortable in our research. Figure 12 illustrates the relationship between the bending radius R_b and the introduced curvature angle α_C (e.g., $\alpha_C = 4°$ corresponds to R_b = 28.7 mm; $\alpha_C = 8°$—R_b = 14.3 mm; $\alpha_C = 12°$—R_b = 9.6 mm, etc.).

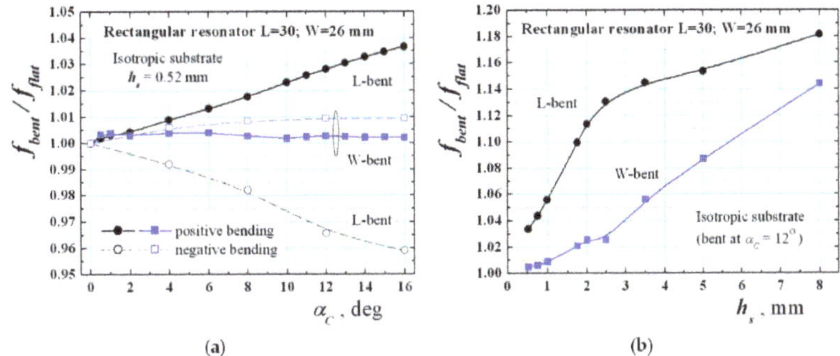

Figure 11. Numerical dependencies of the ratio between the resonance frequencies f_{bent}/f_{flat} of the lowest-order TM_{10} mode for bent and flat rectangular resonators on isotropic substrate versus (**a**) the curvature angle α_C between the substrate slices and (**b**) substrate thickness h_S. The isotropic dielectric constant is chosen to be 3.0, but its concrete value has negligible influence. Positive and negative curvature angles are used.

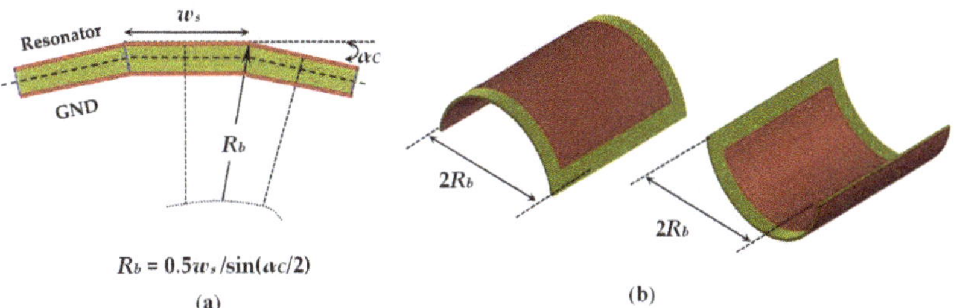

Figure 12. (**a**) Definition of the relation between the curvature angle α_C and bending radius R_b; dashed line: middle line in the resonator substrate, where the effective electrical length of the resonator is formed; (**b**) resonance structures on positive ($+\alpha_C$) and negative ($-\alpha_C$) bent substrate (the bending radius R_b is always determined to the side of the resonator layout).

The presented results show the expected fact (mentioned in [35]) that the resonance frequency of the L-bent resonator increases in comparison to the flat case for pure isotropic substrates. The dependence is not exactly linear. At the same time, the effect on the bending is relatively small for W-bent resonators, which is also an expected result. These dependencies correspond to the classical "positive" bending ($\alpha_C > 0$). What happens during the bending? The material undergoes mechanical deformations, e.g., stretching at the top (to the resonator) and shrinking at the bottom area (to the ground). In our model, we take into account this effect by changing the cross-section shape of the separate slices from rectangular to trapezoidal (illustrated in Figures 9 and 12a). The narrow side of the trapezoid is orientated to the ground of the resonance structure. Thus, the model confirms the assumption that the electrical length L_E of the L-bent resonator decreases in comparison to the geometrical length L (illustrated with the dashed line in Figure 12a). The standing wave of the lowest order TM_{10} mode is located exactly along the curvature in the L-bent structures (see Figure 7a,c) and it explains the increase of the resonance frequency when the curvature angle α_C increase. Contrariwise, during the W-bending the standing wave is located in a perpendicular direction and the influence of the bending is negligible, especially for thin substrates.

Figure 11a presents also the bending effect for the "negative" bending ($\alpha_C < 0$). It is just the opposite and this confirms the origin of the bending effect for the wearable structures. Now, the narrow side of the trapezoid of each slice is orientated to the resonator layout of the resonance structure and in this case, the effective electrical length L_E of the L-bent resonator increases in comparison to the geometrical length L and the corresponding resonance frequency decreases. This type of bending is rarely used and not discussed in detail.

Finally, Figure 11b additionally shows the variations of the bending effect in substrates with different thickness. Now, the effect considerable increases for a thickness interval of 0.5–2.5 mm and then saturation appears for L-bent structures (relatively strong increase is observed also for W-bent structures at bigger thicknesses). However, we cannot observe here the existence of an optimal thickness, where the bending effects are minimized as shown in [30].

The next step is to prove experimentally these tendencies. Figure 13 gives a set of measurement results for the ratio f_{bent}/f_{flat} of the lowest-order TM_{10} mode in bent and flat rectangular resonators on several isotropic substrates versus the bending radius R_b. Three types of dependencies are shown—for L-, W and D-bent resonators. All the results are close to results from the numerical simulations in Figure 11a (D-bent resonators are not simulated). They depend on substrate flexibility and deformations. The best results are got for the well-flexible silicone elastomer (h_s = 0.9 mm), Figure 13c. Good results are obtained by Ro3003 substrate (h_s = 0.52 mm), Figure 13a; however, at small bending radii, this soft substrate undergoes technological stretching and f_{bent} slightly decreases. The harder substrate PC (h_s = 0.5 mm) shows better stability at low R_b. The results for the soft PTFE substrate (h_s = 1.0 mm) deviate from the theoretical dependencies due to the poor adhesion properties of this materials to the metal folio. However, the PTFE-like material with the commercial mark Polyguide®Polyflon (h_s = 1.5 mm) demonstrates better behaviour. In all presented cases, the curves for D-bent substrates (moderate influence) lie between the curves for L-bent (upper curves; stronger influence) and W-bent substrates (lower curves; smaller influence). Thus, we can conclude that the experimental results for the pure bending effect on planar resonators on isotropic substrate fully confirm the numerical simulations, taking into account the possible substrate deformation during the bending on very small radii R_b.

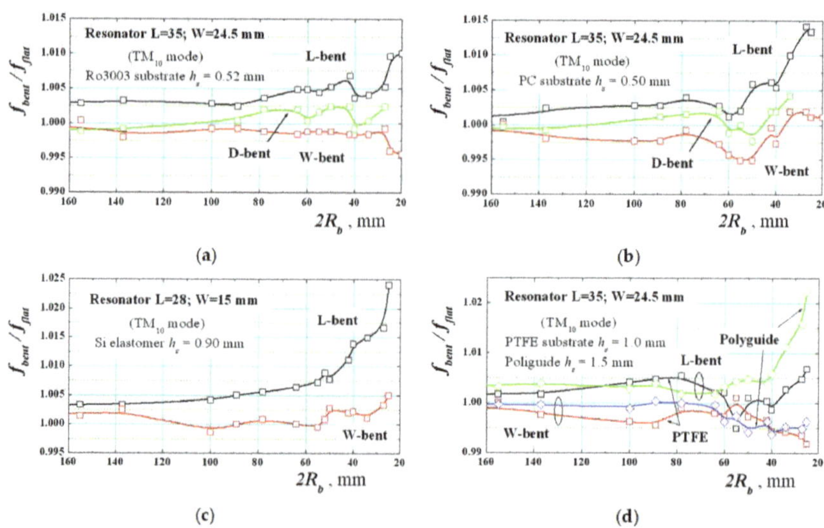

Figure 13. Experimental dependencies of the ratio between the resonance frequencies f_{bent}/f_{flat} of the lowest-order TM_{10} mode for bent and flat rectangular resonators on several isotropic substrates versus the bending radius R_b: **(a)** Ro3003; **(b)** PC; **(c)** commercial silicone elastomer; **(d)** PTFE and Polyguide® Polyflon (http://www.polyflon.com; dielectric parameters 2.05/0.00045).

3.2. Investigation of the Simultaneous Effects of Anisotropy and Bending of Planar Resonators

The main expected results in the research are included in this section. In the beginning, it is important to evaluate the effect of anisotropy in flat resonators. Figure 14a shows the simulated dependencies of the ratio $f_{flat\_aniso}/f_{flat\_iso}$ between the resonance frequencies of modes TM_{10} and TM_{01} for flat rectangular resonators on anisotropic (ΔA_{ε} ~25%) and isotropic substrates versus the substrate thickness h_S. The effect is visibly weak. Only for relatively thick substrates does the resonance frequency shift due to the anisotropy influence with 1–1.5%, which explains why this property is not so popular in the patch antenna design. The explanation is easy—the parallel E fields (to have a noticeable influence of the ε_{par} component) appear only close to the edge of such wide planar structure and the relative effect is practically negligible in comparison to the microstrip line [26].

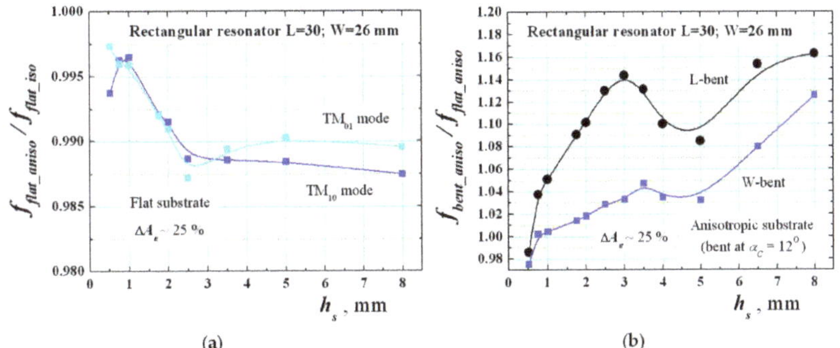

Figure 14. (**a**) Numerical dependencies of the ratio $f_{flat\_aniso}/f_{flat\_iso}$ between the resonance frequencies of modes TM_{10} and TM_{01} for flat rectangular resonators on anisotropic and isotropic substrate versus the substrate thickness h_S. (**b**) Numerical dependencies of the ratio $f_{bent\_aniso}/f_{flat\_aniso}$ of mode TM_{10} for L-/W-bent and flat rectangular resonators on anisotropic substrates versus the substrate thickness h_S.

However, we expect a stronger effect when the resonators are bent. To perform deeper research, a set of bent resonators with different curvature angle are simulated by the help of the 3D models shown in Figures 9 and 10. First, the ratio $f_{bent\_aniso}/f_{bent\_iso}$ is shown in Figure 15a between the resonance frequencies of mode TM_{10} for L-/W-bent rectangular resonators versus the curvature angle α_C. The substrate anisotropy ΔA_{ε} is chosen to be small (~3.5%), moderate (~11%) and big (~25%). This ratio is not measurable, but it shows in a pure form the effect of anisotropy in bent resonators. The results give the useful information, obtained for the first time, that this influence is considerably bigger in comparison with the flat case (up to −5% shifts down). One can see from the presented dependencies that the influence of the substrate anisotropy decreases the resonance frequency in comparison to the hypothetical case of an isotropic bent substrate. Therefore, we can conclude that the effect of the anisotropy of the substrate is just opposite to the effect of bending (as it is shown in Figure 11a). This was our preliminary hypothesis, and it can be considered as proven numerically. Therefore, one can expect that both effects can strongly change the behaviour of these dependencies.

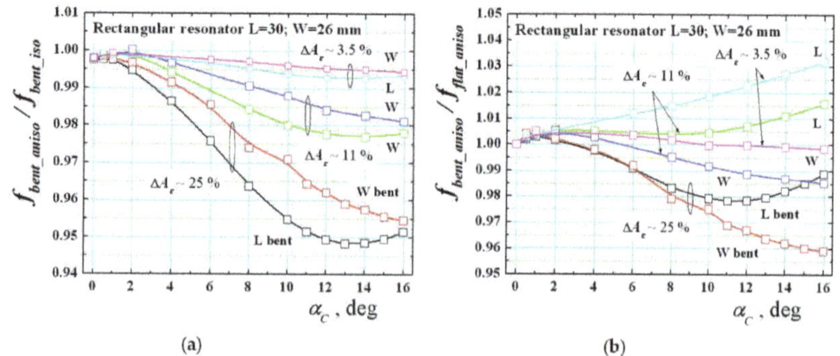

Figure 15. (**a**) Numerical dependencies of the ratio $f_{bent\_aniso}/f_{bent\_iso}$ between the resonance frequencies of mode TM_{10} in L/W-bent resonators on anisotropic and isotropic substrates (h_s = 0.52) versus the curvature angle α_C; (**b**) Numerical dependencies of the ratio $f_{bent\_aniso}/f_{flat\_aniso}$ of mode TM_{10} in L-/W-bent and flat rectangular resonators on anisotropic substrates versus the curvature angle α_C.

Figure 15b presents the ratio $f_{bent\_aniso}/f_{flat\_aniso}$ between the resonance frequencies of mode TM_{10} for L-/W-bent rectangular resonators on anisotropic substrates versus the curvature angle α_C. Now, this ratio is measurable and can be verified experimentally. The new dependencies show that the resonance frequency shift in resonator on realistic (anisotropic) substrates may have as positive, as well as negative signs depending on the actual parameter ΔA_ε, which is impossible for pure isotropic substrates. We also investigate the influence of the substrate thickness h_s on corresponding ratio $f_{bent\_aniso}/f_{flat\_aniso}$. Figure 14b presents curves for L- and W-bent resonators at curvature angle $\alpha_C = 12°$. The results show that the bending effect can compensate the anisotropy influence for thicker substrates to some degree. It is interesting to note that as in [30], we observe the fact that for mediate thicknesses (named "optimal thickness" in [30]) the effect of anisotropy decreases the bending effect; this property probably depends on the curvature angle α_C and not investigated in detail.

Let's now present some experimental dependencies for bent resonators on anisotropic substrates, selected in Section 2.5. The measurement results for the ratio f_{bent}/f_{flat} of the TM_{10} mode in bent and flat rectangular resonators versus the bending radius R_b are presented in Figure 16. They differ from the dependencies shown in Figure 13 for isotropic substrates. In anisotropic case, more or less expressed ripples in the resonance shifts is observed in both L- and W-bent resonators below the resonance frequencies of the corresponding flat resonators (as in paper [35]), which is practically impossible for the isotropic case when accurate measurement procedure has been applied. Therefore, all these cases confirm the simultaneous effects of the anisotropy and bending of used substrates. Very typical are the curves for the textile fabrics denim, linen and commercial multilayer GORE-TEX® and for the flexible polymer PDMS with a small degree of stretching. Similar behaviour is observed for three relatively flexible commercial reinforced substrates: Ro4003; NT9338 and soft ceramic Ro3010. However, the course of dependences here is affected also by the non-plastic deformation in these substrates, which does not allow bending at very small radii. Of course, all presented experimental curves cannot be directly compared with the theoretical ones in Figure 15b due to the difficulties to satisfy the perfect measurement conditions especially at small bending radii, but the trends that reveal the impact of the anisotropy together with the bending effect in wearable structure is obvious.

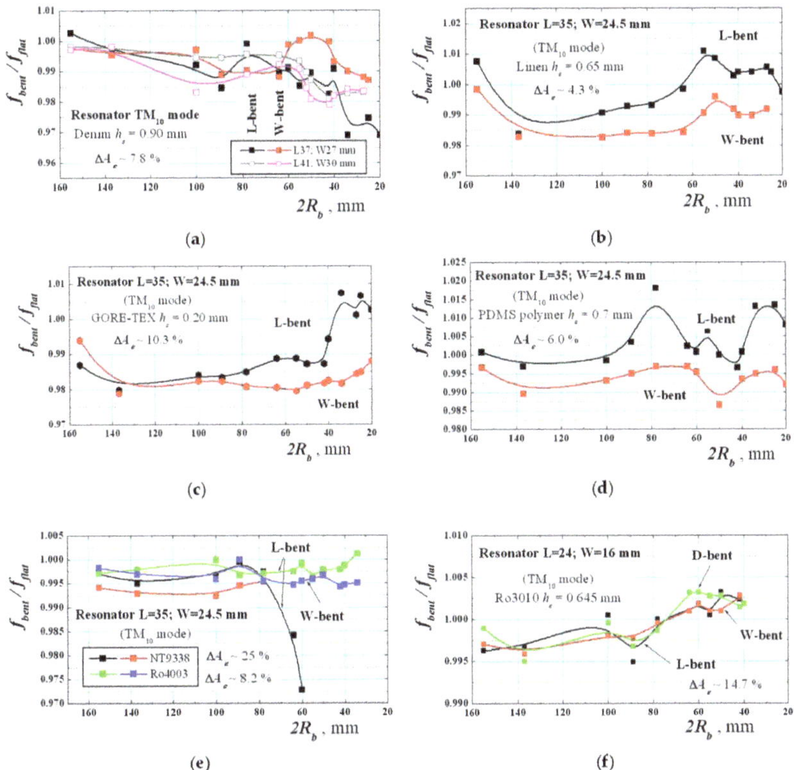

Figure 16. Experimental dependencies of the ratio between the resonance frequencies f_{bent}/f_{flat} of the lowest-order TM_{10} mode for bent and flat rectangular resonators on several anisotropic substrates versus the bending radius R_b: (**a**) Denim; (**b**) Linen; (**c**) commercial textile fabrics GORE-TEX®; (**d**) PDMS; (**e**) NT9338, Ro4003; (**f**) Ro3010.

3.3. Effects of Anisotropy and Bending on More Sophisticated Planar Resonators

The fact, that the substrate anisotropy visible influences together with the bending the resonance behaviour of such simple structure as the rectangular resonator gives us the idea to verify this influence for more complicated planar resonance structures on anisotropic substrates. In this subsection, several resonance structures with slots, defected grounds and Koch fractal contours are numerically investigated to verify the effects of anisotropy and bending in the L and S bands.

Two types of results are presented in Figures 17 and 20. We again investigate the ratio f_{aniso}/f_{iso} between the resonance frequencies of the lowest-order mode for each structure on anisotropic and isotropic substrate (Figure 17). This ratio is a measure of the pure effect of anisotropy. The second type of result is for the ratio $f_{bent\_iso}/f_{flat\_iso}$ between the resonance frequencies of the lowest-order mode in the same planar resonance structures (bent and flat) on isotropic substrates (Figure 20). Now, this ratio is a measure of the pure effect of bending.

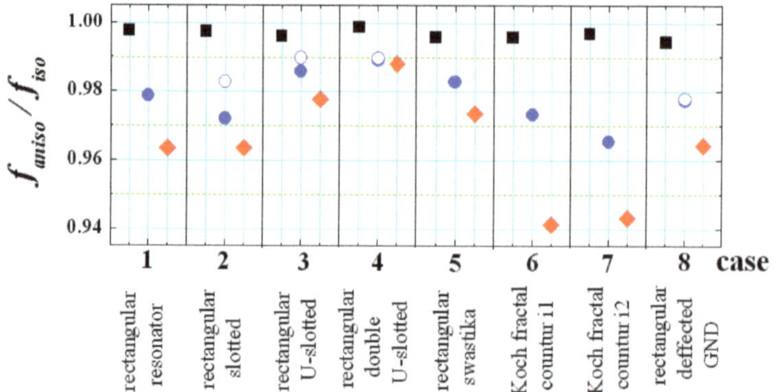

Figure 17. Simulated values of the ratio f_{aniso}/f_{iso} between the resonance frequencies of the lowest-order mode in several planar resonance structures with dimensions 30 × 30 mm on anisotropic and isotropic substrates (h_s = 0.52; ΔA_ε ~ 25%) (this ratio gives the pure effect of anisotropy). The shapes of the considered structures are presented in Figures 18 and 19. The first column for each case corresponds to a flat structure, second—bent at α_C = 8°; third—bent at α_C = 12°; Solid and hollow points correspond to two mutually perpendicular orientations (V & H) of the structure during the bending (when this is possible).

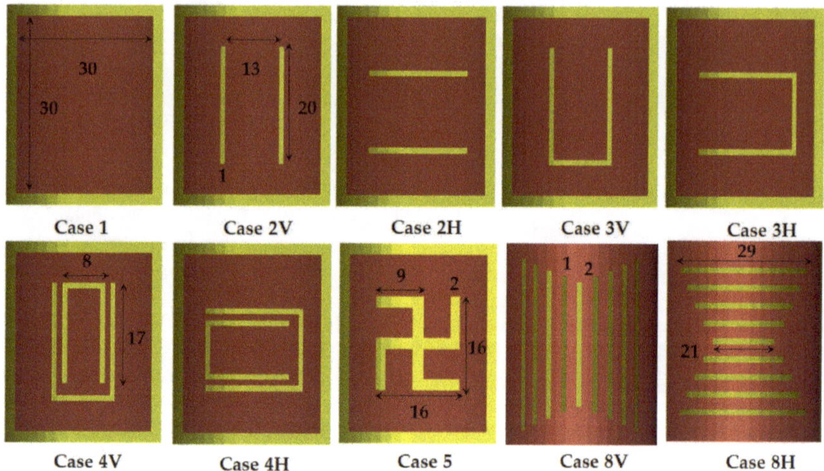

Figure 18. Top view of several resonance structures (bent at α_C = 8°) with dimensions in mm: Case 1—resonator (30 × 30); Case 2—resonator with two slots (V—vertical orientation; H—horizontal orientation); Case 3—resonator with U-shaped slot (V&H); Case 4—resonator with double U-shaped slot (V&H); Case 5—resonator with swastika slot; Case 8—resonator with a defected ground (V&H).

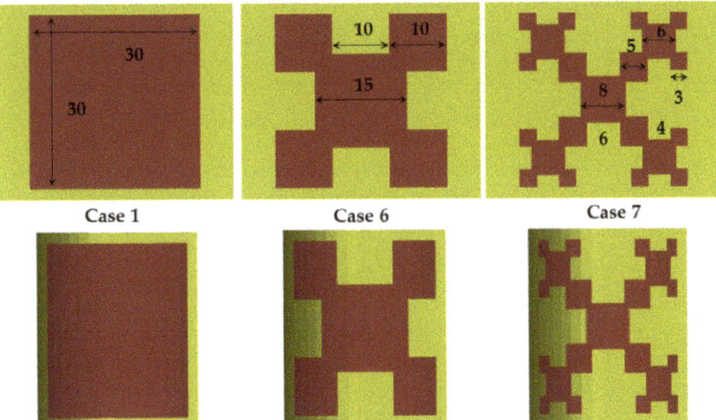

Figure 19. Top view of the first three iterations of Koch fractal contours as a planar resonator (flat—first row and bent at α_C = 8°—second row) with dimensions in mm: Case 1—iteration 0; Case 6—iteration 1; Case 7—iteration 2.

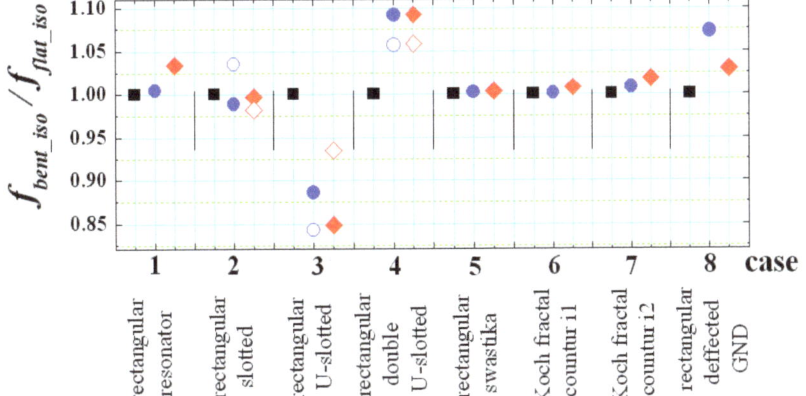

Figure 20. Simulated values of the ratio $f_{bent\_iso}/f_{flat\_iso}$ between the resonance frequencies of the lowest-order mode in several planar resonance structures with dimensions 30 × 30 mm on isotropic substrates (h_s = 0.52; ΔA_ε ~ 25%) (this ratio gives the pure effect of bending). The shapes of the considered structures are presented in Figures 18 and 19. The first column for each case corresponds to a flat structure, second—bent at α_C = 8°; third—bent at α_C = 12°; Solid and hollow points correspond to two mutually perpendicular orientations (V & H) of the structure during the bending (when this is possible).

In the beginning, several rectangular patches with magnetic slots have been considered. These structures are usually applied for a widening of the bandwidth of the corresponding planar patches in comparison to the standard planar patch (Case 1). They include several types of slots (see also Figure 18): Case 2—resonator with two slots [45]; Case 3—resonator with U-shaped slot [30,46]; Case 4—resonator with double U-shaped slot [47]; Case 5—resonator with swastika slot [48]. The structures are not optimized; their dimensions are presented in Figure 18 and are compliant with the used grid of the sliced substrates. The results from Figure 17 show that the effect of the anisotropy decreases (4–1%) with adding the listed slots in the resonator layout. These slots are placed relatively far from the edges, where the parallel E fields exist and the anisotropy cannot change effectively the electrical dimensions of the slotted resonators. At the same time, the pure bending effect is larger, especially for the resonators with U-shaped slots—Figure 20.

The considered defected-ground resonator [49] is also not strongly influenced by the anisotropy (2–3%); the effect is comparable with the effect in the planar resonator with a standard ground. However, the pure bending effect is strong, especially for horizontally-placed slots in the defected ground (case 8H; the increase is more than 25–50%; the values are not shown in Figure 20, because are out of the scale).

Actually, only the considered fractal resonators demonstrate relatively big resonance shifting due to the anisotropy of the substrates—Figure 17 (4–6%), while the pure bending effect is even smaller than in the case of the standard planar resonator (iteration i0)—Figure 20. We investigate the first and second iterations (i1, i2) of classical Koch fractal contours [50], performed on flat and bent substrates—Figure 19. The reason for the increased anisotropy influence is that the portions of the parallel E fields increase considerably with the iteration number of the fractal resonators, which provokes stronger resonance frequency shift down (when $\varepsilon_{par} > \varepsilon_{perp}$). A similar effect can be expected in most of the metamaterial surfaces used in the wearable flat and bent antennas, which is the objective of our future work.

4. Conclusions

The main objective of this study has been accomplished—to prove the opposite influences of the effects of anisotropy and bending on the resonance characteristics of flexible wearable structures. The advantage of this paper is that both effects have been separated in the numerical simulations, which makes it possible to evaluate the degree and sign of the resonance frequency shifts of simple rectangular planar resonators on anisotropic and isotropic substrates in flat and bent states. All simulations and the obtained experimental results show that the pure bending effect, performed only by experimentally-verified isotropic substrates, increases the resonance frequency of the bent rectangular resonators in comparison to the flat ones and proves the origin of this effect in a pure form. Contrariwise, the presented numerical analysis shows that the anisotropy (the existence of direction-dependent dielectric constants ε_{par} and ε_{perp} of the textile materials and similar woven substrates) has just an opposite influence—the resonance frequency of the flat or bent rectangular resonators on anisotropic substrates always decreases (when $\varepsilon_{par} > \varepsilon_{perp}$) in comparison to the same structures on pure isotropic substrates. The last effect is not directly measurable, but it gives the expected pure effect of the substrate anisotropy, which depends on the degree of anisotropy ΔA_ε and the actual bending radius R_b. The combined effects, anisotropy and bending, lead to a more complicated behaviour of the investigated resonance structures when the bent and flat rectangular resonators are considered—as positive, as well as negative resonance frequency shifts.

Now, these combined effects are fully measurable. Applying well-selected flexible anisotropic substrates (including textile fabrics), the resonance shifts in bent and flat resonance structures are measured in the L and S bands. The obtained dependencies for bending radii R_b from 80 up to 10 mm show as increasing (as for the pure bending effect), as well as decreasing of the resonance frequencies (the last phenomenon is theoretically impossible for pure bending effect). Due to the mechanical deformations in the same of the materials during the bending, the obtained dependencies do not fully coincide with the numerical ones, but the tendencies for the opposite influence of the anisotropy and bending are considered as proven. The obtained results explain well the observed dependencies by other authors, even the existence of optimal substrate thicknesses, where the effect of bending (but we add the anisotropy, too), could be minimized. Of course, the last phenomenon depends on the concrete bending radius and anisotropy degree.

Encouraged by the results obtained for such a simple structure as the planar rectangular resonator on wearable substrates, we performed a useful numerical study for the combined effects of anisotropy and bending for more sophisticated structures—planar resonators with slots and defected grounds and fractal resonators. In some of them, the bending effect predominates (resonators with slots and defected grounds), while in the other structures, e.g., the fractal resonators with increased iteration number, the effect of the anisotropy is stronger than the bending effect. These new results determine also the direction of our future research—to investigate complex metamaterial structures for wearable antennas performed on anisotropic substrates at different bending angles. The proposed experimental and numerical methods by the applied TE/TM modes for reliable determination of the direction-dependent equivalent dielectric parameters of different metasurfaces ensure the preliminary determination of the anisotropy of these structures, while the proposed methods for parallel investigation of the effects of bending and anisotropy—the accurate behaviour of metasurfaces with accurate curvature. This is a new scheme of research concerning such meta structures, which will be developed in our future work.

Author Contributions: Conceptualization, P.I.D., P.K.S. and N.G.; methodology, P.I.D.; software and simulations, P.I.D. and P.K.S.; measurements, validation, P.I.D. and P.K.S.; investigation, P.I.D.; resources, P.I.D., P.K.S. and N.G.; writing—original draft preparation P.I.D.; writing—review and editing, P.K.S. and N.G.; visualization, P.I.D.; supervision, P.I.D. and N.G.; project administration, P.I.D. and N.G.; funding acquisition, P.I.D. and N.G. All authors have read and agreed to the published version of the manuscript.

Funding: This research was funded by the Department of Science and Technology, Ministry of Science and Technology, New Delhi, India and the National Science Fund, Ministry of Education and Science, Sofia, Bulgaria under grant number DST INT/BLG/P-01/2019 and KP-06-India-7/2019 under India-Bulgaria Joint Research Projects to the Department of Electrical and Electronics Engineering, Birla Institute of Technology and Science (BITS)—Pilani, Pilani Campus, Rajasthan, India and Faculty of Physics, Sofia University "St. Kliment Ohridski", Sofia, Bulgaria, respectively. The research of anisotropy and methods for its measurement was funded by the National Science Fund, Ministry of Education and Science, Sofia, Bulgaria under grant number DN-07/15 in Faculty of Physics, Sofia University, Sofia, Bulgaria.

Data Availability Statement: Data is contained within the article. More detailed data and data presented in this study are available on request from the corresponding author. Part of them could be included in the Final reports to the corresponding funding organizations.

Conflicts of Interest: The authors declare no conflict of interest.

References

1. Salvado, R.; Loss, C.; Gonçalves, R.; Pinho, P. Textile Materials for the Design of Wearable Antennas: A Survey. *Sensors* **2012**, *12*, 15841–15857. [CrossRef] [PubMed]
2. Priya, A.; Kumar, A.; Chauhan, B. A Review of Textile and Cloth Fabric Wearable Antennas. *Int. J. Comput. Appl.* **2015**, *116*, 1–5. [CrossRef]
3. Yadav, A.; Singh, V.K.; Chaudhary, M.; Mohan, H. A Review on Wearable Textile Antenna. *J. Telecommun. Switch. Syst. Netw.* **2015**, *2*, 37–41.
4. Almohammed, B.; Ismail, A.; Sali, A. Electro-textile wearable antennas in wireless body area networks: Materials, antenna design, manufacturing techniques, and human body consideration—A review. *Text. Res. J.* **2020**, 1–18. [CrossRef]
5. Mohamadzade, B.; Hashmi, R.M.; Simorangkir, R.B.V.B.; Gharaei, R.; Rehman, S.-U.; Abbasi, Q.H. Recent Advances in Fabrication Methods for Flexible Antennas in Wearable Devices: State of the Art. *Sensors* **2019**, *19*, 2312. [CrossRef]
6. Dankov, P.I. Material Characterization in the Microwave Range, When the Materials Become Composite, Reinforced, 3D-Printed, Artificially Mixed, Nanomaterials and Metamaterials. Forum for Electromagnetic Research Methods and Application Technologies (FERMAT Journal). Available online: https://www.e-fermat.org/articles.php (accessed on 20 December 2020).
7. Ouyang, Y.; Chappell, W.J. High Frequency Properties of Electro-Textiles for Wearable Antenna Applications. *IEEE Trans. Antennas Propag.* **2008**, *56*, 381–389. [CrossRef]
8. Sankaralingam, S.; Gupta, B. Determination of Dielectric Constant of Fabric Materials and Their Use as Substrates for Design and Development of Antennas for Wearable Applications. *IEEE Trans. Instrum. Meas.* **2010**, *59*, 3122–3130. [CrossRef]
9. Bal, K.; Kothari, V.K. Measurement of dielectric properties of textile materials and their applications. *Indian J. Fibre Text.* **2009**, *34*, 191–199.
10. Lesnikowski, J. Dielectric permittivity measurement methods of textile substrate of textile transmission lines. *Electr. Rev.* **2012**, *3*, 148–151.
11. Mustata, F.S.C.; Mustata, A. Dielectric Behaviour of Some Woven Fabrics on the Basis of Natural Cellulosic Fibers. *Adv. Mater. Sci. Eng.* **2014**, *2014*, 216548. [CrossRef]

12. Rahim, H.A.; Malek, F.; Soh, P.J.; Romli1, A.; Rani, K.A.; Isa, C.M.N.C.; Fuad, F.A.A. Measurement of Dielectric Properties of Textile Substrates. *J. Teknol. Sci. Eng.* **2015**, *1*, 1–6.
13. Vassiliadis, S.; Kallivretaki, E.; Domvoglou, D.; Provatidis, C. Mechanical analysis of woven fabrics: The state of the art. In *Advances in Modern Woven Fabrics Technology*; Vassiliadis, S., Ed.; In-Tech Publ.: London, UK, 2011; pp. 41–64, ISBN 978-953-307-337-8.
14. Morton, W.E.; Hearle, W.S. *Physical Properties of Textile Fibres*, 4th ed.; Woodhead Publishing: Cambridge, UK, 2008.
15. Hu, J. *Structure and Mechanics of Woven Fabrics*; CRC, Woodhead Publ. Led.: Cambridge, UK, 2004.
16. Raju, G.G. *Dielectrics in Electric Fields*; CRC Press: Boca Raton, FL, USA; Taylor & Francis Group: Abingdon-on-Thames, UK, 2017.
17. Dankov, P.I.; Tsatsova, M.I.; Levcheva, V.P. Investigation of Uniaxial Dielectric Anisotropy of Textile Fabrics and Its Influence over the Wearable Antennas' Behaviour. In Proceedings of the 2017 Progress in Electromagnetics Research Symposium-Fall (PIERS-FALL), Singapore, 19–22 November 2017. [CrossRef]
18. Dankov, P.I. Dielectric anisotropy of modern microwave substrates. In *Microwave and Millimeter Wave Technologies from Photonic Bandgap Devices to Antenna and Applications*; Igor, M., Ed.; In-Tech. Publ.: London, UK, 2010. [CrossRef]
19. Mukai, Y.; Suh, M. Relationships between structure and microwave dielectric properties in cotton fabrics. *Mater. Res. Express* **2020**, *7*, 015105. [CrossRef]
20. Ahmed, M.I.; Ahmed, M.F.; Shaalan, A.A. Investigation and Comparison of 2.4 GHz Wearable Antennas on Three Textile Substrates and Its Performance Characteristics. *Open J. Antennas Propag.* **2017**, *5*, 110–120. [CrossRef]
21. Ibanez-Labiano, I.; Alomainy, A. Dielectric Characterization of Non-Conductive Fabrics for Temperature Sensing through Resonating Antenna Structures. *Materials* **2020**, *13*, 1271. [CrossRef] [PubMed]
22. Zeouga, K.; Osman, L.; Gharsallah, A.; Manar, E.; Gupta, B. Truncated Patch Antenna on Jute Textile for Wireless Power Transmission at 2.45 GHz. *Int. J. Adv. Comput. Sci. Appl.* **2018**, *9*, 301–305. [CrossRef]
23. Ahmed, M.I.; Ahmed, M.F.; Shaalan, A.-E.A. Novel Electrotextile Patch Antenna on Jeans Substrate for Wearable Applications. *Prog. Electromagn. Res. C* **2018**, *83*, 255–265. [CrossRef]
24. Grilo, M.; Correra, F.S. Parametric Study of Rectangular Patch Antenna Using Denim Textile Material. In Proceedings of the SBMO/IEEE MTT-S International Microwave & Optoelectronics Conference (IMOC), Rio de Janeiro, Brazil, 28 October 2013. [CrossRef]
25. Harmer, S.W.; Rezgui, N.; Bowring, N.; Luklinska, Z.; Ren, G. Determination of the Complex Permittivity of Textiles and Leather in the 14-40 GHz, mm wave band using a Free-Wave Transmittance Only Method. *IET Microw. Antennas Propag.* **2008**, *2*, 606–614. [CrossRef]
26. Dankov, P.I. Concept for Equivalent Dielectric Constant of Planar Transmission Lines on Anisotropic Substrates. In Proceedings of the 46th European Microwave Conference (EuMC), London, UK, 3–7 October 2016; pp. 158–161. [CrossRef]
27. Dankov, P. Two-resonator method for measurement of dielectric anisotropy in multilayer samples. *IEEE Trans. Microw. Theory Tech.* **2006**, *54*, 1534–1544. [CrossRef]
28. Dankov, P.I.; Kondeva, M.I.; Baev, S.R. Influence of the Substrate Anisotropy in the Planar Antenna Simulations. In Proceedings of the International Workshop on Antenna Technology (iWAT) Conference, Lisbon, Portugal, 1–3 March 2010. [CrossRef]
29. Wong, K.-L. *Design of Nonplanar Microstrip Antennas and Transmission Lines*; Wiley Series in Microwave and Optical Engineering; John Wiley & Sons: Hoboken, NJ, USA, 2004; ISBN 978-0-471-46390-0.
30. Sanchez-Montero, R.; Lopez-Espi, P.-L.; Alen-Cordero, C.; Martinez-Rojas, J.-A. Bend and Moisture Effects on the Performance of a U-Shaped Slotted Wearable Antenna for Off-Body Communications in an Industrial Scientific Medical (ISM) 2.4 GHz Band. *Sensors* **2019**, *19*, 1804. [CrossRef]
31. Escobedo, P.; de Pablos-Florido, J.; Carvajal, M.A.; Martínez-Olmos, A.; Capitán-Vallvey, L.F.; Palma, A. The effect of bending on laser-cut electro-textile inductors and capacitors attached on denim as wearable structures. *Text. Res. J.* **2020**, *90*, 2355–2366. [CrossRef]
32. Gupta, N.P.; Kumar, M.; Maheshwari, R. Development and performance analysis of conformal UWB wearable antenna under various bending radii. *IOP Conf. Ser. Mater. Sci. Eng.* **2019**, *594*, 012025. [CrossRef]
33. Gharbi, M.E.; Fernández-García, R.; Ahyoud, S.; Gil, I. A Review of Flexible Wearable Antenna Sensors: Design, Fabrication Methods, and Applications. *Materials* **2020**, *13*, 3781. [CrossRef] [PubMed]
34. Jalil, M.E.; Rahim, M.K.A.; Samsuri, N.A.; Murad, N.A.; Majid, H.A.; Kamardin, K.; Abdullah, M.A. Fractal Koch Multiband Textile Antenna Performance with Bending, Wet Conditions and On the Human Body. *Prog. Electromagn. Res.* **2013**, *140*, 633–652. [CrossRef]
35. Ferreira, D.; Pires, P.; Rodrigues, R.; Caldeirinha, R.F.S. Wearable Textile Antennas: Examining the effect of bending on their performance. *Antennas Propag. Mag.* **2017**, *59*, 54–59. [CrossRef]
36. Dankov, P.; Levcheva, V.; Sharma, P. Influence of Dielectric Anisotropy and Bending on Wearable Textile Antenna Properties. In Proceedings of the 2020 International Workshop on Antenna Technology (iWAT), Bucharest, Romania, 25–28 February 2020. [CrossRef]
37. Ye, Y.; Yuan, J.; Su, K. A Volume-Surface Integral Equation Solver for Radiation from Microstrip Antenna on Anisotropic Substrate. *Int. J. Antennas Propag.* **2012**, *2012*, 1–4. [CrossRef]

38. Odabasi, H.; Teixeira, F.L. Analysis of cylindrically conformal patch antennas on isoimpedance anisotropic substrates. In Proceedings of the 2011 30th URSI General Assembly and Scientific Symposium, Istanbul, Turkey, 13–20 August 2011; pp. 1–4. [CrossRef]
39. Levcheva, V.P.; Hadjistamov, B.N.; Dankov, P.I. Two-Resonator Method for Characterization of Dielectric Substrate Anisotropy. *Bulg. J. Phys.* **2008**, *35*, 33–52.
40. Silvestre, L. A characterization of optimal two-phase multifunctional composite designs. *Proc. R. Soc. A Math. Phys. Eng. Sci.* **2007**, *463*, 2543–2556. [CrossRef]
41. Silvestre, L. *Upper Bounds for Multiphase Composites in Any Dimension*; Cornell University Library: Ithaca, NY, USA, 2010.
42. Parnell, W.J.; Calvo-Jurado, C. On the computation of the Hashin–Shtrikman bounds for transversely isotropic two-phase linear elastic fibre-reinforced composites. *J. Eng. Math.* **2015**, *95*, 295–323. [CrossRef]
43. Sihvola, A. *Electromagnetic Mixing Formulas and Applications*; The IEE, Electromagnetic Waves Series 47; IET: London, UK, 1999.
44. Alonso-Gonz'alez, L.; Hoeye, S.V.; Vazquez, C.; Fernandez, M.; Hadarig, A.; Las-Heras, F. Novel Parametric Electromagnetic Modelling to Simulate Textile Integrated Circuits. In Proceedings of the IEEE MTT-S International Conference on Numerical Electromagnetic and Multiphysics Modeling and Optimization for RF, Microwave, and Terahertz Applications (NEMO), Seville, Spain, 17–19 May 2017. [CrossRef]
45. Chen, T.; Chen, Y.; Jian, R. A Wideband Differential-Fed Microstrip Patch Antenna Based on Radiation of Three Resonant Modes. *Int. J. Antennas Propag.* **2019**, *2019*. [CrossRef]
46. Abdelgwad, A.H. Microstrip Patch Antenna Enhancement Techniques. *Int. Sch. Sci. Res. Innov.* **2018**, *12*, 703–710. [CrossRef]
47. V&V Article 3: Modeling Broadband and Circularly Polarized Patch Antennas Using EM. Picasso. Available online: http://www.emagtech.com/wiki/index.php/V%26V_Article_3:_Modeling_Broadband_And_Circularly_Polarized_Patch_Antennas_Using_EM.Picasso (accessed on 20 December 2020).
48. Rathor, V.S.; Saini, J.P. A Design of Swastika Shaped Wideband Microstrip Patch Antenna for GSM/WLAN Application. *J. Electromagn. Anal. Appl.* **2014**, *6*, 31–37. [CrossRef]
49. Khraisat, Y.S.H. Increasing Microstrip Patch Antenna Bandwidth by Inserting Ground Slots. *J. Electromagn. Anal. Appl.* **2018**, *10*, 1–11. [CrossRef]
50. Souza, E.A.M.; Oliveira, P.S.; D'Assunção, A.G.; Mendonça, L.M.; Peixeiro, G. Miniaturization of a Microstrip Patch Antenna with a Koch Fractal Contour Using a Social Spider Algorithm to Optimize Shorting Post Position and Inset Feeding. *Int. J. Antennas Propag.* **2019**, *2019*. [CrossRef]

Article

Near-Field to Far-Field RCS Prediction on Arbitrary Scanning Surfaces Based on Spherical Wave Expansion

Woobin Kim [1], Hyeong-Rae Im [1], Yeong-Hoon Noh [1], Ic-Pyo Hong [2], Hyun-Sung Tae [3], Jung-Kyu Kim [3] and Jong-Gwan Yook [1,*]

1 Department of Electrical and Electronic Engineering, Yonsei University, Seoul 03722, Korea; woobink0203@yonsei.ac.kr (W.K.); ihr3021@yonsei.ac.kr (H.-R.I.); yh.noh@yonsei.ac.kr (Y.-H.N.)
2 Department of Information and Communication Engineering, Kongju National University, Cheonan 31080, Korea; iphong@kongju.ac.kr
3 Aerospace Technology Research Institute, Agency for Defense and Development (ADD), Seosan, Chungnam 32024, Korea; hyunsung2000@add.re.kr (H.-S.T.); jeong806@add.re.kr (J.-K.K.)
* Correspondence: jgyook@yonsei.ac.kr; Tel.: +82-2-2123-4618

Received: 20 November 2020; Accepted: 14 December 2020; Published: 16 December 2020

Abstract: Near-field to far-field transformation (NFFFT) is a frequently-used method in antenna and radar cross section (RCS) measurements for various applications. For weapon systems, most measurements are captured in the near-field area in an anechoic chamber, considering the security requirements for the design process and high spatial costs of far-field measurements. As the theoretical RCS value is the power ratio of the scattered wave to the incident wave in the far-field region, a scattered wave measured in the near-field region needs to be converted into field values in the far-field region. Therefore, this paper proposes a near-field to far-field transformation algorithm based on spherical wave expansion for application in near-field RCS measurement systems. If the distance and angular coordinates of each measurement point are known, the spherical wave functions in an orthogonal relationship can be calculated. If each weight is assumed to be unknown, a system of linear equations as numerous as the number of samples measured in the near electric field can be generated. In this system of linear equations, each weight value can be calculated using the iterative least squares QR-factorization method. Based on this theory, the validity of the proposed NFFFT is verified for several scatterer types, frequencies and measurement distances.

Keywords: radar cross section (RCS) measurement; near-field to far-field transformation (NFFFT); spherical wave expansion (SWE)

---

1. Introduction

In many cases, radar cross section (RCS) measurements for an object under test (OUT) or the radiation patterns of an antenna under test (AUT) are captured in the near-field region inside an anechoic chamber or at an outdoor site. The theoretical RCS value is the power ratio of a scattered wave measured at infinity to an incident plane wave. The far-field region is proportional to the electrical size of the OUT compared to the wavelength. Specifically, it is calculated as $2d^2/\lambda$, where d is the linear size of the OUT and λ is the wavelength. This implies that, as the target expands or the frequency increases, the far-field range criterion will become excessively large for practical measurements. Therefore, in many cases, a specially designed reflector is placed between the feeder and OUT to induce far-field conditions [1]. Another approach is the use of a near-field to far-field transformation (NFFFT) technique in a non-ideal measurement environment. Traditionally, field-transformation algorithms have focused on the processing of field values measured over canonical surfaces (e.g., planes [2], cylinders [3] and spheres [4]) using corresponding modal expansion techniques. To minimize measurement time, studies on non-uniform near-field sampling grids have been conducted continuously for each type of scanning surface [5–7]. In particular, several approaches have been presented for the arbitrary-probe NFFFT of spherical near-field measurements. These approaches include angular scanning [8] and probe modal content tuning [9–12]. However, modal expansion requires every point for measuring a near-field to exist on a surface with a single coordinate system in order to maintain orthogonality. Additionally, the far-field conversion process is unintuitive and complicated, because calculating the integral equation corresponding to each coordinate system is required for the NFFFT process. When measuring within a specific bandwidth rather than at a single frequency, the computational complexity increases significantly because the solution for the new integral equation must be calculated according to the wavenumber corresponding to the sampling frequency within the bandwidth. Additionally, gaps between measurement samples must exist on the canonical surface, which impairs the flexibility of near-field measurements. In a real measurement system, the flexibility of near-field sampling is crucial to avoid limiting the shape of the OUT or the measurement frequency.

To overcome the limitations discussed above, an NFFFT algorithm that can flexibly convert an arbitrary near-field scanning surface measurement value into a far-field RCS within various frequency bands is required. In this paper, a far-field RCS prediction method based on the spherical wave expansion (SWE) method that adapts modal expansion techniques away from canonical surfaces is proposed. SWE-NFFFT has mainly been used to predict far-field radiation patterns based on near-field measurements of AUTs [13–21]. First, the point at which an AUT's radiation field is located is assumed to be the origin of the coordinate system and an arbitrary near-field measurement point can be expressed in spherical coordinates. The electric field at a given location can be expressed as a weighted sum of spherical waves that are orthogonal to each other. Assuming that the weight multiplied by each spherical wave is unknown, a system of linear equations as numerous as the number of samples measured in the near electric field can be generated. The unknown weights can be calculated from the linear system of equations generated in this manner. This process is called an inverse problem. Next, the RCS defined in the far-field domain can be predicted using the calculated weights. This process is called a direct problem, and all processes in the inverse problem and direct problem can be considered as the basic principles of NFFFT, as illustrated in Figure 1.

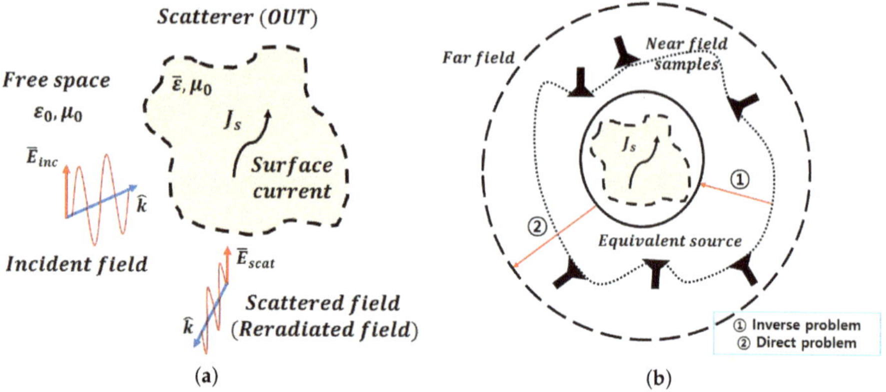

Figure 1. Schematic view of electromagnetic scattering and the process of spherical wave expansion–near-field to far-field transformation (SWE-NFFFT). (**a**) The geometry of electromagnetic scattering. (**b**) Illustration of SWE-NFFFT.

The method described above can be applied to the RCS of an OUT in its current form. Regarding the process of generating scattered waves, when an electric field is incident on an OUT, a surface current density is induced on the OUT surface by the incident waves. This induced surface current density acts as a source and generates another electric field, which is regarded as a scattered wave reflected by the OUT. Therefore, an OUT on which an incident wave generates a surface current density can be modeled as an equivalent current source, such as an AUT emission field.

In this study, an NFFFT algorithm was developed based on the SWE theory to verify the validity of far-field RCS prediction for an arbitr-ry near-field scanning surface. First, NFFFT RCS results for two canonical surfaces (spherical and planar) scans were compared and analyzed using a NASA almond type OUT for which reliable reference measurement data were available. Next, the validity of SWE-based NFFFT for arbitrary scanning surfaces was confirmed based on the extracted far-field RCS results from scanning near-field samples for actual missile type OUTs with a larger electrical size.

2. Theory

2.1. Radar Cross Section

An RCS is utilized as a measure of the reflective strength of a target and is defined as 4π times the ratio of the power per unit solid angle scattered in a specified direction to the power per unit area in a plane wave incident on the scatterer in a specified direction [22]. Specifically, it is defined as the limit value of the distance between the scatterer and the point at which the scattered power is measured at infinity, as follows:

$$\sigma = \lim_{r\to\infty} 4\pi r^2 \frac{|E_{scat}|^2}{|E_{inc}|^2} = \lim_{r\to\infty} 4\pi r^2 \frac{P_{scat}}{P_{inc}} \quad [m^2] \tag{1}$$

$$\sigma_{dBsm} = 10\log_{10}(\frac{\sigma}{\sigma_{ref}}) = 10\log_{10}(\frac{\sigma}{1}) \quad [dBsm] \tag{2}$$

where r is the distance from the OUT to the measurement point, E_{scat} is the scattered electric field and E_{inc} is the incident electric field at the target. The unit for an RCS is square meters. Because objects for which RCSs are measured have various sizes, the logarithm power scale is often adopted, and 1 m^2 is used as a reference value, as shown in Equation (2).

As shown in Figure 2, depending on the direction in which a scattered wave is measured, the RCS can be categorized as monostatic or bistatic. If scattering is observed in the k vector direction, which is the opposite direction of the incident wave, then it is defined as monostatic RCS, while all other cases are defined as bistatic RCS.

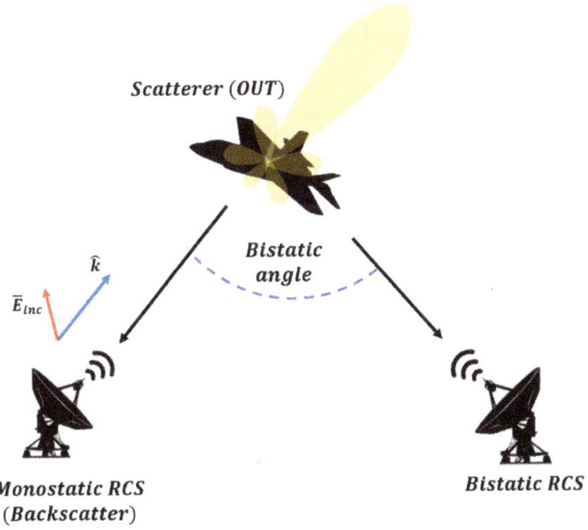

Figure 2. Illustration of monostatic and bistatic radar cross sections (RCSs).

Based on the definition above, the RCS measurement of an OUT can be performed using two methods: outdoor and indoor. According to the theoretical definition of an RCS, the measurement environment that is more suitable for measuring scattered waves at infinite distances is outdoors. However, it is necessary to consider the effects of various types of clutter that can affect RCS measurement values (e.g., weather conditions such as rain or snow or wide-open flat spaces). In most cases, indoor measurements can provide cleaner measurement data with reduced clutter, even though the measurement space is limited in size. In the near-field region, the shape of an electromagnetic wave is closer to the shape of a spherical wave than a plane wave. Therefore, if the power ratio is calculated or measured from a scattered wave in the near-field region, the RCS values will differ from the true values. Therefore, it is necessary to predict a far-field RCS in which a plane wave is reconstructed by applying a post-processing technique to the electric field value measured in the near-field region. The following subsection details the SWE-NFFFT theory for arbitrary scanning surfaces, which is not limited to the canonical scanning surfaces, unlike the conventional NFFFT algorithm.

2.2. Near-Field to Far-Field RCS Prediction Based on SWE

As shown in Figure 1, the scattered wave formed by an OUT can be derived from an equivalent surface current density by utilizing Love's equivalence principle [23]. Specifically, the scattered wave formed by an OUT can be expressed as a weighted sum of spherical wave functions that are orthogonal to each other at an arbitrary measurement point, which matches the form of the following spherical wave expansion:

$$\overrightarrow{E}(r,\theta,\phi) = \sum_{s=1}^{2}\sum_{n=1}^{N}\sum_{m=-n}^{n} Q_{smn}\overrightarrow{F}^{(3)}_{smn} \tag{3}$$

$$N = \lceil ka + 6(ka)^{\frac{1}{3}} \rceil \tag{4}$$

where $s = 1,2$ represent two orthogonal modes of the electromagnetic field, Q_{smn} is the spherical wave coefficient and $\overrightarrow{F}^{(3)}_{smn}$ is the spherical wave basis function. In Equation (4), k is a wave number determined by the measurement frequency and a is the radius of the sphere with the smallest volume containing the entire OUT to be measured. N is the truncation number and is a variable that determines the minimum amount of data required for valid RCS prediction from the limited measurement data. Therefore, the number of m and n are determined to be $N+2$ and N, respectively.

In this case, the spherical wave basis function $\overrightarrow{F}^{(3)}_{smn}$, which takes the form of a vector, is calculated with both theta and phi directional components.

$$\overrightarrow{F}^{(3)}_{1mn}(r,\theta,\phi) = \frac{1}{\sqrt{2\pi}}\frac{1}{\sqrt{n(n+1)}}\left(\frac{-m}{|m|}\right)^{m}\left[z_n^{(3)}(kr)\frac{im\overline{P}_n^{|m|}(\cos\theta)}{\sin\theta}e^{im\phi}\widehat{\theta} - z_n^{(3)}(kr)\frac{d\overline{P}_n^{|m|}(\cos\theta)}{d\theta}e^{im\phi}\widehat{\phi}\right] \tag{5}$$

$$\begin{aligned}\overrightarrow{F}^{(3)}_{2mn}(r,\theta,\phi) = {}& \frac{1}{\sqrt{2\pi}}\frac{1}{\sqrt{n(n+1)}}\left(\frac{-m}{|m|}\right)^{m}\left[\frac{n(n+1)}{kr}z_n^{(3)}(kr)\overline{P}_n^{|m|}(\cos\theta)e^{im\phi}\widehat{r}+\right.\\ &\left.\frac{1}{kr}\frac{d}{d(kr)}\left\{krz_n^{(3)}(kr)\right\}\frac{d\overline{P}_n^{|m|}(\cos\theta)}{d\theta}e^{im\phi}\widehat{\theta} + \frac{1}{kr}\frac{d}{d(kr)}\left\{krz_n^{(3)}(kr)\right\}\frac{im\overline{P}_n^{|m|}(\cos\theta)}{\sin\theta}e^{im\phi}\widehat{\phi}\right]\end{aligned} \tag{6}$$

$$z_n^{(3)}(x) = j_n(x) + iy_n(x) \tag{7}$$

$$j_n(x) = (-x)^n\left(\frac{1}{x}\frac{d}{dx}\right)^n\frac{\sin x}{x} \tag{8}$$

$$y_n(x) = -(-x)^n\left(\frac{1}{x}\frac{d}{dx}\right)^n\frac{\cos x}{x} \tag{9}$$

$$\overline{P}_n^m(x) = (-1)^m\sqrt{\frac{(n+\frac{1}{2})(n-m)!}{(n+m)!}}P_n^m(x) \tag{10}$$

$$P_n^m(x) = (-1)^m(1-x^2)^{m/2}\frac{d^m}{dx^m}(P_n(x)) \tag{11}$$

where $z_n^{(3)}(x)$ is the first kind spherical Hankel function consisting of the sum of the spherical Bessel function $j_n(x)$ and spherical Neumann function $y_n(x)$. $\overline{P}_n^m$ is the normalized associated Legendre polynomial, and P_n^m is the original associated Legendre polynomial.

As shown in Figure 3, if an electric field is expressed by SWE at an arbitrary near-field measurement point $(r_{mn}, \theta_{mn}, \phi_{mn})$, then an $N(N+2)$-weighted sum is generated. This indicates that there are $N(N+2)$ unknown coefficients, which can be derived by constructing a linear system of equations from the near-field measured sample values.

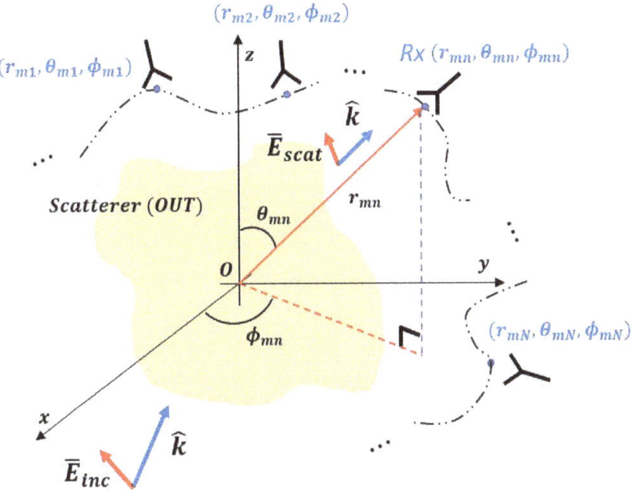

Figure 3. SWE field modeling at the measurement point for a scatterer.

Based on Equations (4)–(11), Equation (3) can be expressed in the form of a matrix-vector product:

$$U = CQ = \begin{bmatrix} F_{\theta,s(-1)1}(r_{m1},\theta_{m1},\phi_{m1}) & F_{\theta,s01}(r_{m1},\theta_{m1},\phi_{m1}) & \cdots & F_{\theta,sNN}(r_{m1},\theta_{m1},\phi_{m1}) \\ F_{\phi,s(-1)1}(r_{m1},\theta_{m1},\phi_{m1}) & F_{\phi,s01}(r_{m1},\theta_{m1},\phi_{m1}) & \cdots & F_{\phi,sNN}(r_{m1},\theta_{m1},\phi_{m1}) \\ F_{\theta,s(-1)1}(r_{m2},\theta_{m2},\phi_{m2}) & F_{\theta,s01}(r_{m2},\theta_{m2},\phi_{m2}) & \cdots & F_{\theta,sNN}(r_{m2},\theta_{m2},\phi_{m2}) \\ \vdots & \vdots & \ddots & \vdots \\ F_{\theta,s(-1)1}(r_{mn},\theta_{mn},\phi_{mn}) & F_{\theta,s01}(r_{mn},\theta_{mn},\phi_{mn}) & \cdots & F_{\theta,sNN}(r_{mn},\theta_{mn},\phi_{mn}) \\ F_{\phi,s(-1)1}(r_{mn},\theta_{mn},\phi_{mn}) & F_{\phi,s01}(r_{mn},\theta_{mn},\phi_{mn}) & \cdots & F_{\phi,sNN}(r_{mn},\theta_{mn},\phi_{mn}) \end{bmatrix} \begin{bmatrix} Q_{s(-1)1} \\ Q_{s01} \\ Q_{s11} \\ \vdots \\ \vdots \\ Q_{sNN} \end{bmatrix} \tag{12}$$

$$Q = (C^H C)^{-1} C^H U \tag{13}$$

In Equations (12) and (13), U is a measured near-field sample vector that includes both θ and ϕ polarization, and C is a matrix that can be calculated from the coordinates of each measurement point. The subscripts θ and ϕ of the elements F constituting the matrix C represent the coefficients of the theta and phi components in Equations (5) and (6), respectively. H denotes the Hermitian operation of a matrix. From U and C, the unknown vector Q can be derived as shown in Equation (13). This process is the inverse problem illustrated in Figure 1b and represents the equivalent current source model for an OUT. The electric field in the far-field region is computed by substituting the r_m value of each measurement coordinate with a significantly large value instead of using the C matrix calculated for the near-field measurement points.

$$E_{far} = C_{far}Q = \begin{bmatrix} F_{\theta,s(-1)1}(\infty,\theta_{m1},\phi_{m1}) & F_{\theta,s01}(\infty,\theta_{m1},\phi_{m1}) & \cdots & F_{\theta,sNN}(\infty,\theta_{m1},\phi_{m1}) \\ F_{\phi,s(-1)1}(\infty,\theta_{m1},\phi_{m1}) & F_{\phi,s01}(\infty,\theta_{m1},\phi_{m1}) & \cdots & F_{\phi,sNN}(\infty,\theta_{m1},\phi_{m1}) \\ F_{\theta,s(-1)1}(\infty,\theta_{m2},\phi_{m2}) & F_{\theta,s01}(\infty,\theta_{m2},\phi_{m2}) & \cdots & F_{\theta,sNN}(\infty,\theta_{m2},\phi_{m2}) \\ \vdots & \vdots & \ddots & \vdots \\ F_{\theta,s(-1)1}(\infty,\theta_{mn},\phi_{mn}) & F_{\theta,s01}(\infty,\theta_{mn},\phi_{mn}) & \cdots & F_{\theta,sNN}(\infty,\theta_{mn},\phi_{mn}) \\ F_{\phi,s(-1)1}(\infty,\theta_{mn},\phi_{mn}) & F_{\phi,s01}(\infty,\theta_{mn},\phi_{mn}) & \cdots & F_{\phi,sNN}(\infty,\theta_{mn},\phi_{mn}) \end{bmatrix} \begin{bmatrix} Q_{s(-1)1} \\ Q_{s01} \\ Q_{s11} \\ \vdots \\ \vdots \\ Q_{sNN} \end{bmatrix} \quad (14)$$

The newly calculated C_{far} matrix with $r \to \infty$ is multiplied by the solved vector Q. This process is the direct problem. By substituting the E_{far} obtained in the same process as Equation (14) into E_{scat} in Equation (1), we can predict the far-field RCS for each near-field measurement point. Each component F in the matrix C_{far} can be computed for arbitrary measurement points that are not limited to specific canonical surfaces. This means that far-field RCS prediction using SWE-based NFFFT is possible if at least $N(N+2)$ measurement samples are secured by the minimum truncation number N, as determined by Equation (4). In this study, in preparation for increases in the matrix size, the least squares QR-factorization (LSQR) method was utilized instead of a direct solver.

3. Numerical Results and Discussion

3.1. SWE-NFFFT Verification

This section describes the validation of the SWE-NFFFT algorithm proposed in Section 2. The far-field reference and near-field data at a specific distance are simulated using a three-dimensional full-wave electromagnetic solver FEKO based on method of moments (MoM) analysis. A perfect electrically conducting NASA almond-type OUT is considered. RCS measurement values have been well documented at specific frequencies for this type of object [24]. As shown in Figure 4, this object can be expressed as a closed surface formed by $(x(t,\psi), y(t,\psi), z(t,\psi))$ according to the positional variables t, ψ.

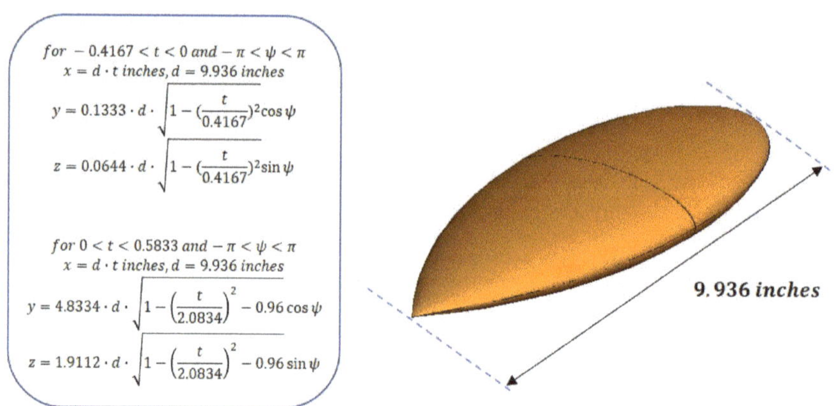

Figure 4. NASA almond-type computer aided design (CAD) model used for simulations.

First, when the SWE-NFFFT procedure is utilized with electrical field data extracted from a distance (other than infinity) that is close to the far-field region, it is necessary to verify whether the OUT is correctly modeled with an equivalent current source. The incident waves are set as θ-polarized plane waves incident from a single direction of $\theta = 90°$ and $\phi = 0°$. The resulting near and far-field scattered wave data are all extracted in a bistatic format. The polarization data used for transformation, based on Equation (12), are θ and ϕ-pol.

To verify the compatibility of SWE-NFFFT with a canonical surface, a spherical scan was performed with a radius of 2 m for the entire azimuth and elevation angle, as shown in Figure 5.

The frequency of the incident wave was set to 100 MHz. Here, 99 near-field samples (N) were required for each polarization, as shown in Equation (4). For the ease of extraction of near-field data, the samples in the θ and ϕ directions were extracted at consistent intervals of $N_\theta = 9$ and $N_\phi = 11$, respectively. Under these simulation conditions, the size of the matrix C in Equation (12), which was required for SWE-NFFFT, was 198×198 and the size of the unknown vector Q was 198×1. Figure 6 presents the far-field RCS prediction results based on the vector Q obtained from SWE-NFFFT. Among the 99 far-field RCS data points predicted for all azimuth angles, the vertical transmit and vertical receive polarization (VV-pol) RCS results are displayed for $\theta = 45°$ and $\theta = 90°$ cut planes. The solid line composed of blue circles represents the reference far-field RCS, whereas the solid line composed of red stars represents the prediction results when using the SWE-NFFFT method. It is clear that the SWE-NFFFT-based RCS values show excellent agreement, except for the extremely small RCS values, which would be negligible a in practical scenario.

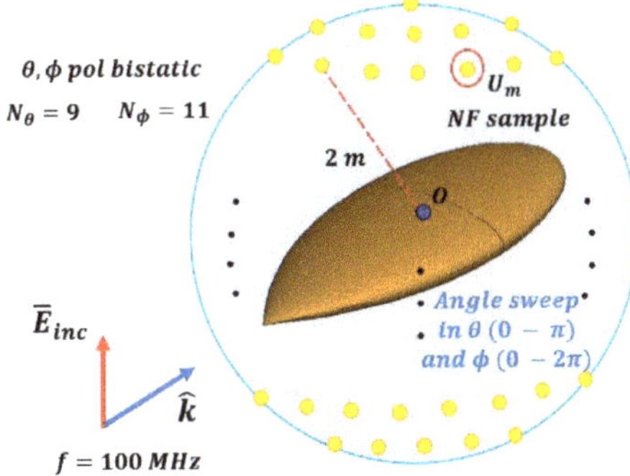

Figure 5. Simulation setup for near-field sample data from a spherical scanning surface.

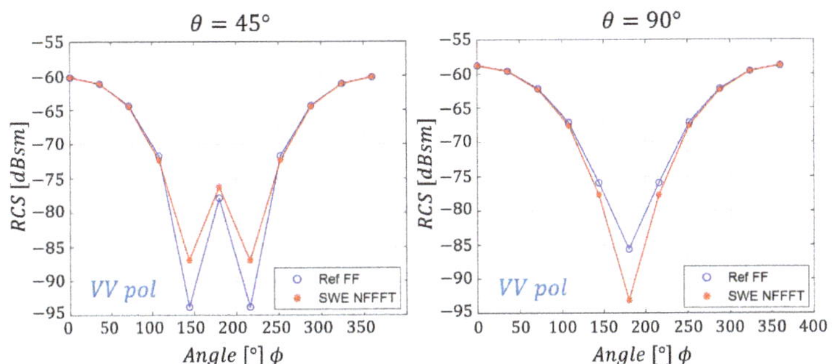

Figure 6. VV-pol SWE-NFFFT results for a 100 MHz incident wave on a $\theta = 45°, 90°$ cut plane.

We then considered a frequency of 300 MHz, and all other simulation conditions were kept the same. Based on this increased operating frequency, the number of near-field samples per polarization was 323. The samples in the θ and ϕ directions were extracted at consistent intervals of $N_\theta = 17$ and $N_\phi = 19$, respectively.

Figure 7 presents the RCS data for various cases with $\theta = 45°$ and $\theta = 90°$ cut planes for VV and VH(-horizontal receive)-pol. Based on the verification processes above, it is confirmed that applying the SWE-NFFFT method is reasonable in the case of spherical scanning at a distance in the far-field region.

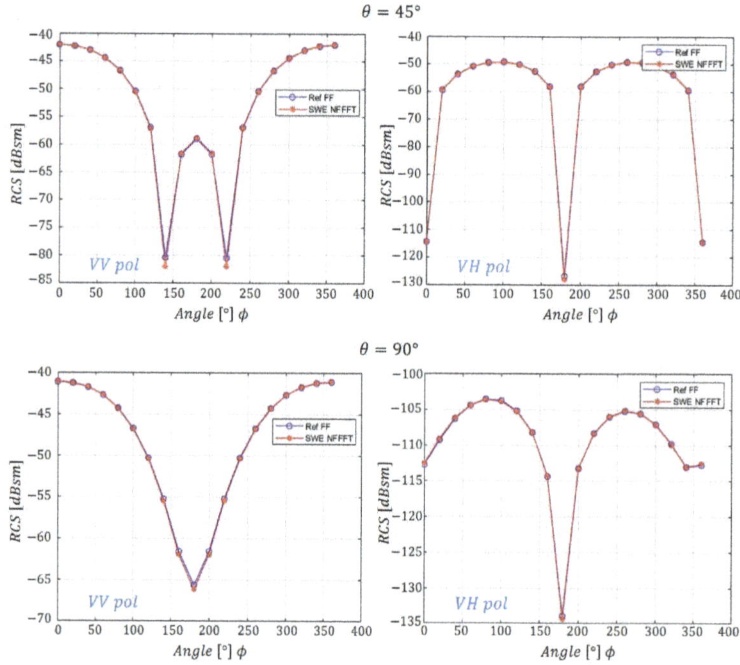

Figure 7. SWE-NFFFT results for a 300 MHz incident wave on a $\theta = 45°, 90°$ cut plane.

We then analyzed the applicability of the SWE-NFFFT method to various scanning surfaces. We also examined tge partial surface scans and compared the results to electric field integral equation-based NFFFT, which is dependent on a canonical scanning surface and requires scanning for all-round surfaces (e.g., spheres and cylinders). Additionally, to verify the validity of RCS prediction in the near-field region, field samples were extracted at a shorter distance compared to the previous simulation.

As shown in Figure 8, verification was performed for two cases: a scan of a spherical surface and a square plane surface containing the same angular region ($\theta = \frac{\pi}{4} \sim \frac{3\pi}{4}, \phi = \frac{-\pi}{4} \sim \frac{\pi}{4}$) with the same frequency of 300 MHz. In the spherical scan, all near-field sample points were extracted at a radius of 0.5 m, whereas in the planar scan, the nearest sample point was located at a distance of 0.3 m on a square plane with a size of 0.6×0.6 m. To compare the far-field RCS results at identical sample points for the two scanning methods, we considered values of $N_\theta = 19$ and $N_\phi = 21$, which were greater than the minimum truncation number required to have the same θ and ϕ values. Based on Equations (5), (6) and (12), SWE-NFFFT was calculated for matrix C under θ polarization and ϕ polarization in the transformation process. In the case of the planar scan, the x, y, z polarization components were extracted and then transformed using Equations (15)–(18);

$$X\hat{\theta} = A\hat{x} + B\hat{y} + C\hat{z} \tag{15}$$

$$Y\hat{\phi} = D\hat{x} + E\hat{y} \tag{16}$$

$$X = A\cos\theta\cos\phi + B\cos\theta\sin\phi - C\sin\theta \tag{17}$$

$$Y = -D\sin\theta + E\sin\phi \tag{18}$$

Figure 9 presents the RCS data from the two different scanning methods and the reference values, which show excellent agreement.

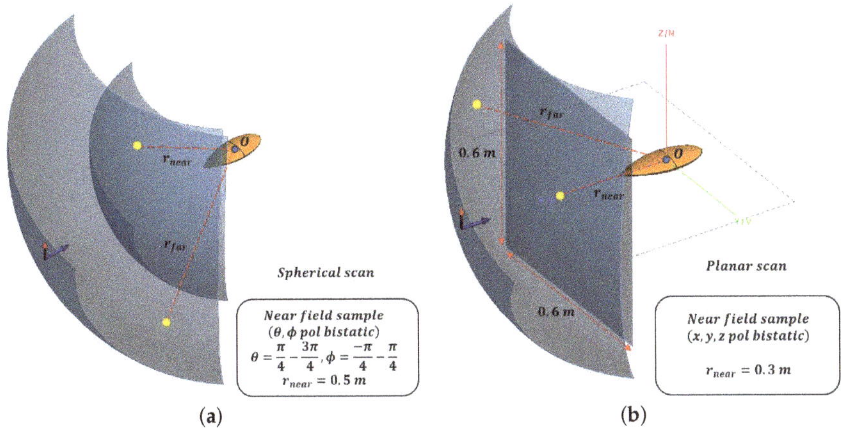

Figure 8. Simulation setup for near-field sample data from two canonical scanning surfaces. (**a**) Spherical near-field scan. (**b**) Planar near-field scan.

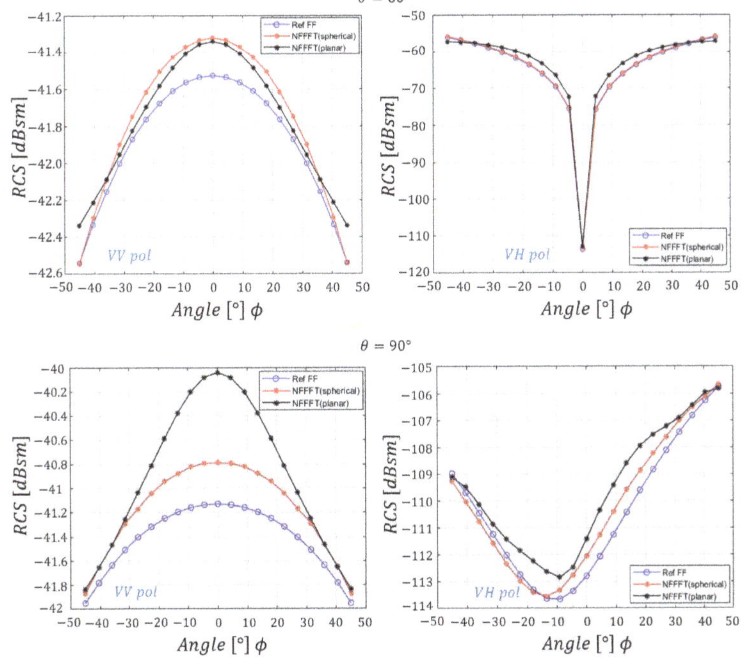

Figure 9. Comparison of SWE-NFFFT results according to the scanning method ($\theta = 60°, 90°$ cut plane).

The VV-pol as well as VH-pol RCS results at the cut planes of $\theta = 60°$ and $\theta = 90°$ are presented in Figure 9, revealing a maximum error of 2 dB, which is almost negligible. The solid line composed of blue circles represents the reference far-field RCS, and the solid line composed of red stars represents the spherical scan NFFFT; the black solid line represents the planar scan NFFFT results. In the $\theta = 90°$ cut plane, for VV-pol, a maximum error of 1 dB occurs at $\phi = 0°$. In the case of the VH-pol, the maximum error is approximately 2 dB around $\phi = 10°$. In the case of a planar scan, because each near-field sample point has a different distance, larger errors occur compared to the spherical scan, but reasonable NFFFT results are obtained in general.

3.2. Far-Field RCS Prediction on Arbitrary Scanning Surfaces

In this section, the proposed SWE-NFFFT algorithm is applied to predict the far-field RCS for an arbitrary scanning surface on a more realistic OUT model.

Compared to the streamlined NASA almond, the OUT model used in this simulation has the shape of a missile with a length of 1 m and width of 0.62 m, as shown in Figure 10.

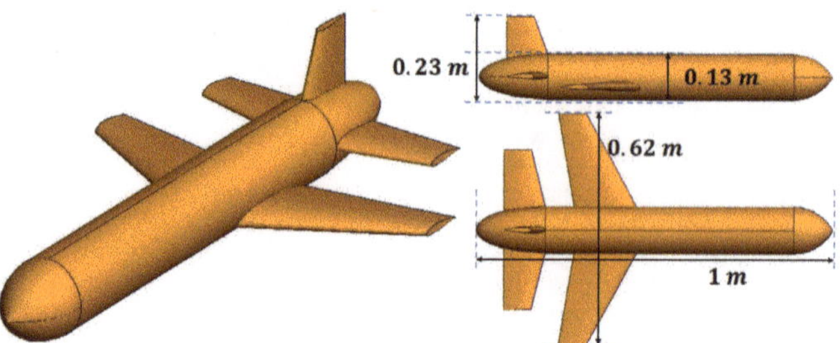

Figure 10. Missile CAD model used for simulation.

The direction and parameters of the incident wave were the same as those mentioned in the previous section and the frequency was increased to 1 GHz. By setting $N = 34$, which is greater than the minimum truncation number determined by Equation (4), the far-field RCS can be calculated from the matrix C (2448 $\times$ 2448). When considering the size of the missile-shaped OUT and the wavelength, the minimum far-field region is 7 m. The near-field measurement data were extracted within a range of 3–5 m, and θ and the distance vector r were set to different random values to ensure that the scanning surface did not correspond to a specific canonical surface. A total of 2448 sample data points were extracted from 12 surfaces (102 points at 1° intervals, with 1224 samples for each polarization).

The SWE-NFFFT-based RCS data are compared with the reference data for 12 different scanning surfaces under VV and VH-pol in Figures 11 and 12, respectively, revealing excellent agreement. Figure 13 presents more detailed data.

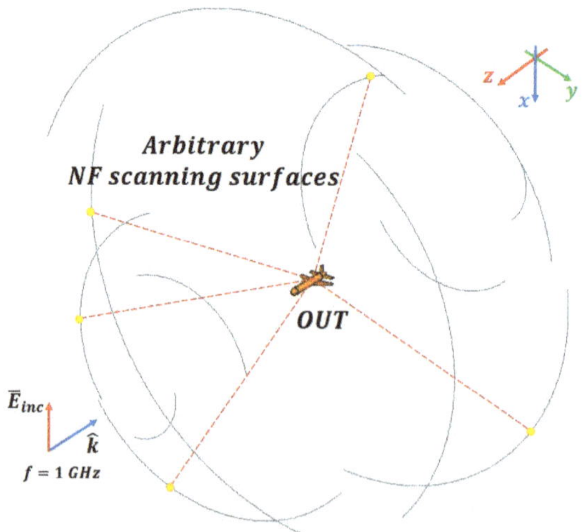

Figure 11. Simulation setup for arbitrary scanning surfaces.

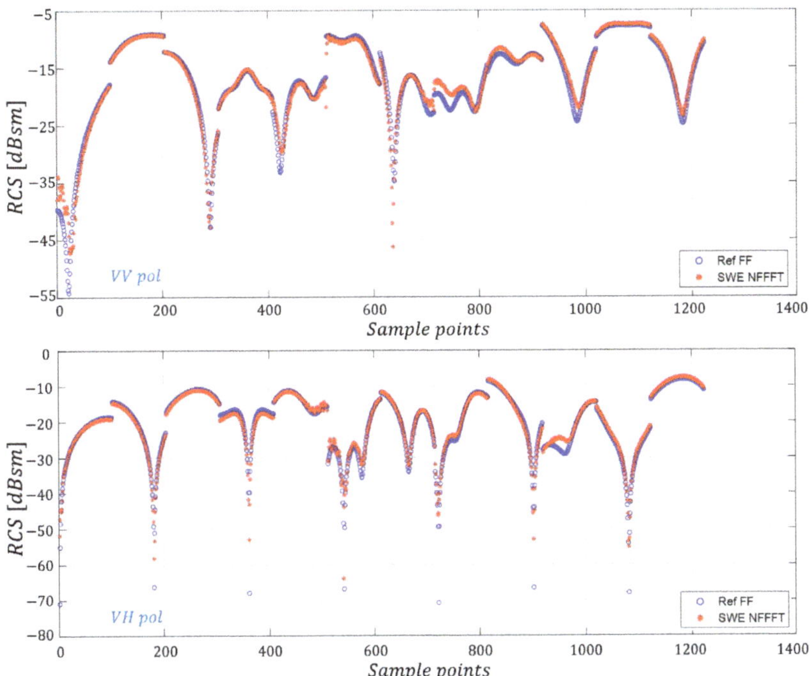

Figure 12. Comparisons of SWE-NFFFT results for all near-field sample points.

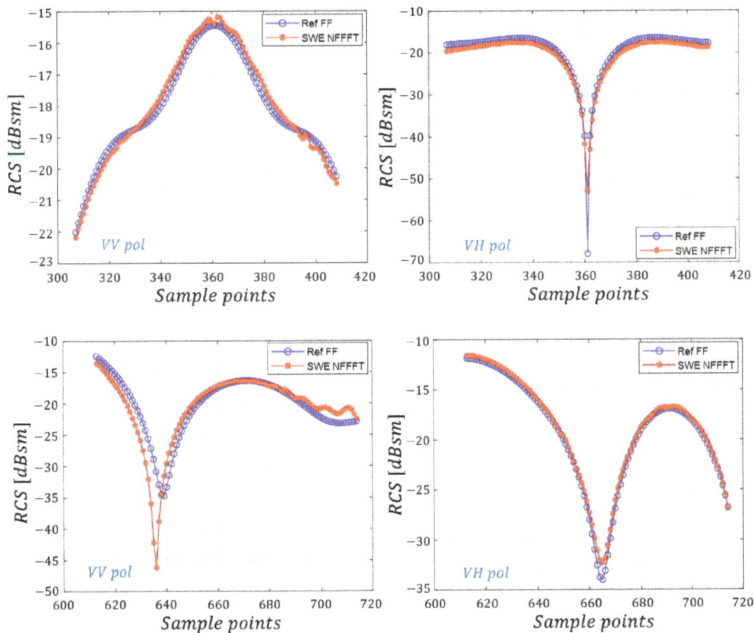

Figure 13. Partial plot of SWE-NFFFT results for all near-field sample points (306∼407, 612∼713).

It is clear that the proposed SWE-NFFFT method enables the prediction of high-accuracy far-field RCS data with arbitrary scanning surfaces for general OUTs.

4. Conclusions

In this paper, a SWE-NFFFT method was proposed for arbitrary measurement points that do not belong to canonical surfaces, and it was proven that the proposed method is applicable to near-field RCS measurement systems. Therefore, any arbitrary near-field data can be utilized to derive far-field RCS data.

First, a system of linear equations is generated from the minimum near-field sample data required for far-field RCS prediction. When the distance and angular coordinates of each measurement point are specified, spherical wave functions that are orthogonal to each other can be calculated. The coefficients multiplied by weights for each function can be derived as determinants by assuming unknown vectors. In the system of linear equations, an unknown vector is calculated as an inverse problem using the LSQR method. It is then modeled as an equivalent current source, and the far-field electric field is derived from the modeled source as a direct problem. Finally, the RCS can be predicted.

In this study, the compatibility of a conventional canonical scanning surface was verified for different frequencies and measurement distances using a NASA almond-type OUT. Next, the proposed SWE-NFFFT was validated on an arbitrary scanning surface for a more general shape of an OUT in the GHz frequency region. Finally, it was verified that RCS prediction can be applied to any arbitrary scanning surface. Therefore, it has been confirmed that RCS prediction using the proposed SWE-based NFFFT algorithm is feasible. In particular, in the final verification process for a missile-shaped OUT, the overall RCS tendency showed high accuracy. It was confirmed that about a 3 dB error occurred at several points, which was about 10 dB at the null point, which would be negligible in a practical scenario.

In future work, we plan to extend the NFFFT method for non-uniform sampling, which extracts near-field sample data more densely from measurement points with large error.

Author Contributions: Conceptualization: W.K., H.-R.I., Y.-H.N. and J.-G.Y.; methodology: W.K., I.-P.H. and J.-G.Y.; software: W.K. and H.-R.I.; validation: W.K., Y.-H.N., I.-P.H. and J.-G.Y.; formal Analysis: W.K., I.-P.H., and J.-G.Y.; investigation: W.K. and H.-R.I.; resources: W.K.; data curation: W.K., I.-P.H. and J.-G.Y.; writing—original draft preparation: W.K. and J.-G.Y.; writing—review and editing: I.-P.H., H.-S.T. and J.-G.Y.; visualization: W.K.; supervision: I.-P.H. and J.-G.Y.; project administration: H.-S.T., J.-K.K. and J.-G.Y.; funding acquisition: H.-S.T. and J.-G.Y. All authors have read and agreed to the published version of the manuscript.

Funding: This research received no external funding.

Acknowledgments: This work was supported by the Aerospace Low Observable Technology Laboratory Program of the Defense Acquisition Program Administration and the Agency for Defense Development of the Republic of Korea.

Conflicts of Interest: The authors declare no conflict of interest.

References

1. Pienaar, M.; Odendaal, J.; Joubert, J.; Pienaar, C.; Smit, J. Bistatic RCS measurements in a compact range. In Proceedings of the 2017 International Conference on Electromagnetics in Advanced Applications (ICEAA), Verona, Italy, 11–15 September 2017; pp. 1199–1202.
2. Hansen, T.; Yaghjian, A.D. *Plane-Wave Theory of Time-Domain Fields: Near-Field Scanning Applications*; IEEE Press: Piscataway Township, NJ, USA, 1999.
3. Leach, W.; Paris, D. Probe compensated near-field measurements on a cylinder. *IEEE Trans. Antennas Propag.* **1973**, *21*, 435–445. [CrossRef]
4. Hansen, J.E. (Ed.) *Measurements, Spherical Near-Field Antenna*; Peter Peregrinus: London, UK, 1988.
5. D'Agostino, F.; Ferrara, F.; Gennarelli, C.; Guerriero, R.; Migliozzi, M. Fast and accurate far-field prediction by using a reduced number of bipolar measurements. *IEEE Antennas Wirel. Propag. Lett.* **2017**, *16*, 2939–2942. [CrossRef]
6. D'Agostino, F.; Ferrara, F.; Gennarelli, C.; Gennarelli, G.; Guerriero, R.; Migliozzi, M. On the direct non-redundant near-field-to-far-field transformation in a cylindrical scanning geometry. *IEEE Antennas Propag. Mag.* **2012**, *54*, 130–138. [CrossRef]

7. D'Agostino, F.; Ferrara, F.; Gennarelli, C.; Guerriero, R.; Migliozzi, M. Non-Redundant Spherical NF-FF Transformations Using Ellipsoidal Antenna Modeling: Experimental Assessments [Measurements Corner]. *IEEE Antennas Propag. Mag.* **2013**, *55*, 166–175. [CrossRef]
8. Laitinen, T. Double ϕ-Step θ-Scanning Technique for Spherical Near-Field Antenna Measurements. *IEEE Trans. Antennas Propag.* **2008**, *56*, 1633–1639. [CrossRef]
9. Laitinen, T.; Breinbjerg, O. A first/third-order probe correction technique for spherical near-field antenna measurements using three probe orientations. *IEEE Trans. Antennas Propag.* **2008**, *56*, 1259–1268. [CrossRef]
10. Laitinen, T.; Pivnenko, S.; Nielsen, J.M.; Breinbjerg, O. Theory and practice of the FFT/matrix inversion technique for probe-corrected spherical near-field antenna measurements with high-order probes. *IEEE Trans. Antennas Propag.* **2010**, *58*, 2623–2631. [CrossRef]
11. Hansen, T.B. Spherical near-field scanning with higher-order probes. *IEEE Trans. Antennas Propag.* **2011**, *59*, 4049–4059. [CrossRef]
12. Saccardi, F.; Giacomini, A.; Foged, L. Probe correction technique of arbitrary order for high accuracy spherical near field antenna measurements. In Proceedings of the 2016 AMTA Proceedings, Austin, TX, USA, 30 October–4 November 2016; pp. 239–244.
13. Culotta-López, C.; Heberling, D.; Bangun, A.; Behboodi, A.; Mathar, R. A compressed sampling for spherical near-field measurements. In Proceedings of the 2018 AMTA Proceedings, Williamsburg, VA, USA, 4–9 November 2018; pp. 1–6.
14. Farouq, M.; Serhir, M.; Picard, D. Antenna far-field assessment from near-field measured over arbitrary surfaces. *IEEE Trans. Antennas Propag.* **2016**, *64*, 5122–5130. [CrossRef]
15. Cornelius, R.; Heberling, D. Spherical near-field scanning with pointwise probe correction. *IEEE Trans. Antennas Propag.* **2016**, *65*, 995–997. [CrossRef]
16. Foged, L.; Saccardi, F.; Mioc, F.; Iversen, P. Spherical near field offset measurements using downsampled acquisition and advanced NF/FF transformation algorithm. In Proceedings of the 2016 10th European Conference on Antennas and Propagation (EuCAP), Davos, Switzerland, 10–15 April 2016; pp. 1–3.
17. Cornelius, R.; Heberling, D. Correction of non-ideal probe orientations for spherical near-field antenna measurements. In Proceedings of the 2017 Antenna Measurement Techniques Association Symposium (AMTA), Atlanta, GA, USA, 15–20 October 2017; pp. 1–5.
18. Chou, H.T.; Pathak, P.H.; Tuan, S.C.; Burkholder, R.J. A novel far-field transformation via complex source beams for antenna near-field measurements on arbitrary surfaces. *IEEE Trans. Antennas Propag.* **2017**, *65*, 7266–7279. [CrossRef]
19. Varela, F.R.; Iragüen, B.G.; Sierra-Castañer, M. Near-Field to Far-Field Transformation on Arbitrary Surfaces via Multi-Level Spherical Wave Expansion. *IEEE Trans. Antennas Propag.* **2019**, *68*, 500–508. [CrossRef]
20. Varela, F.R.; Iragüen, B.G.; Castañer, M.S. Probe-Corrected Near-Field to Far-Field Transformation Using Multiple Spherical Wave Expansions. In Proceedings of the 2019 13th European Conference on Antennas and Propagation (EuCAP), Krakow, Poland, 31 March–5 April 2019; pp. 1–5.
21. Varela, F.R.; SierraCastafier, M.; GalochaIragüen, B. Multi-level spherical wave expansion for fast near-field to far-field transformation. In Proceedings of the 2018 AMTA Proceedings, Williamsburg, VA, USA, 4–9 November 2018; pp. 1–6.
22. Knott, E.F.; Schaeffer, J.F.; Tulley, M.T. *Radar Cross Section*; SciTech Publishing: Raleigh, NC, USA, 2004.
23. Balanis, C.A. *Antenna Theory: Analysis and Design*; John Wiley & Sons: Hoboken, NJ, USA, 2016.
24. Woo, A.C.; Wang, H.T.; Schuh, M.J.; Sanders, M.L. EM programmer's notebook-benchmark radar targets for the validation of computational electromagnetics programs. *IEEE Antennas Propag. Mag.* **1993**, *35*, 84–89. [CrossRef]

Article

Glide-Symmetric Holey Structures Applied to Waveguide Technology: Design Considerations †

Zvonimir Sipus [1,*], Katarina Cavar [1], Marko Bosiljevac [1] and Eva Rajo-Iglesias [2]

[1] Faculty of Electrical Engineering and Computing, University of Zagreb, Unska 3, 10000 Zagreb, Croatia; katarinacavar2@gmail.com (K.C.); marko.bosiljevac@fer.hr (M.B.)

[2] Signal Theory and Communications Department, University Carlos III of Madrid, 28911 Leganés, Spain; eva@tsc.uc3m.es

* Correspondence: zvonimir.sipus@fer.hr

† This paper is an extended version of our paper published in Sipus, Z.; Cavar, K.; Bosiljevac, M. Waveguide Technology Based on Glide-Symmetric Holey Structures: Design Considerations. In Proceedings of the 2020 International Workshop on Antenna Technology (iWAT), Bucharest, Romania, 25–28 February.

Received: 12 November 2020; Accepted: 27 November 2020; Published: 1 December 2020

Abstract: Recently, there has been an increased interest in exploring periodic structures with higher symmetry due to various possibilities of utilizing them in novel electromagnetic applications. The aim of this paper is to discuss design issues related to the implementation of holey glide-symmetric periodic structures in waveguide-based components. In particular, one can implement periodic structures with glide symmetry in one or two directions, which we differentiate as 1D and 2D glide symmetry, respectively. The key differences in the dispersion and bandgap properties of these two realizations are presented and design guidelines are indicated, with special care devoted to practical issues. Focusing on the design of gap waveguide-based components, we demonstrate using simulated and measured results that in practice it is often sufficient to use 1D glide symmetry, which is also simpler to mechanically realize, and if larger attenuation of lateral waves is needed, a diagonally directed 2D glide symmetric structure should be implemented. Finally, an analysis of realistic holes with conical endings is performed using a developed effective hole depth method, which combined with the presented analysis and results can serve as a valuable tool in the process of designing novel electrically-large waveguide-based components.

Keywords: higher symmetries; glide symmetry; periodic structures; mode matching; dispersion analysis

---

1. Introduction

Recent research related to new electromagnetic structures and manufacturing technologies has inspired numerous new developments in the field of periodic structures due to the possibility of realizing components with properties that cannot be obtained using classical materials. Higher symmetries, i.e., periodic structures that are invariant after a translation and a second geometrical operation [1–3], represent a class of periodic structures with additional possibilities in creating new designs, for example in art (medieval Moorish tessellations in the Alhambra, Spain, or the graphic work of M.C. Echer), in numerical geometry (space tessellation and meshing), and in electromagnetics, where their usage has resulted in enhancing the performance of electromagnetic devices [4].

Implementation of higher symmetries in an electromagnetic structure allows various manipulations of the corresponding dispersion diagram. Depending on the desired application and the way in which the periodic structure is used, there are two basic design directions. The first direction is related to the expansion of the non-dispersive part of the dispersion diagram, which is interesting for structures used for guiding waves in desired directions. The second approach is applied when we wish to extend the bandgap in the realized dispersion diagram, for example in structures based on preventing the wave propagation in some directions. Possible applications are numerous—from lenses (e.g., ultra-wideband Luneburg lenses [5,6]), antennas (e.g., leaky-wave antennas with low frequency dependency [7,8] or with high-scanning range [9]), cost-efficient gap waveguide technology [10], phase shifters [11–13], filters [13,14], or contactless flanges in the mm-wave frequency range with low leakage [15].

Although there has been intensive research on periodic structures with higher symmetries over the last five years, there are still many theoretical and practical questions that need to be discussed. Extensive theoretical modeling based on the mode matching approach has shown the advantages and limitations of these structures, and this is in agreement with the results obtained using general numerical solvers [10,16–21]. The focus of these works was on geometries that have glide symmetry in two directions. However, it is possible to have glide symmetry in only one direction, resulting in simpler structures. Thus, it is of interest to analyze how this "simple" approach of implementing glide symmetry (i.e., reduced glide symmetry), which is easier to implement in practice, compares to a full glide symmetry. Furthermore, the base geometrical elements for the realization of these periodic structures are usually cylindrical or rectangular holes. Parametric analyses of the influence that the geometry of these holes has on the dispersion diagram have been extensively reported [16,22–24]; however, in practical realization, drilling of the holes usually means that the actual geometry of the hole is not a perfect cylinder, but rather a cylinder with a conical ending. For this reason we are interested in to what extent the conical ending of the holes influences the characteristics of the designed device. Therefore, the aim of this paper is to make a contribution toward these two practical topics.

The paper is organized as follows. First, holey parallel plate waveguides with implemented one-dimensional (1D) and two-dimensional (2D) glide symmetries are discussed. The focus will be on the differences in obtainable dispersion diagrams. In Section 3 we discuss practical issues related to properties of waveguide components whose design utilizes bandgap properties of holey glide-symmetric structures, and the results of the experimental prototype are presented. Finally, the discussion will focus on another practical issue related to holey waveguide components with a conical shape of the hole bottom in order to also cover this aspect in the realization of components with higher-symmetry periodic structures.

2. Glide-Symmetric Holey Parallel Plate Waveguide

A periodic structure possesses a higher symmetry when more than one geometrical operation is needed for the unit cell to coincide with itself. A common example is glide symmetry, which indicates the invariance of a periodic structure under a translation for half of its period and a mirroring with respect to a symmetry plane defined by the considered structure. Other kinds of higher symmetries are defined by a combination of translation and rotation (twist or screw symmetry) [25] or by time operation (parity-time symmetry) [26].

In parallel-plate waveguide (PPW) applications, a glide symmetry operator should be applied along the waveguide, and there is a degree of freedom in defining the glide-symmetric periodic cell. Two possibilities will be considered, both of them having advantages in particular applications. To define them we introduce two versions of glide operators, 1D and 2D, which can be defined in the following way (the coordinate system is given in Figure 1):

$$\text{1D glide operator}: \begin{cases} (x,y) \longrightarrow (x+\frac{P_x}{2},y) \\ z \longrightarrow -z \end{cases} \tag{1}$$

$$\text{2D glide operator}: \begin{cases} (x,y) \longrightarrow (x+\frac{P_x}{2},y+\frac{P_y}{2}) \\ z \longrightarrow -z \end{cases} \tag{2}$$

A sketch of 1D and 2D glide operators, together with the corresponding PPW unit cells, is given in Figure 1. Note that in the 1D case we have two free parameters: the lateral distance between rows following the 1D glide symmetry Δ_y, and the inclination angle α. Furthermore, there is no requirement for 1D structure to be periodic in the lateral direction (in the example in Figure 1b we have selected a special case, $\alpha = 33.7°$, to ensure periodicity in the lateral direction, which enables calculation of the two-dimensional dispersion diagram). For an inclination angle $\alpha = 45°$ and $\Delta_y = P_x/2$ we get a 2D glide symmetric structure. Note that in the 2D case, the minimum unit cell is positioned in the diagonal direction (see Figure 2b). However, just to be able to compare 1D and 2D cases, we will keep the notation of periodicities in the x- and y-directions, i.e., $P^{2D} = P^{1D}/\sqrt{2} = P_x/\sqrt{2}$.

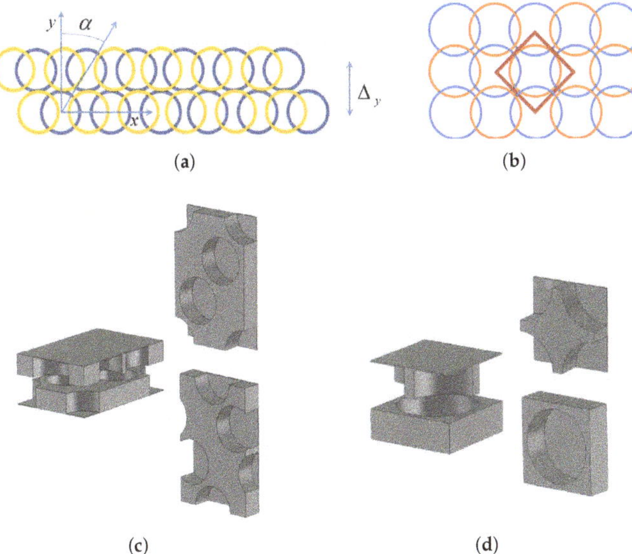

Figure 1. Holey glide-symmetric structures: (**a**) sketch of 1D glide symmetric structure; (**b**) sketch of 2D glide symmetric structure—yellow and blue circles denote holes at the top and bottom plates of parallel-plate waveguide (PPW); (**c**) sketch of periodic unit cell of 1D glide symmetric PPW structure; (**d**) sketch of periodic unit cell of 2D glide symmetric PPW structure.

Differences in the unit cell geometry are also reflected in the corresponding dispersion diagrams. As an example, we considered glide symmetric periodic structures with the stop-band in the K-band. Without loss of generality we fixed the diameter and the depth of the holes ($2r = 7$ mm and $h = 3$ mm, respectively) and the gap between parallel plates ($g = 0.1$ mm). We used the period that maximizes the bandgap, and thus the $P^{2D} = 9$ mm ($r/P^{2D} = 0.39$) for the 2D periodic structure (see also the discussion about the optimum ratio r/P^{2D} given at the end of this section). For 1D glide symmetry we considered two examples, one with zero inclination angle ($P^{1D} = P_x = P^{2D}\sqrt{2} = 13$ mm, $\alpha = 0°$, $\Delta_y = P_x$), and one with inclination angle $\alpha = 33.7°$ ($\Delta_y = P_x/2$, the unit cell of this structure is shown in Figure 1c).

From Figure 2 it can be seen that the 1D holey structure for an inclination angle $\alpha = 0°$ acts as a soft surface, i.e., it prevents the propagation of EM waves in the direction perpendicular to the row of holes (i.e., y-direction) [27,28]. A large stop-band is obtained for the structures where the holes are large enough to overlap, i.e., in the case when there is a non-zero cross-section of the projection of the holes into the symmetry plane (i.e., $2r > P^{1D}/2$). This can be seen in Figure 3a, in which the bandgap cut-off frequencies are shown as a function of periodicity P^{1D} and gap size g. The maximum bandgap is obtained for period $P^{1D} = 9.5$ mm, i.e., for $r/P^{1D} = 0.37$. The stop property is present not only for waves propagating in the y-direction, but the bandgap is also present for EM waves propagating in the cone $\pm 45°$, as seen in Figure 3b. However, the bandgap is positioned at quite low frequencies, between 4 and 18 GHz in this case, which was not our intention in the design.

If we incline the structure in the lateral direction by an angle α, this will enable us to place the lateral walls closer. Furthermore, if the selected diameter of the holes results in mutual overlapping, i.e., if there is a non-zero cross-section of the projection into the symmetry plane of each hole in the upper plate with the projections of the four neighboring holes in the lower plate (and vice versa), then a large bandgap opens at higher frequencies in all directions, see Figure 2b. If we further increase the inclination angle to α = 45°, we will obtain the 2D glide-symmetric structure with maximum bandgap size, see Figure 2c. The dispersion diagram is now quite simple with a small number of modes, mostly because of the presence of a higher symmetry and a smaller unit cell ($P^{2D} = P^{1D}/\sqrt{2}$). One should note that implementing the periodic structure from Figure 1b and not noticing the minimum period of the structure would influence the layout of the dispersion diagram—one would obtain more propagating modes. However, the bandgap would be correctly determined.

The dependency of the bandgap, i.e., the dependency of the lower and upper cut-off frequency of the 2D glide-symmetric structure on the structure period and on the gap size, is given in Figure 4. It can be seen that the maximum bandgap was obtained for P^{2D} = 9 mm, i.e., for r/P^{2D} = 0.39, which is in agreement with results from the literature [23,29]. Furthermore, if there is no mutual overlapping of holes in the lower and upper plates (in the case of periodic structures with $P^{2D} > 9.9$ mm), the bandgap is reduced.

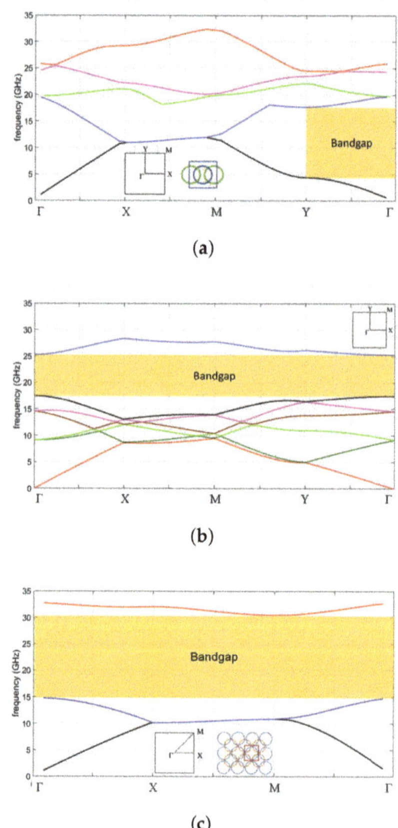

Figure 2. Dispersion diagram of the 1D and 2D glide symmetric structures obtained using CST Microwave Studio: (**a**) 1D case with inclination angle α = 0°; (**b**) 1D case with inclination angle α = 33.7°; (**c**) 2D glide-symmetric case.

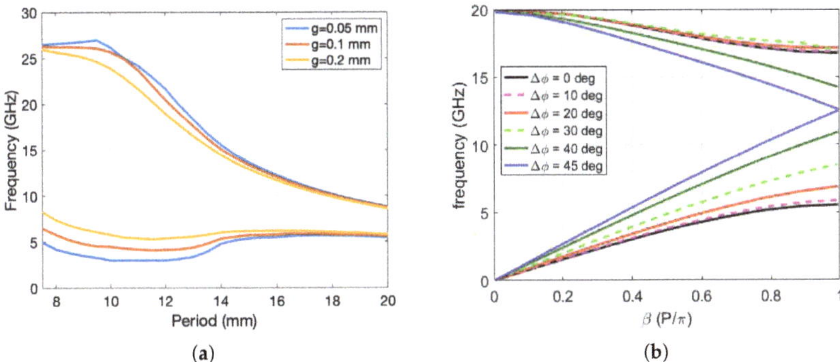

(a) (b)

Figure 3. (**a**) The bandgap cut-off frequencies of a 1D glide-symmetric periodic structure ($\alpha = 0°$, $\Delta_y = P_x$) as a function of periodicity P_x and gap size g; (**b**) dispersion diagram of first two propagating modes of a 1D glide-symmetric periodic structure with $\alpha = 0°$, $\Delta_y = P_x$, and $g = 0.2$ mm for different propagation directions of EM waves.

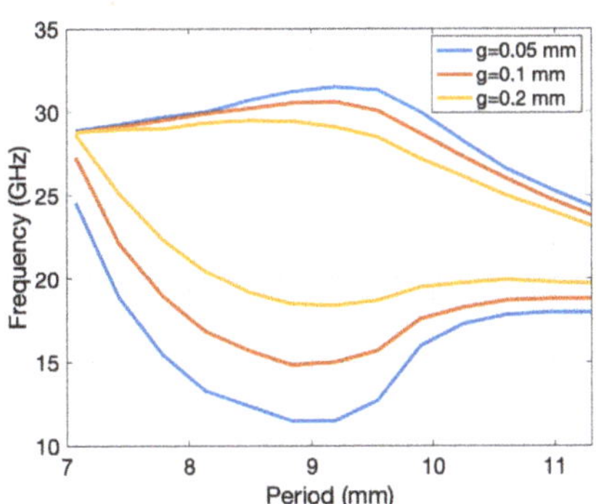

Figure 4. Bandgap cut-off frequencies of 2D glide symmetric structures for different periods and gap sizes; the period of the periodic structure is $P^{2D} = P_x/\sqrt{2}$.

3. Waveguide Components Based on Using Bandgap Properties of Holey Glide-Symmetric Structures

Undesirable EM energy leakage from certain types of waveguides, flanges, shieldings, and similar components in the mm-frequency band is often encountered and is a result of poor manufacturing, small cracks, and material wear. Such waveguide components are typically manufactured in two parts that are joined together, and ensuring very good flatness and good electric contact in these cases is mechanically a very difficult task and other solutions are investigated. Note that even typical production tolerances in surface flatness will cause small gaps when these two parts are mounted together, and in order to avoid leakage of EM energy through the gaps one needs to use a large amount of screws to ensure a firm contact (see, e.g., [30]). One alternative popular solution is based on gap-waveguide technology, which uses the electromagnetic stop-band in order to contain the EM energy inside the structure [28,31,32]. The classical gap-waveguide technology is based on manufacturing a periodic array of pins that form the stop-band in the PPW, and the benefits of using gap waveguide structures to prevent this leakage have been extensively demonstrated. Our aim is to investigate the potential of holey glide symmetric structures in similar applications. Due to its mechanical simplicity (instead of manufacturing the pins one needs only to drill holes, which is easier and cheaper to produce) and good dispersion properties, glide-symmetric structures have great potential in these types of applications. In more detail, the waveguide components are typically fabricated in two separate parts using computer numerical control (CNC) machining or electrical discharge machining (EDM). If the CNC milling process is applied, one should note that drilling the holes is easier than milling the pins (additionally, closely-spaced thin pins are easy to break). If the EDM process is applied, the glide-symmetric holey structure is less sensitive to the fabrication tolerances due to the larger dimensions of the holes, since the periodicity of the holey structure is typically 2–3 times larger than the equivalent pin structure (which is particularly important in the higher mm-wave frequency range) [10]. Understanding these concepts and the benefits of particular realizations is crucial for the development of actual existing devices and the further development of new devices applicable in classical EM applications and also in different sensing applications.

In designing gap waveguide components based on glide-symmetric structures one would like to select dimensions that would maximize the range of the bandgap. The dependency of the lower and upper cut-off frequencies of the 2D glide-symmetric structure on the structure period and on the gap size was already discussed in Figure 4, i.e., the ratio of hole radius and structure period that maximizes the bandgap is $r/P^{2D} = 0.39$ (i.e., $r/P^{1D} = 0.27$). The position of the maximum does not depend on the gap size, making it easy to design components since the sizes of the gaps or cracks are not known in principle. Note that the hole depth for large enough values does not influence the bandgap range since all the modes in the circular waveguides (i.e., in the holes) are evanescent due to the subwavelength radius of the holes, and thus the EM field does not penetrate deep inside the hole.

These results indicate that glide-symmetric holey technology can be efficiently used as a simple way of solving the leakage problem due to the presence of undesired cracks and gaps in waveguide components. Additionally, it is important to highlight that it is very simple to implement holey glide-symmetric structures using the 1D approach. In more detail, from a practical point of view it is often important that a holey glide-symmetric structure occupies as little space as possible in the lateral direction due to the fact that in practical waveguide-based components we would like to put different parts close to each other and in this way to reduce the size of the considered component (a typical example is a waveguide-based feeding network of an antenna array, see, e.g., [30]). Therefore, it is advantageous not to have a half-period shift of the row of holes in the lateral direction, so even when a 2D glide-symmetric structure is implemented, it is oriented in the diagonal direction of the unit cell (i.e., following the 1D approach, see Figure 1).

In order to investigate the optimum glide-symmetric topology, we considered a rectangular waveguide with a gap (i.e., with a PPW) in the lateral walls in which the holes are periodically drilled (without loss of generality the PPWs are placed in the middle of the side waveguide walls). First we need to determine the required number of rows with holes and therefore the following three holey structures were considered (see Figure 5), together with the structure without the holes as a reference case (i.e., a rectangular waveguide with two PPWs in the middle of side walls). By inspecting Figure 6, where the S_{21} parameter of finite-size rectangular waveguides is shown ($L = 143$ mm), we can conclude that even a small gap in the lateral wall causes serious degradation in the transmitting waveguide properties (Figure 6a). Furthermore, it is clear that drilling holes in one plate only does not improve the transmission properties (Figure 6b). However, even one row of 1D glide symmetric holes efficiently prevents the leakage of EM energy, in particular for small sizes of the gap (Figure 6c). For larger gap sizes, two rows of glide-symmetric holes are needed (Figure 6d).

One would expect "perfect" waveguide propagation properties inside the whole bandgap. However the analysis of a finite waveguide structure in Figure 6 shows a strong reflection above 25 GHz. The problem is in the Bragg frequency, i.e., when the period of the holes is equal to the guided wavelength of the waveguide propagating mode a strong reflection occurs since all the small reflections from the holey discontinuities are constructively added in phase. This is also visible if we plot the comparison of the propagation constant of the regular and holey glide-symmetric waveguides, see Figure 7. It can be seen that the difference is negligible until we reach the vicinity of the Bragg frequency.

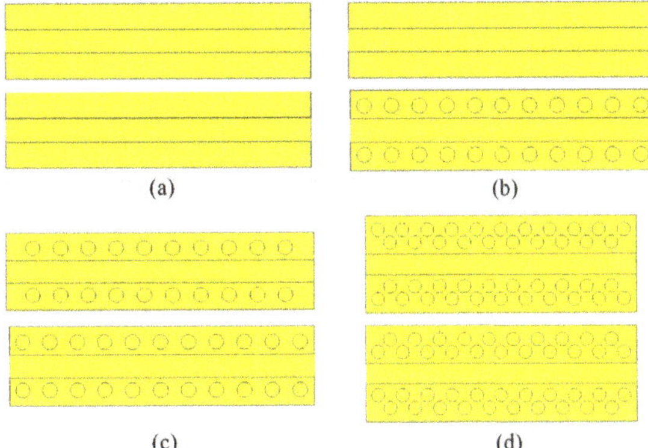

Figure 5. Sketch of the considered holey periodic structures: (**a**) classical waveguide with the PPW in the side walls; (**b**) waveguide with one row of holes in one plate of the PPW only; (**c**) waveguide with one row of holes in both the top and bottom plates of the PPW; (**d**) waveguide with two rows of holes in both the top and bottom plates of the PPW following the 2D glide-symmetry grid.

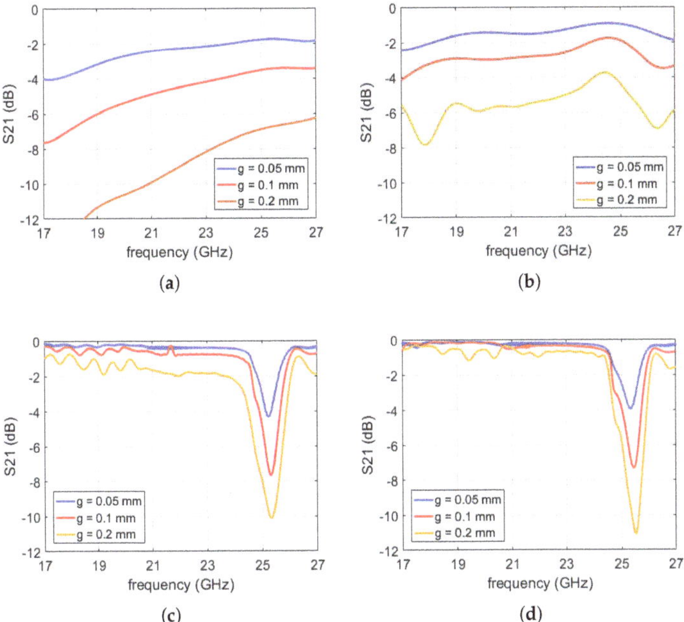

Figure 6. Transmission properties of a finite holey glide-symmetric waveguide (L = 143 mm) as a function of the gap size; the considered structures are shown in Figure 5. The period of the holey glide-symmetric structure is P^{1D} = 13 mm; (**a**) no holes present; (**b**) structure with one row of holes in the bottom; (**c**) structure with one row of holes in the top and bottom plates (shifted half a period versus each other); (**d**) with two rows of holes in the top and bottom plates.

The problem of Bragg frequency reflection can be mitigated by reducing the period of the holes (thus effectively moving the Bragg reflection to a higher frequency). This will reduce the bandgap, but since there is no Bragg reflection, effectively the working frequency range will be extended, as seen in Figure 8. However, the properties of the bandgap are also changed and the ripples in the transmission characteristic are visible since now more than one mode is propagating (e.g., below 22 GHz for g = 0.2 mm). It should be noted that this second propagating mode (due to the fact that the periodic structure is not in the bandgap) is much weaker. This is also visible in Figure 9 in which the field distribution of these two modes is given. It can be seen that the mode that exist outside the bandgap has the E-field maximum in the parallel plate region; thus it is weakly excited by the dominant waveguide mode (excitation of the whole structure). This is the case when we have a geometrically regular parallel plate structure. However, this mode will not be present in reality since the gap will not be regular, i.e., it will be present in places with imperfections of the realized waveguide component.

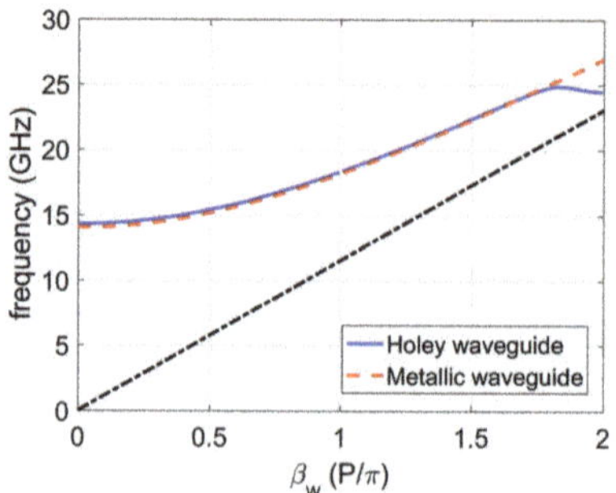

Figure 7. Propagation constant of the regular (fully metallic) and holey glide-symmetric waveguides. The light line is shown with a dash-dotted line.

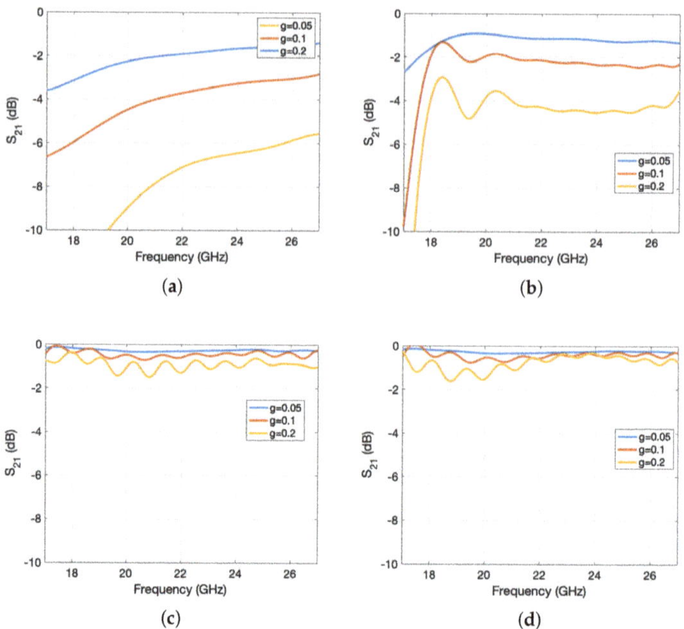

Figure 8. Transmission properties of a finite holey glide-symmetric waveguide (L = 121 mm) as a function of a gap size; the considered structures are shown in Figure 5. The period of the holey glide-symmetric structure is P^{1D} = 11 mm; (**a**) no holes present; (**b**) structure with one row of holes in the bottom; (**c**) structure with one row of holes in the top and bottom plates (shifted half a period versus each other); (**d**) with two rows of holes in the top and bottom plates.

One can conclude that practically there is no difference in the transmission properties of structures with one and two rows of glide symmetric holes, in particular for small sizes of the gap (up to 0.1 mm in the K band). Therefore, it is enough to put one row of holes in different realizations of waveguide components in order to avoid energy leakage due to the presence of undesired cracks and gaps.

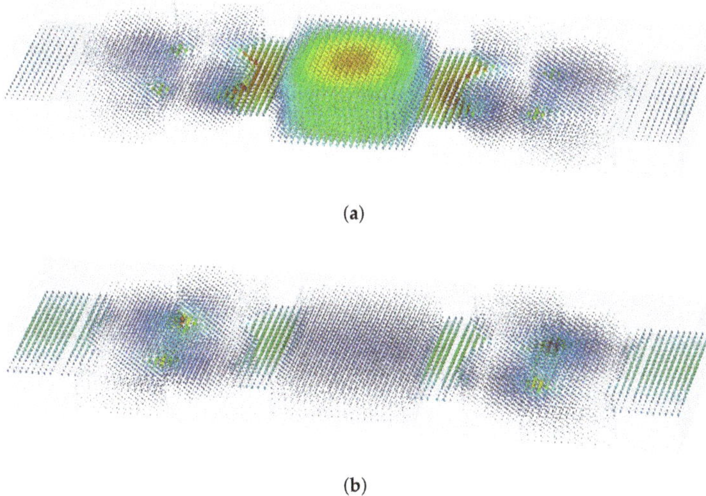

(a)

(b)

Figure 9. E-field distribution within the holey glide-symmetric structure of (**a**) a desired waveguide mode and (**b**) a waveguide mode that exists outside the bandgap.

Practical Realization of Glide-Symmetric Holey Waveguide

To verify the findings from the simulated results we built two waveguide prototypes operating in the K-band. The first waveguide is a classical one, whereas the second one implements the holey gap-waveguide technology in the side walls (Figure 10). The actual waveguide aperture in both cases corresponds to the WG20 waveguide and the dimensions are 10.67×4.32 mm^2 while the length of both waveguides is 205 mm. The applied holey glide-symmetric structure follows the topology from Figure 5d and is analyzed in Figure 8d, i.e., the glide-symmetric structure is made of holes that are 7 mm in diameter and 3 mm in depth, and the periodicity of the holes is 11 mm. The actual manufactured waveguides are shown in Figure 10c,d. The structure was produced using the CNC milling machine with production tolerances around 0.01 mm (the used CNC machine is the three-axis milling machine INGPOS, Laboratory of Machine Tools, Faculty of Mechanical Engineering and Naval Architecture, University of Zagreb). The prototype was made from C45 steel by which we have ensured good flatness of the produced waveguide parts (although the conductivity of steel is smaller comparing to, e.g., aluminum).

The bandgap cut-off frequencies for different sizes of the gap are the ones given in Figure 4, and these were the basis for choosing the parameters of the developed prototype. From a practical perspective we need to cover the whole waveguide band of operation (18.0–26.5 GHz), and as seen from Figure 4, for small gap sizes (below 0.1 mm) the stop-band meets this criterion. Since the actual dimensions of the undesired gap in reality are not known, we used gap sizes of 0.05–0.2 mm in the measurements. These values should give us a good indication whether the proposed holey glide-symmetry technology can solve the abovementioned manufacturing problems with the practical cost being a slightly larger structure in some cases.

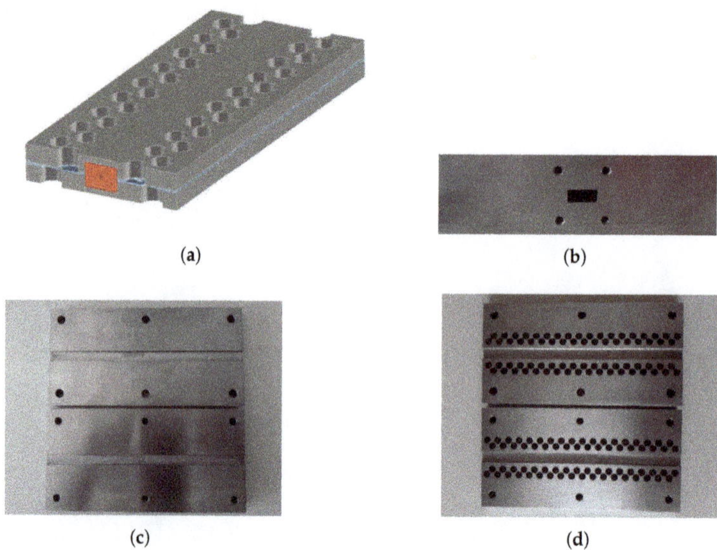

Figure 10. (**a**) Sketch of the developed holey glide-symmetric waveguide; (**b**) picture of the side view of the realized waveguides; (**c**) picture of the realized ordinary waveguide; (**d**) picture of the realized holey glide-symmetric waveguide.

The S-parameters of both waveguides (classical and the one with a holey glide-symmetric structure) were measured and the magnitudes of S_{21} parameters are shown in Figure 11. The results of the classical waveguide show that even a small gap (0.1 mm, which corresponds to the thickness of one 80 g/m^2 paper sheet) causes a significant leakage of energy through the lateral walls of the waveguide. This is not the case for the glide-symmetric version of the waveguide, which efficiently eliminates the leakage problem. Furthermore, the actual experiment revealed that when using the proposed glide-symmetry technology, the required mechanical tolerances can be relaxed and the required number of screws needed to fix the waveguide can be significantly reduced.

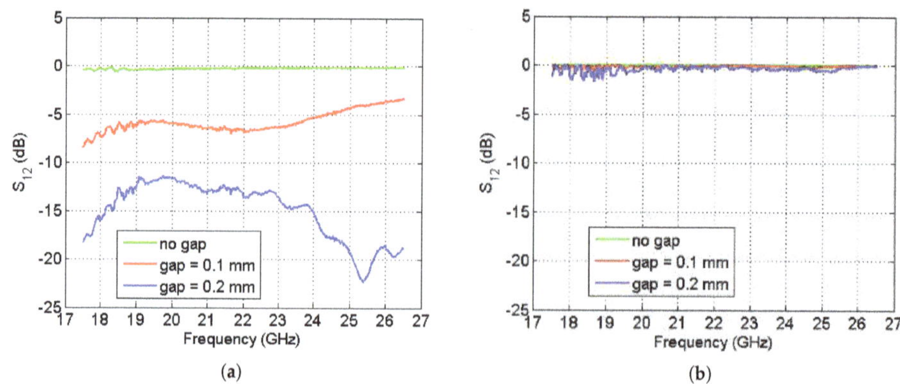

Figure 11. Measured S_{21} parameter of the realized waveguides with different gap sizes: (**a**) ordinary waveguide; (**b**) waveguide realized using the holey glide-symmetric technology.

4. Concept of Effective Hole Depth

When making components using propagation properties of glide-symmetric structures (e.g., lenses in PPW technology), the properties that should be considered first are the obtainable values of the effective refraction index, as well as low-dispersive and isotropic features. Once we select the basic parameters of the glide-symmetric PPW, the obtainable range of refractive indexes is determined by varying the depth of the holes. Note that for larger depths the value of the refractive index will not change due to evanescent field distribution in holes with subwavelength radius, and thus the bottom of the holes is "not visible" for the EM field.

Using a simple drilling manufacturing process for making holey glide-symmetric technology makes this technology cost-effective, in particular in the case when the fast development of proof-of-concept devices is needed. However, this means that the actual conical shape of the hole ending (due to the conical shape of the drill bit) must be taken into account in the design procedure, which is particularly important for lens realizations where the propagation properties of the EM wave in the parallel plate region depend on the hole depth [5–7]. In other words, there is a need for an analysis tool that can efficiently evaluate the effective depth of a circular hole with a conical ending. This is particularly important when analyzing electrically large holey glide-symmetric structures with many elements (i.e., holes), such as lenses and leaky-wave antennas.

The electromagnetic analysis of this problem starts by considering a combination of a circular and conical waveguide (as shown in Figure 12). Two regions are modeled using the appropriate waveguide modes with unknown amplitudes and these are then connected by applying the mode-matching technique from which the mode amplitudes can be obtained. The plane in which the mode-matching is applied is the planar boundary shown in Figure 12, and the details of the complete procedure are given in the Appendix A.

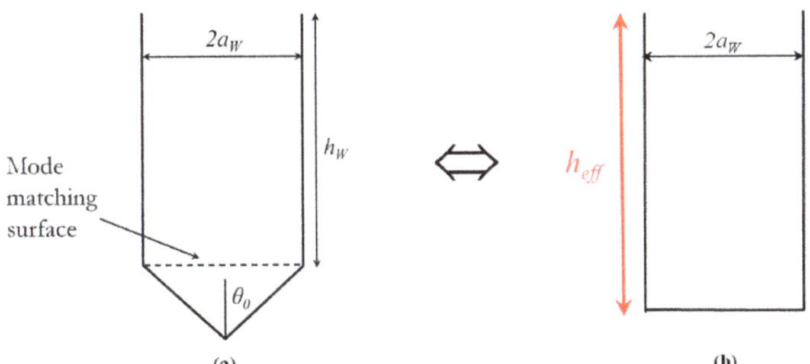

Figure 12. (**a**) Sketch of a hole obtained using the standard drilling procedure. Geometrically, the hole is a combination of a circular and a conical waveguide; (**b**) sketch of the equivalent hole with a straight ending and length h_{eff}.

To determine the effective depth of the conical hole, we defined that the conical hole and the equivalent straight ending hole with depth h_{eff} have the same reflection coefficients when excited with the considered cylindrical waveguide mode (as illustrated in Figure 12). The h_{eff} is then calculated using:

$$\left|\Gamma e^{+2j\beta_w h_{eff}}\right| = \left|\Gamma e^{+2\alpha_w h_{eff}}\right| = 1, \quad (3)$$

where Γ is the reflection coefficient of the considered cylindrical evanescent waveguide mode determined through the mode-matching procedure and $\beta_w = -j\alpha_w$ is its propagation constant.

Figure 13 illustrates the method by showing the effective extension of the hole depth due to the conical bottom. We considered a conical shape defined with $\theta_0 = 60°$ (a typical conical shape related to commercial drilling tools) and a hole diameter of $2r = 7$ mm (the same as in the previous scenarios). The results are shown for both TE_{11} and TM_{11} waveguide mode excitations and were verified by comparison using CST Microwave Studio (for both cases, the mode-matching method and CST MS, the effective extension of the hole depth was calculated using (3)). Although the TE_{11} is the dominant mode, in [33] it was shown that for lower frequencies most of the power traveling along the PPW is coupled to the "superior" TM mode(s), i.e., to the TM_{11} mode in our case. Therefore, in the calculation of the effective depth of holes with conical ending we will consider the TM_{11} mode. When it comes to the mode-matching formulation, that means that the TM_{11} mode excites the structure and the reflection coefficient of the TM_{11} mode is used in determining the effective depths, but other modes are present as well (the reflection coefficients of other modes are much smaller comparing to the one of TM_{11} mode). In total we have considered 4 cylindrical and 12 spherical modes.

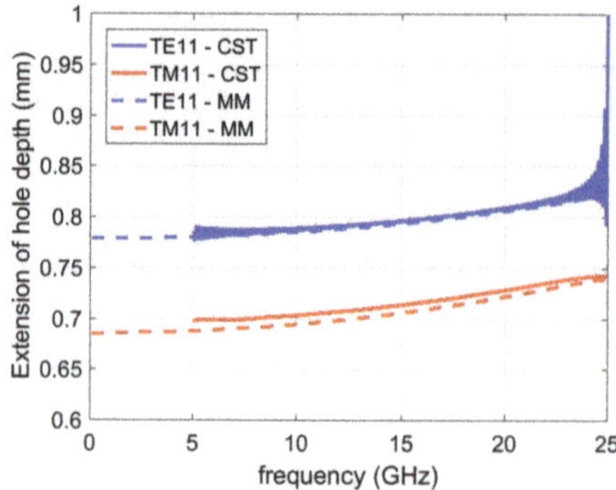

Figure 13. Dependency of the calculated effective extension of the hole depth on the TE_{11} and TM_{11} excitation modes of the cylindrical waveguide section. Solid line—CST Microwave Studio, dashed line—mode matching approach.

The concept of effective hole depth was verified on the 2D glide-symmetric PPW ($P^{2D} = 11$ mm, $2r = 7$ mm, $g = 0.2$ mm). As shown in Figure 14, the effective refraction index was calculated for holes with different depths, all of which have a conical ending with $\theta_0 = 60°$. For comparison, the refractive index of holes with straight endings and a larger depth for the h_{eff} value is also shown ($h_{eff} = 0.7$ mm in our case). The agreement between the obtained effective refractive indexes is very good.

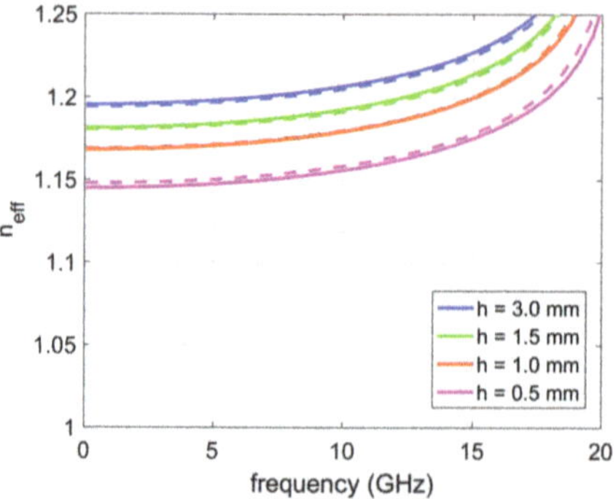

Figure 14. Obtained index of refraction as a function of the hole depth for holes with straight and conical endings (2D glide symmetry case). Straight line—holes with conical ending; dashed line—holes with straight ending (the hole is longer for an effective depth h_{eff} equal to 0.7 mm in the considered case).

5. Conclusions

Periodic structures with higher symmetry have shown a large potential for implementation in novel electromagnetic devices due to their possibility to tailor propagation properties of EM waves. In this paper we discussed some design features of waveguide components containing holey glide-symmetric periodic structures. In particular, we compared periodic structures with a glide symmetry in one and two dimensions. It was shown that it is much simpler to implement a 1D glide-symmetric holey structure, or a diagonally directed 2D structure, into the gap-waveguide-based components, and that in most applications it is enough to have one row of periodic holes. However, in practical realizations it is not enough to consider the stop-band properties of the holey periodic structure only, since one should also avoid Bragg frequencies due to the presence of large reflections around those frequencies. The design of a gap waveguide based on these ideas was practically realized and measured, confirming the simulated results. Finally, we have discussed a design approach of holey structures with circular-cylindrical conical endings (the case of simple production using a standard drilling manufacturing procedure), based on the effective hole depth concept, which can serve as a valuable tool in the analysis of large holey glide-symmetric structures such as lenses and leaky-wave antennas.

Author Contributions: Conceptualization, Z.S., M.B., and E.R.-I.; methodology, Z.S., M.B., and E.R.-I.; software, Z.S.; validation, K.C., M.B., and E.R.-I.; formal analysis, Z.S., M.B., and E.R.-I.; investigation, Z.S., M.B., and E.R.-I.; resources, Z.S. and E.R.-I.; data curation, K.C. and M.B.; writing—original draft preparation, Z.S., M.B., and E.R.-I.; writing—review and editing, Z.S., K.C., M.B., and E.R.-I.; visualization, M.B. and E.R.-I.; supervision, Z.S. and M.B.; project administration, Z.S., M.B., and E.R.-I.; funding acquisition, Z.S., M.B., and E.R.-I. All authors have read and agreed to the published version of the manuscript.

Funding: This work was supported by the Croatian Science Foundation (HRZZ) under the projects IP-2018-01-9753 and IP-2019-04-1064, by the Spanish Government under the projects PID2019-107688RB-C21 and TEC2016-79700-C2-R, and by the European COST Action CA18223-SYMAT.

Acknowledgments: The authors would like to thank Oscar Quevedo-Teruel for helpful discussions.

Conflicts of Interest: The authors declare no conflict of interest.

Appendix A

In this Appendix we will present the analysis procedure for determining the effective hole depth of the cylindrical hole with the conical shape of the hole bottom. Geometrically, such a hole consists of a circular and a conical waveguide sections (see Figure 12). Electromagnetically, each waveguide section can be modeled with inherent waveguide modes, and their amplitudes are determined by applying the mode-matching technique. In more detail, the conical section can be modeled by the following modes:

$$E_r = -j\frac{\eta_0}{k_0}\left(\frac{\partial^2}{\partial r^2} + k_0^2\right) A_r \tag{A1}$$

$$E_\theta = -j\frac{\eta_0}{k_0}\frac{1}{r}\frac{\partial^2 A_r}{\partial r \partial \theta} - \frac{1}{r\sin\theta}\frac{\partial F_r}{\partial \phi} \tag{A2}$$

$$E_\phi = -j\frac{\eta_0}{k_0}\frac{1}{r\sin\theta}\frac{\partial^2 A_r}{\partial r \partial \phi} + \frac{1}{r}\frac{\partial F_r}{\partial \theta} \tag{A3}$$

$$H_r = -j\frac{1}{\eta_0 k_0}\left(\frac{\partial^2}{\partial r^2} + k_0^2\right) F_r \tag{A4}$$

$$H_\theta = -j\frac{1}{\eta_0 k_0}\frac{1}{r}\frac{\partial^2 F_r}{\partial r \partial \theta} + \frac{1}{r\sin\theta}\frac{\partial A_r}{\partial \phi} \tag{A5}$$

$$H_\phi = -j\frac{1}{\eta_0 k_0}\frac{1}{r\sin\theta}\frac{\partial^2 F_r}{\partial r \partial \phi} - \frac{1}{r}\frac{\partial A_r}{\partial \theta} \tag{A6}$$

Here A_r and F_r are the radial components of the magnetic and electric potential, which are related to the considered mode in the form:

$$A_r(r,\theta\phi) = \alpha_{n,m} \cdot \hat{J}_m(k_0 r) P^m_{\nu_n}(\cos\theta) e^{-jm\phi} \tag{A7}$$

$$F_r(r,\theta\phi) = \beta_{n,m} \cdot \hat{J}_m(k_0 r) P^m_{\nu'_n}(\cos\theta) e^{-jm\phi}, \tag{A8}$$

where $\hat{J}_n$ and P^m_ν denote Schelkunoff's spherical Bessel functions and the associated Legendre functions, respectively [34]. As can be seen from Equations (A7)and (A8), the potentials and consequently the EM field distribution depend on the indices ν_n and ν'_n. These two indices of the associated Legendre functions are in general non-integer numbers obtained by imposing the appropriate boundary conditions on the cone metallic surface, i.e., the tangential electric field vanishes when $\theta = \theta_0$. In other words, the indices ν_n and ν'_n are calculated from the following equations:

$$P_{\nu_n}(\cos\theta) = 0 \qquad \text{in TM case} \tag{A9}$$

$$\frac{\partial P_{\nu'_n}(\cos\theta)}{\partial\theta} = 0 \qquad \text{in TE case} \tag{A10}$$

The roots of Equations (A9) and (A10) are numerically determined by implementing the recursive equations for evaluating the associated Legendre functions together with Newton's algorithm for root finding, and the resulting combination enables high computation precision.

The circular waveguide is modeled with "classical" waveguide modes determined by the following form of the z-component of the magnetic and electric vector potentials [34]:

$$A_z(\rho,\phi,z) = \xi_{m,l} \cdot J_m(k_\rho \rho) e^{-jm\phi} e^{\pm jk_z z} \tag{A11}$$

$$F_z(\rho,\phi,z) = \zeta_{m,l} \cdot J_m(k_\rho \rho) e^{-jm\phi} e^{\pm jk_z z}. \tag{A12}$$

The tangential EM fields are determined as:

$$E_\rho = -j\frac{\eta_0}{k_0}\frac{\partial^2 A_z}{\partial\rho\partial z} - \frac{1}{\rho}\frac{\partial F_z}{\partial\phi} \tag{A13}$$

$$E_\phi = -j\frac{\eta_0}{k_0}\frac{1}{\rho}\frac{\partial^2 A_z}{\partial\phi\partial z} + \frac{\partial F_z}{\partial\rho} \tag{A14}$$

$$H_\rho = -j\frac{1}{\eta_0 k_0}\frac{\partial^2 F_z}{\partial\rho\partial z} + \frac{1}{\rho}\frac{\partial A_z}{\partial\phi} \tag{A15}$$

$$H_\phi = -j\frac{1}{\eta_0 k_0}\frac{1}{\rho}\frac{\partial^2 F_z}{\partial\phi\partial z} - \frac{\partial A_z}{\partial\rho}. \tag{A16}$$

Here $k_z = \sqrt{k_0^2 - k_\rho^2}$ and J_m denote the Bessel functions of the first kind. The value of k_ρ depends on the considered waveguide mode and is determined by imposing the boundary condition that the tangential electric field vanishes at the waveguide walls, i.e., $J_m(k_\rho a) = 0$ for the TM modes and $J'_m(k_\rho a) = 0$ for the TE modes. The mode matching is performed over the planar boundary of the cylindrical waveguide (see Figure 12). As testing functions we have selected the H-field distribution of spherical weaveguide modes and E-field distribution of the cylindrical waveguide modes (details about implementation of the mode-matching procedure can be found in [35]).

The effective depth of the hole h_{eff} is determined as follows. Once we determine by the mode-matching procedure the reflection coefficient Γ of the considered cylindrical evanescent waveguide mode (with propagating constant $\beta_w = -j\alpha_w$), the effective depth of the hole h_{eff} is calculated using the following equation:

$$\left|\Gamma e^{+2j\beta_w h_{eff}}\right| = \left|\Gamma e^{+2\alpha_w h_{eff}}\right| = 1 \tag{A17}$$

Note that this equation actually states that the reflection coefficients of the considered hole with the conical ending and of the equivalent hole with the straight ending (of depth h_{eff}) are equal. This concept is also illustrated in Figure 12.

References

1. Crepeau, P.J.; McIsaac, P.R. Consequences of symmetry in periodic structures. *Proc. IEEE* **1964**, *52*, 33–43. [CrossRef]
2. Mittra, R.; Laxpati, S. Propagation in a wave guide with glide reflection symmetry. *Can. J. Phys.* **1965**, *43*, 353–372. [CrossRef]
3. Hessel, A.; Chen, M.H.; Li, R.C.; Oliner, A.A. Propagation in periodically loaded waveguides with higher symmetries. *Proc. IEEE* **1973**, *61*, 183–195. [CrossRef]
4. Quevedo-Teruel, O.; Valerio, G.; Sipus, Z.; Rajo-Iglesias, E. Periodic Structures With Higher Symmetries: Their Applications in Electromagnetic Devices. *IEEE Microw. Mag.* **2020**, *21*, 36–49. [CrossRef]
5. Quevedo-Teruel, O.; Ebrahimpouri, M.; Ng Mou Kehn, M. Ultrawideband Metasurface Lenses Based on Off-Shifted Opposite Layers. *IEEE Antennas Wirel. Propag. Lett.* **2016**, *15*, 484–487. [CrossRef]
6. Quevedo-Teruel, O.; Miao, J.; Mattsson, M.; Algaba-Brazalez, A.; Johansson, M.; Manholm, L. Glide-Symmetric Fully Metallic Luneburg Lens for 5G Communications at Ka-Band. *IEEE Antennas Wirel. Propag. Lett.* **2018**, *17*, 1588–1592. [CrossRef]
7. Chen, Q.; Zetterstrom, O.; Pucci, E.; Palomares-Caballero, A.; Padilla, P.; Quevedo-Teruel, O. Glide-Symmetric Holey Leaky-Wave Antenna With Low Dispersion for 60 GHz Point-to-Point Communications. *IEEE Trans. Antennas Propag.* **2020**, *68*, 1925–1936. [CrossRef]
8. Memeletzoglou, N.; Rajo-Iglesias, E. Holey Metasurface Prism for the Reduction of the Dispersion of Gap Waveguide Leaky-Wave Antennas. *IEEE Antennas Wirel. Propag. Lett.* **2019**, *18*, 2582–2586. [CrossRef]
9. Zhang, G.; Zhang, Q.; Chen, Y.; Murch, R.D. High-Scanning-Rate and Wide-Angle Leaky-Wave Antennas Based on Glide-Symmetry Goubau Line. *IEEE Trans. Antennas Propag.* **2020**, *68*, 2531–2540. [CrossRef]
10. Ebrahimpouri, M.; Rajo-Iglesias, E.; Sipus, Z.; Quevedo-Teruel, O. Cost-Effective Gap Waveguide Technology Based on Glide-Symmetric Holey EBG Structures. *IEEE Trans. Microw. Theory Tech.* **2018**, *66*, 927–934. [CrossRef]
11. Rajo-Iglesias, E.; Ebrahimpouri, M.; Quevedo-Teruel, O. Wideband Phase Shifter in Groove Gap Waveguide Technology Implemented With Glide-Symmetric Holey EBG. *IEEE Microw. Wirel. Compon. Lett.* **2018**, *28*, 476–478. [CrossRef]
12. Palomares-Caballero, A.; Alex-Amor, A.; Padilla, P.; Luna, F.; Valenzuela-Valdes, J. Compact and Low-Loss V-Band Waveguide Phase Shifter Based on Glide-Symmetric Pin Configuration. *IEEE Access* **2019**, *7*, 31297–31304. [CrossRef]
13. Palomares-Caballero, A.; Alex-Amor, A.; Padilla, P.; Valenzuela-Valdés, J.F. Dispersion and Filtering Properties of Rectangular Waveguides Loaded With Holey Structures. *IEEE Trans. Microw. Theory Tech.* **2020**. [CrossRef]
14. Monje-Real, A.; Fonseca, N.J.G.; Zetterstrom, O.; Pucci, E.; Quevedo-Teruel, O. Holey Glide-Symmetric Filters for 5G at Millimeter-Wave Frequencies. *IEEE Microw. Wirel. Compon. Lett.* **2020**, *30*, 31–34. [CrossRef]
15. Ebrahimpouri, M.; Algaba Brazalez, A.; Manholm, L.; Quevedo-Teruel, O. Using Glide-Symmetric Holes to Reduce Leakage Between Waveguide Flanges. *IEEE Microw. Wirel. Compon. Lett.* **2018**, *28*, 473–475. [CrossRef]
16. Ghasemifard, F.; Norgren, M.; Quevedo-Teruel, O.; Valerio, G. Analyzing Glide-Symmetric Holey Metasurfaces Using a Generalized Floquet Theorem. *IEEE Access* **2018**, *6*, 71743–71750. [CrossRef]
17. Mesa, F.; Rodríguez-Berral, R.; Medina, F. On the computation of the dispersion diagram of symmetric one-dimensionally periodic structures. *Symmetry* **2018**, *10*, 307. [CrossRef]
18. Bagheriasl, M.; Quevedo-Teruel, O.; Valerio, G. Bloch Analysis of Artificial Lines and Surfaces Exhibiting Glide Symmetry. *IEEE Trans. Microw. Theory Tech.* **2019**, *67*, 2618–2628. [CrossRef]

19. Chen, Q.; Mesa, F.; Yin, X.; Quevedo-Teruel, O. Accurate Characterization and Design Guidelines of Glide-Symmetric Holey EBG. *IEEE Trans. Microw. Theory Tech.* **2020**, *68*. [CrossRef]
20. Sipus, Z.; Bosiljevac, M. Modelling of glide-symmetric dielectric structures. *Symmetry* **2019**, *11*, 805. [CrossRef]
21. Zetterstrom, O.; Valerio, G.; Mesa, F.; Ghasemifard, F.; Norgren, N.; Quevedo-Teruel, O. Dispersion Analysis of Periodically Loaded Transmission Lines with Twist Symmetry Using the Mode-Matching Technique. *Appl. Sci.* **2020**, *10*, 5990. [CrossRef]
22. Valerio, G.; Ghasemifard, F.; Sipus, Z.; Quevedo-Teruel, O. Glide-Symmetric All-Metal Holey Metasurfaces for Low-Dispersive Artificial Materials: Modeling and Properties. *IEEE Trans. Microw. Theory Tech.* **2018**, *66*, 3210–3223. [CrossRef]
23. Alex-Amor, A.; Valerio, G.; Ghasemifard, F.; Mesa, F.; Padilla, P.; Fernández-González, J.M.; Quevedo-Teruel, O. Wave Propagation in Periodic Metallic Structures with Equilateral Triangular Holes. *Appl. Sci.* **2020**, *10*, 1600. [CrossRef]
24. Alex-Amor, A.; Ghasemifard, F.; Valerio, G.; Ebrahimpouri, M.; Padilla, P.; Fernánd, J.M. Glide-Symmetric Metallic Structures With Elliptical Holes for Lens Compression. *IEEE Trans. Microw. Theory Tech.* **2020**, *68*, 4236–4248. [CrossRef]
25. Ghasemifard, F.; Norgren, M.; Quevedo-Teruel, O. Twist and Polar Glide Symmetries: an Additional Degree of Freedom to Control the Propagation Characteristics of Periodic Structures. *Sci. Rep.* **2018**, *8*, 11266. [CrossRef]
26. El-Ganainy, R.; Makris, K.G.; Khajavikhan, M.; Musslimani, Z.H.; Rotter, S.; Christodoulides, D.N. Non-Hermitian physics and PT symmetry. *Nat. Phys.* **2018**, *14*, 11–19. [CrossRef]
27. Kildal, P.S. Definition of artificially soft and hard surfaces for electromagnetic waves. *Electron. Lett.* **1988**, *24*, 168–170. [CrossRef]
28. Rajo-Iglesias, E.; Sipus, Z.; Zaman, A.U. Gap waveguide technology. In *Surface Electromagnetics: With Applications in Antenna, Microwave, and Optical Engineering*; Yang, F., Rahmat-Sammii, Y., Eds.; Cambridge University Press: Cambridge, UK, 2019.
29. Ebrahimpouri, M.; Quevedo-Teruel, O.; Rajo-Iglesias, E. Design Guidelines for Gap Waveguide Technology Based on Glide-Symmetric Holey Structures. *IEEE Microw. Wirel. Compon. Lett.* **2017**, *27*, 542–544. [CrossRef]
30. Huang, G.; Zhou, S.; Chio, T.; Hui, H.; Yeo, T. A Low Profile and Low Sidelobe Wideband Slot Antenna Array Feb by an Amplitude-Tapering Waveguide Feed-Network. *IEEE Trans. Antennas Propag.* **2015**, *63*, 419–423. [CrossRef]
31. Kildal, P.S.; Alfonso, E.; Valero-Nogueira, A.; Rajo-Iglesias, E. Local Metamaterial-Based Waveguides in Gaps Between Parallel Metal Plates. *IEEE Antennas Wirel. Propag. Lett.* **2009**, *8*, 84–87. [CrossRef]
32. Kildal, P.S.; Zaman, A.U.; Rajo-Iglesias, E.; Alfonso, E.; Valero-Nogueira, A. Design and experimental verification of ridge gap waveguide in bed of nails for parallel-plate mode suppression. *IET Microwaves, Antennas Propag.* **2011**, *5*, 262–270. [CrossRef]
33. Valerio, G.; Sipus, Z.; Grbic, A.; Quevedo-Teruel, O. Nonresonant modes in plasmonic holey metasurfaces for the design of artificial flat lenses. *Opt. Lett.* **2017**, *42*, 2026–2029. [CrossRef] [PubMed]
34. Harrington, R.F. *Time-Harmonic Electromagnetic Fields*; McGraw-Hill Book Company: New York, USA, 1961.
35. Clarricoats, P.J.B.; Slinn, K.R. Numerical method for the solution of waveguide-discontinuity problems. *Electron. Lett.* **1966**, *2*, 226–228. [CrossRef]

Publisher's Note: MDPI stays neutral with regard to jurisdictional claims in published maps and institutional affiliations.

Article

Modified Split Ring Resonators Sensor for Accurate Complex Permittivity Measurements of Solid Dielectrics

Amer Abbood al-Behadili [1,2], Iulia Andreea Mocanu [1,*], Norocel Codreanu [3] and Mihaela Pantazica [3]

[1] Department of Telecommunication, Telecommunications and Information Technology, Faculty of Electronics, University POLITEHNICA of Bucharest, 060042 Bucharest, Romania; amer_osman@uomustansiriyah.edu.iq

[2] Electrical Engineering, College of Engineering, Mustansiriyah University, Baghdad 00964, Iraq

[3] Center for Technological Electronics and Interconnection Techniques, Department of Electronics Technology and Reliability, Telecommunications and Information Technology, Faculty of Electronics, University POLITEHNICA of Bucharest, 060042 Bucharest, Romania; norocel.codreanu@cetti.ro (N.C.); mihaela.pantazica@cetti.ro (M.P.)

* Correspondence: iulia.mocanu@upb.ro

Received: 28 October 2020; Accepted: 29 November 2020; Published: 30 November 2020

Abstract: In this paper, a sensor using modified Split Ring Resonators (SRRs) is designed, simulated, fabricated, and used for advanced investigation and precise measurements of the real part and imaginary part solid dielectrics' permittivity. Adding vertical strips tightly coupled to the outer ring of the SRR leads to the appearance of two resonant frequencies at 1.24 GHz and 2.08 GHz. This modified geometry also assures an improved sensitivity. Using the full wave electromagnetic solver, both the unloaded and loaded sensors are investigated. The numerical simulations are used to develop a mathematical model based on a curve fitting tool for both resonant frequencies, allowing to obtain analytical relations for real and imaginary parts of permittivity as a function of the sample's thickness and quality factor. The sensor is designed and fabricated on 1.6 mm thick FR-4 substrate. The measurements of different samples, such as transparent glass, acrylic glass, plexiglass, and Teflon, confirm that the modified SRR sensor is easy to implement and gives accurate results for all cases, with measurement errors smaller than 4.5%. In addition, the measurements highlight the importance of the second resonant frequency in the cases in which numerical limitations do not allow the usage of the first resonant frequency (1 mm thick sample).

Keywords: metamaterials; planar sensor; non-invasive; Split Ring Resonator; dielectrics measurements; RF absorbing materials

1. Introduction

The electric complex permittivity is one of the most important parameters of material characterization. It is utilized in a large range of applications such as: Material description [1,2], tests of organic tissue [3,4] microfluidics [5–8], bio sensing [9–11], ecological operators [12,13], and quality control in the food industry [14,15]. Accurate determination of the permittivity is an important task and a great challenge for microwave engineering, in general, and therefore many solutions have been investigated lately.

A relatively new option for implementing sensitive planar sensors is to use metamaterials. Metamaterial structures present a major advantage over other conventional options: They can be artificially tailored to achieve better resolution and accuracy. In the last few years, an increased interest for studying the sensors based on resonant metamaterial structures such as Split Ring Resonator (SRR) and Complementary Split Ring Resonator (CSRR) has been noticed due to minimal efforts in sample preparation, nondestructive effect, ability to characterize both low and high losses materials, and higher sensitivity [2,16].

Considering these important advantages, we present a modified SRR sensor for accurate complex permittivity of solid dielectrics. It is implemented in planar technology and it exhibits two resonant frequencies, which are used to overcome numerical limitations that may appear in real-life measurements. The results obtained for our sensor are compared with similar approaches existing in the literature.

There are many SRRs sensors depending on a single resonant frequency, which is produced by the related resonator circuit, and the main focus of the authors is only on the detection of the real part of permittivity.

For example, in reference [5], an SRR sensor working at 2 GHz is proposed. It is implemented in planar technology, on Rogers AD1000 substrate, and it is used to measure the thickness of thin films, as well as the electric permittivity for both dielectrics and liquids, being fully submersible. The extraction of the imaginary part of the permittivity is not rigorous as the authors conclude [5]. They also argue that a more thorough and systematic study investigating many more combinations of real and imaginary parts of the permittivity [5] should be considered for further improvements. Nevertheless, the authors suggest that another version of the sensor should be developed to obtain two resonance frequencies, in order to determine the complex permittivity and the thickness of solids at the same time [5]. The sensor we propose has two resonant frequencies able to measure complex permittivity with errors smaller than 4.5% and for samples with thicknesses from 1 mm to 10 mm.

In ref. [16], an interdigital capacitor based SRR (IDC-SRR) sensor for dielectric testing is investigated. The authors also propose and analyze a meandered line based split ring resonator (ML-SRR) RF sensor for magnetic testing. Both sensors work at 2.45 GHz and are implemented on RT/Duroid 6006 substrate. The accuracy of the real part of permittivity measurements is more than 94% [16], but still, these sensors are not able to measure the complex permittivity as our proposed sensor. Our sensor can measure the imaginary part of permittivity with comparable accuracy.

Another SRR based sensor for magnetodielectric substrates characterization is the one presented in [17]. The device is fabricated using the microstrip technology on a 1.27mm-thick RT/Duroid 6006 substrate and working at 2.5 GHz [17]. The SRRs are magnetically coupled to the microstrip line allowing both electric permittivity and magnetic permeability measurements. Still, only the real parts are measured. Our sensor can measure both real and imaginary parts for the complex permittivity and the errors for real part measurements are comparable for the same samples as in [17].

A different approach for the sensor design is presented in [18]. It uses a two-layer and three-layer magnetic coupled SRR for higher sensitivity, better stability, and stronger anti-jamming ability from the external interface. These sensors have dimensions of $0.052\lambda \times 0.052\lambda$ allowing miniaturization, but no imaginary parts of permittivity measurements are carried out [18]. The influence of the thickness of the sample is not considered in this study. On the other hand, the errors for real part of permittivity measurements are similar to ours.

Another modified SRR sensor is presented in [19], but it is used for thin-film detection, not for thicker dielectrics as ours. Furthermore, for this thin-film sensor, only the frequency shifts are investigated [19], without determining the complex permittivity of the MUT as in our case.

In [6], an SRR-based sensor is presented for measurement of complex permittivity of liquids. The sensitivity of the sensor is improved by overlapping the middle part of the outer ring of the SRR and part of the feeding line. It is best suited for measuring mixed liquids and determining the complex permittivity for each component, but no study regarding its application for solid materials characterization is done.

Regarding the characterization of material under test (MUT), several techniques have been proposed and employed for the permittivity. The most important ones can be categorized as free-space methods, transmission-line methods, and resonant cavity methods [20].

The free-space method commonly employs using extremely directive lens and horn antennas placed on both sides of the MUT. The vector network analyzer (VNA) is connected to the antennas to measure the scattering parameters and phase constant to characterize the sample [21]. This technique has the advantage of being contactless and not wasteful, but it demands the usage of expensive lenses and horn antennas, as well as the need for a large sample.

Another technique for measuring the material's electric permittivity is the transmission-line method. In this method, the MUT is used as a loading material for transmission lines, such as a slice of material that can be incorporated to a waveguide [22] or the deposing materials of a coaxial line that can be replaced by the MUT [23]. The scattering parameters from the MUT-filled region provide the data necessary to extract the material's properties. This method is comparatively lower cost than the free-space method. However, the sensitivity of the scattering parameters approach is not very efficient for low loss samples, and the sample elaboration is also very often a challenging task [24]. The structures of microstrip-line and stripline are also used for this technique [25].

One quite accurate technique is the resonant cavity method [26]. In this method, a cavity resonator is loaded with the MUT, and the shift in the resonance frequency and the variation in the quality factor are determined. Circular resonators and microstrip-line resonators have been also employed for this purpose [27] other than a conventional box resonator. This technique also needs accurate sample elaboration.

In order to overcome the limitations described above and to obtain all the information required to accurately characterize the complex electrical permittivity of a solid material, we propose a modified SRRs planar sensor for noninvasive complex permittivity measurements of solid materials. The resonant structures studied in this paper are based on the well-known SRR structures, with a modified topology, improving the selectivity and assuring two resonant frequencies. The measurement technique adopted in our research is the resonant cavity method.

The sensor is implemented on FR-4 substrate, offering portability, low-cost manufacturing, and easy-to-use and easy-to-interpret results. The structure is designed to be easily manufactured on a single metal layer, while allowing the easiness of integration of resonator elements at the same level.

The propagation phenomena occurring in these modified resonant structures can also be used to create RF absorbing materials, which can lead to designing efficient microwave absorbers for different applications, such as 5G antennas and automotives.

2. Modified SRR Sensor Design

2.1. Resonant Structures for Higher Selectivity

In 1999, J. Pendry proposed a motivating sub wavelength element defined as split ring resonator (SRR) to realize negative permeability [28]. The SRR structure consists of two highly closed concentric metallic split ring resonators etched on a substrate, with two gaps orientated in opposite directions, as shown in Figure 1a. When a magnetic field perpendicular to the ring surface is applied, a current is induced through the rings. These currents go from one ring to another due to the distributed capacitance that appears between them.

The goal of this work is to create an affordable, low-cost manufacturing sensor. Therefore, we choose a FR-4 substrate with relative permittivity $\varepsilon_r = 4.4$ and the dissipation factor, tan δ, approx. 0.02. The thickness of the substrate is equal to 1.6 mm and the cooper metallization electrodeposited on both sides of the substrate has a thickness of 18 μm. Another goal of the work is to characterize materials in the low-GHz band. So, the SRR's dimensions are considered to have measuring applications for frequencies around 2 GHz. In this case, the distance between the rings is c = 1.52 mm, the width of the rings is w = 1.52 mm, the width of the gap is g = 1.22 mm, and the length of the external ring is d = 18.4 mm. Figure 1a shows the geometrical design of the SRR cell.

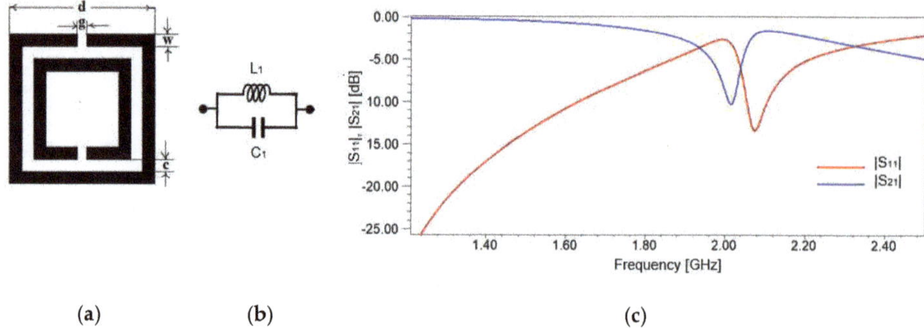

Figure 1. Split Ring Resonator: (**a**) Split Ring Resonator (SRR) with physical dimensions; (**b**) equivalent circuit; (**c**) frequency response of the scattering parameters for SRR.

The material characterization technique used to determine dielectric properties is the resonant one. This process monitors the frequency shift and the variation of the quality factor due to MUT loading the resonator, which is currently represented by the SRRs elements.

The equivalent circuit of the classical SRR in Figure 1a represents a resonant cavity modeled by a LC circuit, Figure 1b, where the inductance L_1 models the effect of the conductive strips of the rings and the capacitance C_1 models the effect of the gap between the two rings [29]. The values for the inductance L_1, the capacitance C_1 and the gap capacitance appearing at the end of each ring, C_g are computed based on the geometrical dimensions of the SRR, the substrate's properties, and the relations given in [29]: L_1 = 38.58 nH, C_1 = 152.49 fF, C_g = 1.69 fF. The gap capacitance, C_g, can be neglected in comparison to the value of the capacitance C_1 [29].

In this case, the total impedance of the resonant equivalent circuit can be written:

$$Z_{T,1} = \frac{j\omega L_1}{1 - \omega^2 L_1 C_1} \quad (1)$$

and the resonant frequency is [29]:

$$f_{r,1} = \frac{1}{2\pi\sqrt{L_1 C_1}} = 2.075\ \text{GHz} \quad (2)$$

The resonant frequency read from Figure 1c is 2.08 GHz, in very good agreement with the one computed using (2). Additionally, Figure 1c shows that the resonant frequency is not so well emphasized, without a sharp response of the SRR.

To obtain a better selectivity, the classical SRR depicted in Figure 1a is modified by adding microstrip vertical strips (VS) of width, w, leaving a gap, s, between the SRR and the vertical strips, as presented in Figure 2a. The gap is set to 0.2 mm to assure a tight coupling effect, but also considering the technological restrains. The dimensions of the resonant structures presented in Figure 1a or Figure 2a are given in Table 1.

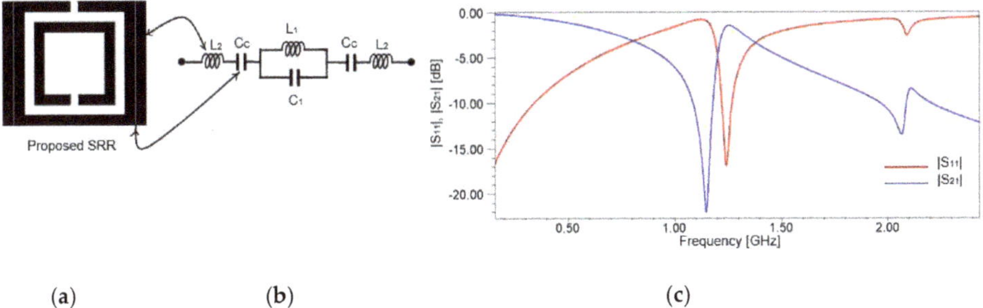

(a) (b) (c)

Figure 2. Proposed SRR: (**a**) SRR with vertical strips, placed at the distance s from the initial SRR in Figure 1a; (**b**) equivalent circuit; (**c**) frequency response of the scattering parameters for proposed SRR.

Table 1. Dimensions of the SRR in Figure 1a or Figure 2a.

Parameters	d [mm]	w [mm]	g [mm]	c [mm]	s [mm]
SRR	18.4	1.52	1.22	1.52	-
Proposed SRR	18.4	1.52	1.22	1.52	0.2

The substrate used for simulations in both figures is FR-4, with a thickness of 1.6 mm and the relative electric permittivity of 4.4. The equivalent circuit for the modified SRR proposed in Figure 2a is the one in Figure 2b, where the effect of the strips is modeled by the inductance L_2 and the coupling effect is modeled by the capacitance C_C. Using our geometrical dimensions and the relations from [29], we obtain: L_2 = 11.04 nH and C_c = 1.49 pF.

Regarding the strong couplings between the vertical strips and the rings of the SRR, the strips themselves lead to the appearance of a second resonant frequency, which assures an improvement in resolution, as shown in Figure 2c, compared to the frequency response from Figure 1c.

The appearance of the second resonant frequency in Figure 2c can be explained by computing the impedance of the resonant equivalent circuit in Figure 2b:

$$Z_{T,2} = \frac{2(1-\omega^2 L_2 C_C)}{\mathrm{j}\omega C_C} + \frac{\mathrm{j}\omega L_1}{1-\omega^2 L_1 C_1} \tag{3}$$

Considering the first resonant frequency given by relation (1), we can rewrite relation (3):

$$Z_{T,2} = \frac{2\left[1-\left(\frac{\omega}{\omega'_{r,1}}\right)^2\right]}{\mathrm{j}\omega C_C} + \frac{\mathrm{j}\omega L_1}{1-\left(\frac{\omega}{\omega_{r,1}}\right)^2} \tag{4}$$

where $\omega'_{r,1} = 2\pi f'_{r,1}$ and

$$f'_{r,1} = \frac{1}{2\pi\sqrt{L_2 C_C}} \tag{5}$$

Using the values determined previously for L_2 and C_C and using relation (5), we can compute the second resonant frequency equal to 1.23 GHz. From Figure 2c, we read that the second resonant frequency is 1.24 GHz. So, the equivalent circuit for the SRR in Figure 1b and for the proposed SRR in Figure 2b used to compute analytically the resonant frequencies proves an accurate modeling of the resonant structures. In addition, we further consider analysis for only the S_{21} parameter because it is more sensitive than S_{11} The two resonant frequencies that occur can be used to obtain a better resolution for measurements. Furthermore, because of an increase in the equivalent capacitance due to adding vertical strips, the resonant frequency decreases to smaller values, around 1.24 GHz, as one can observe in Figure 2c. The other resonant frequency remains the same as in Figure 1c, around 2 GHz.

The next step is to add access transmission lines and practically transform the structure into a planar sensor made of Vertical Strips Split Ring Resonators (VS-SRRs). We calculate the width of the access transmission line corresponding to 50 Ω to be 6.166 mm.

Next, two cases are analyzed: The sensor containing one modified SRR cell and the sensor with two modified SRRs cells. The geometrical dimensions for the SRRs and the vertical strips are the ones given in Table 1. Additionally, we keep the same substrate as in our previous analysis. The overall dimensions of the sensor, including the access transmission lines, are 0.5λ × 0.16λ for one-cell sensor and 0.7λ × 0.16λ, respectively, for the two-cell sensor. Still, the sensors can be easily rescaled and re-designed to be used for other frequency applications or for further miniaturization. The frequency response of both sensors is given in Figure 3.

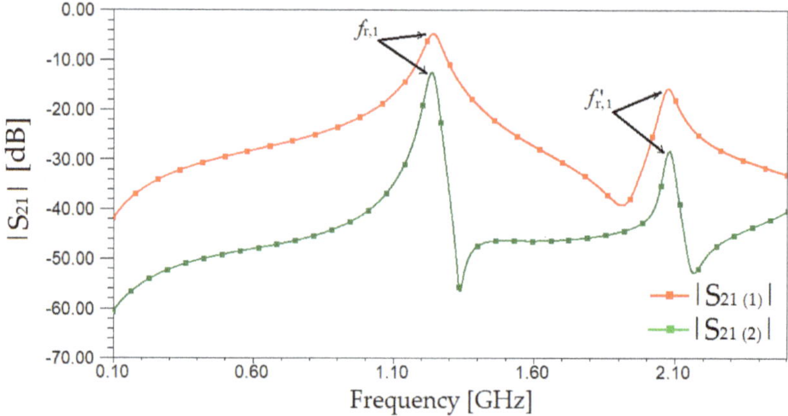

Figure 3. Scattering parameters for sensors made of one modified SRR-$S_{21(1)}$ and two modified SRRs-$S_{21(2)}$.

As it can be observed in Figure 3, selectivity has increased due to adding access transmission lines and increasing the number of cells. In addition, one can observe that both resonant frequencies, $f_{r,1}$ and $f'_{r,1}$ are well emphasized and can be used for further computations. Adding a new modified SRR has led to a better selectivity than using only one. This indicates an improvement in the quality factor of the sensor and thus it will provide a better accuracy for characterizing the dielectric constant of the samples. We consider the results obtained for the two cells vertical strips SRR (VS-SRR) are good enough to further investigate this sensor and not add more SRRs and complicate the structure or increase the manufacturing cost.

2.2. The Resonant Frequencies Analysis

The material under test (transparent box) is placed on the SRR unit cells of the VS-SRR sensor, as depicted in Figure 4a, covering the whole area of the sensor for having an efficient perturbation of the *E*-field and assuring the resonance frequency shift required for precise measurements. When the resonance occurs, the total electric field will be confined to a smaller region of split ring resonator, where the sample is usually placed as shown. This confined electric field is capable of sensing an even smaller change in the dielectric constant of the test sample. The response of the microwave sensor to the change in the effective dielectric constant of the surrounding can be noticed in terms of the change in resonant frequency and the quality factor of the loaded structure [5]. The intensity of the electric field through the sensor, analyzed at the two resonant frequencies, is presented in Figure 4b.

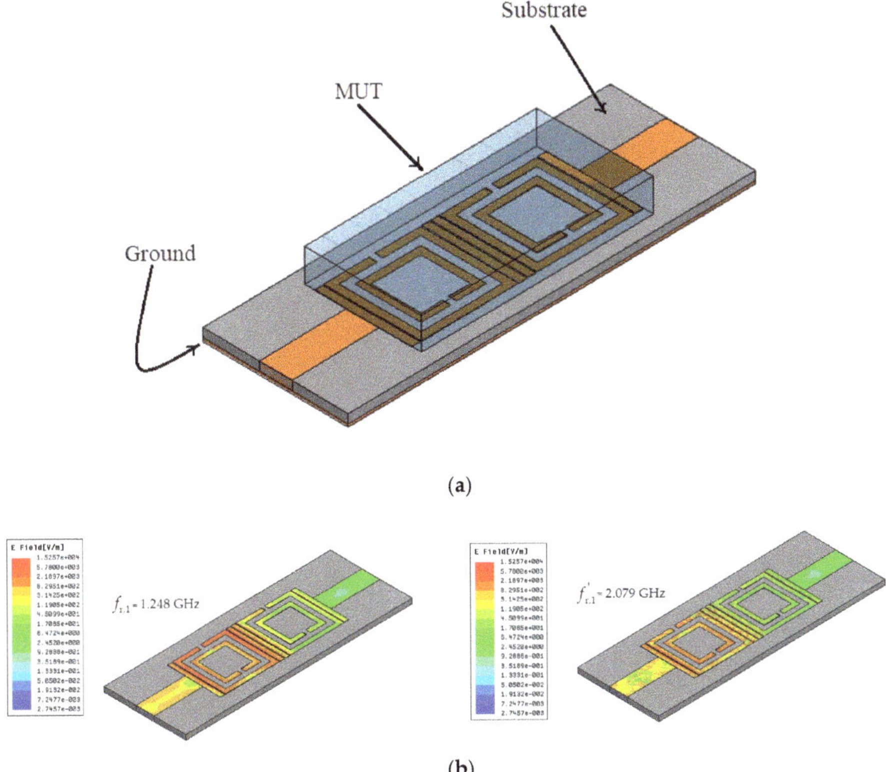

Figure 4. The vertical strips (VS)-SRR sensor: (**a**) Setup of the VS-SRR sensor with material under test (MUT) placed over the two modified SRR unit cells, (**b**) 3D representation of the intensity of the electric field at the first resonant frequency and at the second resonant frequency.

The results in Figure 4b show that when the resonances occur, the total electric field is confined mainly to the first VS-SRR cell of the split ring resonator. This confined electric field is capable of sensing small changes in the dielectric constant of the MUT placed above the sensor. The response of the microwave sensor to the change in the effective dielectric constant is observed as a change in the resonant frequency and the quality factor of the loaded structure. Furthermore, in Figure 4b, one can observe the impact on the electric field distribution of adding vertical strips near the classical SRR. Introducing vertical strips near the SRR, as depicted in Figure 2a, basically increases the effective capacitance of the whole structure. This leads to higher electric field intensity in a small sensing region and, thus, obtaining an improved sensitivity of the sensor.

Moreover, we can see that the electric field is mainly concentrated in the first VS-SRR cell for both frequencies, but through capacitive coupling it propagates to the second cell as well. So, when using the MUT, it is important to place it on the whole sensor, to cover the whole sensing area made of both VS-SRRs.

Next, through full wave electromagnetic simulation in Ansys HFSS, we investigate how the resonant frequencies shift when loading the sensor with different solid dielectrics considered as MUTs. The resonant frequencies in Figure 3 are considered the reference ones for the unloaded sensor ($f_{r,1}$, and $f'_{r,1}$). The proposed sensor is then loaded with various dielectric materials as MUT, with the real part of the relative electric permittivity, ε'_r equal to 2 and to 4 and the loss tangent, tan δ, ranging from 0 to 0.15. The simulated transmission coefficient, S_{21} is depicted in Figure 5.

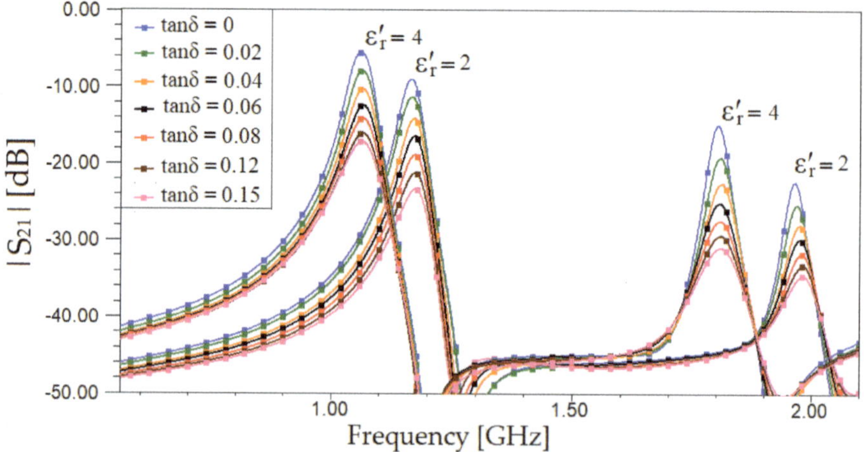

Figure 5. Variation of S_{21} (dB) response of the VS-SRR sensor with loss tangent value varying from 0 to 0.15 and for values of the relative electric permittivity, ε'_r equal to 2 and 4.

From Figure 5, it can be noticed that the shift for the resonant frequency $f'_{r,1}$ is greater than the shift for $f_{r,1}$ in the same conditions: Same variations of ε'_r and tan δ of the material under test. In order to evaluate the sensitivity performance of both resonant frequencies based on the results in Figure 5, a relative frequency shift is defined as:

$$\Delta f_r = \text{unloaded}(f_r) - \text{loaded}(f_r) \tag{6}$$

For our analysis, we consider a broad range of values for the real part of the permittivity, between 0 and 14, and investigate the relative frequency shift for both resonant frequencies. The results of the frequency shift variation, Δf_r with respect to the real part of the relative electric permittivity for both resonant frequencies, are plotted in Figure 6.

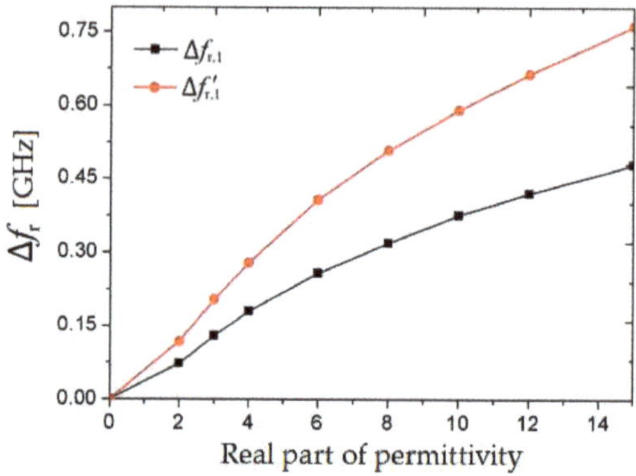

Figure 6. The sensitivity for both resonant frequencies, $f_{r,1}$ and $f'_{r,1}$.

From Figure 6, it can be noticed that the relative frequency shift corresponding to the second resonant frequency, $f'_{r,1}$ is greater than the relative frequency shift produced by the first resonant frequency, $f_{r,1}$. This means that using $f'_{r,1}$ is considered a better option to obtain a higher sensitivity than using the first resonance frequency, $f_{r,1}$. However, in the current work, both resonant frequencies of the VS-SRR sensor will be utilized for MUT characterization in order to add a higher degree of precision, especially if limited by technological or numerical constrains, as proved later when having real measurements.

3. Numerical Analysis

In our further numerical analysis, we investigate both resonant frequencies as argued above. For each of them, we consider the data obtained after full wave electromagnetic simulation, and using a curve fitting tool, we determine analytical expressions for both the real and imaginary parts of the permittivity. These are expressed as a function of the resonant frequencies, the MUT's thickness, and the quality factor of the loaded sensor. A curve fitting tool is often used in the literature [1,16,17] for successfully estimating numerical expressions based on collected data from simulation or measurements.

In the unloaded situation, the simulated resonant frequencies ($f_{r,1}$ and $f'_{r,1}$) of the VS-SRR sensor are 1.24 GHz and 2.08 GHz.

Knowing that the quality factor for general resonators, Q can be written [30]:

$$Q = \frac{f_r}{\Delta f} \tag{7}$$

where fr is the resonant frequency and Δf represents the relative 3dB bandwidth of the resonator's frequency response; we determine the quality factors corresponding to the two resonances as being equal to 35.4 and to 65, respectively.

After loading the sensor with the material under test, a shift in the resonant frequency as well as a change in the magnitude of S_{21} (dB) are recorded as mentioned earlier in Figure 5. The values for the resonant frequencies ($f_{r,1}, f'_{r,1}$) and the quality factors are calculated from the response of the transmission coefficients and are then used to achieve a numerical expression with the aid of a curve fitting tool. A commercially available software OriginPro 8 [16] is used as a curve fitting tool for the data obtained after full wave electromagnetic simulation in Ansys HFSS. The equations are generated using the sets of obtained data. A particular profile curve is chosen based on the least error between the chosen profile and the sets of numerically obtained data [30], as it will be presented in the next sections.

3.1. Deduction of the Real Part of the Permittivity

To determine the type of dependency between the resonant frequency and the real part of the permittivity, we consider the expression for the resonant frequency [31]:

$$f_r = \frac{1}{2\pi\sqrt{L(C + C_{load})}} \tag{8}$$

The capacitance introduced by the load, C_{load} depends directly proportional to the real part of the electrical permittivity [31], so considering relation (8) as well, we obtain $f^{-2}_r \propto \varepsilon'_r$.

So, for the proposed sensor, the affected transmission coefficient due to sample loading can be observed in Figure 7. The inverse squares of the resonant frequencies ($f_{r,1}$ and $f'_{r,1}$) are extracted from the simulated transmission coefficient data and the results with the corresponding real permittivity (ε'_r) of the MUT are plotted and showed in Figure 7a,b.

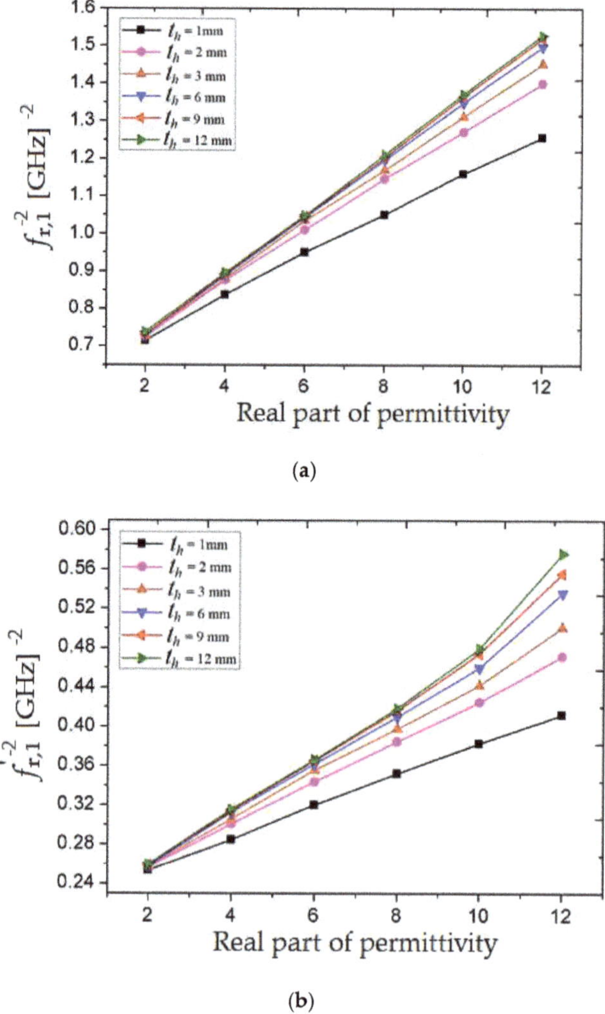

Figure 7. Resonant frequencies in terms of real permittivity (ε'_r) for different thickness of MUT: (**a**) First resonant frequency $(f_{r,1})^{-2}$; (**b**) second resonant frequency $(f'_{r,1})^{-2}$.

In Figure 7a,b, it can be observed that the slope of the plotted curves relies on the thickness of MUT (t_h) and increases as the thickness of the MUT increases. However, in Figure 7a, it can be observed that the slope of the curve residues roughly constant for the sample thickness (t_h) larger than 9 mm. This specific behavior may be noticed from the two lines corresponding to the MUT thickness of 9 mm and 12 mm, where both curves overlap.

Moreover, in Figure 7a, it may be observed that all plotted curves corresponding to the MUT permittivity variation ($\varepsilon'_r = 2$ to $\varepsilon'_r = 12$) have a linear dependency, while in Figure 7b, all plotted curves corresponding to the MUT permittivity ($\varepsilon'_r > 10$) have a roughly exponent dependency while the MUT thickness is increasing. In order to combine all the above effect, the dielectric constant of the specimen is mathematically expressed in terms of the family of straight lines as well as a family of exponential curves, where the freelance parameters are the resonant frequencies ($f_{r,1}$ and $f'_{r,1}$ expressed directly in GHz) and the sample thickness (t_h expressed directly in mm). By taking this aspect into consideration when using the fitting tool practically, the accuracy of the numerical model increases. So, based on the plotted data and using the curve fitting tool, we obtain the expressions for the real permittivity as a function of the MUT's thickness and the two resonant frequencies:

$$\varepsilon_r' = -\frac{1}{1.88 \cdot 10^{-4} \cdot \ln(18472.96 \cdot \ln(t_h))} \cdot \ln\left[\frac{39.6824 - (f_{r,1})^{-2}}{39.095}\right] \tag{9}$$

$$\varepsilon_r' = \exp\left\{2.2607 \cdot \ln\left[\frac{\left(f'_{r,1}\right)^2 - 5.373}{\frac{0.2336}{t_h + 0.1266} - 1.2}\right]\right\} \tag{10}$$

These expressions will be used in our real-life measurements to determine which resonant frequency provides more accurate results. The mathematical limitation of relation (9) is for MUTs with thickness of 1 mm. In this case, relation (9) cannot be used, but we can use relation (10) instead. This case proves the limitation of the sensor, but also the importance of having an alternative analytical formula, based on the second resonant frequency.

3.2. Deduction of the Imaginary Part of the Permittivity

After finding the numerical relations (9) and (10) for determining the real part of the permittivity of the material under test, an identical analysis is completed to find a numerical relation for computing the loss tangent of the tested sample, which will give us information for determining the imaginary part of the permittivity.

The relation between the loss tangent, tan δ, the quality of the proposed sensor after loading the MUT, Q_{MUT}, the real part of the permittivity, ε'_r, and the imaginary part of the permittivity, ε''_r is given by [32]:

$$Q_{MUT} = \frac{1}{\tan\delta} = \frac{\varepsilon_r'}{\varepsilon_r''} \tag{11}$$

where Q_{MUT} states the quality factor of the proposed sensor after loading the MUT, which may be determined applying relation (7). The imaginary part of the complex permittivity is therefore obtained using (7) and (11).

At first, the real part of the permittivity takes values in the range of 2 to 12. For each value, the loss tangent is changed in the range from 0 to 0.12. For each change, the resonant frequencies ($f_{r,1}$ and $f'_{r,1}$) are recorded from the S_{21} parameter's simulation. Then, the quality factor is extracted from the simulation result of S_{21} as depicted earlier in Figure 5, for each resonance frequency. After that, the inverse of the quality factors for each corresponding loss tangent are plotted in Figure 8a,b.

In Figure 8a, it may be observed that all plotted curves corresponding to the MUT loss tangent variation (tan δ = 0 to tan δ = 0.12) have a semi-linear low slope component.

In Figure 8b, all plotted curves remain on a semi-linear high slope component. Therefore, to deduce the tan δ of the MUT, which relies on the loaded quality factor as well as the ε'_r of the MUT, a curve fitting tool is utilized. As in the previous case of the real part of permittivity, a commercially available software OriginPro 8 is used to determine the numerical model for both extracted results in Figure 8a and b as presented in expressions (12) and (13), respectively:

$$\tan\delta = \exp\left\{0.687 \cdot \ln\left[8.165 \cdot 10^{-3} \cdot \left(36.812 \cdot \varepsilon_r'^{-0.338} - Q_{MUT}\right)\right]\right\} \tag{12}$$

$$\tan\delta = 0.1574 \cdot \ln\left[\frac{Q_{MUT}^{-1} + 0.0023}{0.04503 - 0.0318 \cdot (0.94427)^{\varepsilon_r''}}\right] \tag{13}$$

After determining the ε'_r from (9), (10) and tan δ from (12), (13) the imaginary part of the complex permittivity can be obtained using (11).

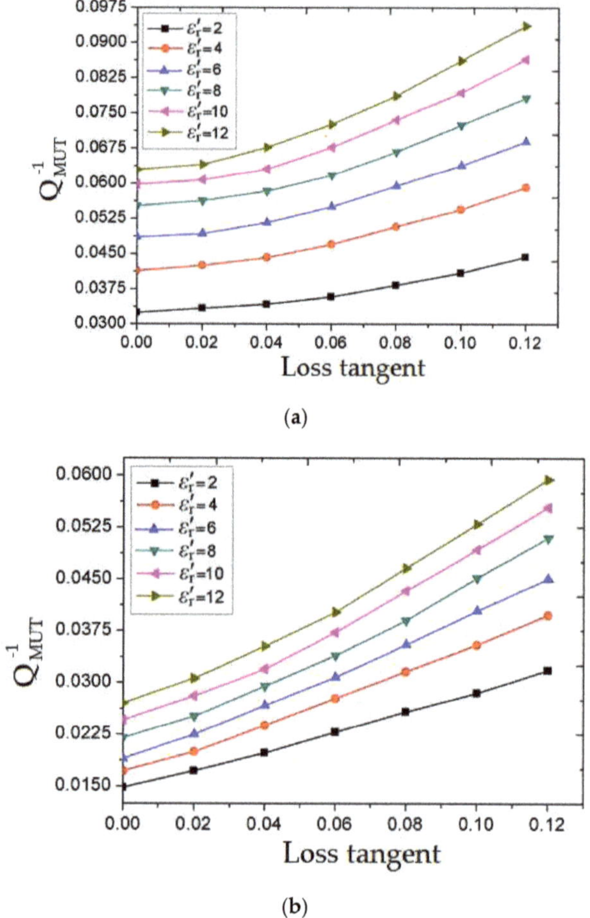

Figure 8. Inverse of Q-factor in terms of tan δ for various values of ε'_r, depending on resonant frequency: (**a**) $f_{r,1}$; (**b**) $f'_{r,1}$.

The mathematical limitation of relations (12) and (13) appear indirectly through the value of the real part of permittivity, ε'_r. If this quantity cannot be determined using the first resonant frequency, as in the case of 1 mm thick MUTs, then automatically neither the loss tangent using relation (12) can be determined. As in the case of real part of permittivity, a second option, one using the second resonant frequency is very useful in practical applications.

4. Results

The sensor proposed in Figure 4 is now implemented and measured. The substrate used is FR-4 (relative permittivity $\varepsilon_r = 4.4$ and the dissipation factor, tan δ, is approximately 0.02), with a thickness of 1.6 mm and cooper metallization electrodeposited on both sides of the substrate, with a thickness of 18 μm.

The technological development and manufacturing of the PCB sensor structure was made using Press-n-Peel Blue transfer foil from Techniks, with the etching process being done in turbulent and warm (approximately 50 °C) ferric chloride ($FeCl_3$), with the concentration of 38%. Press-n-Peel method uses a coated Mylar (Polyester) foil base material in which several layers of release agents and resist coatings are applied, as shown in Figure 9a.

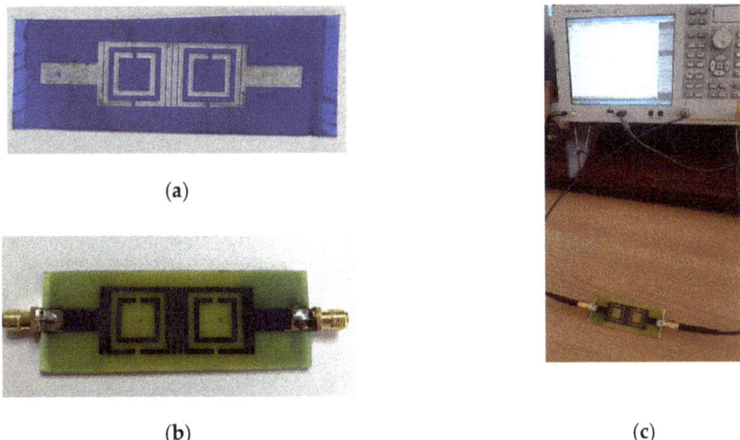

(a)

(b)

(c)

Figure 9. Implementation of the sensor: (**a**) Coated Mylar (Polyester) foil base material; (**b**) the sensor implemented on a FR-4 substrate, with 1.6 mm, with Cooper metallization on both sides; (**c**) measurement setup.

The width of the copper traces has been set to 1.52 mm, with the spacing between two concentric squares also being 1.52 mm, while the spacing between strips is 0.2 mm. The openings in the copper squares are equal to 1.22 mm. The ground plane size is 77 × 18 mm, being developed as a full copper plane onto the bottom side of the PCB. The overall dimensions of the sensor, access transmission lines included, is 76 mm × 26 mm, while the resonant structure itself, where the MUT is placed has a dimension of 44 mm × 18 mm.

The SMA (SubMiniature version A) connecters, which are classical semi-precision coaxial RF connectors used as interface for coaxial cables with screw-type coupling mechanism are mounted on the structure using mechanical welding. The SMA has a 50 Ω characteristic impedance and is designed to work in the range 0–18 GHz, fully matched with the necessities of the current sensing structure. The manufactured sensor is presented in Figure 9b.

The measurement setup consists of the sensor connected to the Agilent E5071C (9 kHz to 6.5 GHz) network analyzer through 50 Ω cables, as shown in Figure 9c. Before starting the measurements, a short-open-load-through (SOLT) calibration was performed using the Agilent calibration Kit. The number of sweep points is chosen 1601.

A set of materials under test: Transparent glass [32], acrylic glass [33], Teflon [32], and Plexiglas [34], with different thicknesses (t_h) of 1 mm, 2 mm, 5 mm, and 10 mm is selected and used for measurements. For each measurement, the sensor is placed on a rough, stable surface and the MUT is carefully placed to cover the whole sensing area. Then, using the Agilent E5071C network analyzer, the magnitude of S_{21} parameter is measured. Further, it is inspected and, using a marker, the resonant frequencies and the relative 3 dB bandwidth of the resonator's frequency response are read. Then, using relation (7), the quality factor of the loaded sensor is determined. The quality factor for the loaded sensor with real MUTs is determined based only on measurements. Both resonant frequencies obtained after measurement for different types of MUT are considered and using relations (9) and (10), two possible values for the real part of the permittivity are obtained. They are compared with reference values [32–34] and the results, including errors, are synthesized in Table 2.

Table 2. Real part of the complex permittivity for different materials under test.

Material	t_h [mm]	ε'_r	f_{r1} [GHz]	ε'_{r1}	error [%]	f_{r2} [GHz]	ε'_{r2}	error [%]
Transparent Glass	5	6	0.9856	5.872	2.12	1.671	6.163	2.72
Acrylic Glass	5	2.7	1.127	2.647	1.93	1.902	2.629	2.59
Acrylic Glass	2	2.7	1.139	2.644	2.04	1.924	2.627	2.7
Teflon	10	2.1	1.152	2.125	1.19	1.941	2.017	3.91
Plexiglas	1	2.597	1.155	-	-	1.97	2.512	3.24

Analyzing the data obtained after measurements, it can be observed that both resonant frequencies can be used to compute the real part of the permittivity, except for the case of 1 mm thickness Plexiglas MUT. In this case, relation (9) cannot be used, because of numerical limitations, but relation (10) gives a value, measured with an error less than 3.5% than the reference value, proving the importance of an extra resonant frequency. The best results are obtained for samples with thicknesses of 2 mm and 5 mm. Another important observation is that when using the second resonant frequency, the errors are slightly larger than those corresponding to using the first resonant frequency. This can be explained because of technological imperfections and placing the probe in direct contact with the sensor. Still, taken into consideration that the errors are quite small for both frequencies, smaller than 4%, it can be concluded that the sensor is suitable for accurate real part of permittivity measurements. The small errors show that the gap between the sensor and the MUT can be ignored.

For the measurement of the imaginary part of permittivity, first, the measured quality factor Q_{MUT} is replaced in relations (12) and (13) and the value of the loss tangent, tan δ is obtained. It is compared to the reference values [32–34] and the results are given in Table 3.

Table 3. Loss tangent for different materials under test.

Material	t_h [mm]	tan δ	Q_{1MUT}	tan δ_1	error [%]	Q_{2MUT}	tan δ_2	error [%]
Transparent Glass	5	0.005	20.18	0.00512	2.39	47.39	0.0051	3.26
Acrylic Glass	5	0.02	26.12	0.019417	2.92	56.1	0.0203	1.88
Acrylic Glass	2	0.02	26.07	0.020532	2.66	56.05	0.0205	2.7
Teflon	10	0.0003	28.5317	0.000308	2.75	69.335	0.0003085	3.24
Plexiglas	1	0.0008	26.648	-	-	65.43	0.00082	3.12

Analyzing the results in Table 3, we notice that for acrylic glass of 5 mm thickness, the error when using the second resonant frequency is smaller than for the first one. Overall, the measurements were done with less than 4% errors. Again, because we could not determine the real part of the permittivity for the first resonant frequency, we could not determine the loss tangent either. A good solution for such cases is to use the alternative, given by the second frequency.

Next, using relation (11), the imaginary part of the permittivity is determined. The results of the measurements are given in Table 4.

Table 4. Imaginary part of the complex permittivity for different materials under test.

Material	t_h [mm]	ε''_r	ε''_{r1}	error [%]	ε''_{r2}	error [%]
Transparent Glass	5	0.03	0.030065	0.215467	0.031431	4.497
Acrylic Glass	5	0.054	0.051397	4.82074	0.053369	1.169
Acrylic Glass	2	0.054	0.054287	0.530756	0.053854	0.27129
Teflon	10	0.00063	0.000655	3.8888	0.00062	1.231
Plexiglas	1	0.002078	-	-	0.00206	0.8548

The errors in Table 4 are smaller when using the second resonant frequency, except for the transparent glass case, proving the importance of the second resonant frequency. Additionally, the errors determined in Table 4 show both the impact of approximations due to computing and the impact of measuring two parameters with different errors: The real part of permittivity and the loss tangent. So, we find cases when the errors are smaller than 1%, even if the corresponding errors for the loss tangent measurements alone are not that small. The observations regarding the technological imperfections and the placing procedure of the probe remain valid. Further, the impact of the air gap over the measurements was not considered and, still, the results are very good, much better than the ones in literature [1]. For example, for Teflon, we have obtained measurement errors of 1.19% and 3.91% for the real permittivity and 3.88% and 1.231% for the imaginary part, proving the accuracy of the results. In reference [1], the errors are 1.9% for the real part of permittivity and 8.6 2% for the imaginary part of permittivity. Nevertheless, it is worth observing that the thickness of the MUT has an impact on the overall response of the sensor. If the thickness of the MUT is increased, the interaction of the electromagnetic field is enhanced, so a change in the sensor's frequency response is more obvious.

The results in the two tables show that the sensor can be used successfully to accurately characterize the dielectric parameters (dielectric constant and loss tangent) for both low-losses and lossy dielectrics, as well as for high dielectric constants dielectrics and small dielectric constants dielectrics.

5. Conclusions

In this paper, we present a modified SRRs planar sensor for noninvasive, accurate complex permittivity measurements of solid dielectrics. Starting from the classical SRR, a modified structure, using vertical strips added at a close distance of 0.2 mm to the SRR is investigated both from the enhanced selectivity perspective and from the overall dimensions. The result is a sensor made of two modified SRRs with lateral vertical strips, exhibiting high sensitivity for two resonant frequencies, at 1.24 GHz and 2.08 GHz.

A simplified equivalent circuit model is used to explain the microwave sensor's design, and a very good agreement between the circuit model and the full electromagnetic simulation results is achieved. After a careful investigation, the two VS-SRRs sensor is selected to be further investigated. For each resonant frequency, we consider the data obtained after full wave electromagnetic simulation and using a curve fitting tool, we determine analytical expressions for both the real and imaginary parts of the permittivity. These are expressed as a function of the resonant frequencies, the MUT's thickness, and the quality factor of the loaded sensor.

The sensor is implemented on an affordable, commercial substrate, FR-4 substrate, with a thickness of 1.6 mm, with reduced dimensions and being able to measure the real and imaginary parts of the permittivity for different solid dielectric samples, with errors less than 4.5% for both resonant frequencies in all analyzed cases. In our work, we have considered a large range of samples, with different thicknesses, different loss tangents, and dielectric constants to better investigate the sensor's capabilities in real-life scenarios. The diversity of the samples helped us to observe the limitations of the numerical model developed in Section 3 and find solutions to overcome them, such as successfully using the second resonant frequency.

Also, we have measured the quality factor both for the unloaded and loaded sensor using the resonant frequency and the relative 3dB bandwidth of the resonator's frequency response. This approach added more practical consistency to our investigation. Still, some improvements can be done with respect to further miniaturization and the possibility to use this sensor for liquid dielectric characterization.

In future, we will investigate if this sensor can be used to measure the permeability for magnetic samples and if a lower losses substrate will improve the results. Last, but not least, the modified structure will be investigated if it is suitable for other resonant applications, which require the usage of similar configurations.

Author Contributions: Conceptualization, A.A.a.-B. and I.A.M.; methodology, A.A.a.-B. and I.A.M.; software, A.A.a.-B., I.A.M., and M.P.; validation, A.A.a.-B., I.A.M., and N.C.; formal analysis, I.A.M. and M.P.; investigation, A.A.a.-B., I.A.M., N.C., and M.P.; resources, N.C.; data curation, M.P.; writing—original draft preparation, I.A.M. and A.A.a.-B.; writing—review and editing, N.C.; visualization, M.P. and I.A.M.; supervision, I.A.M.; project administration, I.A.M.; funding acquisition, N.C. and M.P. All authors have read and agreed to the published version of the manuscript.

Funding: This research received no external funding.

Acknowledgments: The authors would like to acknowledge the Center for Technological Electronics and Interconnection Techniques, from University Politehnica of Bucharest, Department of Electronics Technology and Reliability, (UPB-CETTI) for the materials used to manufacture the sensor and carry the experiments.

Conflicts of Interest: The authors declare no conflict of interest.

References

1. Ansari, M.A.H.; Jha, A.K.; Akhtar, M.J. Design and Application of the CSRR-Based Planar Sensor for Noninvasive Measurement of Complex Permittivity. *IEEE Sens. J.* **2015**, *15*, 7181–7189. [CrossRef]
2. Yang, C.-L.; Lee, C.-S.; Chen, K.-W.; Chen, K.-Z. Noncontact measurement of complex permittivity and thickness by using planar resonators. *IEEE Trans. Microw. Theory Tech.* **2016**, *64*, 247–257. [CrossRef]
3. Puentes, M. *Planar Metamaterial Based Microwave Sensor Arrays for Biomedical Analysis and Treatment*; Springer: Heidelberg, Germany, 2014.
4. Hardinata, S.; Deshours, F.; Alquié, G.; Kokabi, H.; Koskas, F. Miniaturization of Microwave Biosensor for Non-invasive Measurements of Materials and Biological Tissues. *IPTEK J. Proc. Ser.* **2018**, *29*, 90–93. [CrossRef]
5. Galindo-Romera, G.; Herraiz-Martínez, F.J.; Gil, M.; Martinez-Martinez, J.J.; Segovia-Vargas, D. Submersible Printed Split-Ring Resonator-Based Sensor for Thin-Film Detection and Permittivity Characterization. *IEEE Sens. J.* **2016**, *16*, 3587–3596. [CrossRef]
6. Liu, W.; Sun, H.; Xu, L. A Microwave Method for Dielectric Characterization Measurement of Small Liquids Using a Metamaterial-Based Sensor. *Sensors* **2018**, *18*, 1438. [CrossRef]
7. Hao, H.; Wang, D.; Wang, Z.; Yin, B.; Ruan, W. Design of a High Sensitivity Microwave Sensor for Liquid Dielectric Constant Measurement. *Sensors* **2020**, *20*, 5598. [CrossRef]
8. Wei, Z.; Huang, J.; Li, J.; Xu, G.; Ju, Z.; Liu, X.; Ni, X. A High-Sensitivity Microfluidic Sensor Based on a Substrate Integrated Waveguide Re-Entrant Cavity for Complex Permittivity Measurement of Liquids. *Sensors* **2018**, *18*, 4005. [CrossRef]
9. Mondal, D.; Tiwari, N.K.; Akhtar, M.J. Microwave Assisted Non-Invasive Microfluidic Biosensor for Monitoring Glucose Concentration. In Proceedings of the 2018 E-Health and Bioengineering Conference (EHB), New Delhi, India, 28–31 October 2018; pp. 1–4. [CrossRef]
10. Chretiennot, T.; Dubuc, D.; Grenier, K. Double stub resonant biosensor for glucose concentrations quantification of multiple aqueous solutions. In Proceedings of the 2014 IEEE MTT-S International Microwave Symposium (IMS2014), Tampa, FL, USA, 1–6 June 2014; pp. 1–4. [CrossRef]
11. Velez, P.; Grenier, K.; Mata-Contreras, J.; Dubuc, D.; Martín, F. Highly-Sensitive Microwave Sensors based on Open Complementary Split Ring Resonators (OCSRRs) for Dielectric Characterization and Solute Concentration Measurement in Liquids. *IEEE Access* **2018**, *6*, 48324–48338. [CrossRef]
12. Abdolrazzaghi, M.; Khan, S.; Daneshmand, M. A Dual-Mode Split-Ring Resonator to Eliminate Relative Humidity Impact. *IEEE Microw. Wirel. Componen. Lett.* **2018**, *28*, 939–941. [CrossRef]
13. Jafari, F.S.; Ahmadi-Shokouh, J. Reconfigurable microwave SIW sensor based on PBG structure for high accuracy permittivity characterization of industrial liquids. *Sens. Actuators A Phys.* **2018**, *283*, 386–395. [CrossRef]
14. Wessel, J.; Schmalz, K.; Scheytt, J.C.; Cahill, B.; Gastrock, G. Microwave biosensor for characterization of compartments in teflon capillaries. In Proceedings of the 42nd European Microwave Conference, Amsterdam, The Netherlands, 29 October–1 November 2012; pp. 534–537. [CrossRef]
15. Rammah, A.A.; Zakaria, Z.; Ruslan, E.; Isa, A.A. Comparative study of materials characterization using microwave resonators. *Aust. J. Basic Appl. Sci.* **2015**, *9*, 76–85.
16. Shafi, K.T.M.; Jha, A.K.; Akhtar, M.J. Improved Planar Resonant RF Sensor for Retrieval of Permittivity and Permeability of Materials. *IEEE Sens. J.* **2017**, *17*, 5479–5486. [CrossRef]
17. Ansari, M.A.H.; Jha, A.K.; Akhtar, M.J. Design of SRR-Based Microwave Sensor for Characterization of Magnetodielectric Substrates. *IEEE Microw. Wirel. Compon. Lett.* **2017**, *27*, 524–526. [CrossRef]
18. Xu, K.; Liu, Y.; Chen, S.; Zhao, P.; Peng, L.; Dong, L.; Wang, G. Novel Microwave Sensors Based on Split Ring Resonators for Measuring Permittivity. *IEEE Access* **2018**, *6*, 26111–26120. [CrossRef]

19. He, X.-J.; Qiu, L.; Wang, Y.; Geng, Z.-X.; Wang, J.-M.; Gui, T.-L. A Compact Thin-Film Sensor Based on Nested Split-Ring-Resonator (SRR) Metamaterials for Microwave Applications. *J. Infrared Millim. Terahertz Waves* **2011**, *32*, 902–913. [CrossRef]
20. Ghodgaonkar, D.K.; Varadan, V.V.; Varadan, V.K. Free-space measurement of complex permittivity and complex permeability of magnetic materials at microwave frequencies. *IEEE Trans. Instrum. Meas.* **1990**, *39*, 387–394. [CrossRef]
21. Kang, T.; Kim, J.; Lee, D.; Kang, N. Free-space measurement of the complex permittivity of liquid materials at millimeter-wave region. In Proceedings of the 2016 Conference on Precision Electromagnetic Measurements (CPEM 2016), Ottawa, ON, Canada, 10–15 July 2016; pp. 1–2. [CrossRef]
22. Liao, S.; Gao, B.; Tong, L.; Yang, X.; Li, Y.; Li, M. Measuring Complex Permittivity of Soils by Waveguide Transmission/Reflection Method. In Proceedings of the IGARSS 2019—2019 IEEE International Geoscience and Remote Sensing Symposium, Yokohama, Japan, 28 July–2 August 2019; pp. 7144–7147. [CrossRef]
23. Takahashi, K.; Konishi, C.; Loewer, M.; Igel, J. Measuring Complex Permittivity of Soils By Coaxial Transmission Line Method and FDTD. In Proceedings of the IGARSS 2018—2018 IEEE International Geoscience and Remote Sensing Symposium, Valencia, Spain, 22–27 July 2018; pp. 6808–6811. [CrossRef]
24. Narayanan, P.M. Microstrip transmission line method for broadband permittivity measurement of dielectric substrates. *IEEE Trans. Microw. Theory Tech.* **2014**, *62*, 2784–2790. [CrossRef]
25. Cai, Y.; Lu, H.; Fran, J.; Cheng, T.; Lin, T. LTCC Dielectric Constant and Loss Tangent Extraction by Thru-Line Method in Stripline. In Proceedings of the Electrical Design of Advanced Packaging and Systems (EDAPS), KAOHSIUNG, Taiwan, 16–18 December 2019; pp. 1–3. [CrossRef]
26. Kams, D.C.; Weatherall, J.C.; Greca, J.; Smith, P.R.; Yam, K.; Barber, J.; Smith, B.T. Millimeter-Wave Resonant Cavity for Complex Permittivity Measurements of Materials. In Proceedings of the IEEE/MTT-S International Microwave Symposium—IMS, Philadelphia, PA, USA, 10–15 June 2018; pp. 1006–1009. [CrossRef]
27. Petrovic, N.; Risman, P.O. A zirconia cylindrical TM010 cavity for permittivity measurements at 1 GHz. In Proceedings of the IEEE Conference on Antenna Measurements & Applications (CAMA), Vasteras, Sweden, 3–6 September 2018; pp. 1–4. [CrossRef]
28. Pendry, J.B. Holden, A.J.; Robbins, D.J.; Stewart, W.J. Magnetism from conductors and enhanced nonlinear phenomena. *IEEE Trans. Microw. Theory Tech.* **1999**, *47*, 2075–2084. [CrossRef]
29. Naoui, S.; Lassaad, L.; Gharsallah, A. Equivalent Circuit Model of Double Split Ring Resonators. *Int. J. Microw. Opt. Technol.* **2016**, *11*, 1–6.
30. Saadat-Safa, M.; Nayyeri, V.; Khanjarian, M.; Soleimani, M.; Ramahi, O.M. A CSRR-Based Sensor for Full Characterization of Magneto-Dielectric Materials. *IEEE Trans. Microw. Theory Tech.* **2019**, *67*, 806–814. [CrossRef]
31. Lee, C.-S.; Yang, C.-L. Complementary split-ring resonators for measuring dielectric constants and loss tangents. *IEEE Microw. Wireless Compon. Lett.* **2014**, *24*, 563–565. [CrossRef]
32. Rhode and Schwarz, Measurement of Dielectric Material Properties Application Note. Available online: https://cdn.rohde-schwarz.com/pws/dl_downloads/dl_application/00aps_undefined/RAC-0607-0019_1_5E.pdf (accessed on 20 September 2020).
33. Physical Properties of Acrylic Sheets. Available online: https://www.builditsolar.com/References/Glazing/physicalpropertiesAcrylic.pdf (accessed on 20 September 2020).
34. Verma, A.K.; Omar, A.S. Microstrip resonator sensors for determination of complex permittivity of materials in sheet, liquid and paste forms. *Microw. Antennas Propag. IEE Proc.* **2005**, *151*, 47–54. [CrossRef]

Letter

A Wide Frequency Scanning Printed Bruce Array Antenna with Bowtie and Semi-Circular Elements †

Zeeshan Ahmed [1,2,*], Patrick McEvoy [1] and Max J. Ammann [1,2]

[1] Antenna and High-Frequency Research Centre, Technological University Dublin, Dublin, Ireland; patrick.mcevoy@tudublin.ie (P.M.); max.ammann@tudublin.ie (M.J.A.)

[2] CONNECT—Ireland's Research Centre for Future Networks and Communications, Westland Row 34, Dublin, Ireland

* Correspondence: zeeshan.ahmed@ieee.org

† This paper is an extended version of Ahmed, Z.; Hoang, M.H.; McEvoy, P.; Ammann, M.J. Millimetre-wave Planar Bruce Array Antenna. In Proceedings of the 2020 International Workshop on Antenna Technology (iWAT), Bucharest, Romania, 25–28 February 2020.

Received: 26 October 2020; Accepted: 25 November 2020; Published: 27 November 2020

Abstract: A printed edge-fed counterpart of the wire Bruce array antenna, for frequency scanning applications, is presented in this paper. The unit-cell of the proposed antenna consists of bowtie and semi-circular elements to achieve wide bandwidth from below 22 GHz to above 38 GHz with open-stopband suppression. The open-stopband suppression enables a wide seamless scanning range from backward, through broadside, to forward endfire. A sidelobe threshold level of −10 dB is maintained to evaluate efficient scanning performance of the antenna. The antenna peak realized gain is 15.30 dBi, and, due to its compact size, has the ability to scan from −64° to 76°.

Keywords: meander line antenna; periodic structure; millimeter-wave antenna; frequency scanning antenna; leaky-wave antenna

1. Introduction

In 1931, Edmond Bruce patented the idea of the Bruce array antenna (BAA) in which a long wire antenna was bent in equal and periodic meandered intervals. The antenna was designed for amateur radio applications in which bi-directional broadside radiation and high gain are required. Figure 1 shows a typical BAA fed from the center of the structure using a twin-line feed mechanism. In the figure, the lengths and directions of the arrows are representations of the magnitudes and flow of the current, respectively. The horizontal and vertical segments of the meander line were both kept equal in size, i.e., approximately λ/4 for ham radio applications, except for the last two inward bent segments, which are half the length of the other segments. The currents in the horizontal segments, represented by light grey colored arrows in Figure 1, flow in opposite directions so as to add together destructively, thus cancelling out radiation in ideal circumstances. These horizontal elements are, therefore, considered interconnecting segments. The currents in the vertical segments flow in the same direction, adding constructively in phase to give broadside radiation, which is why these segments are termed radiating elements. The half-length segments, which are bent inward on either ends of the structure, have little to no magnitude of current; therefore, they maintain reasonably low cross-polarized radiation [1]. As the number of radiating elements are added to the structure, the half power beamwidth (HPBW) becomes sharp with the increase in peak realized gain and the radiation pattern in the broadside becomes so compressed and narrow that it can be classified as a highly directive fan-beam radiation pattern. The BAA offers reduced complexity, substantially greater bandwidth than other wire antennas (such as the bobtail curtain and half-square antennas), and, for a relatively low height requirement, it can achieve the maximum possible gain in a given area [2]. Suited to a particular installation, the wire BAA can also be fed at points other than the center, the lengths can be varied to tune the resonant frequency, and while it usually does not require a ground system, an extensive ground system can be deployed under the BAA to mitigate the losses if there is adequate space [2].

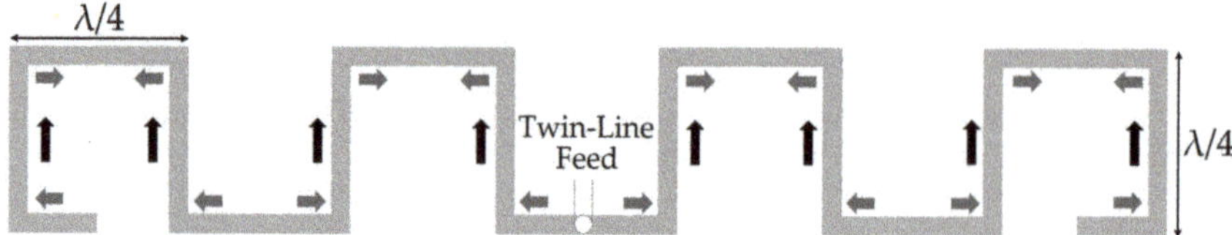

Figure 1. A typical twin-line fed 8-element wire Bruce array antenna (BAA).

The BAA has been around for over a century, but, in spite of its simplicity, modern researchers have overlooked its development and utilization in modern day antenna applications. There are only a handful of ideas proposed to make use of the structure, some of which include Nakano et al.'s concatenation of the Bruce and Franklin antennas' performance at 12.50 GHz [3], Chen's twin-line fed slot-type BAA planar equivalent [4], a tri-band mm-wave printed counterpart of BAA [5], and an edge-fed printed BAA [6].

In 1940, W. W. Hansen patented the first waveguide-based leaky-wave antenna (LWA) [7]. Several other researchers later elaborated the concept in their research [8–11], but it was A. A. Oliner who streamlined the working mechanism in 1984 [12]. In IEEE standard 145-2013, an LWA is defined as "An antenna that couples power in small increments per unit length, either continuously or discretely, from a travelling-wave structure to free space" [13]. LWAs are generally divided into two categories, namely uniform and periodic LWAs [14], the latter of which are widely used in mm-wave frequency regions as well as other scanning applications because of their ability to scan a wider area than the uniform LWAs [15]. Planar periodic LWAs are low-profile, relatively easier to fabricate, and can scan in the backward or the forward endfire direction with a fan-beam radiation pattern with frequency tuning. Several types of LWAs based on a range of technologies have been proposed in the scientific literature, including periodically meandered rampart array [16], sharpening the bends [17], squarely modulated reactance surface (SquMRS) [18], composite right/left-handed structures (CRLH) [19,20], slot or coplanar lines [21], substrate integrated waveguide (SIW) structures [22–27], Goubau line structures [24], spoof plasmon transmission line (SSP-TL) structures [25], and periodically loaded microstrip structures [26].

In the case of periodic LWAs, a steep gain-drop is usually observed around the broadside when scanning from the backward to forward endfire, because of which the antenna suffers from pattern degradation. This is because of the presence of the so-called open-stopband (OSB) at which the LWA, which usually supports a traveling-wave, exhibits standing-wave characteristics with equal excitation of the unit-cells. At OSB frequency, the incident power from the unit-cell that is supposed to radiate outwards instead reflects into the source due to the coupling of a contra-directional pair of space harmonics (Floquet modes) [28]. There are numerous periodic LWAs that have either overcome or suppressed this problem. Balanced transmission lines are used in Metamaterial LWAs to enable seamless scanning through the broadside [29,30]. Other than that, SIW structures use shorting vias [31], unequal width in transversal elements of meander lines [16], and non-identical elements in their unit-cells [32]. A lattice-network based TL model [33] has also been reported to suppress the OSB.

This paper presents a modification of planar, edge-fed, periodic array using meandering concept of wire BAA geometry and the suppression of the OSB around the broadside by replacing horizontal and vertical segments with semi-circular and novel bowtie elements, respectively. The unit-cell is the repetitive part of the structure designed at the broadside frequency. The optimizations and simulations were performed using CST Microwave Studio, in which the dielectric and metallization losses were considered. Finally, the prototype antenna was fabricated and measured responses were compared against the simulated results, from which a satisfactory agreement was attained.

2. Unit-Cell and Antenna Geometry

The configuration of the unit-cell of modified printed BAA is shown in Figure 2. The vertical and horizontal segments of the meander-line BAA antenna, shown in Figure 1, were replaced with bowtie and semi-circular elements, respectively. Either ends of the bowtie had the same width as the semi-circular segment, i.e., w_c. Compared to the BAA, the meandered segments' lengths are approximately λ/4, but in the mm-wave region, this corresponds to a very small size which gives rise to coupling between the vertical segments; therefore, these lengths are varied. However, the length and diameter of the unit-cell, l_v and l_c, respectively, are kept the same, as shown in Table 1. The periodic unit-cell's dominant mode does not radiate on its own because of its slow-wave characteristics; the free space wavenumber, k_0, is less than the phase constant ($\beta > k_0$). Floquet's theorem states that as the unit-cells are combined in series, the periodicity introduces an infinite number of Floquet modes in a leaky-wave structure. These Floquet modes are represented by phase constant β_n [14].

$$\beta_n p_{unit} = \beta_0 p_{unit} + 2\pi n; \; n = 0, \pm 1, \pm 2, \pm 3 \tag{1}$$

where p_{unit} is the period of the unit-cell defined by $4 \times (l_c - w_c/2)$, n is the nth number space harmonic, and β_0 is the phased constant of the dominant mode of the now modulated and uniform waveguide. From Equation (1) the β_0 is slow-wave, but the structure is designed in a way that the other modes are fast. In order to scan a single beam in a directive manner, the first space harmonic, i.e., $n = -1$, is substituted into Equation (1) and is written as

$$\beta_{-1} p_{unit} = \beta_0 p_{unit} - 2\pi \tag{2}$$

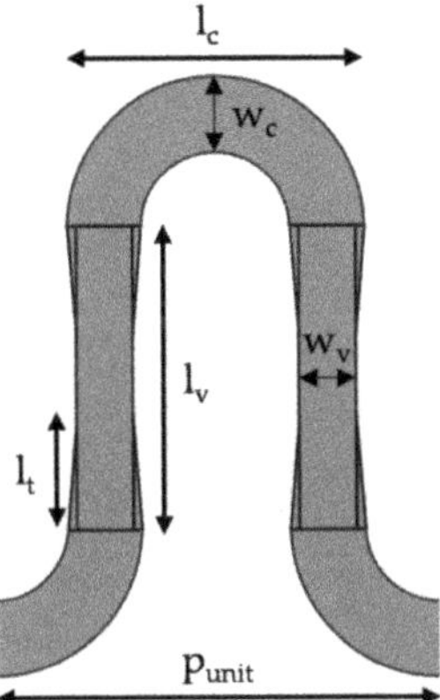

Figure 2. Geometry of the modified BAA unit-cell with vertical bowtie and horizontal semi-circular segments printed on a 0.254 mm thick grounded Arlon DiClad 880.

Table 1. Table of parameters of unit-cell.

Parameter	Size (mm)
$l_c = l_v$	3.50
l_t	1.68
w_c	0.76
w_v	0.65
p_{unit}	5.48

The scanning direction of the periodic LWA can be expressed using [14]

$$\sin\theta_m = \frac{\beta_{-1}}{k_0} \tag{3}$$

where θ_m is the maximum beam angle deviation from the broadside and k_0 is the free space wave number. Subsequently, the beamwidth is given by

$$\Delta\theta \approx \frac{1}{\left(\frac{L}{\lambda_0}\right)\cos\theta_m} \tag{4}$$

where L is the overall length of the leaky-wave antenna structure.

The geometry of the periodic modified BAA antenna, with vertical bowtie and horizontal semi-circular segments, is presented in Figure 3a. As is the case with linear arrays, the addition of unit-cells in series increases the gain and decreases the beamwidth along the length of the antenna, but a large number of unit-cells prevents the increase in gain due to the lower power delivered to the last unit-cells. Thirteen unit-cell elements, presented in Figure 2, were connected in series, and the configuration presented in Table 1 was used for simulation and prototyping of the structure. The periodic modulation of the geometry assisted with radiation along the length of the antenna. The structure was fed using transmission line of length 6.18 mm and width 0.76 mm; the last unit-cell element was terminated using a similar transmission line and another 50 Ω port that acted like a resistor to avoid reflections. Arlon DiClad 880 substrate, with a thickness of 0.254 mm, ε_r = 2.2 and tan δ = 0.0009, was used to fabricate the prototype presented in Figure 3b; the measurements were performed using a Rhode and Schwarz Vector Network Analyzer (ZVA40). The overall dimensions of the antenna were 83.60 × 18.0 × 0.254 mm^3.

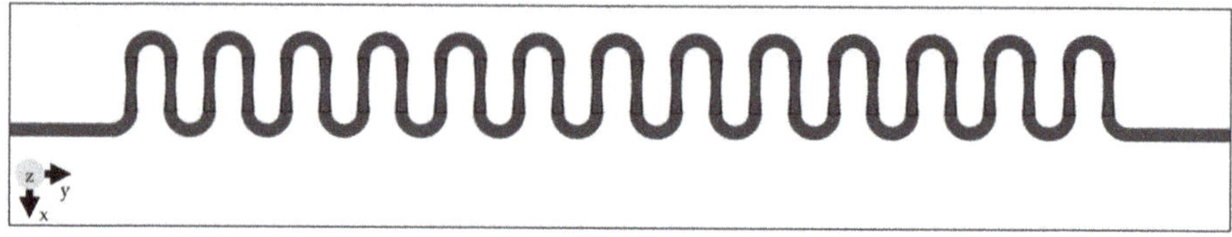

(**a**)

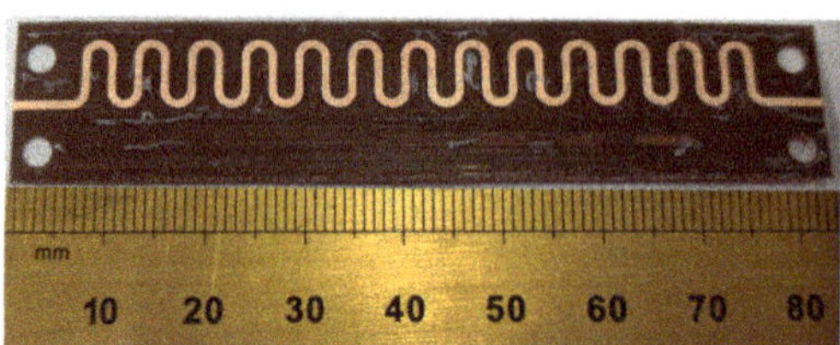

(**b**)

Figure 3. Top view of the proposed periodic 13 unit-cell antenna fabricated on 0.254 mm thick grounded Arlon DiClad 880 substrate. (**a**) Geometry. (**b**) Prototype.

3. Parametric Analysis

Figure 4 shows the effects on $|S_{11}|$ for the structure with 13 unit-cell elements presented in Figure 3 by simultaneously varying l_v and l_c with w_v = 0.76 mm, without the vertical bowtie element. The multiplying factor of the wire BAAs, λ/4, was increased to avoid unwanted resonances in the mm-wave region, due to close separation distance at λ/4 between vertical elements; it varied from 1.69 × λ/4 to 1.89 × λ/4 (3.30 mm to 3.70 mm). With this configuration, mismatching was observed around the broadside frequencies, for which $|S_{11}| > -10$ dB indicated the presence of OSB. Additionally, an increase in l_v and l_c by 0.1 mm tuned down the OSB mismatched region by approximately 1 GHz. The $|S_{11}|$ above and below the OSB, between 20 GHz and 40 GHz, remained less than −10 dB.

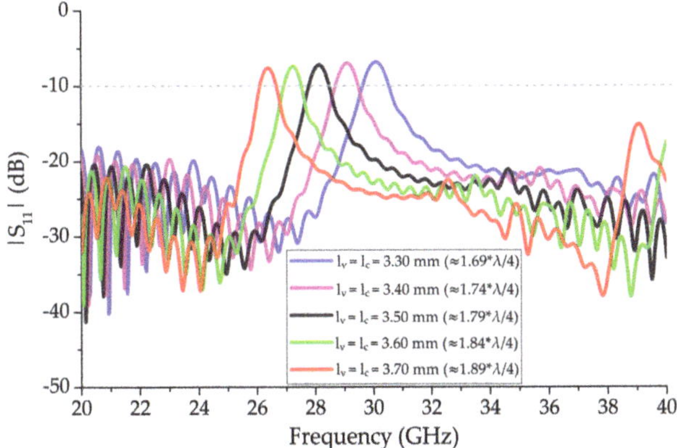

Figure 4. Effect of varying length, l_v, and diameter, l_c, simultaneously for a 13 unit-cell element periodic structure with $w_v = 0.76$ mm unit-cell segments on $|S_{11}|$.

The effect of parameters associated with the vertical bowtie segments, w_v and l_t, of the 13 unit-cell structure, on $|S_{11}|$, is shown in Figure 5. The other parameters, presented in Table 1, remain unchanged. A noticeable improvement in impedance matching, around the OSB region, was observed as the width, w_v, was increased from 0.50 mm to 0.60 mm, with l_t fixed at 1.68 mm, in which $|S_{11}|$ improved from −7.24 dB to −3.54 dB at 28.0 GHz. In the other case, where w_v was kept constant at 0.65 mm, the variation in l_t from 1.45 mm to 1.65 mm improved the impedance matching significantly without any upward or downward tuning of frequency.

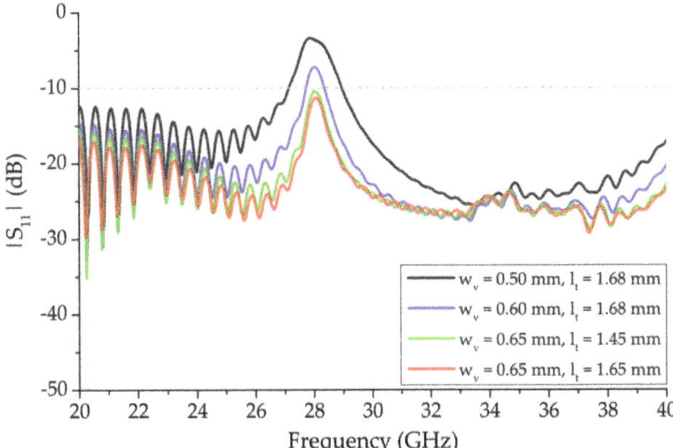

Figure 5. Effect of varying vertical bowtie segment parameters, w_v and l_t, of 13 unit-cell element periodic structure independently on $|S_{11}|$ while the other parameters remain the same as Table 1.

Figure 6 shows the effect of independently varying l_v and l_c with the fixed $w_v = 0.65$ mm bowtie vertical unit-cell segment on $|S_{11}|$. The frequency was tuned down by approximately 1.15 GHz and the impedance matching deteriorated when l_v was increased by 0.20 mm, between 3.40 mm and 3.60 mm, and l_c was kept constant at 3.50 mm. A downward shift in frequency of 0.70 GHz was observed when l_v was fixed at 3.50 mm and l_c was increased from 3.40 mm to 3.60 mm, with a minute effect on impedance matching.

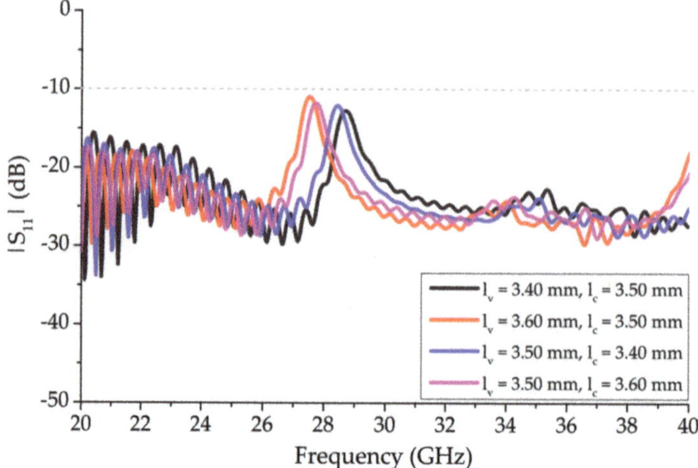

Figure 6. Effect of varying length, l_v, and diameter, l_c, independently of 13 unit-cell element periodic structure with fixed w_v = 0.65 mm bowtie segments on $|S_{11}|$.

Figure 7 shows the realized gain of the 13 unit-cell antenna structure without the vertical bowtie element at between 22 GHz and 38 GHz with $l_v = l_c$ = 3.50 mm and w_v = 0.76 mm. The gain rose gradually between 23.0 GHz and 26.0 GHz and, while still under 12.0 dBi at 26.0 GHz, a sharp gain drop was observed around 28.0 GHz. The realized gain around 28.0 GHz was considerably less than the realized gain in the rest of the forward endfire region. This is consistent with the OSB region, identified in Figure 4, where $|S_{11}| > -10$ dB.

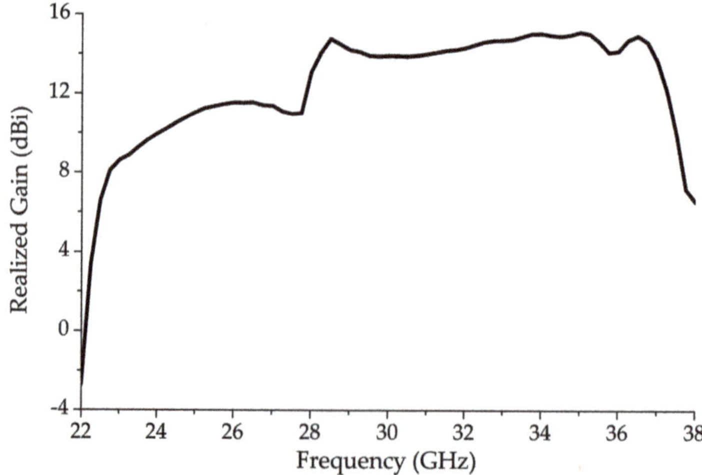

Figure 7. Realized gain of 1-D periodic BAA-LWA with $l_v = l_c$ = 3.50 mm and w_v = 0.76 mm unit-cell showing gain degradation around 28.0 GHz.

4. Results and Discussion

Figure 8 shows the simulated and measured response of the $|S_{11}|$ of the 13-element structure with the bowtie vertical segment unit-cells that were shown in Figure 2 and the parameters presented in Table 1. The mismatched OSB frequency range, for when l_v and l_c is 3.50 mm (~1.79 × λ/4) and w_v = 0.76 mm as presented in Figure 4, improved without any frequency tuning, resulting in the mitigation of the OSB. The $|S_{11}|$, at 28.0 GHz, improved to −12.01 dB with the vertical bowtie element, compared to −7.73 dB without the bowtie element. With the bowtie element, the $|S_{11}|$was ≤ −10 dB from below 22.0 GHz and above 40.0 GHz with a fractional bandwidth of more than 67%.

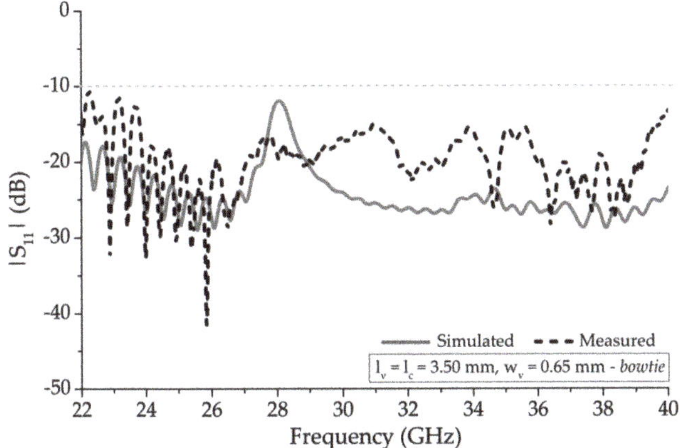

Figure 8. Simulated and measured $|S_{11}|$ of 13 unit-cell element periodic structure with bowtie vertical unit-cell segments.

Figure 9 shows the realized gain plot of the antenna structure with the modified bowtie unit-cell shown in Figure 2. The sharp decline in realized gain around 28.0 GHz region, shown in Figure 7, considerably improved with this arrangement using the parameters presented in Table 1. The gain profile gradually increased around 23.0 GHz, and onwards, with peak realized gain of 15.30 dBi at 35.0 GHz.

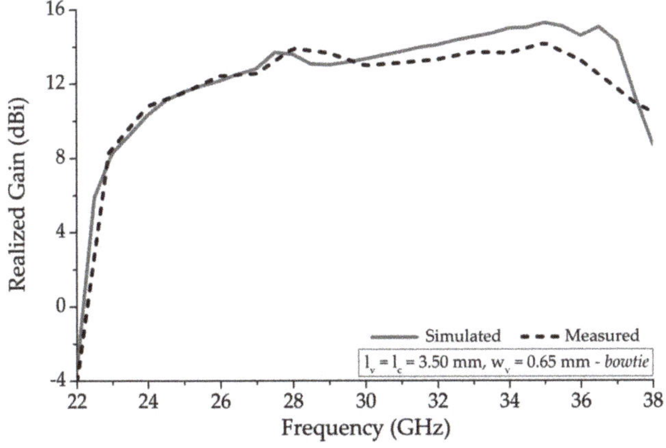

Figure 9. Simulated and measured realized gain comparison of a 13 unit-cell element periodic structure with w_v = 0.76 mm and a modified bowtie vertical unit-cell segments.

The 3D radiation patterns for the simulation of the 13-element structure, with bowtie and semi-circular unit-cell, are presented for backward endfire, broadside and forward endfire regions in Figure 10a–c, respectively. The patterns, at 24.0 GHz, 28.0 GHz and 35.0 GHz, showed a fan-beam scanning with an increase in frequency and radiation angles of −42°, 0° and 56°, respectively. The scanning range of the proposed antenna is presented in Figure 11. Figure 11a shows the scanning range from backward endfire approaching towards the broadside. The antenna scanned from −64° at 22.87 GHz. Figure 11b shows the scanning range from the broadside to the forward endfire. The antenna scanned seamlessly through the broadside, due to the mitigation of OSB, until 76°, i.e., 37.0 GHz.

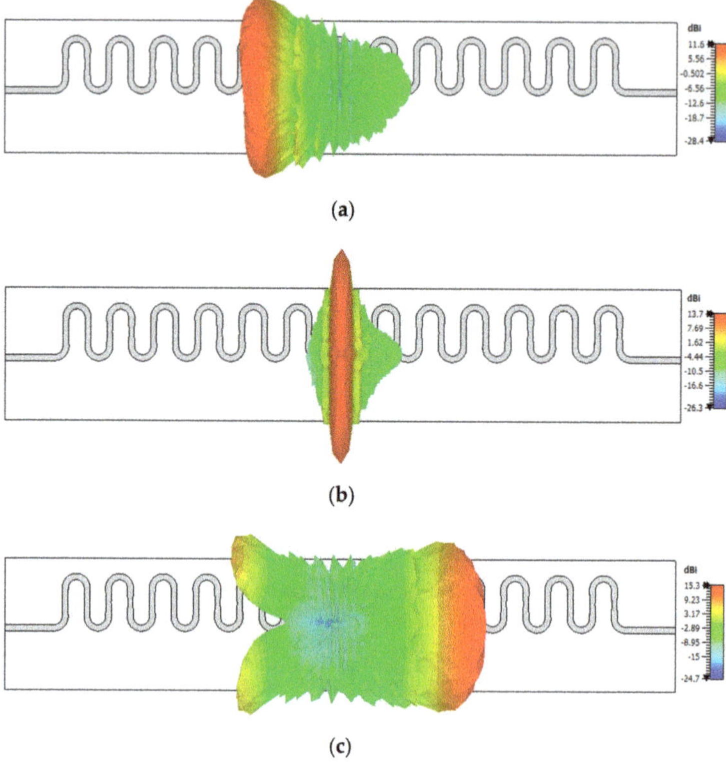

Figure 10. Three-dimensional radiation pattern visualizing scanning at (**a**) 24.0 GHz, (**b**) 28.0 GHz, and (**c**) 35.0 GHz.

Figure 12 shows the main beam direction and sidelobe level (SLL), in the yz-plane, of the proposed antenna. From Figure 8, it can be seen that the antenna has wide bandwidth below 22.87 GHz and above 37.0 GHz, but these frequencies are not considered as part of the scanning range in Figure 11 because an SLL threshold of −10 dB is maintained to efficiently define the scanning region which. As the mainlobe of the scanning range approaches forward endfire after 76°, the rise in the antenna's backlobe and the increase in SLL makes it unsuitable to scan in a single direction as efficient as the rest of the considered bandwidth.

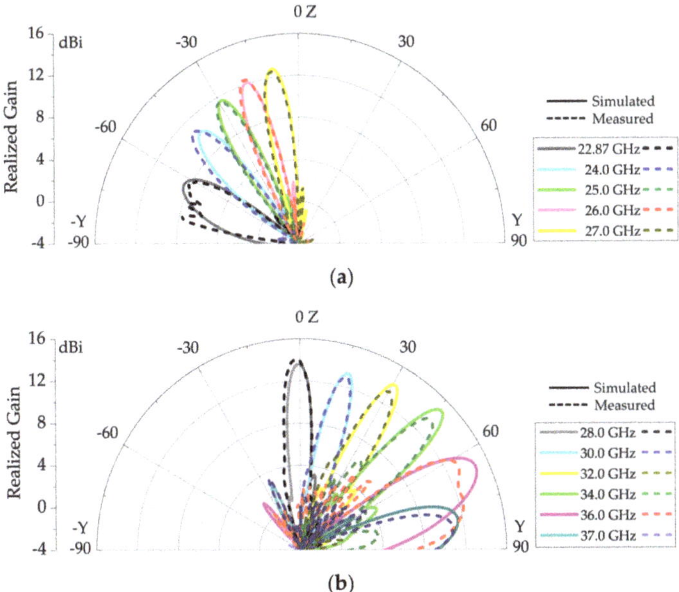

Figure 11. Scanning range of the proposed 1-D periodic modified BAA-LWA with bowtie and semi-circular unit-cell (**a**) backward quadrant and (**b**) forward quadrant.

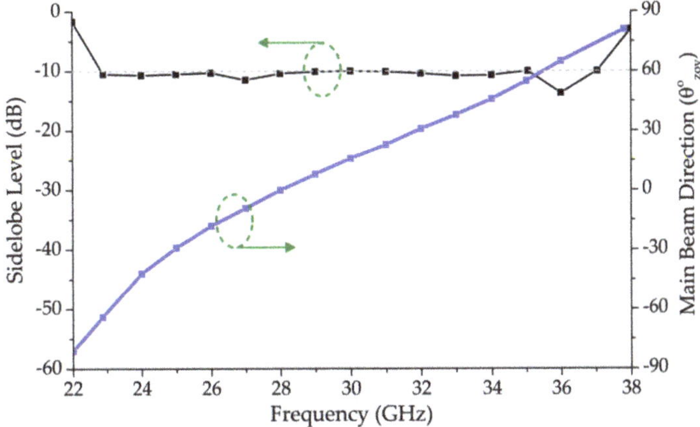

Figure 12. Sidelobe level and main beam direction in the yz-plane of the proposed LWA.

Figure 13 shows the radiation efficiency of the proposed antenna array. The antenna had more than 60% radiation efficiency throughout the scanning range, and, between 25.0 GHz and 37.0 GHz, the efficiency was more than 80%. The HPBW in both xz and yz-planes, across the entire scanning range, is also shown in the figure. As the beam approaches broadside from backward endfire, the HPBW in xz-plane increased and stabilized before dropping again as it approached forward endfire shown in Figure 10c. From the yz-plane, it can be seen that the antenna had a narrow radiation beam throughout the scanning range which can be classified as fan-beam radiation pattern.

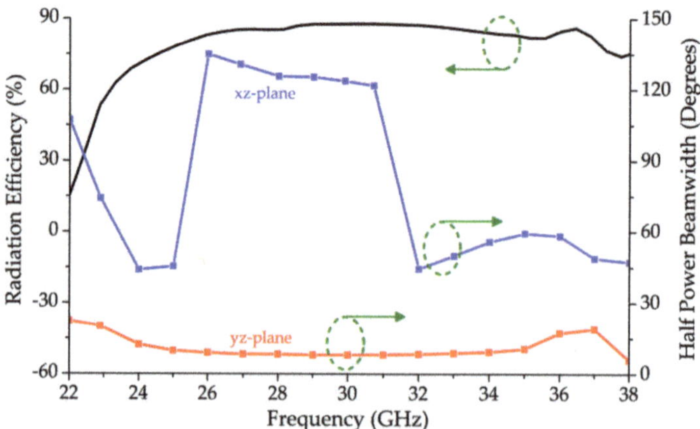

Figure 13. Radiation efficiency and half power beamwidth (HPBW) plots of the proposed LWA.

The performance of the proposed antenna's scanning characteristics was compared with other antennas in the scientific literature and is presented in Table 2. The proposed antenna had wider bandwidth and scanning range along with the peak realized gain than [34–37] and in [35], only the SLL at the broadside is mentioned. There is no mention of SLL threshold for efficient scanning in any articles except [6,27,38,39]. The presented antenna had better scanning range than the first continuous scanning range of [6]. In [39], although the realized gain and bandwidth are better, it has overall dimensions of $133 \times 93 \times 21$ mm^3, with a narrower scanning range; it is also difficult to fabricate geometry compared to the presented structure. If not for the SLL ≤ -10 dB threshold maintained throughout this work, the proposed antenna may have had a wider scanning range due to the available bandwidth below 22.87 GHz in the backward endfire and above 37.0 GHz in the forward endfire. The proposed antenna can thus be used for 28 GHz 5G and Ka-Band millimeter-wave imaging applications [40].

Table 2. Table of performance comparison between proposed LWA and LWAs in literature.

Ref.	Antenna Type	Total Length (λ_0)	Bandwidth (GHz)	Scanning Range	Realized Gain (dBi)	SLL Threshold (dB)
[6]	Periodic dual BAA	6.80	18.0–38.63	−38° to 54° −32° to−5°	10.60–16.44 16.20–18.61	10
[16]	Periodic microstrip	3.68	3.70–6.80	55° to 63°	10.0 (Peak)	N/A
[20]	Multilayered CRLH TL	11.58	20.0–30.0	−25° to 50°	10.0–14.0	N/A
[27]	Half-mode SIW	≈9.30	13.50–16.50	−30° to 30°	10.0 (Peak)	−10
[34]	Periodic CRE	N/A	13.0–19.45	−48° to 35°	≈12.0–14.0	N/A
[35]	Periodic microstrip	7.48	20.0–29.0	−50° to 45°	12.20 (Peak)	−13 (Broadside)
[36]	Metasurface LWA	15.85	15.50–16.20	47° to 7°	15.10 (Peak)	N/A
[37]	* Periodic HW-MLWA	3.99	4.40–8.80	144° to 41°	≈1.0–8.0	N/A
[38]	Periodic combline	29.49	8.0–11.40	−25° to 10°	21.0 (Peak)	−10
[39]	WG-CTS	15.08	26.0–42.0	−56° to 2°	≈22.90–29.20	−12.6
This Work	Bowtie and semi-circular periodic BAA	7.80	22.87–37.0	−64° to 76°	8.57–15.30	−10

CRE—complimentary radiation elements; TL—transmission line; HW-MLWA—half width microstrip LWA; WG—waveguide; CTS—continuous transverse stub; N/A—not available; * OSB from 5.30 GHz to 6.20 GHz.

5. Conclusions

A wide backward to forward endfire scanning leaky-wave antenna is proposed in this paper, as well as a discussion of the results. The initial concept to design the unit-cell of the antenna is taken from meandered wire Bruce array antenna and transformed to printed geometry. The horizontal and vertical segments of the meandered unit-cell were replaced with semi-circular and bowtie segments, respectively, of which the latter assists in the mitigation of the open-stopband at broadside. The length and diameter of both vertical and horizontal segments, respectively, are kept equal at 3.50 mm. The proposed antenna has a wide operational bandwidth from below 22.0 GHz to above 38.0 GHz; however, an SLL threshold of −10 dB was enforced to define an efficient scanning range between 22.87 GHz and 37.0 GHz. The 13 unit-cell periodic antenna has a compact size, offers a scanning range between −64° to 76°, and has peak gain of 15.30 dBi.

Author Contributions: Conceptualization, Z.A.; methodology, Z.A.; software, Z.A.; validation, Z.A.; formal analysis, Z.A.; investigation, Z.A.; resources, Z.A., P.M., and M.J.A.; data curation, Z.A.; writing—original draft preparation, Z.A.; writing—review and editing, Z.A., P.M., and M.J.A.; visualization, Z.A.; supervision, P.M. and M.J.A.; project administration, Z.A. and M.J.A.; funding acquisition, M.J.A. All authors have read and agreed to the published version of the manuscript.

Funding: This publication has emanated from research conducted with the financial support of Science Foundation Ireland (SFI) and is co-funded under the European Regional Development Fund under Grant Number 13/RC/2077.

Conflicts of Interest: The authors declare no conflict of interest.

References

1. Bruce, E. Aerial System. U.S. Patent 1813143, 7 July 1931.
2. Straw, R.D. Multielement Arrays. In *The ARRL Antenna Book*, 19th ed.; The American Radio Relay League: Newington, CT, USA, 2000; pp. 8–47.
3. Nakano, H.; Odachi, N.; Mimaki, H.; Yamauchi, J. An array of Franklin and Bruce antennas. In Proceedings of the IEEE Antennas and Propagation Society International Symposium: 1996 Digest, Baltimore, MD, USA, 21–26 July 1996; IEEE: Piscataway, NJ, USA, 1996; Volume 2, pp. 1130–1133.
4. Chen, S. Broadband Slot-Type Bruce Array Fed by a Microstrip-to-Slotline T-Junction. *IEEE Antennas Wirel. Propag. Lett.* **2009**, *8*, 116–119. [CrossRef]
5. Ahmed, Z.; Hoang, M.H.; McEvoy, P.; Ammann, M.J. Millimetre-wave Planar Bruce Array Antenna. In Proceedings of the 2020 International Workshop on Antenna Technology (iWAT), Bucharest, Romania, 25–28 February 2020; pp. 1–3.
6. Ahmed, Z.; John, M.; McEvoy, P.; Ammann, M.J. Investigation of Frequency Scanning Printed Bruce Array Antenna. *IEEE Access* **2020**, *8*, 189003–189012. [CrossRef]
7. Hansen, W.W. Radiating Electromagnetic Wave Guide. U.S. Patent 2402622, 25 June 1946.
8. Hines, J.N.; Upson, J.R. *A Wide Aperture Tapered-Depth Scanning Antenna*; Report 667-7; Ohio State Univ. Res. Found.: Columbus, OH, USA, 1957.
9. Trentini, G.V. Partially reflecting sheet arrays. *IRE Trans. Antennas Propag.* **1956**, *4*, 666–671. [CrossRef]
10. Rotman, W.; Karas, N. The sandwich wire antenna: A new type of microwave line source radiator. In Proceedings of the 1958 IRE International Convention Record, New York, NY, USA, 21–25 March 1957; pp. 166–172.
11. Rotman, W.; Oliner, A. Asymmetrical trough waveguide antennas. *IRE Trans. Antennas Propag.* **1959**, *7*, 153–162. [CrossRef]
12. Oliner, A.A. Historical Perspectives on Microwave Field Theory. *IEEE Trans. Microw. Theory Tech.* **1984**, *32*, 1022–1045. [CrossRef]
13. Roederer, A.; Farr, E.; Foged, L.J.; Francis, M.; Hansen, R.; Haupt, R.; Warnick, K. *IEEE Standard for Definitions of Terms for Antennas. IEEE Std 145-2013 (Revision of IEEE Std 145-1993)*; IEEE: New York, NY, USA, 2014; pp. 1–50.
14. Xu, F.; Wu, K. Understanding Leaky-Wave Structures: A Special Form of Guided-Wave Structure. *IEEE Microw. Mag.* **2013**, *14*, 87–96. [CrossRef]

15. Huo, X.; Wang, J.; Li, Z.; Li, Y.; Chen, M.; Zhang, Z. Periodic Leaky-Wave Antenna with Circular Polarization and Low-SLL Properties. *IEEE Antennas Wirel. Propag. Lett.* **2018**, *17*, 1195–1198. [CrossRef]
16. Cheng, S.; Li, Y.; Liang, Z.; Zheng, S.; Long, Y. An Approximate Circuit Model to Analyze Microstrip Rampart Line in OSB Suppressing. *IEEE Access* **2019**, *7*, 90412–90417. [CrossRef]
17. Kandwal, A.; Nie, Z.; Wang, L.; Liu, L.W.Y.; Das, R. Realization of low profile leaky wave antennas using the bending technique for frequency scanning and sensor applications. *Sensors* **2019**, *19*, 2265. [CrossRef]
18. Yu, H.; Zhang, K.; Ding, X.; Wu, Q. A Dual-Beam Leaky-Wave Antenna Based on Squarely Modulated Reactance Surface. *Appl. Sci.* **2020**, *10*, 962. [CrossRef]
19. Xie, S.; Li, J.; Deng, G.; Feng, J.; Xiao, S. A Wide-Angle Scanning Leaky-Wave Antenna Based on a Composite Right/Left-Handed Transmission Line. *Appl. Sci.* **2020**, *10*, 1927. [CrossRef]
20. Jiang, W.; Liu, C.; Zhang, B.; Menzel, W. K-Band Frequency-Scanned Leaky-Wave Antenna Based on Composite Right/Left-Handed Transmission Lines. *IEEE Antennas Wirel. Propag. Lett.* **2013**, *12*, 1133–1136. [CrossRef]
21. Zhu, L. Guided-wave characteristics of periodic coplanar waveguides with inductive loading - unit-length transmission parameters, *IEEE Trans. Microw. Theory Tech.* **2003**, *51*, 2133–2138.
22. Zheng, D.; Lyu, Y.; Wu, K. Longitudinally Slotted SIW Leaky-Wave Antenna for Low Cross-Polarization Millimeter-Wave Applications. *IEEE Trans. Antennas Propag.* **2020**, *68*, 656–664. [CrossRef]
23. Lyu, Y.; Liu, X.; Wang, P.; Erni, D.; Wu, Q.; Wang, C.; Kim, N.; Meng, F. Leaky-Wave Antennas Based on Noncutoff Substrate Integrated Waveguide Supporting Beam Scanning From Backward to Forward. *IEEE Trans. Antennas Propag.* **2016**, *64*, 2155–2164. [CrossRef]
24. Zhang, G.; Zhang, Q.; Chen, Y.; Murch, R.D. High-Scanning-Rate and Wide-Angle Leaky-Wave Antennas Based on Glide-Symmetry Goubau Line. *IEEE Trans. Antennas Propag.* **2020**, *68*, 2531–2540. [CrossRef]
25. Yang, Z.; Guan, D.; Zhang, Q.; You, P.; Huang, X.; Hou, X.; Xu, S.; Yong, S. Low-Loss Spoof Surface Plasmon Polariton Based on Folded Substrate Integrated Waveguide. *IEEE Antennas Wirel. Propag. Lett.* **2019**, *18*, 222–225. [CrossRef]
26. Baccarelli, P.; Paulotto, S.; Jackson, D.R.; Oliner, A.A. Analysis of printed periodic structures on a grounded substrate: A new Brillouin dispersion diagram. In Proceedings of the IEEE MTT-S International Microwave Symposium Digest 2005, Long Beach, CA, USA, 17 June 2005; p. 4.
27. Rezaee, S.; Memarian, M. Analytical Study of Open-Stopband Suppression in Leaky-Wave Antennas. *IEEE Antennas Wirel. Propag. Lett.* **2020**, *19*, 363–367. [CrossRef]
28. Jackson, D.R.; Caloz, C.; Itoh, T. Leaky-Wave Antennas. *Proc. IEEE* **2012**, *100*, 2194–2206. [CrossRef]
29. Cao, W.; Chen, Z.N.; Hong, W.; Zhang, B.; Liu, A. A Beam Scanning Leaky-Wave Slot Antenna with Enhanced Scanning Angle Range and Flat Gain Characteristic Using Composite Phase-Shifting Transmission Line. *IEEE Trans. Antennas Propag.* **2014**, *62*, 5871–5875. [CrossRef]
30. Yang, Q.; Zhao, X.; Zhang, Y. Design of CRLH Leaky-Wave Antenna with Low Sidelobe Level. *IEEE Access* **2019**, *7*, 178224–178234. [CrossRef]
31. Zhou, W.; Liu, J.; Long, Y. Investigation of Shorting Vias for Suppressing the Open Stopband in an SIW Periodic Leaky-Wave Structure. *IEEE Trans. Microw. Theory Tech.* **2018**, *66*, 2936–2945. [CrossRef]
32. Liu, J.; Zhou, W.; Long, Y. A Simple Technique for Open-Stopband Suppression in Periodic Leaky-Wave Antennas Using Two Nonidentical Elements Per Unit Cell. *IEEE Trans. Antennas Propag.* **2018**, *66*, 2741–2751. [CrossRef]
33. Otto, S.; Rennings, A.; Solbach, K.; Caloz, C. Transmission Line Modeling and Asymptotic Formulas for Periodic Leaky-Wave Antennas Scanning Through Broadside. *IEEE Trans. Antennas Propag.* **2011**, *59*, 3695–3709. [CrossRef]
34. Lyu, Y.; Meng, F.; Yang, G.; Wu, Q.; Wu, K. Leaky-Wave Antenna with Alternately Loaded Complementary Radiation Elements. *IEEE Antennas Wirel. Propag. Lett.* **2018**, *17*, 679–683. [CrossRef]
35. Rahmani, M.H.; Deslandes, D. Backward to Forward Scanning Periodic Leaky-Wave Antenna with Wide Scanning Range. *IEEE Trans. Antennas Propag.* **2017**, *65*, 3326–3335. [CrossRef]
36. Zhang, A.; Yang, R.; Li, D.; Hu, B.; Lei, Z.; Jiao, Y. Metasurface-Based Tapered Waveguide Slot Array Antennas for Wide Angular Scanning in a Narrow Frequency Band. *IEEE Trans. Antennas Propag.* **2018**, *66*, 4052–4059. [CrossRef]
37. Li, Y.; Xue, Q.; Tan, H.; Long, Y. The Half-Width Microstrip Leaky Wave Antenna with the Periodic Short Circuits. *IEEE Trans. Antennas Propag.* **2011**, *59*, 3421–3423. [CrossRef]

38. Williams, J.T.; Baccarelli, P.; Paulotto, S.; Jackson, D.R. 1-D Combline Leaky-Wave Antenna with the Open-Stopband Suppressed: Design Considerations and Comparisons with Measurements. *IEEE Trans. Antennas Propag.* **2013**, *61*, 4484–4492. [CrossRef]
39. You, Y.; Lu, Y.; You, Q.; Wang, Y.; Huang, J.; Lancaster, M.J. Millimeter-Wave High-Gain Frequency-Scanned Antenna Based on Waveguide Continuous Transverse Stubs. *IEEE Trans. Antennas Propag.* **2018**, *66*, 6370–6375. [CrossRef]
40. Sheen, D.M.; McMakin, D.L.; Hall, T.E. Three-dimensional millimeter-wave imaging for concealed weapon detection. *IEEE Trans. Microw. Theory Tech.* **2001**, *49*, 1581–1592. [CrossRef]

Article

Compact N-Band Tree-Shaped Multiplexer-Based Antenna Structures for 5G/IoT Mobile Devices

Amélia Ramos [1,2,*], Tiago Varum [1,2] and João N. Matos [1,2]

1 Instituto de Telecomunicações, Campus Universitário de Santiago, 3810-193 Aveiro, Portugal; tiago.varum@ua.pt (T.V.); matos@ua.pt (J.N.M.)

2 Campus Universitário de Santiago, Universidade de Aveiro, 3810-193 Aveiro, Portugal

* Correspondence: ameliaramos@ua.pt; Tel.: +351-234-377-900

Received: 9 October 2020; Accepted: 7 November 2020; Published: 8 November 2020 

Abstract: This paper presents a simple, compact and low-cost design method that allows one to obtain low-profile multi-band antennas for the overcrowded future generation networks, which are widely versatile and very heterogeneous in the K/Ka bands. The proposed antennas comprise n radiating monopoles, one for each of the desired operating frequencies, along with a frequency selective feeding network fed at a single point. This concept enables a single antenna to be shared with different radio-frequency (RF) frontends, potentially saving space. Typically, with *n-band* structures the biggest challenge is to make them highly efficient and here this is assured by multiplexing the frequency, and thus isolating each of the monopoles, allowing the design of scalable structures which fit the 5G applications. Based on the vision proposed here, a dual-band and a tri-band structures were built and characterized by their main parameters. Both prototypes achieved peak efficiencies around 80%, with adequate bandwidths and gains, as well as great compactness.

Keywords: frequency multiplexed; IoT; millimeter-waves; multi-band; n-band antenna; antenna as a sensor

1. Introduction

It is clear that the global tendency of exchanged data traffic is growing exponentially and, with it, the number of interconnected devices, pushing existing systems to their limits. In addition to the increasing traffic on each network, the variety of the services supported is increasing too, which naturally gives rise to great concerns about energy consumption. In this sense, 5G wireless systems must fulfil three main requirements: (i) have high throughput; (ii) serve many users simultaneously; and (iii) have less energy consumption [1], the latter being probably the biggest driving force of 5G. Associated with this, the high user mobility will force new antenna designs and new concepts to be implemented [2].

In addition, migration to mmWaves becomes mandatory, because despite implying higher propagation issues, it is probably the most effective way to achieve the necessary bandwidth [3]. In these frequencies, new challenges arise, given the reduced dimensions, however, there is an inherent opportunity to produce compact solutions and an increasing demand for robust multipurpose dual-band or multi-band antenna systems for 5G applications. Today, mobile devices comprise a wide range of applications and features, most of them involving several communication frontends, and also multiple antennas, requiring space, which can create problems such as the coupling between them.

A possible solution to combat the lack of available space resulting from this growing incorporation of communication systems, is to develop a single antenna, operating in each frequency band, and shared by the various communication systems. Figure 1 clarifies the scenario referring to a mobile device where a single antenna structure interacts with multiple radio-frequency frontends, which can be one of the main characteristic nowadays, since even a common smartphone gathers multiple antennas, enabling the connection with the most varied services.

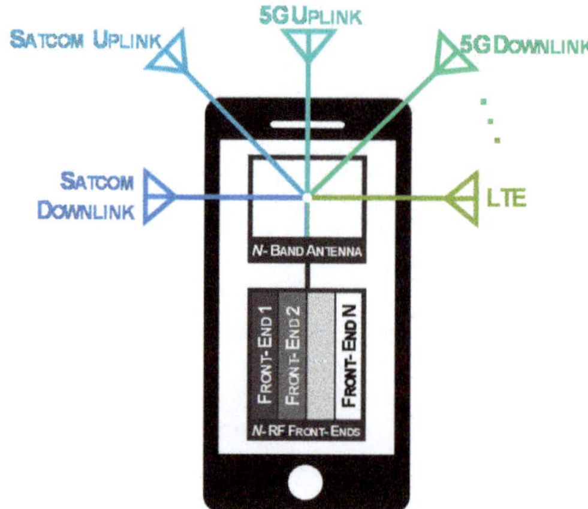

Figure 1. Single N-Band Antenna Shared by Multiple Frontends.

There are several ways to design an antenna that meets these requirements. One is through the bandwidth, i.e. producing an ultra-wideband antenna, however, it this (obtaining a band that includes all the necessary frequencies) is an almost impossible condition, especially given the disparity of frequency bands used. In addition, isolation between bands and frontends would not be guaranteed, decreasing the efficiency.

Another possibility is to develop antennas resonating in different bands, which are the commonly named multi-band antennas. This option consists of a process with a high degree of randomness, since in most cases this is done at the expense of deformations of the radiating structure, presenting little freedom to the control of the antenna's properties. There is also the group of reconfigurable antennas (in this case, in terms of frequency), although these antennas are much more complex because they are active and require additional control to select the operating frequency. In addition, these antennas use switches in the form of discrete elements, which consume energy. Due to all this, an alternative design is proposed.

This paper is divided into 6 sections, starting with this introduction. Section 2 presents a careful revision of the related works, focusing on multi-band antennas and in frequency-selective feeding structures. Then, in Section 3 the design principle is presented, clarifying the concept proposed. Section 4 contains all methods and presents the structures developed, each in a different sub-section. The results, both measured and simulated can be found in Section 5, as well as the prototypes built. Finally, major conclusions can be found in Section 6.

2. Related Work

In this way, several authors have tested the feasibility of reducing the number of antennas in each equipment and thereby saving space. In [2] a structure based on a dual-band slot is proposed and is designed to operate at 28/38 GHz. Along with the operating frequencies, this antenna exhibits adequate bandwidth values, as well as adequate gains for incorporation in a 5G mobile device. Despite the interesting results, this work lacks measures and the authors end up defending their proposal by comparing CST with HFSS.

In [4] another dual-band antenna in a massive MIMO arrangement is described. This structure is based on a series-fed antenna array and one can denote high gains (of more than 12 dB at each band) and good radiation patterns, however, once again, only simulation values are shown.

On the contrary, in [5] the measurement results support the work on a substrate integrated waveguide (SIW) antenna array for the Ka-band. There, authors implemented a linear array to improve the overall antenna's performance and they were able to achieve satisfactory bandwidths (lower than 5%) and proper gain measurements (around 5 dBi). The authors claim that by using the presented structure, multi-band antennas can be designed, but no implementation is shown to prove this statement.

Additionally, a dual-band dual-circularly-polarized antenna operating in the Ka-band is presented in [6]. The work is supported by simulated and measured results, however, a mediocre correspondence can be denoted between simulations and measurements. The gain, both in the uplink and downlink bands, is improved when the radiating element is implemented into an array, however measurements regarding axial ratio are poor, the complexity is huge and both the antenna's scalability and the robustness are questionable.

When it comes to tri-band antennas, neither the microstrip patch antenna presented in [7] nor the antenna array with defected ground seen in ref. [8] show any measures confirming their respective simulations. Both structures have resonances in the mmWaves region, along with suitable bandwidths and gains. In [7] the efficiency values shown are quite high, whereas in [8] these values are not mentioned.

Additionally, a scalable structure can be found in [9]. It proposes a compact dual-band and small slot antenna without compromising its performance. Measurements show good correspondence with simulations, as well as good gain, considering the typical values for slot antennas. Apart from these good results, one could argue that the major contribution of the work is its scalability, since authors defend that by adding more slots to the structure the desired multi-band behavior can be achieved. Nevertheless, the structure proposed operates at 2.4/5.2 GHz, very low frequencies for the 5G context.

Recently, an aperture-sharing integration methodology implementing a 3.5/28 GHz antenna with mmWave beam steering capability was proposed [10]. The main concept is to share the aperture of a linear 28 GHz array, comprised by four separately fed dipoles, with a 3.5 GHz dipole antenna. Favorable results were obtained regarding a stable mmWave beam at different scanning angles, meanwhile with broad impedance bandwidths in both operating bands (over 20%). Nevertheless, only one of the resonating frequencies suits the mmWave spectrum region and the overall size of the structure is close to the size of a single 3.5 GHz dipole antenna, which, in many of the future applications may be inappropriate.

In [11] an interesting approach is conducted on using the half-mode substrate integrated waveguide (HMSIW) technique to design low-profile cavity-backed multi-band antennas. Authors designed single, dual and triple band structures to validate the multi band responses which indicated favorable results on the radiation patterns' stability and the front-to-back ratio. Moreover, the radiation efficiency being higher than 80% at the operating frequencies is a quite encouraging result to test this concept of introducing U-shaped strips outside the aperture of an HMSIW cavity in the mmWaves region of the electromagnetic spectrum, since the tests made were at the C-band.

In [12], an architecture whose main objective is to solve the bandwidth limitations of phased arrays was proposed. The suggested design includes five printed quasi-Yagi antennas, which should be placed in the upper edge of a mobile device. Their placement, and the orientation of the active element and the directors are crucial to solve such bandwidth limitations. With the suggested configuration and by not using phase shifters and simply switching the feeding to one of the quasi-Yagi elements it is possible to scan the desired areas. However, this switchable antenna system results in a physically larger setup, since only one antenna is used at a time, which when compared to the phased array is a disadvantage, as in these structures, antenna area can be saved, as the whole aperture is exploited.

Another structure which explores alternative designs is presented in [13]. Considering the advantages of omnidirectional radiation patterns in communicating regardless of direction, a modified fork-shaped microstrip monopole antenna with a probe feed line shows wideband and multi-band characteristics. Here, the impedance bandwidth is improved by designing a dual-triangle portion of the ground plane, yet the resonant frequencies are quite low, suitable for example for the GSM band.

Regarding the emerging MIMO systems, in [14] a MIMO antenna system for multi-band 5G (mmWave) and wideband 4G application is shown. This structure works at triple bands (28, 37 and 39 GHz) for 5G and the wideband (1.8–2.6 GHz) for 4G. Each one of the MIMO elements consists of a slot in the ground plane and two microstrip feeding ports in the top layer (the isolation is also enhanced by using a low pass filter). Indeed, this design can work as a tapered slot antenna for 5G, covering 27.5–40 GHz or as an open-ended slot antenna for 4G covering 1.8–2.6 GHz. However, as noted, in the 5G band one denotes wideband operation, instead of multiple resonances at the frequencies of interest.

One of the major challenges of a multi-band antenna structures is its efficiency, a specification that gains further importance when operating in the mmWave region. The major contribution of this work is the microstrip feeding arrangement, which allows to section the antenna into n frequency-selective parts. In the literature, other techniques can be found to improve the feeding network's efficiency.

In [15], an antenna array consisting of five radiating elements is designed and measured, operating at the 2–4 GHz range, achieving a beamwidth around 24 degrees. The frequency selective feeding network delivers the signal to the selected elements at $2f_0$ and gradually switches the signal between elements as the frequency decreases to f_0. The major advantage of this strategy is an almost constant beamwidth over a broad frequency range.

A similar concept is presented in [16] where a six-element antenna array operating in 1.75–3.5 GHz frequency range is seen. In this report, the feeding network uses a directional filter where the adjustment of coupled-line sections allows for the flexible selection of transmission coefficients. This structure permits a constant beamwidth in an octave frequency range the signal is redirected from the center elements to the outer elements. In the end, this feeding network's achievements are reached at the expense of a single directional filter and equal split power dividers.

More recently in [17] an interesting approach on multi-band filtering slot antennas is proposed. There, authors realize a duplexing and filtering antenna by integrating a multi-band antenna and the multimode resonator. The work is validated since three antennas were designed, fabricated and measured. Although the correspondence between simulations and measurements is quite satisfactory, neither the antenna's scalability is a priority nor the operating frequency is suitable for the future generation of mobile communications. Above all, the concept proposed in [17] requires the usage of two feeding ports, which differs from the goal of this work which lies on having a single antenna, fed in a single point, capable of interacting with multiple RF frontends, isolating each frequency.

All of these suggested frequency selective networks are associated with higher complexity in the structure's design and fabrication. They demand for additional components and the antennas were designed for lower frequencies than what is expected within the 5G context. The work presented here defends an innovative concept where neither the manufacturing is compromised nor the feeding network imposes the usage of any additional components.

In this sense a new concept to design a structure for a multi-band antenna operating in the K/Ka bands is proposed. The design idea lies on sectioning a n-band antenna into n parts, having each section properly isolated (in frequency) in order to maximize the efficiency, and that is achieved by applying the adequate impedance matching as it will be explained later. This frequency multiplexing concept represents the main difference between the state-of-the-art studied and the structures proposed in this report. With this alternative it is possible to several antennas in a single structure with a single feeding point. Two prototypes were built and tested representing the cases of a dual and a tri-band antenna, ensuring the scalability to other resonances.

3. Design Principle

Devices which operate in multiple bands are highly interesting, given the wide and varied protocols that they are equipped to operate. To exemplify, the architecture of a current smartphone includes capabilities to operate with different communication technologies such as Wi-Fi, Bluetooth, GSM, 3G/4G (LTE), GPS, NFC, among others. In addition to multiple radios, these devices require several antennas, increasing interferences and couplings, and it is an ever more difficult task to accommodate them due to the reduced space available. It is in this context that a new concept for designing a multi-band antenna, is inserted.

Figure 2 presents a schematic example of a multi-band antenna for modern terminals. As noted, the antenna structure in the example operates in the downlink and uplink bands of 5G, SATCOM and LTE. In the illustration, instead of five antennas, with this concept it is possible to design a single antenna operating in all desired bands, with only one feeding point.

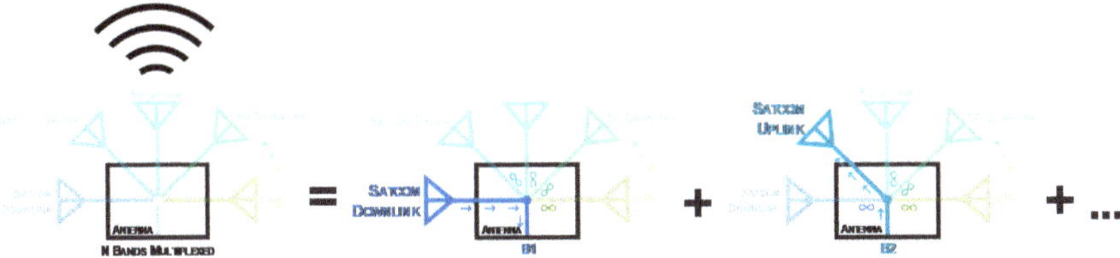

Figure 2. Schematic of the Main Design Concept.

Being an antenna shared by different radios, it is important to assure isolation between all bands, that is, that each RF frontend receives/transmits information using only its respective resonating element. The project begins with the design of all n resonant antenna elements, one for each operating frequency. After having the n radiating elements, the feeding structure ensures the high global efficiency of the antenna, forcing theoretically infinite impedances in the operating bands of neighboring elements, clearly conducting the signal towards/from the respective resonant element.

As an example, Figure 2 can be analyzed in depth, where in the middle schematic the user wants to receive data in the SATCOM downlink band, and thus the path is clear for the respective frequency band B1, and for all the other operating bands the impedance seen will be infinite. This concept is scalable for the n different bands of the antenna. In short, this concept allows to design a *n-band* structure as efficient as the summation of several resonant antennas in each of the n bands.

4. Methods and Structures

The great advantage of having the prototype composed of n sections is to guarantee that for each operation frequency, the input impedance is only imposed by the respective resonant element. This leads to a very efficient feeding network, in line with the challenging requirements of 5G, massive IoT and the future satellite communications.

Nowadays, wireless communications systems are mainly equipped using printed antennas [18,19] mostly due to their low cost and ease of fabrication. Another important aspect is the ability to produce various structures, with unlimited design shapes and typologies. Bearing in mind the need to ensure communication regardless of direction, monopole and dipole-based structures present interesting advantages.

In order to verify the practicability of this concept, it was decided to demonstrate it, designing two prototypes, a dual band and a tri-band antenna. Starting with the dual-band model, it is composed by two printed monopoles, each resonant at a different frequency, 28 GHz and 38 GHz respectively. Additionally, confirming the scalability of the concept, a tri-band antenna was developed, and for this, a radiator at 20 GHz was added. Both schemes are presented in Figure 3, clarifying the arrangement and the impedances expected in different parts of the structure.

The key aspect of these antennas is the feeding structure. The sectioning of this antenna into different parts lies on the existence of the connection points presented in Figure 3, thus, to scale this concept to a *n-band* antenna, *n-1* connection points are required. These connection points represent the division of the feeding network. In the example, for a dual-band antenna, there is one single connection point, however, for the tri-band structure, two connection points are required, and so on.

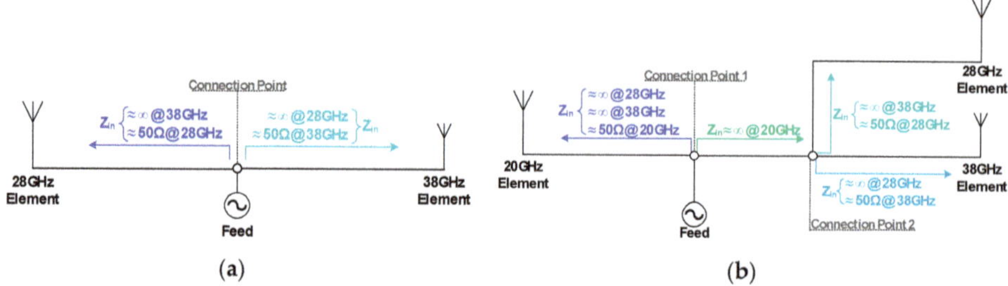

Figure 3. Schemes Proposed for the (**a**) Dual-Band and (**b**) Tri-Band antenna.

The resulting parts of the overall antenna have an adequate matching at the resonant frequency of the respective radiating element, while simultaneously showing a close to infinite impedance at the frequency of the opposite radiating element. Therefore, at each resonant frequency, a parallel association of infinite impedances with a characteristic impedance of interest is observable.

Clarifying with Figure 3a, at the connection point and looking at the 28 GHz element: at the resonant frequency, 28 GHz, an adequate matching is verified ensuring that it works properly, and, at the same time, the feeding network warrants that at 38 GHz a very high impedance appears. These principles are applied symmetrically to the opposite side of the structure.

Both antennas proposed are designed over a single layer of the dielectric substrate Rogers RO4350B, which main properties include a dielectric constant $\varepsilon_r = 3.48$, thickness $h = 0.254$ mm and dissipation factor of $\tan(\delta) = 0.0037$ @ 10 GHz. All simulations have been carried out using the electromagnetic simulator software which is the Computer Simulation Technology Microwave Studio Suite (CST-MWS).

4.1. Dual-Band Antenna Design

Figure 4 identifies the main design parameters of the dual-band structure. Two monopoles with dimensions *LMp28* × *WMp28* and *LMp38* × *WMp38* were designed. A 50 Ω impedance microstrip line, along with a quarter-wavelength transformer lead the input signal to the connection point referred above. Additionally, in order to maximize the structure's compactness, meander lines were used for the greater stubs.

From the connection point up to the monopoles, there is a group of microstrip lines which guarantees the adequate matching (at the desired frequency) and the close to infinite impedance at the adjacent operating band.

The acceptable matching is achieved by the colored identified lines highlighted in Figure 4a. These lines' dimensions and placement were implemented bearing the main concepts of microwave propagation. Starting with the open-ended stub named *Stub 1*, which has a 90° electrical length of 90° at 38 GHz, it turns infinite impedance into a short circuit. Naturally, *Line 2*, having the same electrical length at 38 GHz, converts the above-mentioned short circuit once again into a very high impedance at 38 GHz.

Through this pair of lines, the first requirement at the connection point is guaranteed: an (close to) infinite impedance at 38 GHz. However, these lines surely also influence the impedance in this part of the antenna, at 28 GHz. Therefore, the third line, *Stub 3*, was placed to compensate for this undesired influence, ensuring the impedance matching at 28 GHz.

Symmetrically, on the other side of the connection point, there is an open-ended stub, immediately followed by another stub, both with an electrical length of 90° at 28 GHz, enabling the open circuit in the connection point at 28 GHz. Subsequently, a third stub once again compensates the influence in the antenna matching observed at 38 GHz.

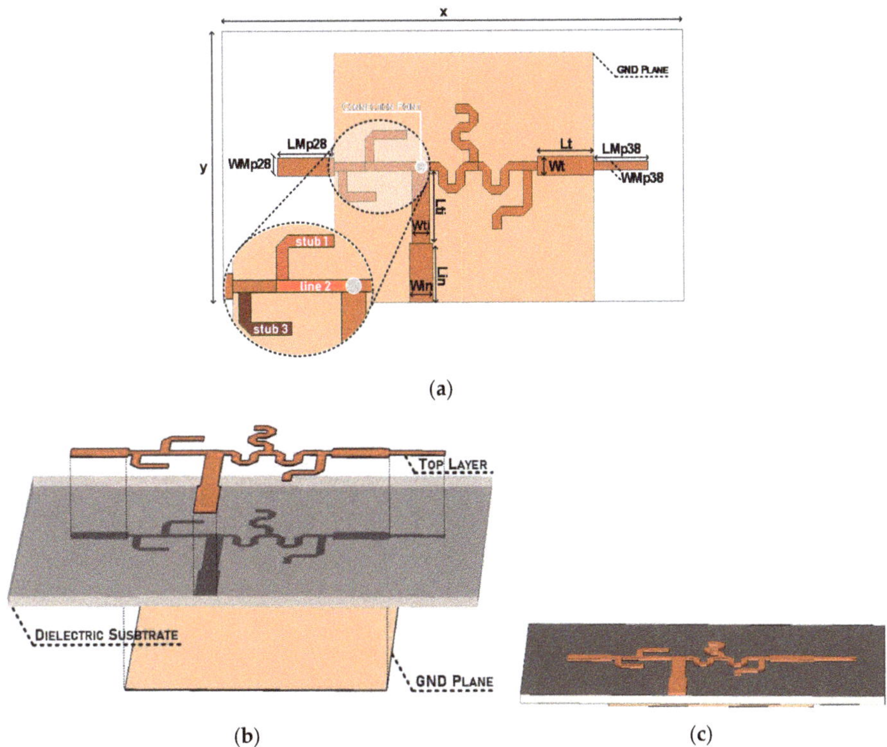

Figure 4. Dual-Band Antenna: (**a**) Design Parameters, (**b**) Layers Identification and (**c**) Final Structure.

After bearing all theoretical propagation considerations, the dual-band antenna optimized design parameters are presented in Table 1 and the feeding network line width is constant throughout the antenna (0.2 mm), with the exception of the quarter-wavelength transformer (*Lti* × *Wti)* and the 50 Ω input line (*Lin* × *Win*), connected to the single feeding point.

Table 1. Dual-Band Antenna's Dimensions.

Parameter	Dimension
LMp28	1.30
WMp28	0.40
LMp38	1.24
WMp38	0.20
Lt	1.28
Wt	0.4115
Lti	1.60
Wti	0.40
Lin	1.30
Win	0.5228
x	10.50
y	6.00

4.2. Tri-Band Antenna Design

Using the same method as before, and considering both the impedance requirements showed in Figure 3 and the design principle of Section 3, the tri-band antenna was designed over the same dielectric substrate used for the dual-band structure and its main design parameters are presented in Figure 5.

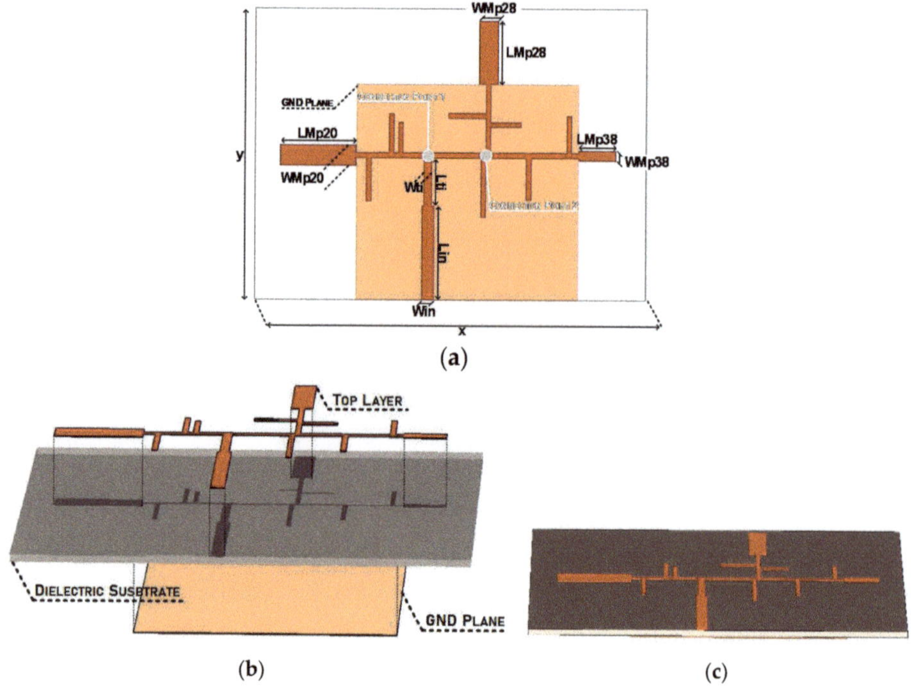

Figure 5. Tri-Band Antenna: (**a**) Design Parameters, (**b**) Layers Identification and (**c**) Final Structure.

The three resonant elements are easily denoted in Figure 5 through their associated design parameters *LMpx* × *WMpx*. The main design parameters are listed in Table 2. All microstrip lines designed have 0.2 mm width, apart from the 50 Ω input line (*Lin* × *Win*) and the quarter-wavelength transformer (*Lti* × *Wti*), as in the dual-band example. In this case, the final version of the monopoles has lengths equivalent to $0.4\lambda_d$, $0.45\lambda_d$ and $0.37\lambda_d$, regarding the 20, the 28 and the 38 GHz monopoles, respectively.

Table 2. Tri-Band Antenna's Dimensions.

Parameter	Dimension
LMp20	3.25
WMp20	0.86
LMp28	2.60
WMp28	0.80
LMp38	1.557
WMp38	0.43
Lti	2.00
Wti	0.33
Lin	3.90
Win	0.51
x	16.50
y	12.00

5. Results and Measurements

The previously described prototypes were simulated and built and they are shown in Figure 6. It is important to mention that regarding the tri-band structure, Figure 6b presents a connectorized version of the antenna, where the length of the 50 Ω input feed line is excessive and just needed to place the connector. The antennas were measured, and the main results obtained are presented throughout this section.

Figure 6. Prototypes Built: (**a**) Dual-Band and (**b**) Tri-Band Antennas.

Figure 7 exhibits the simulated and measured results of the reflection coefficient for the dual-band structure.

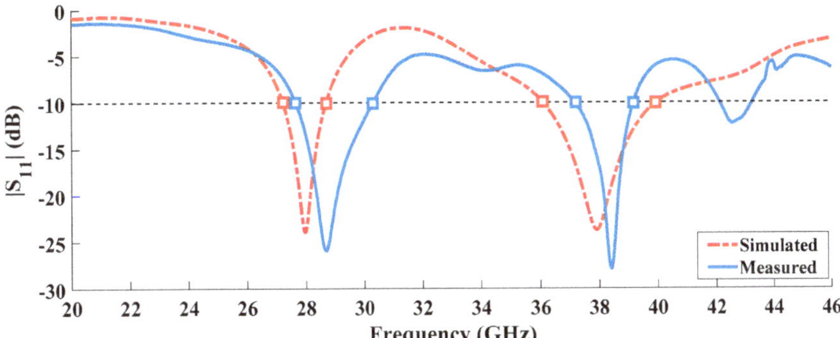

Figure 7. Simulated and Measured Reflection Coefficient of the Dual-Band Antenna.

For this prototype, a good impedance matching was achieved, with both curves presenting a similar behavior despite some frequency shifts of 680 MHz and 410 MHz from the desired 28 and 38 GHz, respectively. These small discrepancies seen might be justified either by slight construction imperfections (which is something natural given the antennas' small size). Nevertheless, the measured values of S_{11} at the operation frequencies are −13.64 dB and −16.94 dB, respectively, making of these results quite satisfactory.

Regarding the bandwidth, and considering the $S_{11} < 10$ dB criteria, the prototype achieved an operating band of 2.55 GHz [27.54–30.09 GHz], representing 8.8% bandwidth around 28 GHz. On the other hand, around 38 GHz, a 1.98 GHz [37.2–39.18 GHz] band was reached, meaning 5.2%. Table 3 summarizes the measured and simulated results of the dual-band radiating element.

Table 3. Dual-Band S_{11} Parameter Characteristics.

	Measured Frequency Shift	Simulated Bandwidth	Measured Bandwidth
Around 28 GHz	680 MHz	1.48 GHz [27.22–28.7 GHz] 5.3%	2.55 GHz [27.54–30.09 GHz] 8.8%
Around 38 GHz	410 MHz	3.84 GHz [36.08–39.92 GHz] 10.1%	1.98 GHz [37.2–39.18 GHz] 5.2%

The second prototype built proves the concept here proposed in a tri-band antenna. Figure 8 identifies a reasonable impedance matching between simulated and measured results in respect to the reflection coefficient, since the minimum value for S_{11} occurs at the frequencies of interest, despite the slight frequency shifts around 20 GHz and 38 GHz.

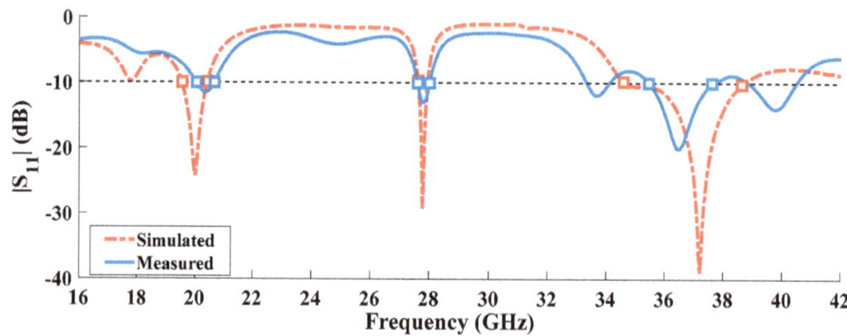

Figure 8. Simulated and Measured Reflection Coefficient of the Tri-Band Antenna.

These minor mismatches can be once again justified by the same reasons presented for the previous prototype: construction imperfections, or dielectric permittivity variations. Regarding the first operation band (at 20 GHz), the antenna shows a measured bandwidth of 0.62 GHz, equivalent to 3%. At the other operation frequencies, 0.4 GHz and 2.16 GHz are seen at 28 and 38 GHz respectively. Table 4 summarizes the measured and simulated results of the tri-band radiating structure.

Table 4. Tri-Band S_{11} Parameter Characteristics.

	Measured Frequency Shift	Simulated Bandwidth	Measured Bandwidth
Around 20 GHz	330 MHz	0.88 GHz [19.56–20.44 GHz] 4.4%	0.62 GHz [20.05–20.67 GHz] 3%
Around 28 GHz	190 MHz	0.29 GHz [27.64–27.93 GHz] 1.1%	0.4 GHz [27.61–28.01 GHz] 1.4%
Around 38 GHz	1.5 GHz	4.03 GHz [34.60–38.62 GHz] 11%	2.16 GHz [35.45–37.61GHz] 5.9%

Through the reflection coefficient of the latest prototype analyzed it is possible to confirm the adequacy of the prototypes built and more importantly the principle of design proposed seems to guarantee the impedance matching.

Other aspect that sustains the appropriate performance of these antennas in 5G and the massive IoT scenario is the information about the antennas' efficiency, which is presented in Figures 9 and 10, regarding the dual and the tri-band radiating structures, respectively.

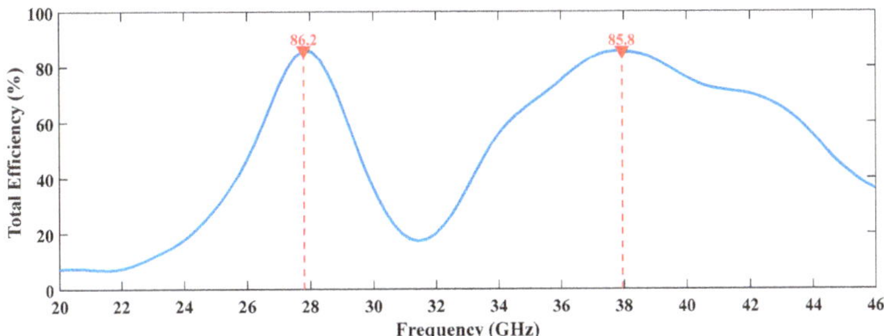

Figure 9. Efficiency Variation Over Frequency of the Dual-Band Antenna.

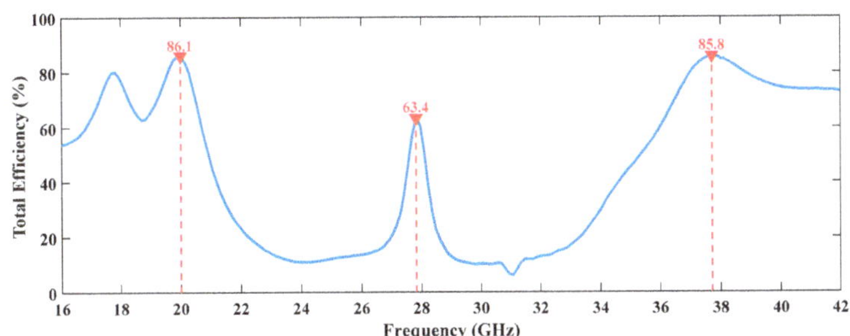

Figure 10. Efficiency Variation Over Frequency of the Tri-Band Antenna.

Figure 9 presents the simulated efficiency of the dual-band antenna over the frequency. Naturally two local maximums are highlighted. These maximums are according to expected since they appear adjacent to the frequencies of operation. Their magnitude is higher than 85% in both operating frequencies, an important value given the structure's compactness and size.

Similarly, for the tri-band antenna, there are clearly three local maximums in Figure 10, which are once again, near the frequencies of interest, in this case 20, 28 and 38 GHz, confirming the proper functioning of the antenna.

Besides the efficiency peaks, the surface current distribution also confirms the proper functioning of the feeding network, verifying once again the frequency multiplexed structure achieved. Figure 11 starts by presenting the surface current distribution at the operating frequencies of the dual-band antenna, 28 GHz and 38 GHz.

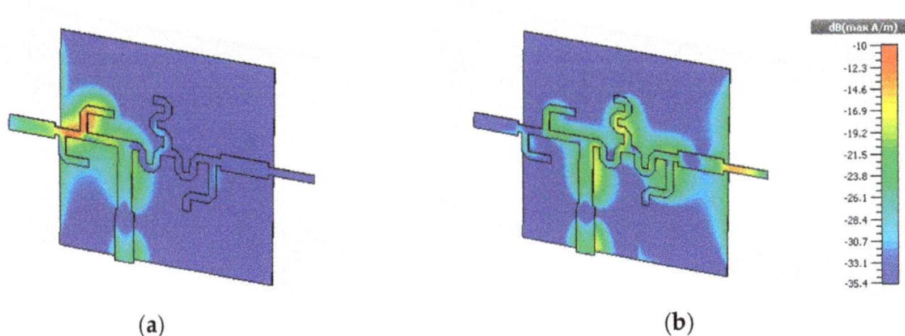

Figure 11. Surface Current Distribution at (**a**) 28 GHz and (**b**) 38 GHz.

It can be easily observed that, for both frequencies, only the respective resonant monopole is being fed. The opposite monopole has no current flowing at all.

Figure 12 exhibits the respective results for the tri-band structure, denoting the surface current distribution for the three operating frequencies.

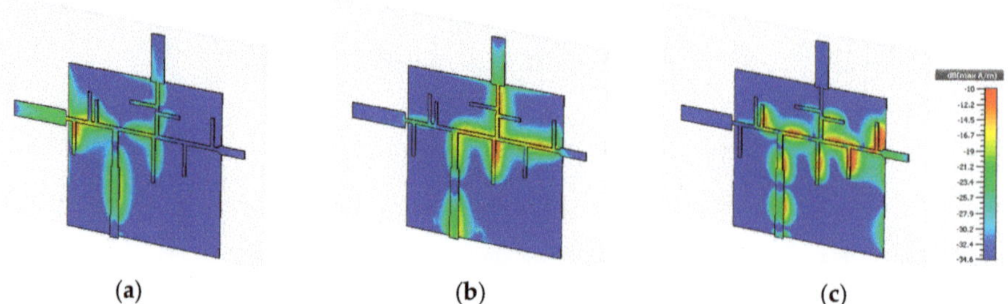

Figure 12. Surface Current Distribution at (a) 20 GHz, (b) 28 GHz and (c) 38 GHz.

Once again at each one of the resonant frequencies, only the respective monopole has any current through it, all the others remain not fed. With this, it is possible to state that the frequency multiplexing concept proposed allowed to isolate, in frequency, each radiating element, maintaining a single input feed and without requiring any additional filters.

Figures 13 and 14 show the gain variation of both prototypes over frequency, highlighting the respective values at the operating frequencies. It is possible to verify that, in the tri-band case, the gain values vary between 2.5 dBi and 4.5 dBi, in the frequencies of interest and they are mainly dependent on the antenna structure.

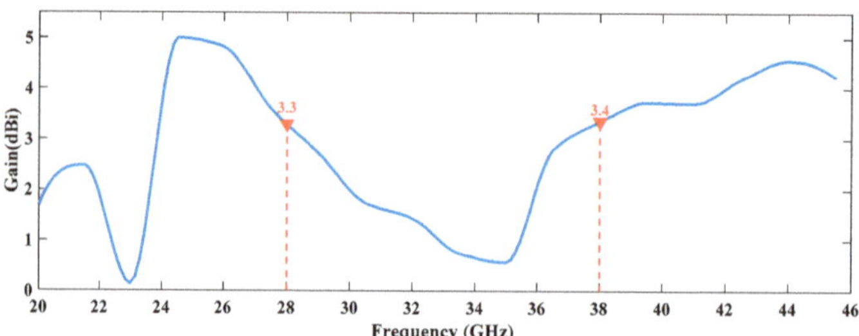

Figure 13. Gain Variation Over Frequency of the Dual-Band Antenna.

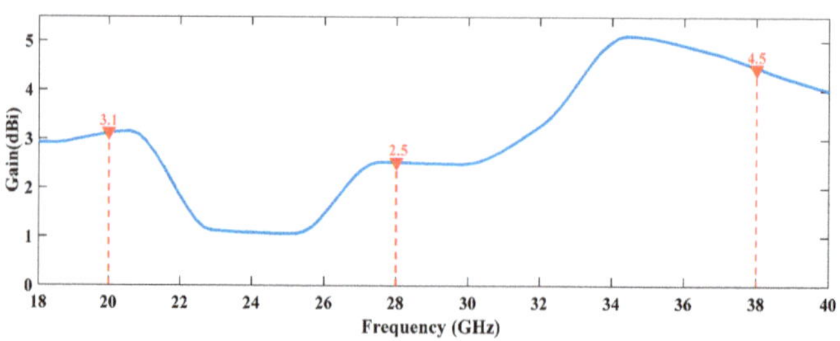

Figure 14. Gain Variation Over Frequency of the Tri-Band Antenna.

Regarding the gain of the antennas, in Figure 15 it is possible to observe the simulation and measured results of the two main radiation planes at both frequencies of interest, 28/38 GHz, for the dual band antenna.

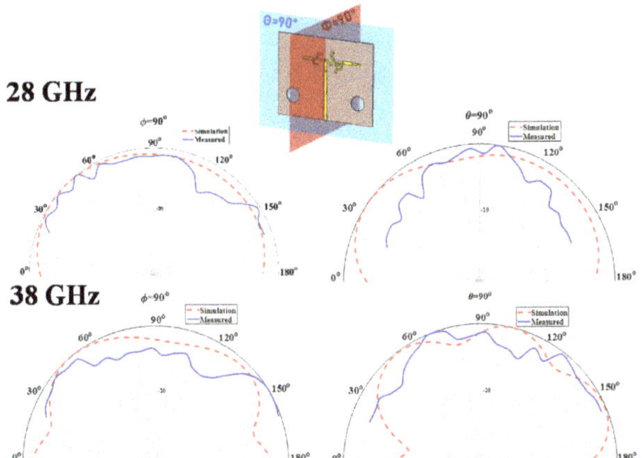

Figure 15. Simulated and Measured Normalized Radiation Pattern of the Dual-Band Antenna at 28 GHz and 38 GHz.

As it is natural in very compact structures, particularly in multi-band antennas, the radiation patterns obtained seem a bit distorted from what was expected to be seen in monopoles. Apart from that, the dual-band antenna presents a maximum simulated gain of 3.3 dBi at 28 GHz, while at 38 GHz that value is 3.4 dBi.

When it comes to the tri-band antenna, the same two radiation planes are shown, but now for the three frequencies of interest. These results are presented in Figure 16 and a similar distortion in the radiation pattern shapes is observed. This radiating structure achieved a maximum simulated gain of 3.1, 2.5 and 4.5 dBi at 20, 28 and 38 GHz.

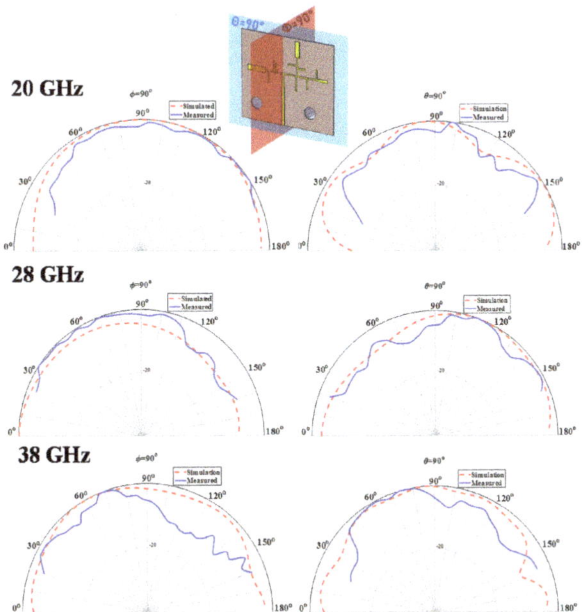

Figure 16. Simulated and Measured Normalized Radiation Pattern of the Tri-Band Antenna at 20/28/38 GHz.

The main achievements of these structures are summarized and compared with the current state-of-the-art of dual and tri-band antennas in Table 5, analyzing the publication year (PY), the operating frequencies, bandwidth, size, scalability and the isolation in frequency factor.

Table 5. State-of-the-Art of Multi-Band Antennas: Main Results.

[Ref.] PY	Antenna's Description	Operating Frequencies	Bandwidth	Measurements	Scalability	Frequency Isolation
[2] 2015	Dual-Band Slot-Based	28/38 GHz	24–30 GHz 34–41 GHz	No	No	No
[4] 2016	Series-Fed Antenna Array	28/38 GHz	27.8–28.6 GHz 36–38.6 GHz	No	No	No
[5] 2015	SIW Antenna Array	28/38 GHz	0.32 GHz 1.9 GHz	No	No	No
[6] 2017	Annular Ring Slots	20/30 GHz	20.13–20.8 GHz 30–30.85 GHz	Yes	Yes	No
[7] 2018	Microstrip Patch	24.4/28/38 GHz	23.6–24.4 GHz 27.2–28.4 GHz 37.3–38.1 GHz	No	No	No
[8] 2019	Antenna Array with Defected Ground	29/37.4/41 GHz	28.5–29.7 GHz 36.7–38 GHz 40–42.1 GHz	No	No	No
[9] 2008	Printed Slot-Based Structures	2.4/5.2 GHz	2.48–3.02 GHz 3.81–5.65 GHz	Yes	Yes	No
[10] 2020	Aperture-Sharing Technique with a 4 Unit Linear Array and a Dipole	3.5/28 GHz	3.12–3.84 GHz 24.9–30.6 GHz	Yes	No	No
[11] 2019	Cavity-Backed Antennas using HMSIW Technique	4.6/5.48 GHz	91 MHz 20 MHz	Yes	Yes	No
		4.5/5/5.5 GHz	70 MHz 20 MHz 20 MHz	Yes	Yes	No
[13] 2004	Fork-Shaped Microstrip Monopole Antenna	860/2280 MHz	731–987 MHz 1498–3080 MHz	No	No	No
This Work	Monopole-Based with Frequency Selective Feeding Network	28/38 GHz	27.54–30.09 GHz 37.2–39.18 GHz	Yes	Yes	Yes
		20/28/38 GHz	20.05–20.67 GHz 27.61–28.01 GHz 35.45–37.61 GHz	Yes	Yes	Yes

Knowing that one of the main concerns of this letter is the microstrip frequency-selective feeding network able to maximize the efficiency as well as easing he interaction of a single antenna with multiple RF Frontends, Table 6 compiles some of the state-of-the-art works related to the frequency multiplexing argued, showing how it was obtained and the main characteristics of the antenna structure.

Table 6. State-of-the-Art in Frequency Selective Antenna Structures: Main Results.

[Ref.] PY	Antenna's Description	Central Frequencies	Frequency Isolation Technique	Feeding Network Complexity	Number of Ports/Element	Does it Require Additional Components?
[14] 2019	Ground Tapered Slot and Microstrip Feeders	2/32.5 GHz	Low-Pass Filter	High	2	Yes
[15] 2016	Antenna Array	3 GHz	3 dB directional couplers + Schiffman C sections + Filters and Power Dividers	Extreme	1	Yes

Table 6. *Cont.*

[Ref.] PY	Antenna's Description	Central Frequencies	Frequency Isolation Technique	Feeding Network Complexity	Number of Ports/Element	Does it Require Additional Components?
[16] 2017	Antenna Array	2.6 GHz	Directional Filter and Power Dividers	High	1	Yes
[17] 2020	Filtering Slot Antennas	3.5/5.2 GHz	*n*-mode resonators	Medium	2	No
		2.5/3.5/5.1 GHz				
This Work	Monopole-Based with Frequency Selective Feeding Network	28/38 GHz	Impedance Matching	Low	1	No
		20/28/38 GHz				

6. Conclusions

In this paper, a new multi-band antenna design concept is proposed, explained and proved through some practical prototypes. The main basis is the use of a feeding structure which ensure the impedance matching at a given operating frequency, isolating the structure at the remaining frequencies. When compared with an eventual alternative of a radiator with dual resonance, the option here presented, besides maximizing the efficiency, it avoids the usage of any bulky filters at the input.

This strategy was implemented in a dual-band and a tri-band prototypes, both suitable to integrate in IoT sensors. Individually these antennas present good impedance bandwidths for all frequencies of interest, since the smallest measured value is 400 MHz.

Regarding the antennas' gain, and considering that these structures are based in monopoles, these values are also satisfactory, especially when allied to the efficiency values witnessed. More importantly, the proposed concept of producing a *n-band* antenna has proven not only to work, but it also assures higher efficiency values than what is commonly seen in the state-of-the-art for multi-band antennas.

Author Contributions: All the authors have contributed to this paper. A.R. has designed the proposed antennas, performed the simulations and the measurements, and wrote the paper. T.V. and J.N.M. have supervised the entire research, the approach used, the results analysis, and discussion. Additionally, both have strongly contributed to the writing of the paper. All authors have read and agreed to the published version of the manuscript.

Funding: This work is funded by FCT/MCTES through national funds and when applicable co-funded EU funds under the project UIDB/50008/2020-UIDP/50008/2020This work was partially supported by FCT/MCTES through national funds and when applicable cofunded EU funds under the project UIDB/50008/2020-UIDP/50008/2020 and the European Regional Development Fund through the Competitiveness and Internationalization Operational Program, Regional Operational Program of Lisbon, Regional Operational Program of the Algarve, in component FEDER, and the Foundation for Science and Technology, Project RETIOT, POCI-01-0145-FEDER-016432.

Conflicts of Interest: Authors declare no conflict of interest.

References

1. Ngo, H. *Massive MIMO: Fundamentals and System Designs*; Linköping University Electronic Press: Linköping, Sweden, 2015.
2. Haraz, O.; Ali, M.; Alshebeili, S.; Sebak, A. Design of a 28/38 GHz dual-band printed slot antenna for the future 5G mobile communication Networks. In Proceedings of the 2015 IEEE International Symposium on Antennas and Propagation & USNC/URSI National Radio Science Meeting, Vancouver, BC, Canada, 19–25 July 2015. [CrossRef]
3. Samsung. *5G Vision*. February 2015. Available online: https://www.samsung.com/global/business/networks/insights/whitepaper/?white-paper (accessed on 9 October 2018).
4. Ali, M.; Sebak, A. Design of compact millimeter wave massive MIMO dual-band (28/38 GHz) antenna array for future 5G communication systems. In Proceedings of the 2016 17th International Symposium on Antenna Technology and Applied Electromagnetics (ANTEM), Montreal, QC, Canada, 10–13 July 2016. [CrossRef]

5. Ashraf, N.; Haraz, O.; Ashraf, M.; Alshebeili, S. 28/38-GHz dual-band millimeter wave SIW array antenna with EBG structures for 5G applications. In Proceedings of the 2015 International Conference on Information and Communication Technology Research (ICTRC), Abu Dhabi, UAE, 17–19 May 2015. [CrossRef]
6. Mener, S.; Gillard, R.; Roy, L. A Dual-Band Dual-Circular-Polarization Antenna for Ka-Band Satellite Communications. *IEEE Antennas Wirel. Propag. Lett.* **2017**, *16*, 274–277. [CrossRef]
7. Kamal, M.; Islam, M.; Uddin, M.; Imran, A. Design of a Tri-Band Microstrip Patch Antenna for 5G Application. In Proceedings of the 2018 International Conference on Computer, Communication, Chemical, Material and Electronic Engineering (IC4ME2), Rajshahi, Bangladesh, 8–9 February 2018. [CrossRef]
8. Naqvi, S.N.; Azam, M.; Amin, Y.; Arshad, F.; Loo, J.; Tenhunen, H. Tri-Band Antenna Array with Defected Ground Structure for mm-Wave 5G Applications. In Proceedings of the 2019 IEEE 4th International Conference on Computer and Communication Systems (ICCCS), Singapore, 23–25 February 2019. [CrossRef]
9. Lee, Y.; Sun, J. Compact Printed Slot Antennas for Wireless Dual- and Multi-Band Operations. *Prog. Electromagn. Res.* **2008**, *88*, 289–305. [CrossRef]
10. Lan, J.; Yu, Z.; Zhou, J.; Hong, W. An Aperture-Sharing Array for (3.5, 28) GHz Terminals with Steerable Beam in Millimeter-Wave Band. *IEEE Trans. Antennas Propag.* **2020**, *68*, 4114–4119. [CrossRef]
11. Yang, X.; Ge, L.; Ji, Y.; Zeng, X.; Luk, K. Design of Low-Profile Multi-Band Half-Mode Substrate-Integrated Waveguide Antennas. *IEEE Trans. Antennas Propag.* **2019**, *67*, 6639–6644. [CrossRef]
12. di Paola, C.; Zhang, S.; Zhao, K.; Ying, Z.; Bolin, T.; Pedersen, G. Wideband Beam-Switchable 28 GHz Quasi-Yagi Array for Mobile Devices. *IEEE Trans. Antennas Propag.* **2019**, *67*, 6870–6882. [CrossRef]
13. Go, H.; Jang, Y. Multi-band modified fork-shaped microstrip monopole antenna with ground plane including dual-triangle portion. *Electron. Lett.* **2004**, *40*, 575. [CrossRef]
14. al Abbas, E.; Ikram, M.; Mobashsher, A.; Abbosh, A. MIMO Antenna System for Multi-Band Millimeter-Wave 5G and Wideband 4G Mobile Communications. *IEEE Access* **2019**, *7*, 181916–181923. [CrossRef]
15. Wincza, K.; Gruszczynski, S. Broadband Scalable Antenna Arrays with Constant Beamwidths Fed by Frequency-Selective Networks. *IEEE Trans. Antennas Propag.* **2016**, *64*, 2936–2943. [CrossRef]
16. Wincza, K.; Gruszczynski, S. Frequency-Selective Feeding Network Based on Directional Filter for Constant-Beamwidth Scalable Antenna Arrays. *IEEE Trans. Antennas Propag.* **2017**, *65*, 4346–4350. [CrossRef]
17. Xie, Y.; Chen, F.; Qian, J. Design of Integrated Duplexing and Multi-Band Filtering Slot Antennas. *IEEE Access* **2020**, *8*, 126119–126126. [CrossRef]
18. Ávila-Navarro, E.; Carrasco, J.; Reig, C. Design of Yagi-like printed antennas for WLAN applications. *Microw. Opt. Technol. Lett.* **2007**, *49*, 2174–2178. [CrossRef]
19. Balanis, C.A. *Antenna Theory Analysis and Design*, 4th ed.; John Wiley & Sons: New York, NY, USA, 2005.

Letter

A Single-Fed Multiband Antenna for WLAN and 5G Applications

Zakir Khan [1], Muhammad Hunain Memon [1], Saeed Ur Rahman [2], Muhammad Sajjad [2], Fujiang Lin [1] and Liguo Sun [1,*]

[1] Micro-/Nano-Electronic System Integration Center, University of Science and Technology of China, Hefei 230027, China; zakirkhan@mail.ustc.edu.cn (Z.K.); Mhm121@mail.ustc.edu.cn (M.H.M.); Linfj@ustc.edu.cn (F.L.)

[2] College of Electronic and Information Engineering, Nanjing University of Aeronautics & Astronautic, Nanjing 211106, China; saeed@nuaa.edu.cn (S.U.R.); sajjadwazir@nuaa.edu.cn (M.S.)

* Correspondence: liguos@ustc.edu.cn

Received: 22 September 2020; Accepted: 20 October 2020; Published: 6 November 2020

Abstract: In this paper, a slotted conical patch connected to a small triangular patch multiband antenna for both microwave and millimeter-wave applications is presented. The designed antenna has three characteristics. The proposed antenna is a multiband, having a compact size of $0.35\lambda_0 \times 0.35\lambda_0 \times 0.004\lambda_0$ at its lowest operational frequency, i.e., 2.4 GHz, and more importantly, it can cover both the microwave and millimeter-wave bands with a single feeding. According to the −10 dB matching bandwidth, experimental results show that the antenna operates at (2.450–2.495) GHz, (5.0–6.3) GHz, and (23–28) GHz. The reduced size, simple design, and multiband large bandwidth are some of the advantages over the reported designs in the latest literature. Both simulated and experimental results show a good agreement, and the proposed antenna can be used for wireless local area network (WLAN) applications and fifth-generation (5G) wireless communication devices.

Keywords: multibandoperation; slotted antenna; microwave; millimeter-wave band; WLAN; 5G

1. Introduction

A wireless local area network (WLAN) is a widely used network for short-range wireless communication applications. According to 802.11b/g and 802.11a standards, the bands used for WLAN are (2.400–2.484) GHz, (5.15–5.35) GHz, and (5.725–5.825) GHz [1]. Moreover, the need for high quality videos and other high data rate applications require wider bandwidth. Therefore, to fulfill the requirement of wider bandwidth, 28/38 GHz frequencies are seen as the most promising choices for fifth-generation (5G) technology [2].

Numerous researches have been exerted over the last few years pertaining to the evolution of existing standards of the wireless communication system to future 5G wireless communication standards, which is likely going to be implemented in the 2020s [3,4]. For that reason, more and more requirements have been made in the design of an antenna, in terms of size, multiband operation, and radiation pattern [5]. Many researchers are focusing on the advancement of an antenna system to operate in both current and future standards of wireless communication systems. Therefore, the acceptable way to be considered is the designing of a multiband antenna that can be integrated as a single element in many standards [6], such as WLAN, global positioning system (GPS), and other wireless communication applications.

For designing multiband antennas, different techniques were used previously to achieve multiband operating frequency standards [7–17]. The following study, of different multiband antennas covering the microwave band for WLAN applications, has been conducted in [1,7–14]. A defected ground structure (DGS) monopole antenna operating at triple frequencies for WLAN applications is presented in [6]. The radiating patch and ground of the antenna were etched on both sides of a printed-circuit board (PCB). The ground plane was modified by two equal-shaped slots on the right and left sides. Similarly, a multiband characteristic of the antenna in [9] was generated by a rectangular slot on the upper side of the antenna substrate loaded with differently shaped stubs on each side of the slot. In [10], a slotted monopole antenna, having a C-shaped patch introduced by a G-shaped parasitic strip and a partial ground plane, is used to obtain a larger bandwidth of 3.5 GHz at (3.92–7.52 GHz). Two elements of a multiple-input–multiple-output (MIMO) antenna etched with a different slot is reported in [13]. Similarly, a triple-band antenna for 2.4, 5.2, and 5.8 GHz applications in [1] and a dual-band antenna operating at 2.4 GHz and 5.2 GHz in [14] are presented. Meandering slots etched in the patch and a slotted ground DGS is used, respectively, to obtain the triple and dual-band characteristics. In [15], a 28-GHz mm-wave antenna of size 30 mm × 20 mm for 5G is reported, which is the combination of a waveguide aperture and several microstrip patches. Further, the study of antennas covering both microwave and mm-wave bands simultaneously were performed in [16,17]. In [16], a multi-layer antenna system having a dual-element MIMO on the top layer operating in the microwave band, and an antenna array at the bottom layer for the 5G band, is presented. A multiband antenna operating in both microwave and mm-wave is introduced in [17], which consists of a monopole antenna operating at 2.4/5.5 GHz and a rectangular patch covering the mm-wave 5G band. The comparison of the proposed work with the available designs of [7–17] in terms of bandwidth, multiband operation, substrate availability, design complexity, the number of layers, and feeding used is shown in Table 1. It was observed from the comparison table that the available designs have large size, complex geometry, multi-ports, and they can only cover the microwave band, or only mm-wave band, but cannot cover both bands with one feeding. Thus, the challenging part of this work is to design an antenna that can cover both the microwave band and mm-wave band with a single feeding and a compact size.

Table 1. Comparison of different multiband designs.

Ref.	Dim (mm^2)	Frequency (GHz)	Remarks
[6]	20 × 30	(2.14–2.52) (2.8–3.7) (5.15–6.02)	Cover only microwave band.
[9]	56 × 44	(1.57–1.66), (2.40–2.54), (3.27–3.97), (5.17–5.93)	Cover only microwave band.
[10]	32 × 30	(3.92–7.52)	Cannot cover 2.4 and the Mm-wave band.
[11]	24 × 16	(5.15–5.35), (5.72–5.82)	Cannot cover 2.4 and the Mm-wave band.
[15]	20 × 30	(26–29)	Only the mm-wave band.
[16]	60 × 100	(1.870–2.530) (26–28)	Complex geometry, Dual port and cannot cover 5.2, 5.8 GHz.
[17]	45 × 40	(2.16–2.53), (4.58–5.80), (26.8–30.0).	Large size, Complex geometry.
This work	30 × 30	(2.46–2.49), (5.0–6.3), (23–28)	Covers microwave, mm-wave band, with simple geometry and compact size.

To solve these problems (of large size and complex geometry), a compact, multiband antenna covering the microwave band and the mm-wave band is proposed. Rogers RT/Duroid 5880, a widely available and inexpensive substrate was used to design the proposed antenna. We modified the radiating patch by truncating the corners of the two rectangles to form a cone and a triangle. We etched different slots in the conical patch, which increased the electrical length of the antenna and made it more compact. Additionally, corners of the ground plane were truncated and cut by different slots to form a DGS, unlike the conventional solid ground plane. The mentioned design techniques were applied to make the antenna resonate at about (2.4, 5.2, 5.8) GHz and (28 GHz). To endorse the concept and validate the simulated results, a prototype is fabricated and results are measured. The simulated and measured results suggest that the designed antenna is the best candidate for various wireless communication applications in terms of multiband operations, compactness, large bandwidth, ease of design, and low cost. In Section 3, an explanation of a parametric study has been discussed to properly select optimized dimensions of the proposed design and achieve good results of multiband. In Section 4, different types of results were discussed followed by the comparison and conclusion of the paper in Sections 5 and 6.

2. Geometry of the Antenna

The detailed geometry of the proposed antenna will be discussed in this section. The three-dimensional (3D) electromagnetic wave solver, computer simulation technology (CST) microwave studio [18] was used for numerically investigating and optimizing the configuration of the proposed designed antenna. The front and back view of the antenna is depicted in Figure 1e. From figure, the blue color is the copper used at the top and bottom layer and the brown color is the substrate. Rogers RT5880 (ε_r= 2.2, tan δ = 0.0009) is used as a substrate to design the antenna. The overall dimension is $30 \times 30 \times 0.508$ mm^3. The final geometry consists of a slotted conical patch connected to a small triangle by narrow lines. The conical patch covers the WLAN band and the triangular patch covers the 5G band. Two meandering slots and a triangular slot were etched on a conical patch. Feeding is given through a 50 Ω microstrip line. A defected ground plane etched with six slots and truncated corners are used at the bottom layer of the antenna.

The final geometry of the designed antenna was obtained by different modifications in two rectangular patches antenna (Antenna1) shown in Figure 1a. The Antenna1, patch dimensions such as width and length were obtained by Equations (1) and (2) [19]. From Figure 2a,b the Antenna1 is resonating at (24.32–24.54) GHz and (25.5–25.7) GHz. It also gives resonance at (5.3–6.0) GHz but that's not below −10 dB. We used corners truncation and meandering slits for compactness and multiband operation [6,9]. These two techniques lower the frequency of operation of the antenna by increasing the electrical length that results in the compactness of the antenna [6,9,20,21]. The ground plane used at the bottom layer is defected with different meandering slots and truncated corners to get higher bandwidth in both operating bands of the antenna. More details of truncating the corners of the square patches to form a conical and triangular patch are also discussed in Section 3.1 of this article. The next stage (Antenna2) in Figure 1b had its corners truncated from rectangular patches and the ground plane to form a cone-like and a small triangle. From Figure 2a, the resonating frequency of Antenna2 is lowered to (21.80–22.42) GHz and (26.2–27.6) GHz. Moreover, resonating at 4.7 and 7.0 GHz, but the resonance was not below −10 dB. In the next stage, Antenna3, a slot of optimized value in the conical patch and two slots in the truncated ground were etched as shown in Figure 1c. From Figure 2a, the higher resonating frequency turns into a broad band, i.e., (22–26) GHz. While at the lower band, the antenna resonates at 3.9 GHz, which is below −10 dB. It also resonates at 2.60 and 5.05 GHz, but that is above −10 dB. Moving to the next stage, Antenna4, one more slot in the patch and two more slots in the ground plane were etched, shown in Figure 1d. The antenna resonates at 2.3 GHz and (5.0–6.3) GHz, but there is a mismatch at 5.2 GHz. Thus, further improvement is needed. The final stage is to etch a triangular slot in the patch along with two more slots in the ground plane, shown in Figure 1e. From Figure 2a,b it can be seen that the antenna is exactly resonating at (2.46–2.49) GHz, (5–6.3) GHz, and (23–28) GHz.

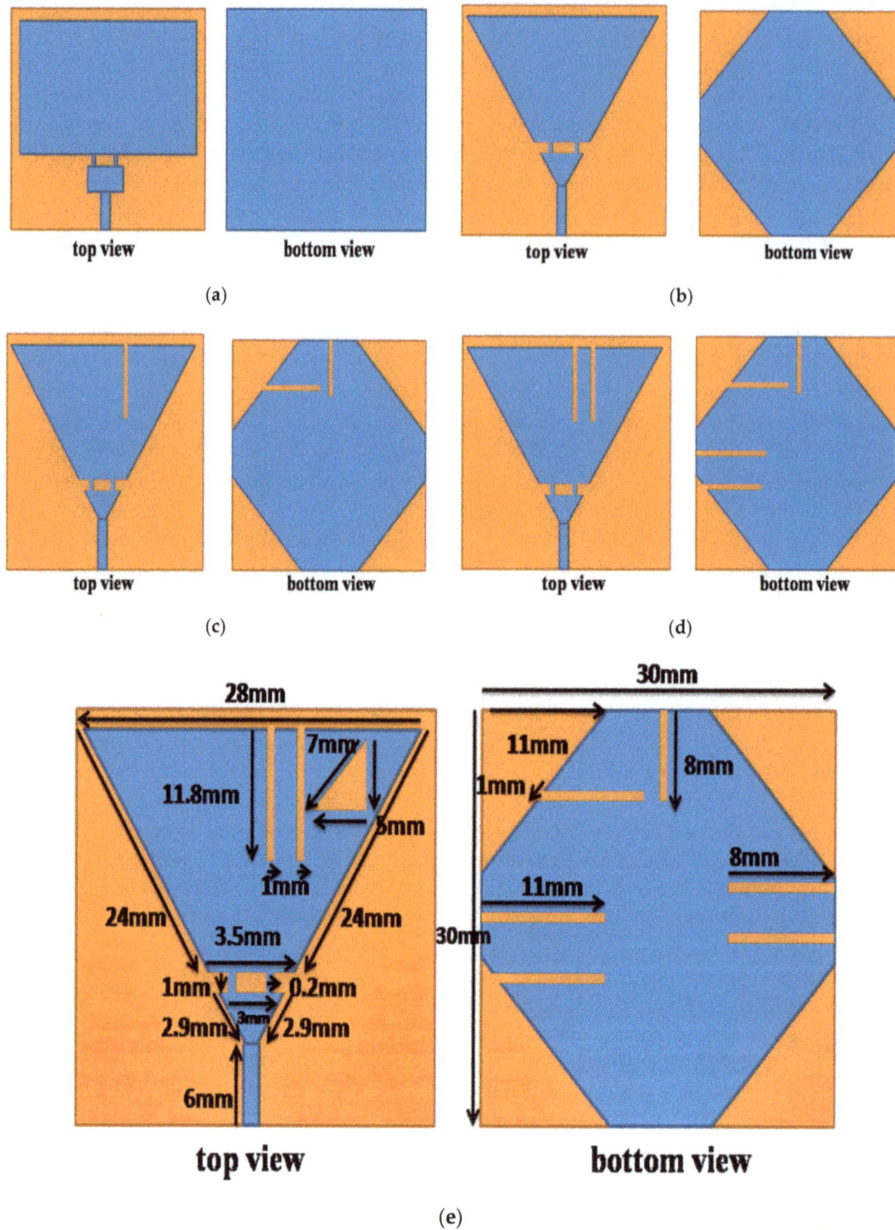

Figure 1. Evolution of proposed antenna front and back view (**a**) Antenna1, (**b**) Antenna2, (**c**) Antenna3, (**d**) Antenna4, and (**e**) proposed antenna.

$$W = \frac{(2N+1)}{\sqrt{(\varepsilon_r + 0.5)}} \times \frac{\lambda o}{2} \tag{1}$$

$$L = \frac{(2N+1)}{\sqrt{\varepsilon_{eff}}} \times \frac{\lambda o}{2} - 2\Delta L \tag{2}$$

ε_r = dielectric constant, λ_o = free space wavelength.

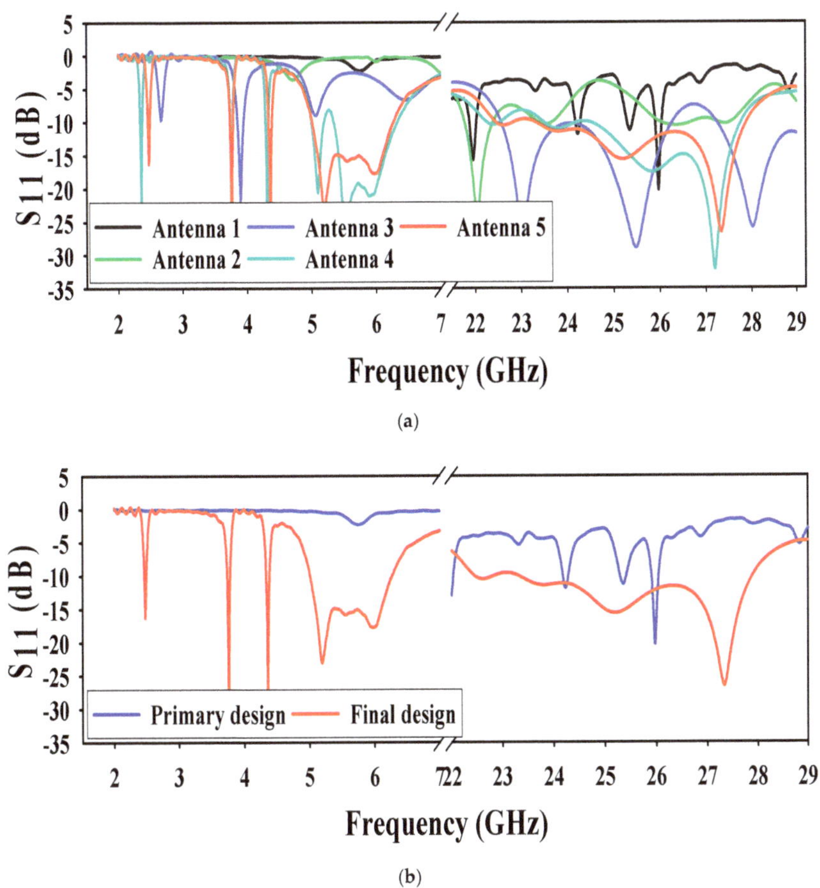

Figure 2. Simulated S_{11} (**a**) antenna 1 to 5, (**b**) primary and final design.

ΔL is the effective length and can be found by the Equation (3)

$$\frac{\Delta L}{h} = 0.412\left[\frac{\varepsilon_{eff}+0.3}{\varepsilon_{eff}-0.258}\right]\left[\frac{\frac{w}{h}+0.264}{\frac{w}{h}+0.813}\right] \tag{3}$$

where ε_{eff} = effective relative permittivity of the substrate

$$\varepsilon_{eff} = \frac{\varepsilon_r+1}{2} + \frac{\varepsilon_r+1}{2}\sqrt{(1+12h/2w)} \tag{4}$$

3. Parametric Study

To understand the impact of various parameters on different results and to achieve the best optimized dimensions of the final design, a parametric analysis has been done on different parameters of the antenna. All other parameters were kept at their final value during the parametric study.

3.1. Effect of Truncating Corners of the Patch

The corners of the rectangular patches shown in Figure 3 were truncated at three different values to form a conical and triangular shape patch. A visible effect at both operating frequency bands i.e., microwave and mm-wave band, was observed. At first value, the antenna is only resonating at 2.4 GHz in the microwave band while at (21 GHz–22.5 GHz) in the mm-wave band. At the second value, the antenna resonating frequencies are 2.4 GHz, (4 GHz–5 GHz), and (22 GHz–23.5 GHz). At the third value, the antenna gives resonance at 2.4 GHz, 5.8 GHz, and (21 GHz–24.8 GHz). The effect of different values along with the optimized value results are shown in Figure 3.

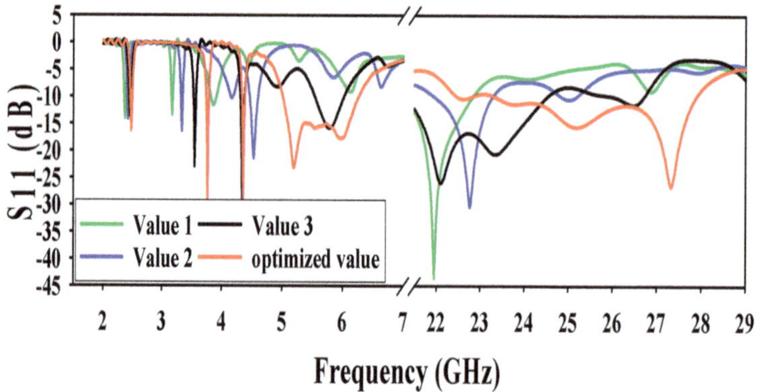

Figure 3. Effect of truncating the corners of patches.

3.2. Effect of Truncating Corners of the Ground

The DGS also has a very high impact on both microwave and mm-wave bands of operation of the antenna. Two types of techniques were used in this paper for defected ground. The first method was a truncation of the ground at three different values at all corners. At first value, the antenna resonated at 2.3 GHz, whereas the other resonating frequency shifted to 6.5 GHz. Moreover, there was a mismatch at (22.5–24.3) GHz. At the second value, the resonance frequency shifted to 2.4 GHz and 6.4 GHz, and a mismatch at (25.5–26.5) GHz. Finally, when the value increased from its optimized value, the resonance frequency moved to 4 GHz and 5.8 GHz. Whereas, at mm-wave, there was mismatching at (25.7–27.1) GHz. All of the results (of truncating the corners of the ground at different values along with its optimized values) are shown in Figure 4.

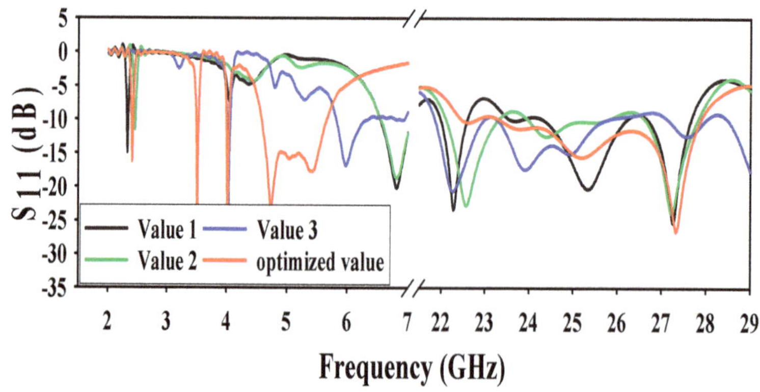

Figure 4. Effect of truncating corners of the ground.

3.3. Effect of Different Values of Slots in the Ground

The second technique used for DGS was to etch different values of meandering slots at the bottom layer. Widths of the ground slots were varied at different values to analyze its performance at all the operating frequencies. Each slot width decreased to 0.5 mm from its optimized value (1 mm) and it was observed (from the result shown in Figure 5) that the antenna was not resonating at 2.4, 5.2, and 5.8 GHz. Again, when width of the slots increased from 0.5 to 0.8 mm, the resonance was above −10 dB. Finally, when the width of each slot increased to 1.2 mm, the antenna resonated only at 5.8 GHz, with the maximum resonance of −12 dB. There is no clear effect on the mm-wave operation band apart from a little mismatch at 0.5 mm on (21–25) GHz. The effect of variation of slots in the ground layer is shown in Figure 5.

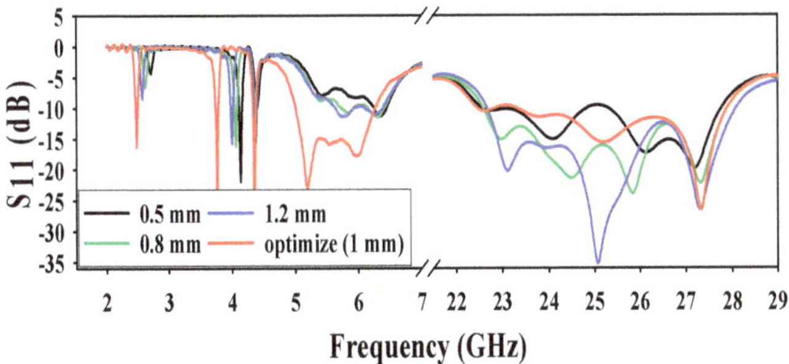

Figure 5. Effect of different values of slots in the ground.

3.4. Effect of Distance between Two Patches

The antenna performance was analyzed by different values of spacing between a triangle and a conical patch. As illustrated in Figure 6, by increasing the distance between the patches, the resonance at the mm-wave band deteriorated with every variation, and had almost no effect on the lower frequencies. Initially, the distance between two patches was kept at 0.7 mm and a mismatch was observed at (25.5–27.8) GHz. When the distance further increased to 1.1 mm, the resonating frequency emanated to (22–24.2) GHz. Finally, when the distance was kept at 1.5 mm, then the resonance came further down to (21–23.5) GHz.

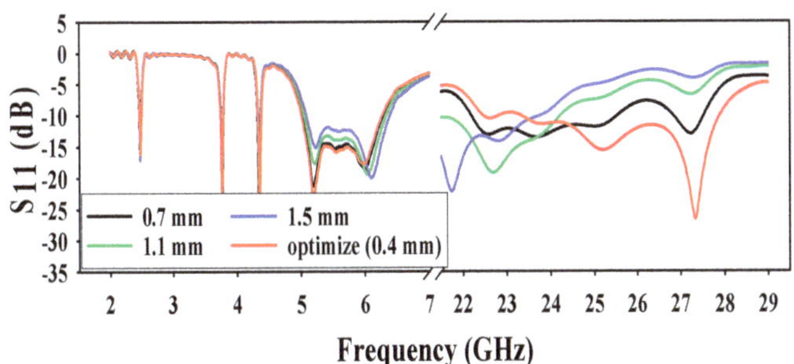

Figure 6. Effect of distance between patches.

3.5. Effect of Different Values of Slots in the Patch

The conical patch slots width were varied at three different values and the results were analyzed. In the first step, the widths of the slots were kept 0.5 mm, and it was noted that the antenna was resonating at 5.2 GHz and 5.8 GHz, and the resonance frequency of 2.4 GHz shifted to 2.6 GHz.

When widths of the slots increased to 1.3 mm, it was observed that the antenna was only resonating at 5.8 GHz, whereas there was no resonance at 2.4 GHz and 5.2 GHz. Finally, when the width of the slots further increased to 1.7 mm, again, the antenna only resonated at 5.8 GHz, and there was no resonance at 2.4 GHz and 5.2 GHz. From Figure 7, there is no effect of patch slot variation at the mm-wave operating frequency.

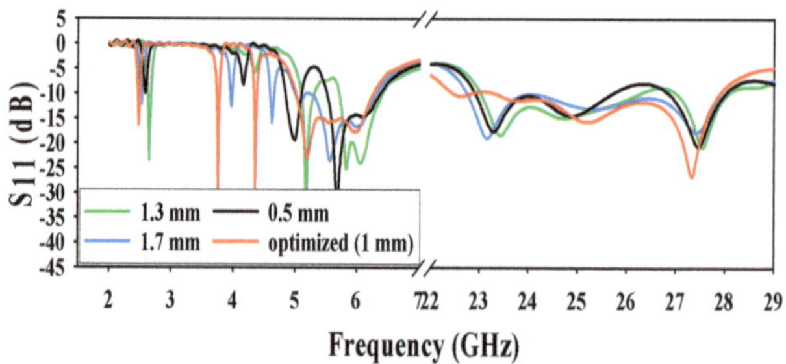

Figure 7. Effect of different values of slots in the patch.

4. Simulated and Measured Results Discussion

The prototype antenna, shown in Figure 8a, is fabricated and measured to confirm simulated results. Different results of the proposed antenna, such as reflection coefficient, radiation pattern, current density, and antenna gain will be discussed in the subsections below.

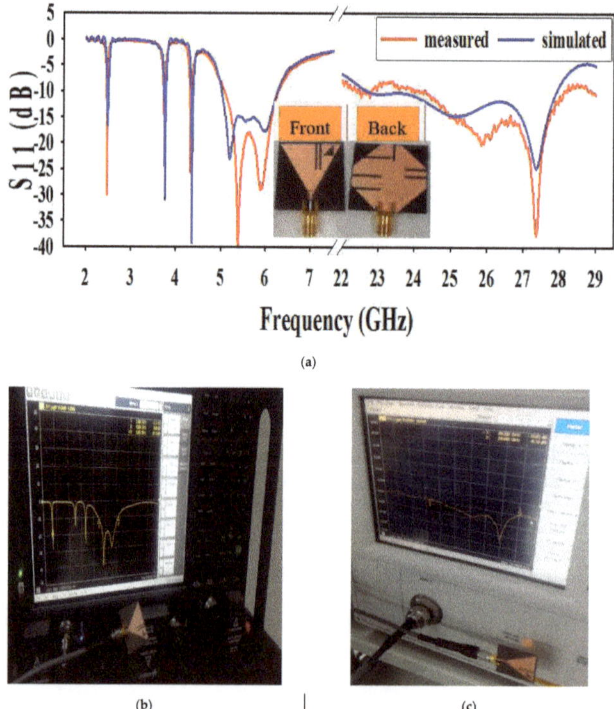

Figure 8. (**a**) Simulated and measured S-parameter, (**b**) measurement setup 2–9 GHz, and (**c**) measurement setup 21 to 29 GHz.

4.1. Reflection Coefficient

The proposed antenna simulated along with measured S_{11} results are depicted in Figure 8a. The S_{11} result from 2 to 7 GHz was measured by the subminiature version A (SMA)-1 connector (D550B51H01-118) and the S_{11} result from 23GHz to 28 GHz was measured by SMA-2 2.92 (D360B50H01-118). The S_{11} value is below −10 dB at all frequencies of operation. The antenna operates at different frequency bands, i.e., in microwave band at (2.45–2.495) GHz, (5.0–6.3) GHz and in mm-wave band at (23–28) GHz. The demonstrated measurement setup for S_{11} is shown in Figure 8b,c, measuring the microwave and mm-wave band respectively. In Figure 8a–c, a very good agreement can be seen between the results of simulated and measurement. However, the slight dissimilarity between the two results, especially at the mm-wave band of operation, can be noticed, and it could be possible because of the practical factors, which include SMA connector loss and hand soldering of the SMA-D360B50H01-118 to the antenna.

4.2. Current Density

To understand further explanation of the multiband operation, the surface current distribution of the designed antenna was analyzed at 2.4, 5.2, 5.8, and 28 GHz. As shown in Figure 9a–c, the maximum current density is along the different parts of the conical shape patch at 2.4 GHz, 5.2 GHz, and 5.8 GHz. Whereas in Figure 9d, it can be realized that the maximum current strength of the antenna is mainly associated with the smaller triangle at 27.5 GHz.

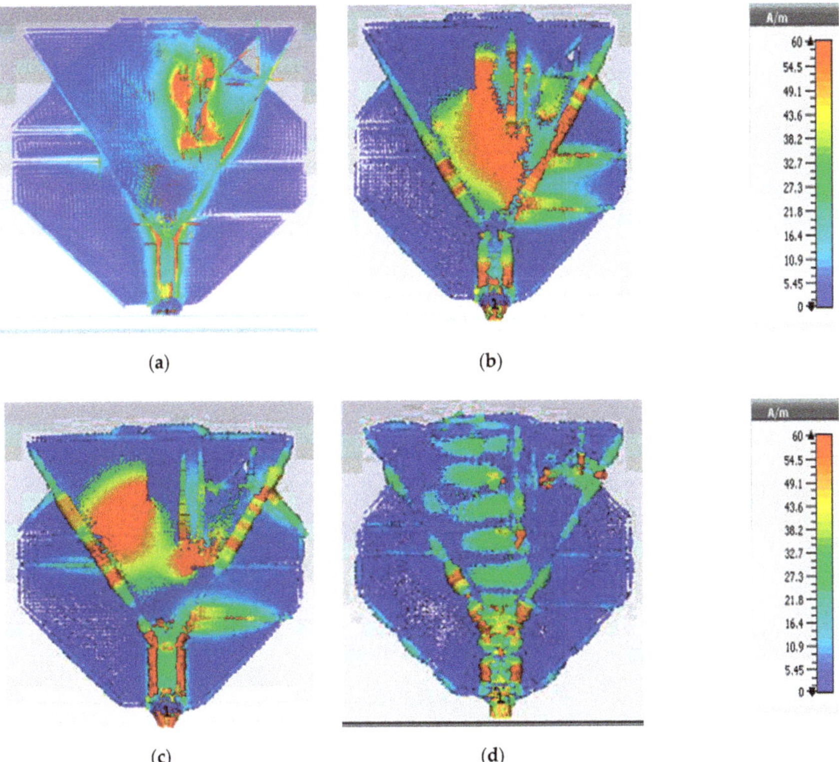

Figure 9. Surface current density (**a**) 2.4 GHz, (**b**) 5.2 GHz, (**c**) 5.8 GHz, and (**d**) 27.5 GHz.

4.3. Radiation Pattern

The far-field radiation pattern and gain were measured at 2.4 GHz, 5.2 GHz, 5.8 GHz, and 26.5 GHz. Both the E-plane and H-plane radiation patterns were given in Figure 10a–d. The radiation pattern in E-plane at 2.4 GHz, 5.2 GHz, and 5.8 GHz is nearly a dumbbell shape, whereas it is nearly omnidirectional in H-plane, which makes it suitable for multiple wireless systems. Similarly, the E-plane and H-plane radiation pattern at 26.5 GHz nearly has a directional pattern, as shown in Figure 10c. The measured radiation pattern results have a good agreement with the simulated results. However, again, a minor discrepancy can be noticed, and it could be possible as a result of the hand soldering of the 2.92 mm SMA D360B50H01-118 connector and measurement errors.

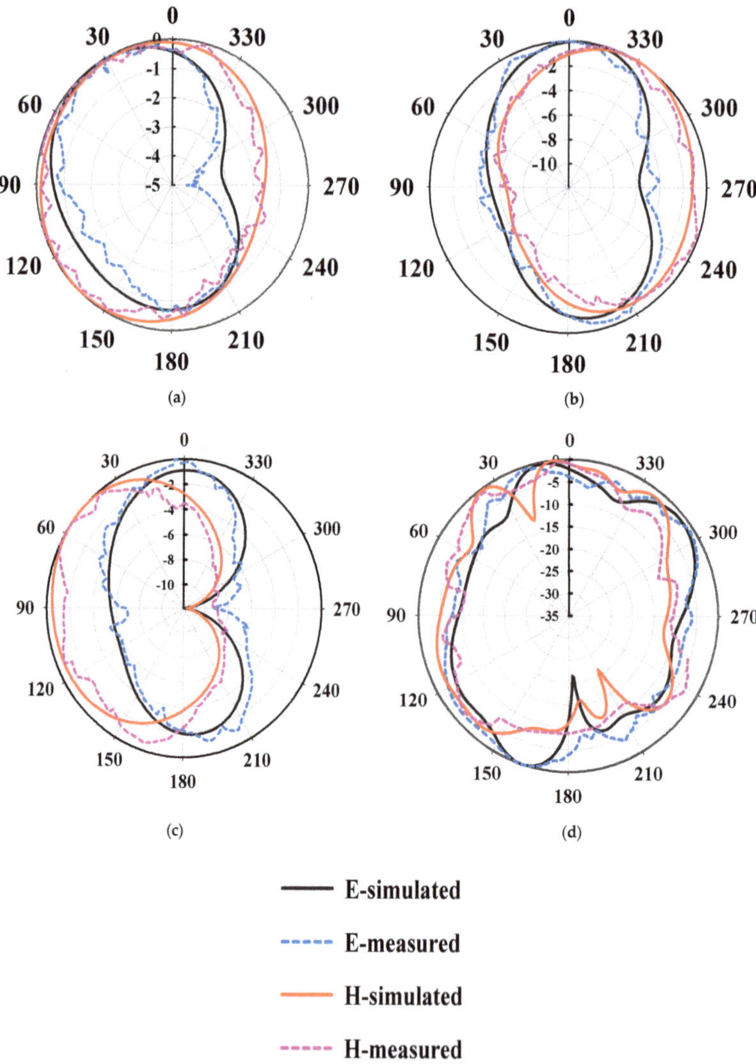

Figure 10. Radiation pattern in E-plane and H-plane at (**a**) 2.4 GHz, (**b**) 5.2 GHz, (**c**) 5.8 GHz, and (**d**) 26.5 GHz.

4.4. Antenna Gain

The antenna gain was calculated using an anechoic chamber at different frequencies of operation in both microwave and the mm-wave bands are depicted in Figure 11. A horn antenna was used as a reference antenna and the measurement setup in the anechoic chamber can be seen in Figure 12. From Figure 11, it can be seen that the antenna gives a maximum gain of 3.55 dB at 5.2 GHz, 4.72 dB at 5.8 GHz, and 5.85 dB at 26.5 GHz, respectively.

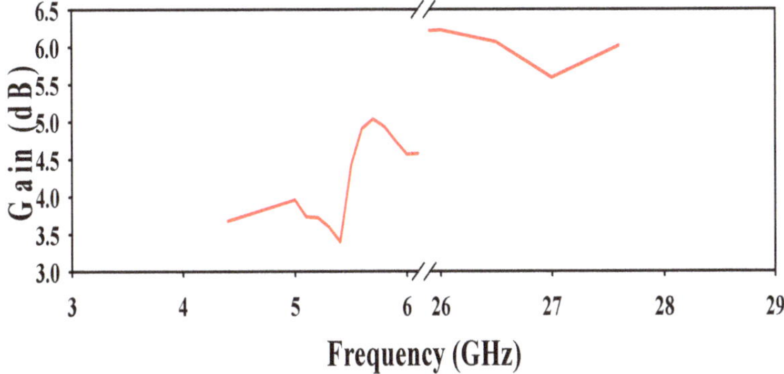

Figure 11. Proposed antenna gain over frequency.

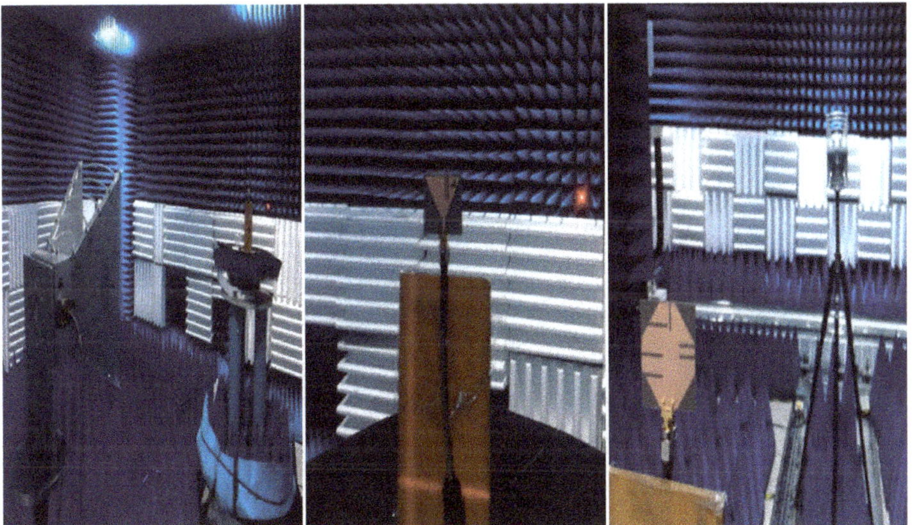

Figure 12. Measurement setup for radiation pattern.

5. Comparison

The comparison of the proposed design with other state-of-the-art designs is presented in Table 1. It can be seen in Table 1 that the antenna designs in the literature can only operate in microwave band for WLAN applications or in mm-wave band for 5G applications while the proposed designed antenna in this paper can operate in both microwave and mm-wave bands. The proposed antenna can be useful for two different communication technologies. Moreover, the proposed design can cover a large bandwidth as compared to the available designs. Further, the proposed antenna designed in this paper gives better performance in terms of multiband operation, light weight, low profile, low fabrication cost, simple geometry, and compactness.

6. Conclusions

In this paper, a DGS slotted double patch antenna, having a compact size, single feeding, simple design, and multiband characteristics, was designed and measured. The designed multiband antenna consists of (2.4, 5.2, and 5.8) GHz slotted conical patch antenna and 28 GHz triangular patch antenna. Good results were achieved in both the microwave band and mm-wave band. High-quality results compared to the latest literature were obtained by optimizing different parameters of the antenna. Measured results confirmed the proposed antenna is a suitable candidate for WLAN (5.0–6.3) GHz and 5G (23–28) GHz applications.

Author Contributions: Conceptualization, Z.K.; methodology, Z.K.; software, Z.K.; supervision, F.L., L.S.; validation, S.U.R., M.S.; Resources, S.U.R.; formal analysis, Z.K.; writing—original draft, Z.K.; writing—review and editing, Z.K., M.H.M., L.S., F.L., S.U.R.; Funding acquisition, Z.K. All authors have read and agreed to the published version of the manuscript.

Funding: This work was supported by the National Key R&D Program of China under Grant 2019YFB2204601.

Acknowledgments: The authors would like to thank the Information Science Laboratory Center, University of Science and Technology of China (USTC), for hardware and software services. The authors would also like to thanks Jalil Ul Rehman Kazim for providing technical help.

Conflicts of Interest: The authors declare no conflict of interest.

References

1. Varma, R.; Ghosh, J. Analysis and design of compact triple-band meandered PIFA for 2.4/5.2/5.8 GHz WLAN. *IET Microw. Antennas Propag.* **2019**, *13*, 505–509. [CrossRef]
2. Rappaport, T.S.; Sun, S.; Mayzus, R.; Zhao, H.; Azar, Y.; Wang, K.; Wong, G.N.; Schulz, J.K.; Samimi, M.; Gutierrez, F. Millimeter wave mobile communications for 5G cellular: It will work. *IEEE Access* **2013**, *1*, 335–349. [CrossRef]
3. Roh, W.; Seol, J.-Y.; Park, J.; Lee, B.; Lee, J.; Kim, Y.; Cho, J.; Cheun, K.; Aryanfar, F. Millimeter-wave beamforming as an enabling technology for 5G cellular communications: Theoretical feasibility and prototype results. *IEEE Commun. Mag.* **2014**, *52*, 106–113. [CrossRef]
4. Tan, J.Y.; Jiang, W.; Gong, S.X.; Cheng, T.; Ren, J.Y.; Zhang, K.Z. Design of a dual-beam cavity-backed patch antenna for future fifth generation wireless networks. *IET Microw. Antennas Propag.* **2018**, *12*, 1700–1703. [CrossRef]
5. Quevedo-Teruel, S.; Rajo-Iglesias, E. Inverted mode patch antenna for dual-band communications. *IEEE Antennas Wirel. Propag. Lett.* **2008**, *7*, 792–794. [CrossRef]
6. Liu, W.-C.; Wu, C.-M.; Dai, Y. Design of triple-frequency microstrip-fed monopole antenna using defected ground structure. *IEEE Trans. Antennas Propag.* **2011**, *10*, 2457–2463. [CrossRef]
7. Wang, H.; Zheng, M. An internal triple-band WLAN antenna. *IEEE Antennas Wirel. Propag. Lett.* **2011**, *10*, 569–572. [CrossRef]
8. Zhai, H.; Zhang, K.; Yang, S.; Feng, D. A low-profile dual-band dual-polarized antenna with an AMC surface for WLAN applications. *IEEE Antennas Wirel. Propag. Lett.* **2017**, *16*, 2692–2695. [CrossRef]
9. Cao, Y.F.; Cheung, S.W.; Yuk, T.I. A multiband slot antenna for GPS/WiMAX/WLAN systems. *IEEE Trans. Antennas Propag.* **2015**, *63*, 952–958. [CrossRef]
10. Midya, M.; Bhattacharjee, S.; Mitra, M. Broadband circularly polarized planar monopole antenna with G-shaped parasitic strip. *IEEE Antennas Wirel. Propag. Lett.* **2019**, *18*, 581–585. [CrossRef]
11. Chakraborty, U.; Kundu, A.; Chowdhury, S.K.; Bhattacharjee, A.K. Compact dual-band microstrip antenna for IEEE 802.11 a WLAN application. *IEEE Antennas Wirel. Propag. Lett.* **2014**, *13*, 407–410. [CrossRef]
12. Wu, C.M.; Chiu, C.N.; Hsu, C.K. A new non-uniform meandered and fork-type grounded antenna for triple-band WLAN applications. *IEEE Antennas Wirel. Propag. Lett.* **2006**, *5*, 346–348. [CrossRef]
13. Nandi, S.; Mohan, A. A compact dual-band MIMO slot antenna for WLAN applications. *IEEE Antennas Wirel. Propag. Lett.* **2017**, *16*, 2457–2460. [CrossRef]

14. Salih, A.A.; Sharawi, M.S. A dual-band highly miniaturized patch antenna. *IEEE Antennas Wirel. Propag. Lett.* **2016**, *15*, 1783–1786. [CrossRef]
15. Park, J.S.; Ko, J.B.; Kwon, H.K.; Kang, B.S.; Park, B.; Kim, D. A tilted combined beam antenna for 5G communications using a 28-GHz band. *IEEE Antennas Wirel. Propag. Lett.* **2016**, *15*, 1685–1688. [CrossRef]
16. Aleshia, A.T.; Hussain, R.; Podilchak, S.K.; Sharawi, M.S. A dual-element MIMO antenna system with a mm-wave antenna array. In Proceedings of the 2016IEEE 10th European Conference on Antennas and Propagation (EuCAP), Davos, Switzerland, 10–15 April 2016; pp. 1–4.
17. Yassin, M.E.; Hesham, M.A.; Abdallah, E.A.; El-Hennawy, H.S. Single-fed 4G/5G multiband 2.4/5.5/28 GHz antenna. *IET Microw. Antennas Propag.* **2018**, *13*, 286–290. [CrossRef]
18. CST Microwave Studio. Available online: http://www.cst.com (accessed on 10 September 2020).
19. Balanis, C.A. *Antenna Theory: Analysis and Design*; A John Wiley and Sons: Hoboken, NJ, USA, 2016.
20. Chiang, K.H.; Tam, K.W. Microstrip monopole antenna with enhanced bandwidth using defected ground structures. *IEEE Antennas Wirel. Propag. Lett.* **2008**, *7*, 532–535. [CrossRef]
21. Nouri, A.; Dadashzadeh, G.R. A compact UWBband-notched printed monopole antenna with defected ground structure. *IEEE Antennas Wirel. Propag. Lett.* **2011**, *10*, 1178–1181.

Letter

Integration of Microstrip Slot Array Antenna with Dye-Sensitized Solar Cells

Bowen Bai *, Zheng Zhang, Xiaoping Li, Chao Sun and Yanming Liu

The Key Laboratory of Information and Structure Efficiency in Extreme Environment, The Ministry of Education of China, and The School of Aerospace Science and Technology, Xidian University, Xi'an 710071, China; zzhang_1992@stu.xidian.edu.cn (Z.Z.); xpli@xidian.edu.cn (X.L.); sunc@xidian.edu.cn (C.S.); ymliu@xidian.edu.cn (Y.L.)

* Correspondence: bwbai@xidian.edu.cn; Tel.: +86-13572025103

Received: 15 September 2020; Accepted: 31 October 2020; Published: 2 November 2020

Abstract: This paper describes the integration of microstrip slot array antennas with dye-sensitized solar cells that can power array antennas at 5.8 GHz, ensuring normal communications. To appraise the antennas, a 2 × 2 circularly polarized microstrip slot array antenna integrated with dye-sensitized solar cells using a stacked design method was analyzed, fabricated and measured. The size of the entire array is 140 mm × 140 mm, where the size of each solar cell is 35 mm × 35 mm. The results show that the effect of the antenna has a slight influence on the output performance of the solar cells, and the interference of the output current of the solar cells to the antenna feeding system is negligible. The gain of the array antenna increases by 0.12 dB and the axial ratio is reduced to 1.50 dB after the integration of dye-sensitized solar cells. The integration saves a lot of space, and has the ability of self-sustaining power generation, thus providing reliable and long-term communication for various communication systems.

Keywords: dye-sensitized solar cells; integration; antenna array; solar antenna

---

1. Introduction

Recently, solar energy has received more and more attention as a clean renewable energy source, and the solar antenna (solant) has drawn a large amount of concern because it can not only transmit and receive electromagnetic waves, but also generate electricity [1]. The research on the fusion technology of solar cells and antennas can be traced back to 1995. Tanaka et al. [2] took the lead in designing a fusion device of solar cells and patch antenna, and the device was successfully applied to a microsatellite. Compared with the simple juxtaposition of the antenna and solar cell, the integration of the antenna and the solar cell has certain advantages in volume, weight, appearance and electrical performance. Both amorphous (a-Si) and crystalline (c-Si)-type silicon solar cells with integrated antennas have been reported [3–8]. A single crystal silicon solant is proposed [9]; several solar cells are placed 5 mm above the microstrip slot antenna, indicating a poor combination between the solar cells and the antenna. S. V. Shynu et al. [10] integrated a double-slot antenna with an amorphous silicon solar cell by covering a dual-band WLAN. K Yang et al. [11] replaced the copper grounding plane of the vivaldi antenna with an amorphous silicon solar cell; while the solant achieves a high degree of integration, there is a certain interference between the solar cell and the antenna. M. Danesh et al. [12] used a monopole antenna in combination with an amorphous silicon solar cell, and placed only the solar cell in the radiating portion of the monopole antenna, resulting in low space utilization. Therefore, it is still a challenge to integrate antennas with the solar cell to the greatest extent and eliminate the interference between antennas and solar cells.

Dye-sensitized solar cells have been widely studied and applied [13–15] due to their lower processing cost compared with crystalline silicon cells. With the development of research, the conversion efficiency of dye-sensitized solar cells has been improved [16], and can be printed on flexible conductive plastic layers for enhanced integration [17,18]. However, there is little literature on the integration of antennas with dye-sensitised solar cells. The first dye-sensitised solar cell antenna in a proof-of-concept dipole configuration was studied in [19]; series-connected dye-sensitised cells could produce 1.49 V and 15.5 mA, which meets design requirements. However, the interference caused by the integration of antennas with dye-sensitised solar cells still needs to be analyzed. A compatible integration of a circularly polarized omnidirectional metasurface antenna with solar cells has been reported in [20]. While the antenna gain of type IV is 4.1 dBi, it can be predicted that the output power generated by the solar cell array is not high due to the fact that the solar cells are not connected to each other, which limits its practical application.

In this paper, an integration of a 2 × 2 circularly polarized microstrip slot array antenna with dye-sensitized solar cells is designed. A novel stack design method makes the solar cell and array antenna well integrated and the simulation and measurement results show that the gain of the array antenna increased by 0.12 dB, reaching 6.60 dBi, and the axial ratio was reduced to 1.50 dB after the integration of dye-sensitized solar cells. The solar cells and the microstrip slot array antenna are perfectly integrated. The integration saves a lot of space, especially when the proposed antenna is used in satellite communication. Compared with the existing circularly polarized microstrip slot array antenna, the proposed antenna adds the output of dye-sensitized solar cells into the voltage regulation circuit to form a stable power supply for the radio frequency system, which ensures the operation of the microstrip slot array antenna. In other words, the antenna has the ability of self-sustaining power generation capabilities, so as to provide reliable and long-term communication for the communication system when the power is not easy to obtain.

2. Integrated Design Array Antenna and Solar Cells

Due to the influence of climate, environment and other factors, the linear polarization wave easily causes polarization deflection loss. These factors have little effect on the polarization deflection of circularly polarized waves, and, concerning circularly polarized antennas, polarization mismatch does not easily occur. In order to suit practical applications, a circularly polarized microstrip slot antenna is used in this paper. The geometry of a 2 × 2 circularly polarized microstrip slot array antenna integrated with dye-sensitized solar cells is illustrated in Figure 1a. The size of the entire array is 140 mm × 140 mm, where the size of each solar cell is 35 mm × 35 mm. The circularly polarized waves are excited by arranging the four slots crosswise and designing the microstrip feeder network reasonably.

The dielectric substrate of the microstrip slot antenna uses a Rogers R4350b plane, a dielectric constant of 3.48, a dielectric loss angle of 0.0037 and a thickness of 0.5 mm. The equal division Wilkinson power divider is used in the feed network to obtain the excitation signal with equal amplitude and phase difference of 90°, as shown in Figure 1b. The isolation resistance is 100 Ω, and the impedance of the feed port is set to 50 Ω. Through optimization analysis, the width of microstrip lines is W1 = 1.15 mm (50-Ω) and W2 = 0.63 mm (70.71-Ω). The slot size is 14.5 mm × 1.8 mm, and the eccentric distance of the feed point is 3 mm. The microstrip slot array antenna gain and the available solar cell area on the antenna surface need to be taken into consideration, concerning a slot spacing of 44 mm.

Dye-sensitized solar cells are a new type of photovoltaic technology developed by simulating the principle of plants in nature using solar energy for photosynthesis. DSSCs are based on dye sensitizers and nano-TiO_2, which can make the photoelectric conversion efficiency reach a better level. At the same time, dye-sensitized solar cells (DSSCs) are rich in raw materials, non-polluting and low in cost; the manufacturing cost is only one fifth to one tenth that of silicon solar cells [21]. Therefore, dye-sensitized solar cells were chosen for this paper. Dye-sensitized solar cells adopt a stacked structure similar to the microstrip slot antenna, so the metal ground plane of the microstrip slot antenna can be used as the substrate of the solar cell.

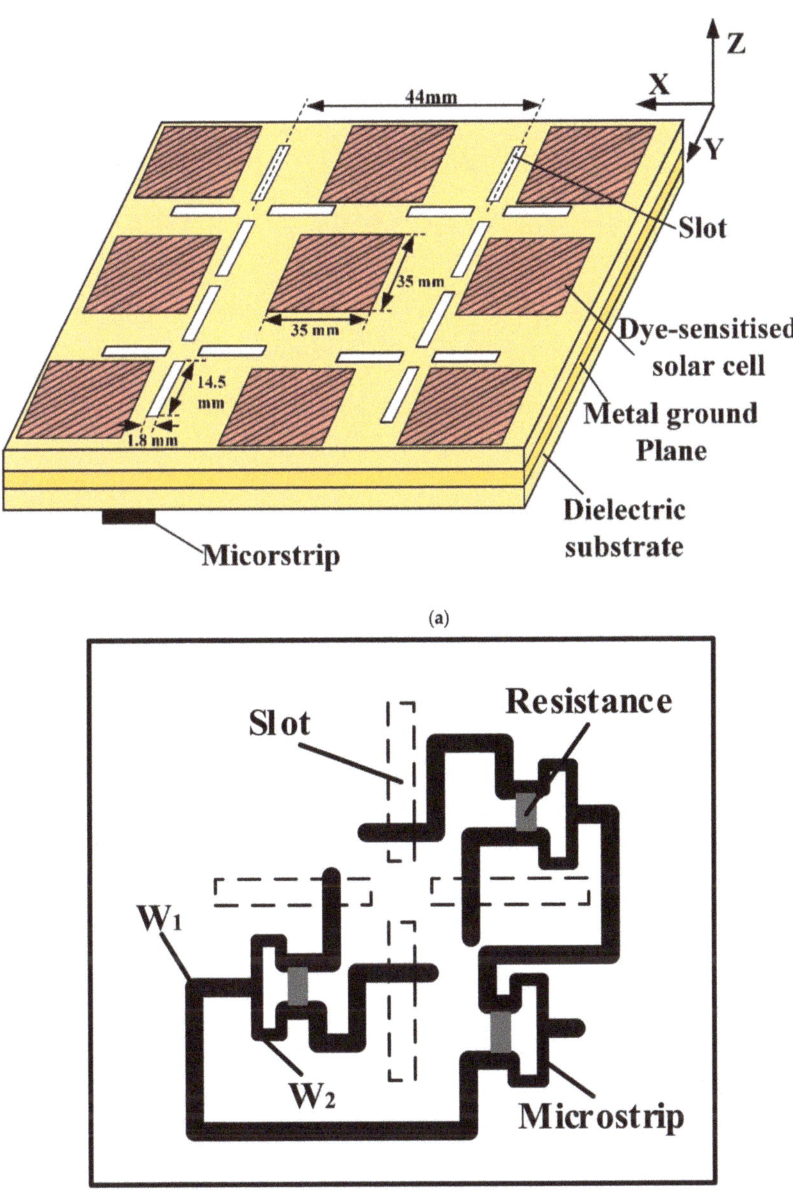

Figure 1. (**a**) The geometry of a 2 × 2 circularly polarized microstrip slot array antenna integrated with dye-sensitized solar cells; (**b**) microstrip feeder network.

The integrated design does not change the intrinsic structure of the antenna, and will minimize the impact of the antenna. Due to the conductive material in the dye-sensitized solar cell, it will block the radiation of electromagnetic waves, so it is necessary to retain the slot of the antenna. The dye-sensitized solar cell and the microstrip slot antenna share a metal ground plane, and the cell structure is as described in [22]. The Components of the integration of a circularly polarized microstrip slot antenna with a dye-sensitized solar cell are shown in Figure 2. The dielectric constant and conductivity of the dye-sensitized solar cell materials at room temperature are shown in Table 1. The dye-sensitized layer is an electrolyte containing I^- and I^{-3}, mixed with sensitizers, potassium chloride, etc.

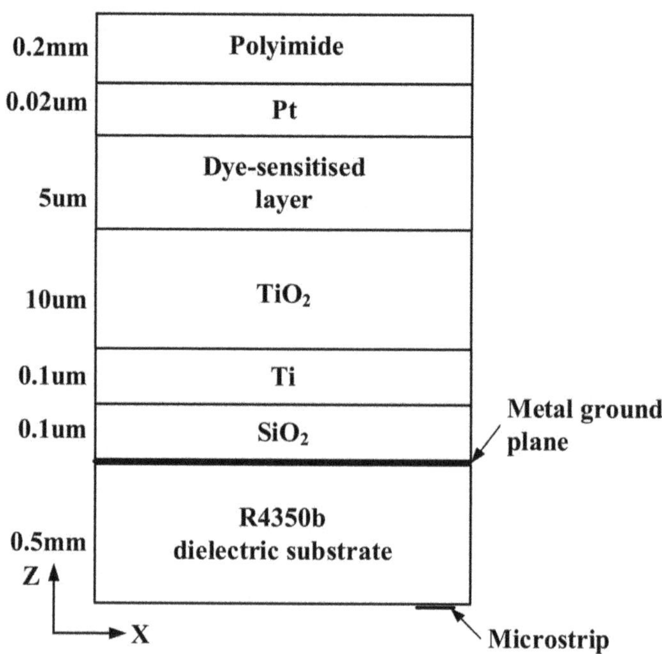

Figure 2. Component of the integration of a circularly polarized microstrip slot antenna with a dye-sensitized solar cell.

Table 1. Solar cell material properties.

Material	Dielectric Constant	Conductivity
Polyimide	3.4	3.92 × 10–15 S/m
Pt	1	9.3 × 106 S/m
Dye-sensitized layer	90	1.51 × 10–2 S/m
TiO_2	86	100 S/m
Ti	1	1.82 × 106 S/m
SiO_2	4	0 S/m

According to Figure 2 and Table 1, for the antenna structure shown in Figure 1, the following two spectral domain integral equations can be obtained by using the boundary condition that the total electric field tangential direction of the antenna surface is 0:

$$\iint \left[\tilde{G}_{xx}\tilde{J}_x + \tilde{G}_{xy}\tilde{J}_y\right] \exp\{-j(k_x x + k_y y)\} dk_x dk_y = \iint \tilde{G}_{xz}\tilde{J}_z \exp\{-j(k_x x + k_y y)\} dk_x dk_y \tag{1}$$

$$\iint \left[\tilde{G}_{yx}\tilde{J}_x + \tilde{G}_{yy}\tilde{J}_y\right] \exp\{-j(k_x x + k_y y)\} dk_x dk_y = \iint \tilde{G}_{yz}\tilde{J}_z \exp\{-j(k_x x + k_y y)\} dk_x dk_y \tag{2}$$

where $\tilde{G}_{xx}$, $\tilde{G}_{xy}$, $\tilde{G}_{xz}$, $\tilde{G}_{yx}$, $\tilde{G}_{yz}$, $\tilde{G}_{yy}$ are the components of Green's function in the electric field spectral domain, $\tilde{J}_x$ and $\tilde{J}_y$ are the x and y spectral components of the unknown current on the surface, and $\tilde{J}_z$ are the spectral components of the feed current of the coaxial probe. The antenna is covered with multi-layer dielectric plates, where each layer of the dielectric plate is a lossy medium, their relative dielectric constant is ε_{ri}, the thickness of each covering layer is h_i and the permeability of each layer of the medium is μ_0. According to reference [23], the analytical calculation formula of spectral domain Green's functions of the electric field on the antenna surface can be obtained as follows:

$$\tilde{G}_{xxi}(k_x, k_y) = k_0\eta_0[B^h k_y^2/k_t^2 + B^e k_x^2/k_t^2] \tag{3}$$

$$\tilde{G}_{xyi}(k_x, k_y) = k_0\eta_0 k_x k_y/k_t^2[B^e - B^h] \tag{4}$$

$$\tilde{G}_{yxi}(k_x, k_y) = \tilde{G}_{xyi}(k_x, k_y) \tag{5}$$

$$\tilde{G}_{yyi}(k_x, k_y) = k_0\eta_0[B^h k_x^2/k_t^2 + B^e k_y^2/k_t^2] \tag{6}$$

$$\tilde{G}_{xzi}(k_x, k_y) = jk_0\eta_0 B^e k_x/\gamma^2 \tag{7}$$

$$\tilde{G}_{yzi}(k_x, k_y) = (k_y/k_x)\tilde{G}_{xzi}(k_x, k_y) \tag{8}$$

where:

$$B^h = (A_{11}^h + A_{21}^h)/U^h \tag{9}$$

$$\begin{bmatrix} (A_{11}^h) & (A_{12}^h) \\ (A_{21}^h) & (A_{22}^h) \end{bmatrix} = \prod_{i=1}^{N} \frac{1}{1+R_i^h} \begin{bmatrix} e^{j\gamma_i h_i} & R_i^h e^{j\gamma_i h_i} \\ R_i^h e^{-j\gamma_i h_i} & e^{-j\gamma_i h_i} \end{bmatrix} \tag{10}$$

$$R_i^h = (\gamma_i - \gamma_{i-1})/(\gamma_i + \gamma_{i-1}) \tag{11}$$

$$R_i^e = (\varepsilon_{ri}\gamma_{i+1} - \varepsilon_{ri+1}\gamma_i)/(\varepsilon_{ri}\gamma_{i+1} + \varepsilon_{ri+1}\gamma_i) \tag{12}$$

$$U^h = \left\{ j\gamma \cot y(\gamma h)\left[(A_{11}^h) + (A_{21}^h)\right] - \gamma_1[(A_{11}^h) - (A_{21}^h)]\right\} \tag{13}$$

$$U^e = \left\{ j(k_0^2\varepsilon_r/\gamma)\cot y(\gamma h)\left[(A_{11}^e) + (A_{21}^e)\right] - (k_0^2\varepsilon_r/\gamma)[(A_{11}^e) + (A_{21}^e)]\right\} \tag{14}$$

$$\gamma = \sqrt{\varepsilon_r k_0^2 - k_t^2} \tag{15}$$

$$\gamma_i = \sqrt{\varepsilon_r k_0^2 - k_t^2} \tag{16}$$

$$k_t = \sqrt{k_x^2 + k_y^2} \tag{17}$$

Replace the superscript 'h' in Equations (9) and (10) with 'e' to obtain the expressions for each component of B^e, A_{11}^e, A_{12}^e, A_{21}^e and A_{22}^e. k_0 is the free space wave number, and η_0 is the free space wave impedance. Equations (3)–(8) are the new analytical calculation formulas for the spectral domain Green's function of the microstrip slot antenna structure covered by the multilayer dielectric. After obtaining the calculation formula of Green's function in the spectral domain, the solution of integral Equations (1) and (2) can be discussed, as below. Assuming that the coaxial probe is fed at point (x_p, y_p) and there is a constant current I_0 on the probe, the formula for calculating the spectral domain of the current on the probe can be obtained as follows:

$$\tilde{J}_z = I_0 \exp[j(k_x x_p + k_y y_p)] \tag{18}$$

Let the unknown current on the antenna surface be:

$$\tilde{J}_s(x, y) = J_x(x, y)\tilde{x} + J_y(x, y)\tilde{y} \tag{19}$$

$J_x(x, y)$ and $J_y(x, y)$ are expanded by a set of basis functions, and then Fourier transform is used to obtain the spectral domain expression:

$$\tilde{J}_x(k_x, k_y) = \sum_{n=1}^{N_x} C_{xn}\tilde{J}_{xn}(k_x, k_y) \tag{20}$$

$$\tilde{J}_y(k_x, k_y) = \sum_{n=1}^{N_y} C_{yn}\tilde{J}_{yn}(k_x, k_y) \tag{21}$$

$\tilde{J}_{xn}(k_x, k_y)$ and $\tilde{J}_{yn}(k_x, k_y)$ in Equations (20) and (21) are the spectral domain expressions of the selected basis functions. By introducing Equations (18), (20) and (21) into Equations (1) and (2), the integral Equations (1) and (2) can be solved using the Galerkin method to obtain the current coefficients C_{xn} and C_{yn}. Then the relevant characteristic parameters of the antenna can be further calculated.

The electromagnetic simulation software ANSYS HFSS is used to obtain the relevant electromagnetic parameters of the antenna [24]. The center frequency of the antenna is set to 5.8 GHz and the polarization mode is right-handed circular polarization. The microstrip slot antenna and the microstrip slot antenna integrated with dye-sensitized solar cell are simulated separately, and the related reflection coefficient and radiation efficiency are shown in Figure 3. It can be seen that the solar cell has little influence on the impedance matching of the antenna, and the microstrip slot antenna integrated with dye-sensitized solar cell has an impedance bandwidth of 2.33 GHz, from 4.23 to 6.56 GHz. After the integration of the dye-sensitized solar cell, the radiation efficiency of the antenna decreases from 76.1% to 68.9% when the operation frequency is 5.8 GHz. The radiation efficiency of the antenna decreases slightly, which indicates that the dye-sensitized solar cell has little influence on the circular polarization radiation characteristics of the antenna.

The simulation results show that the presence of the solar cell has little effect on the performance of the antenna, but the solar cell generates a corresponding current when receiving visible light irradiation. Therefore, when the cell is in working condition, the influence on the antenna performance needs to be measured.

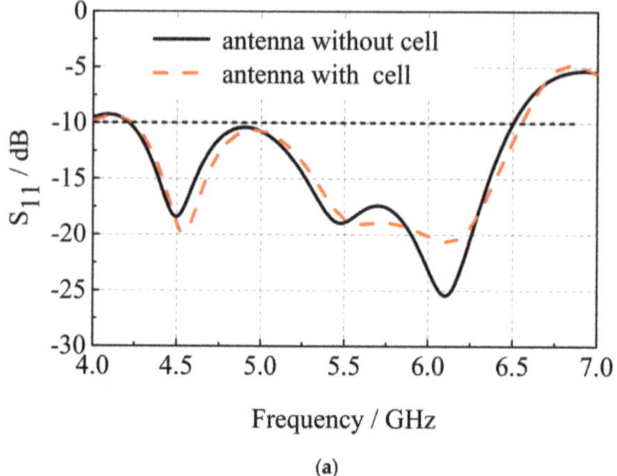

(a)

Figure 3. *Cont.*

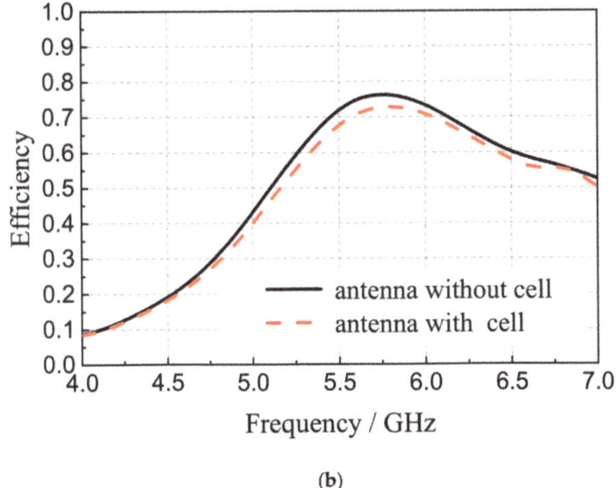

(**b**)

Figure 3. (**a**) Reflection coefficient; (**b**) radiation efficiency of the microstrip slot with or without solar cell.

3. Results and Discussion

The photograph of the 2 × 2 circularly polarized microstrip slot array antenna integrated with the dye-sensitized solar cells is presented in Figure 4. It can be seen that the slot divides the entire antenna into nine parts, and each part is tightly connected to a dye-sensitized solar cell. Among them, three independent dye-sensitized solar cells are connected in series to improve the output voltage of the solar cells, and three series connected solar cells are connected in parallel to increase the output current of the solar cells. The dye-sensitized solar cell uses glass as the substrate, the cell and the antenna are grounded together by welding the negative electrode of the cell to the metal ground plant of the microstrip slot antenna.

Figure 4. Photograph of the 2 × 2 circularly polarized microstrip slot array antenna integrated with dye-sensitized solar cells.

The high-illuminance xenon lamp is used as the light source to illuminate the microstrip slot array antenna integrated with dye-sensitized solar cells at a suitable distance. Figure 5a shows the related experiments on the energy output characteristics of the microstrip slot array antenna integrated with the dye-sensitized solar cells. The open-circuit voltage and short circuit current of dye-sensitized solar cells are 1.94 V and 99 mA, respectively, whether the antenna works or not. The dye-sensitized solar cells are externally connected with a sliding rheostat whose resistance varies from 0 to 100 Ω. The relationship between the output voltage and the output power of the solar cell under the two conditions of antenna operation and non-operation is measured by changing the value of the sliding rheostat, as shown in Figure 5b. It can be seen that the curves coincide basically when the antenna works or does not work, which indicates that, whether it works or not, the antenna has little effect on the performance of dye-sensitized solar cells. The 2 × 2 circularly polarized microstrip slot array antenna integrated with dye-sensitized solar cells is measured in an anechoic chamber to further validate its design; the illumination intensity was maintained at 150 W/m^2 and the relevant measurement environment is shown in Figure 6.

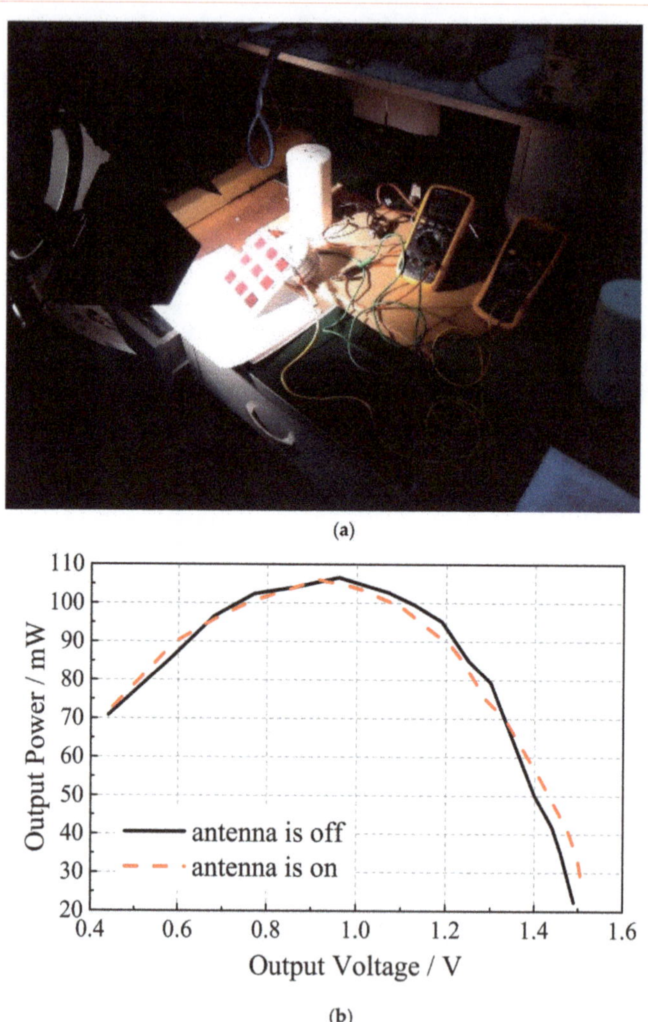

Figure 5. (**a**) The related experiments on the energy output characteristics; (**b**) output power versus output voltage.

Figure 6. The experimental setup and the measurement environment.

Figure 7 shows the measured reflection coefficient and radiation pattern of the microstrip slot array antenna and the array antenna integrated with dye-sensitized solar cells. From the measurement results in Figure 7a, when the array antenna is integrated with dye-sensitized solar cells, the current generated when the solar cells work will have a certain influence on the reflection coefficient of the antenna, but the reflection coefficient performance of the array antenna integrated with dye-sensitized solar cells is still good on the whole. From the measurement results of Figure 7b, the gain of the microstrip slot array antenna is 6.48 dBi at 5.8 GHz, and the gain of the array antenna integrated with dye-sensitized solar cells is 0.12 dB higher than that of the microstrip slot array antenna, reaching 6.60 dBi. Generally, the existence of dye-sensitized solar cells has little effect on the gain performance of microstrip slot array antenna.

Figure 8 shows the measurement results of the influence of the axial ratio parameters of the microstrip slot array antenna integrated with dye-sensitized solar cells. It can be found from Figure 8a that the axial ratio of the microstrip slot array antenna is 1.65 dB at 5.8 GHz; the axis ratio of the array antenna integrated with dye-sensitized solar cells is 1.50 dB. The existence of dye-sensitized solar cells has little effect on the circular polarization radiation performance of the microstrip slot array antenna. The axial ratio radiation patterns in Figure 8b show that the working solar cells have almost no effect on the antenna axis ratio.

The results of normalized radiation patterns of the array antenna integrated with dye-sensitized solar cells are shown in Figure 9. It can be seen that the existence of dye-sensitized solar cells has little influence on the directivity of the microstrip slot array antenna. The above results show that, when the antenna is integrated with dye-sensitized solar cells, the measurement results in the figures will show slight changes, which are mainly caused by the interference between the solar cells and the mutual coupling between the wires. To sum up, the dye-sensitized solar cells and the microstrip slot array antenna are perfectly combined; the interference between the solar cells and the antenna is minimal. When solar cells with higher cost but better photoelectric conversion efficiency are used, the antenna size can be reduced to further improve the performance.

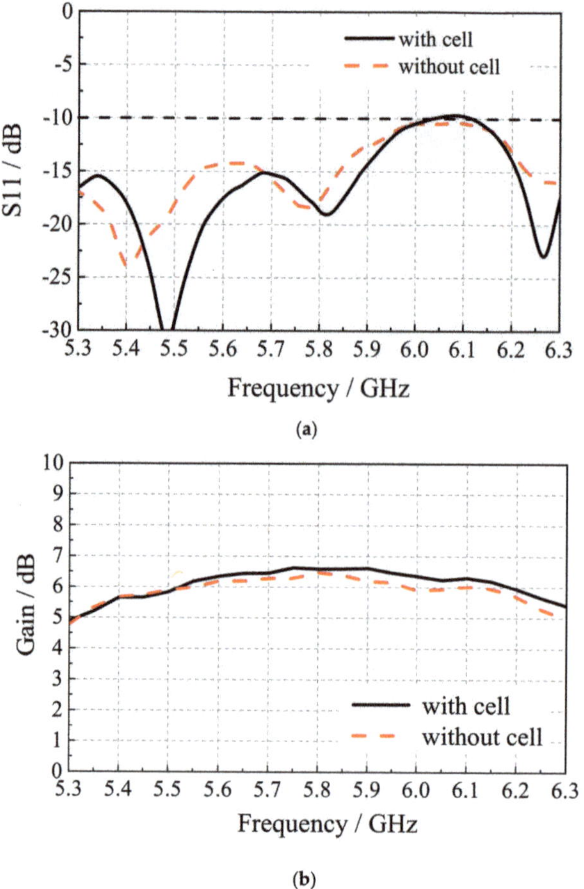

(b)

Figure 7. (**a**) Reflection coefficient; (**b**) gain of the microstrip slot antenna integrated with dye-sensitized solar cells varies with frequency.

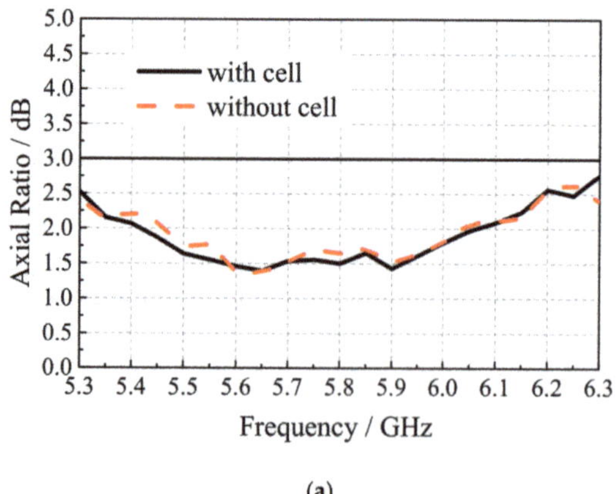

(**a**)

Figure 8. *Cont.*

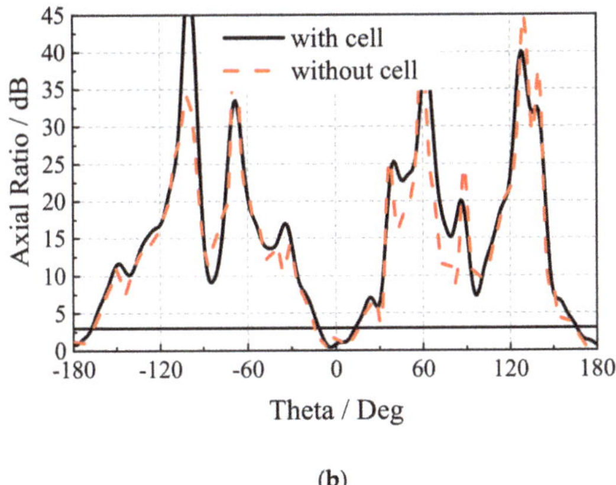

(b)

Figure 8. (a) The axial ratio varies with frequency; (b) axial ratio radiation pattern of the proposed antenna.

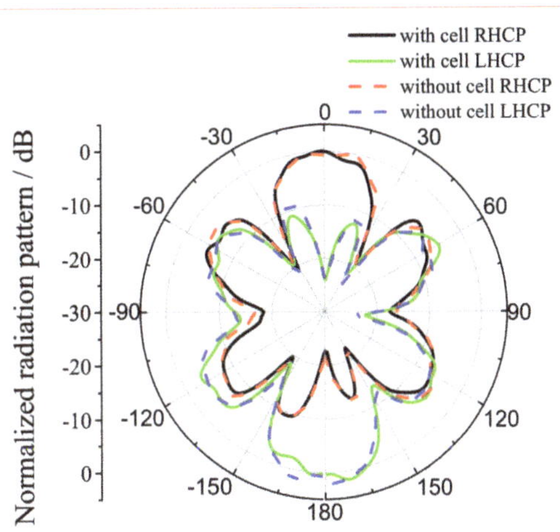

Figure 9. Normalized radiation patterns of the proposed antenna.

4. Conclusions

In this paper, a 2 × 2 circularly polarized microstrip slot array antenna integrated with dye-sensitized solar cells is designed. A novel stack design method makes the solar cells and the array antenna well integrated. The simulation and measurement results show that the gain of the array antenna increases by 0.12 dB and the axial ratio decreases to 1.50 dB after the integration of the dye-sensitized solar cells. Whether the antenna works or not has little influence on the performance of the dye-sensitized solar cells. The array antenna integrated with dye-sensitized solar cells has a similar radiation performance to the traditional microstrip slot array antenna, and can also provide electricity. Compared with the existing circularly polarized microstrip slot array antenna, the proposed antenna adds the output of dye-sensitized solar cells into the voltage regulation circuit to form a stable power supply to the radio frequency system, which ensures the operation of the microstrip slot array antenna. In other words, the antenna has the ability of self-sustaining power generation capabilities, so as to provide reliable and long-term communication for the communication system when the power is not easy to obtain.

Author Contributions: Conceptualization, B.B.; methodology, B.B., C.S. and Z.Z.; software, B.B., Z.Z.; validation, B.B., Z.Z. and C.S.; formal analysis, Z.Z. and C.S.; investigation, B.B., Z.Z. and X.L.; resources, B.B., X.L. and Y.L.; writing—original draft preparation, Z.Z.; writing—review and editing, B.B. and C.S.; visualization, Z.Z.; supervision, X.L., Y.L.; project administration, B.B., X.L.; funding acquisition, B.B. All authors have read and agreed to the published version of the manuscript.

Funding: This work was supported in part by the National Natural Science Foundation of China under Grant 61801343 Grant 61701381, Grant 61431010, and Grant 61627901, in part by the Natural Science Basic Research Plan in Shaanxi Province of China under Grant 2019JM-177, and in part by the Chinese Postdoctoral Science Foundation.

Conflicts of Interest: The authors declare no conflict of interest.

References

1. Szlufcik, J.; Sivoththaman, S.; Nlis, J.F. Low-cost Industrial Technologies of Crystalline Silicon Solar Cells. *Proc. IEEE* **1997**, *85*, 711–730.
2. Tanaka, M.; Suzuki, Y.; Arak, K.; Suzuki, R. Microstrip Antenna with Solar Cells for Microsatellites. *Electron. Lett.* **1995**, *31*, 5–6.
3. Georgiadis, A.; Collado, A.; Kim, S.; Lee, H.; Tentzeris, M.M. UHF solar powered active oscillator antenna on low cost flexible substrate for wireless identification applications. In Proceedings of the 2012 IEEE/MTT-S International Microwave Symposium Digest, Montreal, QC, Canada, 17–22 June 2012; pp. 1–3. [CrossRef]
4. O'Conchubhair, O.; Mc Evoy, P.; Ammann, M.J. Integration of Inverted-F antenna with solar cell substitute. In Proceedings of the 2012 Loughborough Antennas & Propagation Conference (LAPC), Institute of Electrical and Electronics Engineers (IEEE), Loughborough, UK, 12–13 November 2012; pp. 1–4.
5. Huang, J.; Zawadzki, M. Antennas integrated with solar arrays for space vehicle applications. In Proceedings of the 2000 5th International Symposium on Antennas Propagation and EM Theory, Beijing, China, 15–18 August 2000; pp. 86–89.
6. Yurduseven, O.; Smith, D. A solar cell stacked multi-slot quadband PIFA for GSM, WLAN and WiMAX networks. *IEEE Microw. Wirel. Compon. Lett.* **2013**, *23*, 285–287. [CrossRef]
7. Shynu, S.V. Integration of microstrip patch antenna with polycrystalline silicon solar cell. *IEEE Trans. Antennas Propag.* **2009**, *57*, 3969–3972. [CrossRef]
8. Shynu, S.V.; Ammann, M.J.; McCormack, S.J.; Norton, B. Emitter-wrap-through photovoltaic dipole antenna with solar concentrator. *Electron. Lett.* **2009**, *45*, 241–242.
9. Henze, N.; Giere, A.; Fruchting, H. GPS patch antenna with photovoltaic solar cells for vehicular applications. *Veh. Technol. Conf.* **2003**, *1*, 50–54.
10. Shynu, S.V.; Roo~Ons, M.J.; Ammann, M.J.; McCormack, S.J.; Norton, B. Dual band a-Si:H solar-slot antenna for 2.4/5.2 GHz WLAN applications. In Proceedings of the 2009 3rd European Conference on Antennas and Propagation, Berlin, Germany, 23–27 March 2009; pp. 408–410.
11. O'Conchubhair, O.; Yang, K.; Mcevoy, P. Dye-sensitized Solar Vivaldi Antenna. *IEEE Antennas Wirel. Propag. Lett.* **2016**, *15*, 893–896. [CrossRef]
12. Danesh, M.; Long, J.R.; Simeoni, M. Small-area solar antenna for low-power UWB transceivers. In Proceedings of the 4th European Conference on Antennas and Propagation, Berlin, Germany, 12–16 April 2020; pp. 1–4.
13. Gurung, A.; Elbohy, H.; Khatiwada, D.; Mitul, A.F. A simple cost-effective approach to enhance performance of bifacial dye-sensitized soalr cells. *IEEE J. Photovolt.* **2016**, *6*, 912–917. [CrossRef]
14. Chiang, C.; Chien, L. A pitaya dye-sensitized solar cell monitor for environmental sunlight intensity detection. *IEEE Sens J.* **2019**, *19*, 4229–4236. [CrossRef]
15. Chou, J.C. IGZO/TIO2 composited film as a ptotoelectrode with reduced graphene oxide/Pt counter lelctrode for a dye-sensitized soalr cell. *IEEE J. Photovolt.* **2018**, *8*, 769–776.
16. Cardoso, B.N.; Kohlrausch, E.C.; Laranjo, M.T. Tuning anatase-rutile phase transition temperature: TIO_2/SIO_2 nanoparticles applied in dye-sensitized solar cells. *Int. J. Photoenergy* **2019**, *2019*, 1–9. [CrossRef]
17. Bomben, P.G.; Robson, K.C.D.; Koivisto, B.D.; Berlinguette, C.P. Cyclometalated ruthenium chromophores for the dye-sensitized solar cell. *Coord. Chem. Rev.* **2012**, *256*, 1438–1450. [CrossRef]
18. Yu, M.; Wang, C. The Research Progress of Dye-sensitized Solar Cells. *Shangdong Chem. Ind.* **2016**, *9*, 45–47.
19. O'Conchubhair, O.; McEvoy, P.; Ammann, M.J. Dye-Sensitized Solar Cell Antenna. *IEEE Antennas Wirel. Propag. Lett.* **2016**, *16*, 352–355. [CrossRef]

20. Liu, S.; Yang, D.; Chen, Y.; Zhang, X.; Xiang, Y. Compatible integration of circularly polarized omnidirectional metasurface antenna with solar cells. *IEEE Trans. Antennas Propag.* **2019**, *68*, 4155–4160. [CrossRef]
21. Zhou, W.; Fu, L. Research on Evaporation-induced Sdlf-assembly Method and Photoelectric Properties of TIO2 Films. *Mater. Rep.* **2016**, *30*, 10–13.
22. Tao, J.; Tang, Y.; Yu, J.; Deng, J. Flexible Dye Sensitized Solar Cell Based on Stainless Steel and Its Preparation Method. CN Patent CN101447341 A, 9 February 2011.
23. Li, K. General full-wave Green's functions in spectral domain for arbitrarily multilayered dielectric media. *IEEE Int. Microw. Symp. Dig.* **1997**, *3*, 1571–1574.
24. Ansoft HFSS Online Help. Available online: https://ansyshelp.a-nsys.com/ (accessed on 15 October 2019).

Letter

Timestamp Estimation in P802.15.4z Amendment †

Ioan Domuta *, Tudor Petru Palade, Emanuel Puschita and Andra Pastrav

Communication Department, Technical University of Cluj-Napoca, 400027 Cluj-Napoca, Romania; tudor.palade@com.utcluj.ro (T.P.P.); Emanuel.Puschita@com.utcluj.ro (E.P.); Andra.PASTRAV@com.utcluj.ro (A.P.)

* Correspondence: ioan.domuta@2dd.ro; Tel.: +40-264-440000

† This paper is an extended version of our paper: Domuta, I.; Palade, T.P.; Puschita, E.; Pastrav, A. Localization in 802. 15.4z Standard. In Proceedings of the 2020 International Workshop on Antenna Technology (iWAT), Bucharest, Romania, 25–28 February 2020; pp. 1–4, doi:10.1109/iWAT48004.2020.1570615511.

Received: 9 July 2020; Accepted: 20 September 2020; Published: 22 September 2020 

Abstract: Due to the known issue that the ranging in the 802.15.4™-2015 standard is prone to external attacks, the enhanced impulse radio (EiR), a new amendment still under development, advances the secure ranging protocol by encryption of physical layer (PHY) timestamp sequence using the AES-128 encryption algorithm. This new amendment brings many changes and enhancements which affect the impulse-radio ultra-wideband (IR-UWB) ranging procedures. The timestamp detection is the base factor in the accuracy of range estimation and inherently in the localization precision. This paper analyses the key parts of PHY which have a great contribution in timestamp estimation precision, particularly: UWB pulse, channel sounding and timestamp estimation using ciphered sequence and frequency selective fading. Unlike EiR, where the UWB pulse is defined in the time domain, in this article, the UWB pulse is synthesized from the power spectral density mask, and it is shown that the use of the entire allocated spectrum results in a decrease in risetime, an increase in pulse amplitude, and an attenuation of lateral lobes. The paper proposes a random spreading of the scrambled timestamp sequence (STS), resulting in an improvement in timestamp estimation by the attenuation lateral lobes of the correlation. The timestamp estimation in the noisy channels with non-line-of-sight and multipath propagation is achieved by cross-correlation of the received STS with the locally generated replica of STS. The propagation in the UWB channel with frequency selective fading results in small errors in the timestamp detection.

Keywords: UWB; 802.15.4z; timestamp detection; ranging; multipath; frequency fading

1. Introduction

The wireless localization is a key part of many emerging technologies: internet of things (IoT), intelligent transportation systems (ITS), autonomous robots, or unmanned aerial vehicles. For many critical applications, localization accuracy is a basic requirement of localization systems. Due to its large bandwidth, the impulse radio ultra-wideband (IR-UWB) technology provides the best precision in range measurement by time-of-flight (ToF) estimation.

The basic feature in the accuracy of estimating the ToF is the shape of the pulse, more precisely the speed of increase of the pulse front. Nowadays the IEEE Task Group 4z, (TG4z), [1] is working on the enhanced impulse radio (EiR) project focused on localization safety improvement. This new amendment proposes a new UWB reference pulse and a time domain mask. In this paper, the UWB pulse is synthesized from power spectral density (psd) specified by the 802.15.4-2015 standard [2]. Two shapes of pulses were synthesized and compared, the first pulse being synthesized using only the central lobe of the psd mask and the second one being synthesized from the entire allocated spectrum.

The EiR amendment proposes that the timestamp estimation is validated by the cross-correlation of locally generated STS replica with received STS sequence. To avoid the interferences, the pulses are spread out on a symbol. This article proposes a supplementary spreading by a bit position modulation with a randomly generated sequence and shows, by simulated experiments, that lateral lobes of cross-correlation are mitigated by this modulation.

The behavior of the proposed methods is analyzed in a noisy radio channel with non line-of-sight (NLOS) and multipath propagation. The channel impulse response is estimated and subsequently used for the generation of a local replica of STS.

The main contributions of this article are that it:

- demonstrates that the inclusion of lateral lobes with very low power spectral density (−51.3 dBm and −59.3 dBm) in pulse synthesis leads to a pulse with a tighter shape and a steeper rising edge than the pulse recommended by EiR;
- shows that, by a random spreading of STS, that it results in easier extraction of the main lobe by mitigating lateral lobes of cross-correlation.

This paper is organized as follows. Section 2 presents state of the art research in the field. In Section 3, the UWB pulses are synthesized. Section 4 presents the random spreading of STS sequence and timestamp estimation in noisy channels and NLOS propagation. All sections incorporate simulated experiments, and because the simulation results from a subsection are used in the subsequent ones, they will be presented along with the theoretical aspects in the corresponding subsection.

2. Literature Review

The UWB radio holds a large bandwidth, but the harmonized standards [3] impose upper limits for power spectral density (−41.3 dBm/MHz, which is under the noise floor), resulting in great difficulty in the extraction of signal from noise. The research in [4,5] shows the presence of intra-symbol interference (IASI), inter-symbol interference (ISI), and multipath interference (MUI). In order to minimize the interferences, the pulse should have a small duration of the leading lobe and a high attenuation of the side lobes. It must be noted that this small duration of the main lobe can lead to an excess in bandwidth. The pulse shape has to be a compromise between the regulatory compliance, the need of low voltage and low power supply, low duration for maximization of data rate, ranging accuracy, and minimalization of interferences. In most cases, the UWB pulse is synthesized from Gaussian impulse [6], its derivatives [7], or a linear combination of Gaussian pulses [8]. Keshavarz et al. [9] infer the weights of the derivatives in the impulse structure by particle swarm optimization (PSO) algorithm, and PSO algorithm is used for optimization of the architecture of a UWB transmitter [10]. A linear combination of Gaussian monocycles with weight optimization by semidefinite programming is used for pulse synthesis [11]. Baranauskas and Zelenin present a direct waveform synthesis of UWB pulse by high speed DAC [12].

The EiR amendment specifies two pulses [13] as boundaries for the UWB pulse and a time domain mask [14] as a constraint for pulse shape. The UWB pulse synthesized from the entire allocated spectrum falls into the time domain mask, has high energy, and can be used as a reference in UWB pulse design.

In the 802.15.4 standard [2], the UWB PHY is specified in detail and, the transceivers manufactured in this technology are widely used in localization. However, several researches show that the range measurement in the current technology is prone to external attacks. Francillon et al. [15] present a relay attack, Taponecco et al. [16] show a delay attack and Singh et al. [17] propose a modulation scheme that secures the distance measurement against relay attack. The EiR project [1] brings a lot of improvements, including UWB reference pulse shape, preamble symbols revision, addition of scrambled timestamp sequence for secure ranging, an increase in data rate and PHY payload length, and the modification and addition of a new MAC primitive for key management. An overview of the EiR standard is presented in the work of Sedlacek et al. [18].

This article is limited to estimating the timestamp, without going into detail regarding ranging or location methods. Alarifi et al. perform a deep analysis of ultrawide band indoor positioning Technologies [19]. Several works deal with the wireless localization in internet of things (IoT) [20–22] and many research depict the localization in vehicular technologies [23–25].

3. UWB Pulse Synthesis

The pulse shape plays an essential role in the ranging accuracy and in reaching the maximum distance, while maintaining regulatory compliance. The 802.15.4-2015 standard, hereinafter called 'old standard' has been defined as a root raised cosine reference pulse. As this pulse has a precursor, it can mask the attenuated first path signal. The TG4 proposes [13] that the transmitted pulse shape p(t) to be constrained by the time domain mask, specified by the standard. The EiR specifies that the pulse risetime, 10–90%, for 500 MHz channels has to be maximum 2 ns.

This paper proposes the synthesis of UWB pulse from the compliant power spectral density mask (psd) [2] by Kolmogorov factorization (detailed in Appendix A) [26], because this method provides a minimum phase pulse, as it is specified by the EiR. The old standard specifies the psd mask as it is depicted in Figure 1a, namely the trace 802.15.4a mask. The power is expressed in Watts, in order to get the pulse amplitude in volts (1 Ω load). For pulse synthesis, a new spectral mask is designed, trace 1.5 ns risetime impulse, using a raised cosine profile

$$H = \begin{cases} psd_H & \left|f - f_c\right| < (1-\beta)F_c \\ \left(\frac{psd_H + psd_L}{1}\right)\left(1 + \cos\left[\frac{\pi}{2\beta}\left(\frac{f-f_c}{F_c} - 1 + \beta\right)\right]\right) + psd_L & (1-\beta)F_c \le \left|f - f_c\right| \le (1-\beta)F_c \\ 0 & otherwise \end{cases} \tag{1}$$

where central frequency $f_c = 0$, cutoff frequency $F_c = 315$ MHz, roll-off factor $\beta = 0.25$ and psd_H *and* psd_L are the high and low value of psd mask. The sampling frequency is $f_s = 10$ GHz and the window length is $N = 10^4$ samples for 1 MHz frequency resolution.

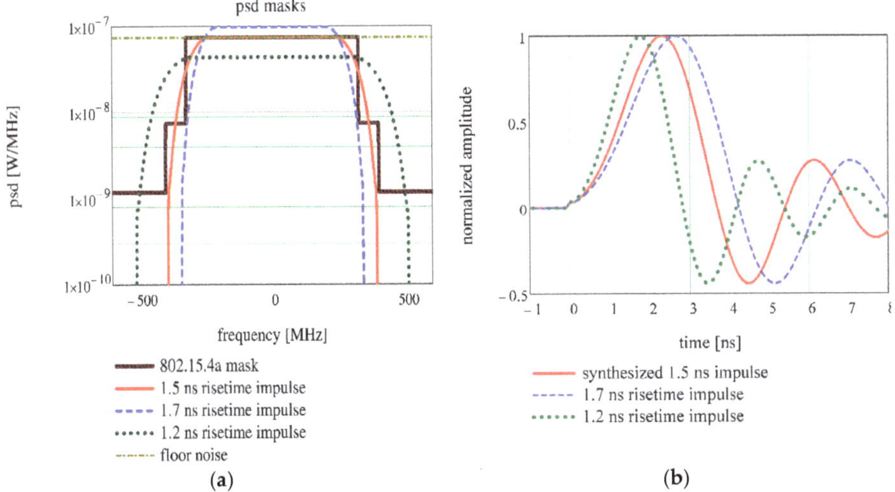

Figure 1. UWB impulse synthesis: (**a**) Spectral masks; (**b**) The pulses in time domain.

Different from standard [3], where the pulse energy is averaged on a 1 ms interval, in the paper, the mean power is computed on a 1 μs interval, considering that the pulse repetition frequency $PRF = 1$ MHz. Accordingly, the pulse amplitude is inferred based on this PRF. The EiR defines many mean PRFs and, in order to comply with the regulations, the determined amplitude has to be scaled with $\sqrt{PRF}$ [MHz]. Furthermore, the regulation imposes the pulse peak power to a value that shall not exceed 0 dBm on 50 MHz bandwidth, and the impulse has to respect this restriction too.

Figure 1b shows the synthesized pulse, trace synthesized 1.5 ns impulse, compared to EiR compliant pulses for the 499.2 MHz bandwidth (i.e., 1.7 ns risetime impulse and 1.2 ns risetime impulse traces). The synthesized pulse is situated between recommended pulse limits, so it respects the specifications. The traces 1.7 ns risetime and 1.2 ns risetime in Figure 1a show that the pulses suggested by the EiR do not fit exactly in the standard psd mask, the first exceeding the maximum psd and the second exceeding the bandwidth.

Recently, TG4z has defined a time domain mask for UWB impulse [12], as illustrated in Figure 2b. Based on this mask, it is appropriate to search for a new UWB pulse shape which falls in this mask for impulse energy maximization and risetime reduction. The Cramer–Rao lower bound in ToF estimation is inversely proportional to effective bandwidth [27], so it is convenient to use the lateral lobes of low power for pulse synthesis. In order to do this, a new psd mask is designed using a sum of raised cosine profiles, Figure 2, having the following parameters: ($F_{c1} = 315$ MHz, $\beta_1 = 0.05$); ($F_{c2} = 400$ MHz, $\beta_2 = 0.05$); ($F_{c3} = 500$ MHz, $\beta_3 = 0.05$).

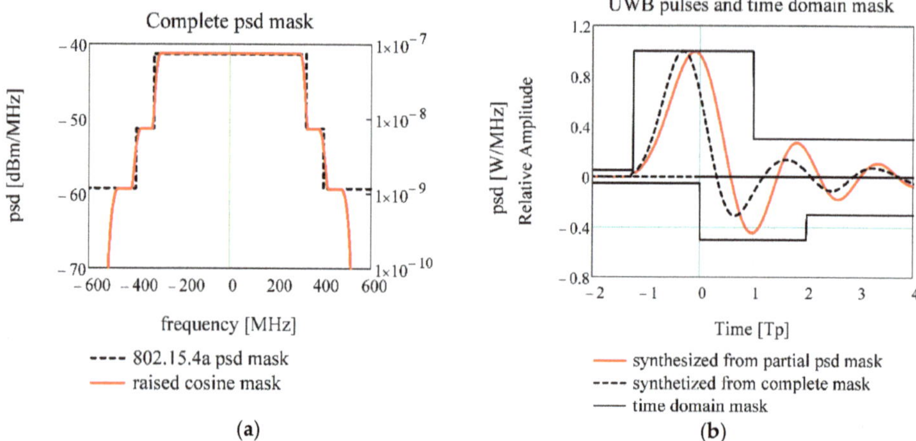

Figure 2. UWB impulse synthesis from complete spectral mask: (**a**) Spectral mask from multiple raised cosine profiles; (**b**) Time domain mask and synthesized pulses.

Figure 2b shows that the pulse synthesized from complete mask has a smaller risetime and a stronger attenuation of lateral lobes compared to the pulse synthesized from partial psd mask. As such, the pulse synthesized from complete mask allows for more precise timestamp estimation and less interferences. The pulse synthesized from complete mask has the amplitude 0.187 V/1 Ω/1 µs, higher energy, low risetime $T_{10\%\rightarrow 90\%} = 1.2$ ns and smaller lateral lobes than the pulse synthesized from partial psd mask which has an amplitude of 0.164 V/1 Ω/1 µs and a risetime of $T_{10\%\rightarrow 90\%} = 1.5$ ns. The pulse risetime plays a key role in timestamp estimation precision. Therefore, it is preferable to choose the pulse synthesized from complete mask as UWB reference pulse.

The degree of spectrum usage is evaluated by normalized effective signal power (NESP) [6]

$$\text{NESP} = \frac{\int |S(f)|^2 df}{\int M(f) df} \quad (2)$$

where $|S(f)|^2$ is spectral density of the impulse and $M(f)$ is allocated spectral mask.

Because the signal does not have a uniform distribution on the entire spectrum, the effective bandwidth β is defined

$$\beta = \sqrt{\frac{\int f^2 |S(f)|^2 df}{\int |S(f)|^2 df}} \quad (3)$$

Usually, the quality of the UWB pulse synthesis is defined in terms of occupied bandwidth and pulse duration. Table 1 compares the quality parameters of proposed pulse with P802.15.4z reference pulse.

Table 1. The synthesis quality parameters.

Synthesis Method	Bandwidth β [MHz]	NESP [%]	Risetime [ns]	Amplitude [V]
802.15.4z reference pulse [11]	533	94.5	1.5	0.164
Pulse synthesized from entire spectral mask (Figure 2a)	535	98.5	1.2	0.187

The minimum uncertainty, σ, for range estimation in the time of arrival method is quantified by Cramér–Rao lower bound (CRLB) [28]

$$\sigma = \frac{c}{\beta \sqrt{8\pi^2 SNR}} \quad (4)$$

where β is effective bandwidth, c is speed of the light, and SNR is signal to noise ratio.

Relation (100) clearly shows that for minimizing the uncertainty the effective bandwidth has to be maximized, but within the limit of regulations.

4. Timestamp Estimation in the 802.15.4z Standard

4.1. Timestamp Estimation by Random Spreading of STS

The new standard brings in a new physical protocol data unit (PPDU) structure by incorporating the STS for secure ranging. The STS is encrypted using the AES-128 algorithm, the time of arrival estimation is achieved on STS, and the range measurement is validated only if the received STS cross correlated with the locally generated reference exceeds the "match level" [29] threshold.

The default PHY frame format proposed by EiR [29] is depicted in Figure 3, where SHR is the synchronization header (preamble), STS is the scrambled timestamp sequence, and PHR is the PHY header.

Figure 3. PHY frame format in the 802.15.4z amendment.

To avoid the inter-pulse interferences, [30] stipulates that every component B_k of Ipatov ternary symbol (ITS), or STS, is spread out on a symbol of length δ_L by $\sum_{n=0}^{\delta_L-1} \delta(n)\, p(t - nT_{ch})$; where $\delta(n)$ is Kronecker delta, $p(t)$ is UWB pulse and $T_{ch} = 2\ ns$ is chip duration. To mitigate the side lobes of correlation, this paper proposes a supplementary spreading by a randomly generated sequence S_k. The sequence of length N is:

$$s(t) = \sum_{k=0}^{N-1} B_k \sum_{n=0}^{\delta_L-1} (\delta(n) \cdot p(t - (k \cdot \delta_L + S_k n_s + n) T_{ch})). \quad (5)$$

The STS sequence of length $N = 128$, *s*(*t*), is generated by taking $B_k = 1 - 2A_k$, A being the result of AES encryption. The peak pulse repetition frequency (PRF) is 499.2 MHz, the mean PRF is 62.4 MHz, resulting $\delta_L = 8$ chips. Figure 4a depicts the STS sequences: STS standard spread out is generated according to the standard specifications, $S_k = 0$, STS reference signal is the signal needed to achieve correlation and for STS randomly spread out, the S_k is generated based on a linear feedback shift register with the characteristic polynomial $x^3 + x + 1$. The cross-correlation of the reference signal and STSs are shown in Figure 4b. The ratio between the main lobe and maximum side lobe is $\eta \cong 10$ for cross-correlation randomly spread out and $\eta \cong 3.33$ for cross-correlation standard spread out, which proves the efficiency of random spreading.

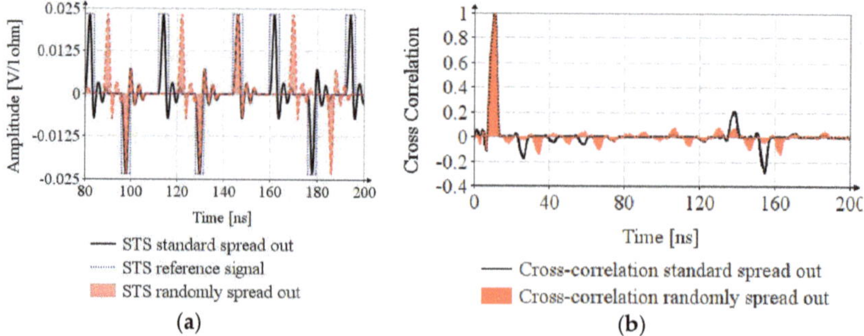

Figure 4. Timestamp estimation by STS: (**a**) UWB pulse position in STS sequence; (**b**) cross-correlation of STS sequences with reference signal.

Figure 4a shows that the random spreading increases the risk of IASI. Therefore, it is necessary to perform the analysis of timestamp estimation for propagation in noisy channel and for multipath propagation.

4.2. Sounding the Channel with Multipath Propagation

The channel sounding is the estimation of the channel impulse response (CIR) using preamble sequence, in order to remove the noise and design the channel equalizer.

The frequency-dependent path gain $G(d, f)$ is modeled considering an isotropic radiation pattern, with the "antenna attenuation factor" [31] of $\frac{1}{2}$:

$$G(d,f) = \frac{P_{Rx}(d,f)}{P_{Tx}(f)} = \frac{1}{2}\, G_0\eta_{Tx}\eta_{Rx}\frac{(f/f_c)^{-2(\kappa+1)}}{(d/d_0)^n}, \tag{6}$$

where $P_{Rx}(d,f), P_{Tx}(f)$ are received and transmitted power, η_{Tx}, η_{Rx} are the transmission and reception antenna gains, G_0 is the path gain at reference distance d_0, d is the distance between transmitter and receiver, n is the path gain exponent, f_c is the carrier frequency, f is the frequency and κ is the frequency decaying factor.

The frequency decaying factor follows the Friis equation and, in Equation (6), it has the value $\kappa = 0$. The path loss varies from $n = 1.2$ in industrial LOS, $n = 1.76$ in outdoor LOS, to $n = 4.58$ in residential NLOS. It is noted that multipath propagation in the industrial environment leads to an increase in the path gain.

Using the Saleh-Valenzuela statistical model, the propagation paths are designed as the sum of clusters, every cluster having multiple rays. The impulse response $h^m(t)$ is

$$h^m(t) = \Sigma_{l=0}^{L}\Sigma_{r=0}^{K}a_{r,l}exp\left(j\varnothing_{r,l}\right)\delta\left(t - T_l - \tau_{r,l}\right), \tag{7}$$

where $a_{r,l}$ is the tap weight of r ray in cluster l, $\varnothing_{r,l}$ is the ray phase, T_l is the delay of l cluster and $\tau_{r,l}$ is the delay of r ray relative to cluster l front. The intervals between the time arrivals of the clusters is modeled as Poisson process with the arrival rate Λ, and the ray delays inside the cluster are modeled as a mixture of two Poisson processes with arrival rates λ_1, λ_2 and mixing weight β. The mean power of arriving clusters follows an exponential decay with time constant Γ, having a normal distribution around the mean value $\sigma_{cluster}$, and the cluster shape also bears to an exponential decay with time constant γ.

In the above CIR model, the first path has the highest energy. In non-line-of-sight propagation, there are cases when the first path is strongly attenuated. For such situations, [31] proposes a new modeling for the first path.

$$E\{a_{k,0}\} \propto \left(1 - \chi exp\left(-\frac{\tau_{k,l}}{\gamma_{rise}}\right)\right) exp(-\frac{\tau_{k,l}}{\gamma_1}), \tag{8}$$

where χ describes the attenuation of the first path, γ_{rise} determines how fast the power delay profile (PDP) increases, and γ_1 determines the profile decay. By joining Equations (7) and (8), the CIR, $h(t)$ can be found.

The preamble sequence consists of a string of 32 or 64 Ipatov ternary symbols, every symbol having 91 elements with 81 non-zero elements [30]. An Ipatov symbol has "perfect" periodic autocorrelation, i.e., all side lobes of autocorrelation are zero, and using Wiener-Hopf equation, by cross-correlation of the received signal $y(t)$ with the input Ipatov sequence $I(t)$ results immediately the CIR, $h(t)$

$$y(t) \star I(t) = h(t) * I(t) \star I(t) = h(t) * R_{I,I}(t), \tag{9}$$

where autocorrelation of ITS is: $R_{I,I}(0) = \text{N}; R_{I,I}(t) = 0$ for $t \neq 0$.

For outdoor, NLOS and multipath propagation environment, the channel PDP, $h(t)$, is modeled based on Equations (7) and (8), with the parameters retrieved from [31], and synthesized in Table 2.

Table 2. The UWB channel parameters for outdoor NLOS propagation.

Name [unit]	Symbol	Value
Path gain:		
Reference distance [m]	d_0	1
Gain at reference distance [dB]	G_0	−73
Path gain exponent	n	2.5
Frequency decaying factor	κ	0.13
Power delay profile:		
Expected number of clusters	$\overline{L}$	10.5
Clusters arrival rate [1/ns]	Λ	0.0243
Clusters decay time constant [ns]	Γ	104.7
Rays arrival rate [1/ns]	λ	0.223
Intracluster decay time constant [ns]	γ	9.3
First path attenuation	χ	0.65
Path increase time constant [ns]	γ_{rise}	5

For channel sounding, the EiR standard specifies a mandatory preamble sequence of 32 or 64 ITSs, the Ipatov symbol having a number of 91 elements with 10 zero elements. In this paper, the CIR is estimated only using one ITS with a total of 57 elements including 8 zero elements [32]. The transmitted sequence of pulses, $i(t)$, is generated based on Equation (5), with $B_k = I_k$, where I_k is the kth element of ITS. The received signal $y_i(t) = h(t) * i(t)$ is the convolution of PDP with the emitted sequence. The CIR estimation, $\hat{h(t)}$, is achieved by the cross-correlation of the received signal $y(t)$ with the Ipatov symbol I. Figure 5 shows that the CIR estimation, estimated PDP, is close enough to the true PDP. By successive transmission of ITSs contained in the preamble, the estimations are accumulated and averaged, resulting in SNR reduction and CIR estimation improvement.

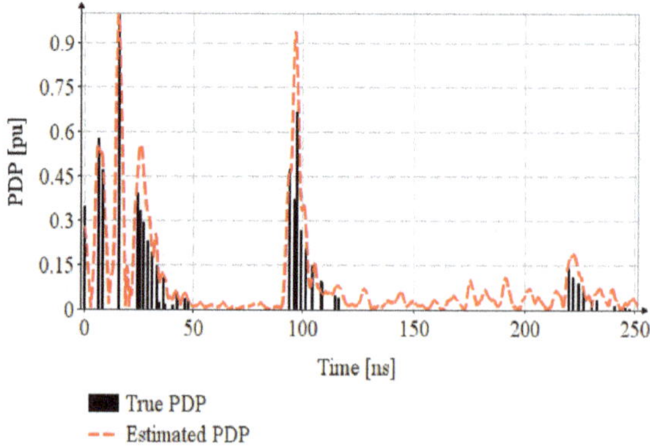

Figure 5. CIR estimation for channel with multipath propagation.

This CIR estimate will be used in the following paragraphs, for analysis of timestamp estimation in channels with multipath propagation. This is a classical channel model, more detailed models are being published in the recent research [33,34].

4.3. Timestamp Estimation in Channel with NLOS and Multipath Propagation

The EiR specifies that STS consists of 32, 64, 128 AES-128 sequences, successively transmitted, but in this section only one AES sequence is considered for analysis the impact of multipath propagation.

The STS is generated based on Equation (5) with random spread out, and for propagation simulation it is convoluted with $h(t)$.

In the usual way, the received signal is passed through an equalizer filter and cross-corelated with the STS reference $s_r(t)$. In this article, an easier way for timestamp detection is proposed, that is, to generate a virtual propagated STS reference by convolution of the STS reference with the estimated CIR, $h\hat{(t)}$, and cross-correlation of received sequence $y_s(t)$ with locally generated replicas $y_r\hat{(t)}$ as depicted in Algorithm 1.

Algorithm 1: Timestamp estimation in multipath propagation

Inputs: STS reference (AES-128 sequence), $s_r(t)$; Received STS, $y_s(t)$

1. Generate locally replica of STS:
$$y_r\hat{(t)} = h\hat{(t)} * s_r(t)$$
2. Compute the received sequence:
$$y_s(t) = h(t) * s(t)$$
3. Timestamp estimation:
$$r_{rs}(t) = y_r\hat{(t)} \star y_s(t)$$

Output: return $r_{rs}(t)$

Figure 6 shows that by cross-correlating the received STS, $y_s(t)$, with the STS reference, $s_r(t)$, the timestamp is not detectable, (see the correlation with STS reference trace), and that the cross-correlation of received signal with the locally generated replica, $y_r\hat{(t)}$, the side lobes are strongly attenuated (see the correlation with STS reference convoluted with PDP trace).

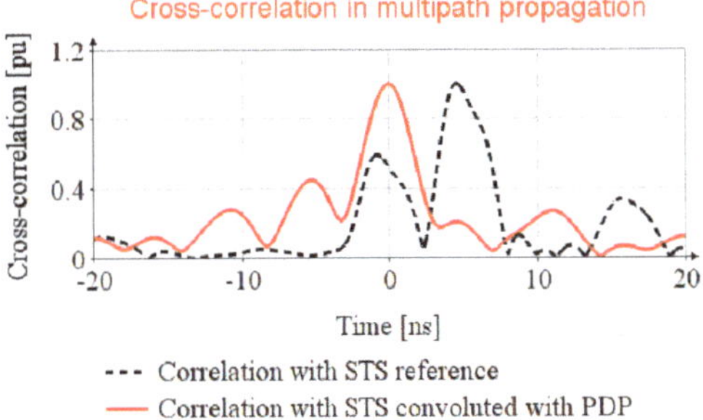

Figure 6. Timestamp estimation in channel with NLOS and multipath propagation.

4.4. Timestamp Estimation in Noisy Channel with NLOS and Multipath Propagation

The EiR standard specifies that the typical range of radio is 100 m. For timestamp estimation in noisy channel we consider that the transmitter is situated at 22 m and the noise source is situated at reference distance of 1 m.

The transmitter emits with maximum compliant power (Figure 2a) and the noise source emits with floor noise level (i.e., −41 dBm/MHz) [35]. In order to mitigate the effect of noise, consider that the receiver has a 1 GHz passband filter on the input. The standard deviation of noise is $\sigma = 0.0228$ V and the signal to noise ratio on the receiver is $SNR = -38$ dB. In this situation, the timestamp is undetectable from only a single AES-128 sequence (Figure 7b, trace 1 STS), so STS will consist of multiple AES sequences as the EiR standard specifies.

The timestamp detection is detailed in Algorithm 2 and the results are shown in Figure 7.

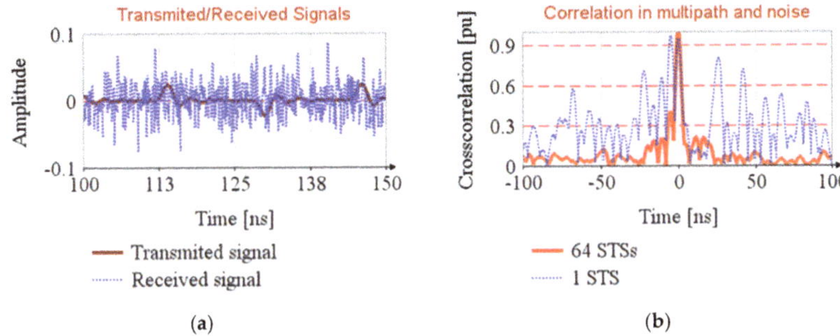

Figure 7. Timestamp estimation in noisy channel with multipath propagation.

Algorithm 2: Timestamp estimation in noisy channel

Inputs: STS reference (AES-128 sequence), $s_r(t)$; Received STSs, $y_s^i(t)$; Number of STS, N;

1. Generate locally replica of STS:
$$\hat{y_r(t)} = \hat{h(t)} * s_r(t)$$
2. Compute the received sequences:
$$for\ i \in [0, N-1]\ \ do:$$
$$y_s^i(t) = h(t) * s(t) + n^i(t)$$
3. Average the received sequences:
$$y_s(t) = \frac{1}{N} \sum_{i=0}^{N-1} y_s^i(t)$$
4. Timestamp estimation:
$$r_{rs}(t) = \hat{y_r(t)} \star y_s(t)$$

Where $n^i(t)$ is the noise with standard deviation σ

Output: return $r_{rs}(t)$

4.5. UWB Channel with Frequency Selective Fading

Depending on antenna design, or antenna position relative to external objects, it is possible to encounter frequency selective fading [35,36]. To analyze the impact of frequency fading in timestamp detection, consider UWB channel 9 with the spectral mask profile power spectral density, illustrated in Figure 8, having a selective fading of 12 dBm at carrier frequency and frequency decaying factor of $\kappa = 0.13$. The channel frequency response, $H(f)$, is achieved by Kolmogorov factorization (Appendix A).

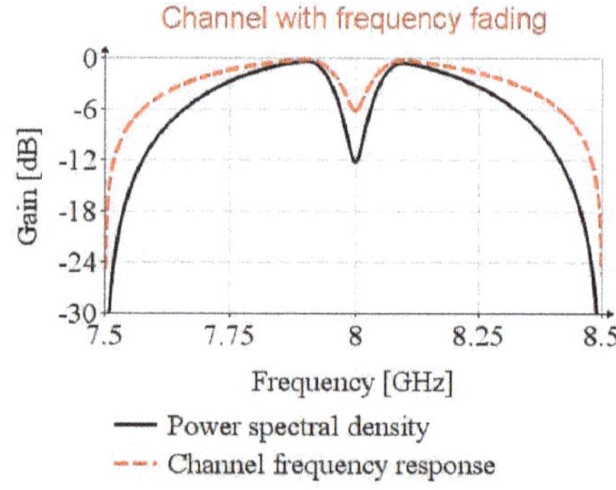

Figure 8. Psd mask and channel response.

To estimate the CIR, $H\hat{(f)}$, the Ipatov symbol, $i(t)$, is shifted on to the carrier, f_c, by complex modulation with $\exp(j2\pi f_c t)$. The cross-correlation is performed in the frequency domain, and to avoid the circular correlation, the series is padded with zeros. The propagation of STS in the faded channel is simulated by the shifting of the STS on to the carrier and by the convolution with $H(f)$. The timestamp estimation is detailed in Algorithm 3.

Algorithm 3: Timestamp estimation in channel with frequency selective fading

---

Inputs: Ipatov symbol, $i(t)$; STS, $s(t)$;

CIR estimation:

1. Lift up the Ipatov symbol on the carrier:
$$i_c(t) = e^{j2\pi f_c t} \cdot i(t)$$
2. Complete the series with zero and perform FFT.
$$I_c(f) = FFT(i_c(t))$$
3. Convolve the Ipatov symbol with CIR:
$$Y_I(f) = H(f) \cdot I_c(f)$$
4. CIR estimation by cross-correlation:
$$H\hat{(f)} = \frac{Y_I(f) \cdot \overline{I_c(f)}}{I_c(f) \cdot \overline{I_c(f)}}$$

Timstamp estimation from STS:

5. Shift STS to carrier, complete with zero and perform FFT:
$$S_c(f) = FFT\left(e^{j2\pi f_c t} \cdot s(t)\right)$$
6. Generate locally replica of STS:
$$S_r\hat{(f)} = H\hat{(f)} \cdot S_c(f)$$
7. Simulate the received STS by convolution:
$$Y_S(f) = H(f) \cdot S_c(f)$$
8. Cross-correlation in the frequency domain:
$$R_{rs}(f) = S_r\hat{(f)} \cdot \overline{Y_S(f)}$$
9. Correlation in time domain:
$$r_{rs}(t) = IFFT(R_{rs}(f))$$

Output: return $r_{rs}(t)$

Figure 9a, psd at the emission, shows that Ipatov symbol has a uniform distribution over the entire bandwidth, that it follows the compliant spectral mask (Figure 2a), and that after propagation, psd after fading, it borrows CIR spectral mask.

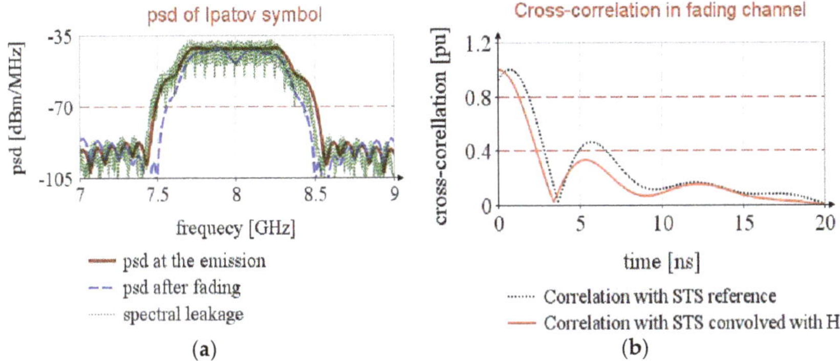

Figure 9. Timestamp estimation in channel with selective fading.

The propagation in a channel with frequency fading leads to a small error in timestamp estimation, as shown in Figure 9b, correlation with STS reference, and this error is cancelled if the received STS is correlated with the locally generated replica, correlation with STS convolved with H.

It should be highlighted that the spectral leakage in the FFT leads to the introduction of nonuniformity in the estimated CIR spectrum, Figure 10a, resulting in incorrect results regarding timestamp estimation, Figure 10b.

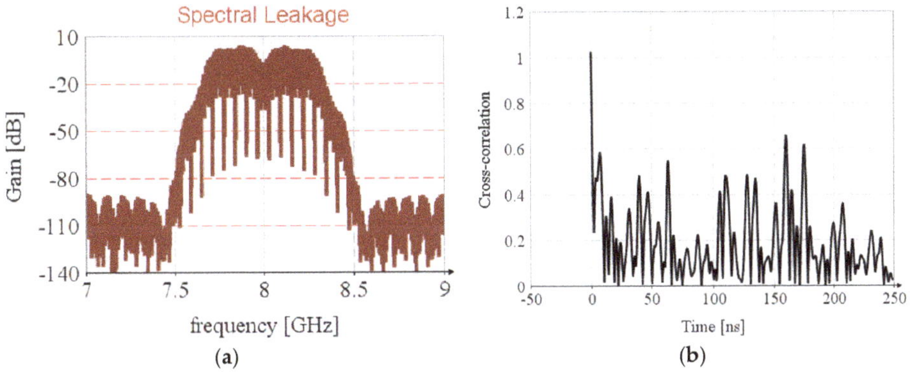

Figure 10. Spectral leakage in timstamp estimation: (**a**) Nonuniformity in CIR estimation; (**b**) Erroneous estimation of timestamp.

5. Results

The NESP shows the usage efficiency of the allocated bandwidth. Table 3 displays the occupied bandwidth and NESP for the pulse synthesized from entire spectral mask which is compared with previously published related works.

Table 3. UWB pulse parameters.

Synthesis Method	Bandwidth β [GHz]	NESP [%]
Pulse synthesized from entire spectral mask (Figure 2a)	0.537	98.5
Gaussian with NESP maximization [6]	7.160	98.7
Linear combination of Gaussian pulses [8]	7.200	91
VC oscillator with PSO optimization [10]	1.332	NA
Gaussian with semidefinite programing [11]	12	85.5

The uncertainty in range estimation for noisy channel with multipath propagation is computed based on Equation (4). The effective bandwidth is computed based on Equation (3). The leading lobe in Figure 7b is extracted, the samples series is completed with zero and then transformed in frequency domain. The effective bandwidth is nearly constant $\beta = 234$ MHz.

The propagation loss is determined considering outdoor LOS channel with path loss exponent $n = 1.76$,

$$L = 10 \cdot n \cdot \log\left(\frac{d}{d_o}\right) \tag{10}$$

where $d_0 = 1$ m is reference distance.

The noise has psd at floor level $P_n - 41$ dBm and 7 GHz bandwidth.

The SNR and minimum uncertainty are depicted in Table 4.

Table 4. CRLB for timestamp estimation by STS in noisy channels and multipath propagation.

Distance [m]	1	10	20	40	80	160
SNR [dB]	−10.2	−27.8	−33.1	−38.4	−43.7	−49
CRLB [m]	0.1665	1.25	2.31	4.25	7.78	14.33

6. Discussion

The synthesis of UWB impulse from a complete psd mask (Figure 2a) results in a risetime reduction, amplitude increasing, and the attenuation of side lobes relative to the impulse proposed by EiR (Table 1). Furthermore, the pulse psd fully complies with the standard spectral mask. Therefore, this impulse is advisable as a UWB reference pulse.

The random spreading of STS leads to an attenuation of lateral lobes of cross-correlation, but also leads to an increase in the probability of intra-symbol interferences. Thus, more investigations should be performed for testing these interferences in noisy channels with multipath propagation.

The cross-correlation in a noisy channel with multipath propagation displays a wide main lobe, which leads to a decrease in the accuracy of the timestamp estimation. This result seems to be due to the fact that the AES-120 sequence does not have uniform spectral distribution in the entire frequency band. The whitening of the AES sequence or of the entire STS by additional randomization could lead to the main lobe narrowing, and this is a subject of interest in our research.

Author Contributions: Conceptualization, I.D. and T.P.P.; methodology, E.P.; software, I.D.; investigation and writing, I.D., E.P. and A.P.; visualization, A.P.; supervision, T.P.P.; funding acquisition, T.P.P. All authors have read and agreed to the published version of the manuscript.

Funding: This research received no external funding.

Acknowledgments: This study was accomplished within the Centre of competence for wireless intrasatellite technologies (IntraSAT-Tech), Technical University of Cluj-Napoca.

Conflicts of Interest: The authors declare no conflict of interest.

Appendix A

Kolmogorov Factorization

Algorithm A: Kolmogorov Factorization

Inputs: correlation $\{r_t\}$, psd $S(\omega)$

1. Inputs test:
$$\{r_t\}\ ?\ \text{Continue: Goto 3}$$
2. Compute psd:
$$S(\omega) = FFT(\{r_t\})$$
3. Compute cepstrum coefficients:
$$ReH(\omega) = \tfrac{1}{2}\ln S(\omega)$$
4. Determine imaginary part of log filter by Hilbert transform:
$$REH(\omega) = FFT(ReH(\omega))$$
$$Hl_i = \begin{cases} 0 & if\ i \equiv 0 \\ -j\,REH_i & if\ i > 0 \\ j\,REH_i & if\ i < 0 \end{cases}$$
$$ImH(\omega) = IFFT(Hl)$$
5. Transfer function in frequency domain:
$$H(\omega) = e^{ReH + j\,ImH}$$
6. Transfer function in time domain of moving average type:
$$h = IFFT(H(\omega))$$
7. Autoregressive function:
$$a = \frac{h_0}{h}$$

Outputs: return $H(\omega)$, h, a

References

1. IEEE 802.15 WPAN™ Task Group 4z Enhanced Impulse Radio. Available online: http://www.ieee802.org/15/pub/TG4z.html (accessed on 19 October 2019).
2. IEEE Computer Society. *IEEE Standard for Low-Rate Wireless Networks Amendment 2: Ultra-Low Power Physical Layer*; IEEE Computer Society: Washington, DC, USA, 2016; Volume April.
3. ETSI. *Electromagnetic Compatibility and Radio Spectrum Matters (ERM); Short Range Devices (SRD) using Ultra Wide Band (UWB); Transmission Characteristics; Part 2: UWB Mitigation Techniques*; ETSI: Sophia Technology Park, France, 2014.
4. Luo, C.; Wu, X.; Cao, Y. Multipath interference analysis of IR-UWB systems in indoor office LOS environment. In Proceedings of the 2011 6th International ICST Conference on Communications and Networking in China, CHINACOM 2011, Harbin, China, 17–19 August 2011; pp. 846–850.
5. Stanciu, M.I.; Azou, S.; Rədoi, E.; Serbənescu, A. A statistical analysis of multipath interference for impulse radio UWB systems. *J. Frankl. Inst.* **2015**, *352*, 5952–5967. [CrossRef]
6. Li, B.; Zhou, Z.; Zou, W.; Li, D.; Zhao, C. Optimal waveforms design for ultra-wideband impulse radio sensors. *Sensors* **2010**, *10*, 11038–11063. [CrossRef] [PubMed]
7. Sharma, A.; Sharma, S.K. Spectral efficient pulse shape design for UWB communication with reduced ringing effect and performance evaluation for IEEE 802.15.4a channel. *Wirel. Netw.* **2019**, *25*, 2723–2740. [CrossRef]
8. Liu, X.; Premkumar, A.B.; Madhukumar, A.S. Pulse shaping functions for UWB systems. *IEEE Trans. Wirel. Commun.* **2008**, *7*, 1512–1516. [CrossRef]
9. Keshavarz, S.N.; Hamidi, M.; Khoshbin, H. A PSO-Based UWB Pulse Waveform Design Method. In Proceedings of the 2010 Second International Conference on Computer and Network Technology, Bangkok, Thailand, 23–25 April 2010; pp. 249–253. [CrossRef]
10. Ben Issa, D.; Samet, M. Design and Optimization of Dual-Band Energy-Efficient OOK UWB Transmitter Via PSO Algorithm. *J. Circuits Syst. Comput.* 2019. [CrossRef]
11. Wu, X.; Tian, Z.; Davidson, T.N.; Giannakis, G.B. Optimal waveform design for UWB radios. *IEEE Trans. Signal Process.* **2006**, *54*, 2009–2021. [CrossRef]

12. Baranauskas, D.; Zelenin, D. A 0.36W 6b up to 20GS/s DAC for UWB Wave Formation. In Proceedings of the 2006 IEEE International Solid State Circuits Conference—Digest of Technical Papers, San Francisco, CA, USA, 8–10 February 2006; pp. 2380–2389. [CrossRef]
13. McLaughin, M.; Niewczas, J.; Verso, B. IEEE P802.15-19-0443-01-004z, Text to Address Comment id r1-0820. 2018. Available online: http://www.ieee802.org/15/pub/TG4z.html (accessed on 19 October 2019).
14. McLaughin, M.; Hammerschmidt, J.; Ibrahim, B.; Verso, B. IEEE P802.15-20-0089-01-004z, Pulse Shape Text Changes for HRP UWB PHY. 2020. Available online: http://www.ieee802.org/15/pub/TG4z.html (accessed on 1 June 2020).
15. Francillon, A.; Danev, B.; Capkun, S. Relay Attacks on Passive Keyless Entry and Start Systems in Modern Cars. *Netw. Distrib. Syst. Secur. Symp.* **2011**, *431–439*, 8529521.
16. Taponecco, L.; Perazzo, P.; D'Amico, A.A.; Dini, G. On the feasibility of overshadow enlargement attack on IEEE 802.15.4a distance bounding. *IEEE Commun. Lett.* **2014**, *18*, 257–260. [CrossRef]
17. Singh, M.; Leu, P.; Capkun, S. UWB with Pulse Reordering: Securing Ranging against Relay and Physical-Layer Attacks. *NDSS* **2019**. [CrossRef]
18. Sedlacek, P.; Slanina, M.; Masek, P. An overview of the IEEE 802.15.4z standard its comparison and to the existing UWB standards. In Proceedings of the 2019 29th International Conference Radioelektronika, RADIOELEKTRONIKA 2019—Microwave and Radio Electronics Week, MAREW, Pardubice, Czech Republic, 16–18 April 2019; pp. 1–6. [CrossRef]
19. Alarifi, A.; Al-Salman, A.; Alsaleh, M.; Alnafessah, A.; Al-Hadhrami, S.; Al-Ammar, M.A.; Al-Khalifa, H.S. Ultra wideband indoor positioning technologies: Analysis and recent advances. *Sensors (Switzerland)* **2016**, *16*, 707. [CrossRef]
20. Khelifi, F.; Bradai, A.; Benslimane, A.; Rawat, P.; Atri, M. A Survey of Localization Systems in Internet of Things. *Mob. Netw. Appl.* **2019**, *24*, 761–785. [CrossRef]
21. Shit, R.C.; Sharma, S.; Puthal, D.; Zomaya, A.Y. Location of Things (LoT): A review and taxonomy of sensors localization in IoT infrastructure. *IEEE Commun. Surv. Tutor.* **2018**, *20*, 2028–2061. [CrossRef]
22. E Silva, P.F.; Kaseva, V.; Lohan, E.S. Wireless positioning in IoT: A look at current and future trends. *Sensors (Switzerland)* **2018**, *18*, 2470. [CrossRef] [PubMed]
23. Kuutti, S.; Fallah, S.; Katsaros, K.; Dianati, M.; Mccullough, F.; Mouzakitis, A. A Survey of the State-of-the-Art Localization Techniques and Their Potentials for Autonomous Vehicle Applications. *IEEE Internet Things J.* **2018**, *5*, 829–846. [CrossRef]
24. Balico, L.N.; Loureiro, A.A.F.; Nakamura, E.F.; Barreto, R.S.; Pazzi, R.W.; Oliveira, H.A.B.F. Localization prediction in vehicular Ad hoc networks. *IEEE Commun. Surv. Tutor.* **2018**, *20*, 2784–2803. [CrossRef]
25. Del Peral-Rosado, J.A.; Seco-Granados, G.; Kim, S.; López-Salcedo, J.A. Network Design for Accurate Vehicle Localization. *IEEE Trans. Veh. Technol.* **2019**, *68*, 4316–4327. [CrossRef]
26. Scharf, L. *Statistical Signal Processing: Detection, Estimation and Time Series Analysis*, 1st ed.; Addison-Wesley Publishing Company, Inc.: Boston, MA, USA, 1991; pp. 424–429. ISBN 0-201-19038-9.
27. Qi, Y.; Kobayashi, H.; Suda, H. On time-of-arrival positioning in a multipath environment. *IEEE Trans. Veh. Technol.* **2006**, *55*, 1516–1526. [CrossRef]
28. Álvarez, R.; Díez-González, J.; Alonso, E.; Fernández-Robles, L.; Castejón-Limas, M.; Perez, H. Accuracy Analysis in Sensor Networks for Asynchronous Positioning Methods. *Sensors* **2019**, *19*, 3024.
29. Niewczas, J.; Verso, B.; Fagan, T.; Leong, F.; Hammerschmidt, J.; Ibrahim, B.; Sasoglu, E.; Knobloch, D.; Reisinger, T. IEEE P802.15-19-0134-00-004z, Security vs. Sequence Length Considerations. 2019. Available online: http://www.ieee802.org/15/pub/TG4z.html (accessed on 21 October 2019).
30. Verso, B.; Leong, F.; Hammerschmidt, J.; Niewczas, J.; Ibrahim, B.; Shah, T.; Reisinger, T.; Daniel Knobloch, D. IEEE P802.15-18-0477-00-004z, HRP UWB PHY Enhanced Mode Converged Consensus. 2018. Available online: http://www.ieee802.org/15/pub/TG4z.html (accessed on 19 October 2019).
31. Molisch, A.F.; Cassioli, D.; Chong, C.C.; Emami, S.; Fort, A.; Kannan, B.; Karedal, J.; Kunisch, J.; Schantz, H.G.; Siwiak, K.; et al. A comprehensive standardized model for ultrawideband propagation channels. *IEEE Trans. Antennas Propag.* **2006**, *54*, 3151–3166. [CrossRef]
32. Li, X.; Fan, P.; Mow, W.H. Existence of ternary perfect sequences with a few zero elements. In Proceedings of the 5th International Workshop on Signal Design and Its Applications in Communications, IWSDA'11, Guilin, China, 10–14 October 2011; pp. 88–91.

33. Kram, S.; Stahlke, M.; Feigl, T.; Seitz, J.; Thielecke, J. UWB channel impulse responses for positioning in complex environments: A detailed feature analysis. *Sensors (Switzerland)* **2019**, *19*, 5547. [CrossRef]
34. Nwadiugwu, W.P.; Kim, D.S. Ultrawideband Network Channel Models for Next-Generation Wireless Avionic System. *IEEE Trans. Aerosp. Electron. Syst.* **2020**, *56*, 113–129. [CrossRef]
35. International Telecommunication Union (ITU). Presentation: Advanced wireless technologies and spectrum management. In *Radio Spectrum Management for a Converging World*; ITU: Geneva, Switzerland, 2004; Available online: http://handle.itu.int/11.1002/pub/800c7e3d-07f64a9d-en (accessed on 25 October 2019).
36. Otim, T.; Díez, L.E.; Bahillo, A.; Lopez-Iturri, P.; Falcone, F. Effects of the body wearable sensor position on the UWB localization accuracy. *Electronics* **2019**, *8*, 1351. [CrossRef]

Article

Small Antennas for Wearable Sensor Networks: Impact of the Electromagnetic Properties of the Textiles on Antenna Performance

Gabriela Atanasova [1,*] and Nikolay Atanasov [1,2]

1 Department of Communication and Computer Engineering, South-West University "Neofit Rilski", 2700 Blagoevgrad, Bulgaria; natanasov@swu.bg
2 Electromagnetic Compatibility Laboratory, Bulgarian Institute of Metrology, 1040 Sofia, Bulgaria
* Correspondence: gatanasova@windowslive.com

Received: 24 August 2020; Accepted: 8 September 2020; Published: 10 September 2020

Abstract: The rapid development of wearable wireless sensor networks (W-WSNs) has created high demand for small and flexible antennas. In this paper, we present small, flexible, low-profile, light-weight all-textile antennas for application in W-WSNs and investigate the impact of the textile materials on the antenna performance. A step-by-step procedure for design, fabrication and measurement of small wearable backed antennas for application in W-WSNs is also suggested. Based on the procedure, an antenna on a denim substrate is designed as a benchmark. It demonstrates very small dimensions and a low-profile, all while achieving a bandwidth ($|S_{11}| < -6$ dB) of 285 MHz from 2.266 to 2.551 GHz, radiation efficiency more than 12% in free space and more than 6% on the phantom. Also, the peak 10 g average SAR is 0.15 W/kg. The performance of the prototype of the proposed antenna was also evaluated using an active test. To investigate the impact of the textile materials on the antenna performance, the antenna geometry was studied on cotton, polyamide-elastane and polyester substrates. It has been observed that the lower the loss tangent of the substrate material, the narrower the bandwidth. Moreover, the higher the loss tangent of the substrate, the lower the radiation efficiency and SAR.

Keywords: small antenna; textile antenna; wearable antenna; SAR; flexible antenna; low-profile antenna; sensor network; active test

1. Introduction

Wearable wireless sensor networks (W-WSNs) can be applied in diverse areas, including health care (clinical diagnostics, rehabilitation), sports (athlete activity profile, energy expenditure during training) and work safety (monitors for the temperature, humidity, CO_2) [1,2]. These networks are a particular case where sensors are deployed on the user clothing and/or directly on the body to measure physiological signals of a human and/or to monitor its environment [3–5]. Hence, each W-WSN consists of multiple wearable sensor nodes which are capable of communicating with each other (on-body communications) or with external devices (off-body communications) allowing a connection with a monitoring centre (smartphone, local or cloud webserver). Several frequency bands (such as 2.36–2.38 GHz medical body area network (MBAN), 2.4–2.48 GHz and 5.75–5.82 GHz industrial, scientific, and medical (ISM) bands) and wireless technologies such as IEEE 802.11, IEEE 802.15, LTE, LoRA, etc. are used to connect wearable sensor nodes. These wireless technologies require each wearable sensor node to be equipped with a sensing element, processor, memory, power module, transceiver and an antenna [2,4,6].

The antenna plays a key role in the link performance of each wearable wireless sensor node because it determines the reliability of the wireless link and directly influences the power consumption of the node [2]. Moreover, small, flexible, low-profile, and light-weight wearable antennas based on materials which are deformable, twistable and stretchable are needed because the sensor node needs to be seamlessly worn [7]. Most of the proposed flexible wearable antennas are based on polymers [5,8,9], textiles [1,7,10–17] or flexible ceramics [18].

Among flexible materials, textiles are the most widely employed materials for wearable antennas due to their ease of integration on the clothes. The operating frequency bands and radiation efficiency of such a kind of antenna structure can be controlled by proper selection of the antenna topology, substrate thickness and substrate electromagnetic (EM) properties [1,19,20]. Consequently, design of the all-textile wearable antennas requires precise knowledge of the EM properties of textiles used for the antenna substrate at the frequencies of interest.

The second major challenge when designing wearable antennas is the performance reduction (such as shifting of the resonant frequency, changing of the input impedance, reducing the radiation efficiency of the antenna) caused by the specific environment, in which wearable antennas operate (close to the human body) [5,17,21,22]. Hence, antenna topologies with high body-antenna isolation are needed to guarantee a satisfactory performance in varying operating conditions and to reduce the specific absorption rate (SAR) [2].

In the literature, a diverse range of techniques have been reported for reducing interaction between the antenna and human tissue. One popular technique to reduce electromagnetic coupling between the antenna and human body is to use metamaterials such as electromagnetic bandgap [10–14] and artificial magnetic conducting surfaces [16,17]. Another technique is to use a reflector [1,5,8,15] or a full ground plane [7].

Also an essential aspect would be considered when designing antennas for wearable sensor nodes is miniaturization. Most of the proposed antenna designs still suffer from relatively large size [1,9,14–17] or high profile [5,10–12] and do not meet the requirements (for low-profile and small size) of the antennas for wearable wireless sensor nodes. Consequently, the design of antennas for wearable wireless sensor nodes is a complex task. Generally, in wearable antenna design, electrical, mechanical, and safety requirements described in [2] should be taken into account.

In this paper, we extend our conference paper [1] where the characterization of the EM properties of the regular textiles and study of their effects on performance (radiation efficiency, reflection coefficient magnitude, bandwidth, and maximum gain) of wearable backed antennas have been presented. Now, details about the design, manufacturing and performance measurements of small wearable backed antennas for applications in sensor nodes are provided. The proposed step-by-step methodology allows us to design and experimentally realize new small, low-profile, lightweight and flexible all-textile antennas with high body-antenna isolation. Compared with [1], additional contents of this expansion paper are as follows: (1) a step-by-step design, fabrication and measurement procedure of small wearable backed antennas for application in sensor nodes; (2) new small all-textile antennas for potential integration into everyday clothing; (3) a study (by simulations and measurements) of the antenna performance in free space and on a phantom of the human body.

2. Design, Fabrication and Measurement of Small Antennas for W-WSNs

Figure 1 shows the generic flowchart of the design, optimization, fabrication and measurement processes of small antennas for W-WSNs used in this work. As shown in Figure 1, the antenna design process starts with the definition of the design goal and target antenna specifications. The design goal of this work is to create new small low-profile all-textile antennas which provide sufficient radiation efficiency and appropriate isolation to the human body for off-body communications in W-WSNs, for potential integration into everyday clothing. The target antenna specifications and required performance associated with this goal are set out below in Table 1.

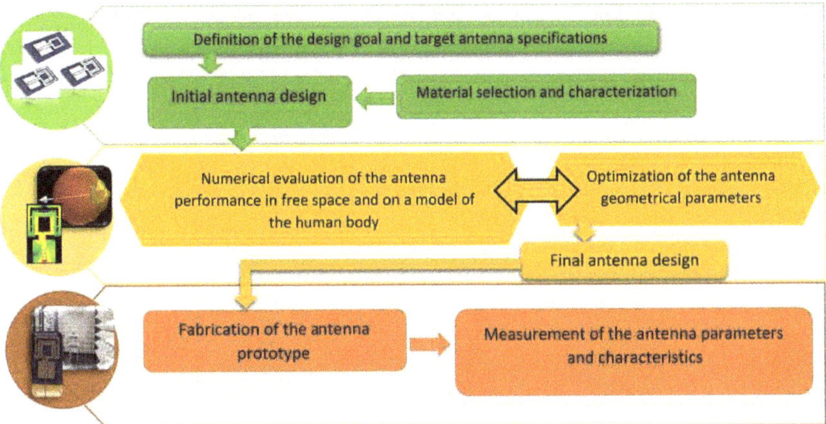

Figure 1. Block diagram of the design, optimization, fabrication and measurement process of small antennas for W-WSNs.

Table 1. Target antenna parameters and characteristics.

Antenna Specification	
Frequency range (GHz)	2.36–2.48
$\|S_{11}\|$ in operating frequency range (dB)	≤−6
Minimum radiation efficiency on the human body (%)	≥5
Radiation pattern	unidirectional
Minimum front to back ratio (dB)	≥10
Peak 10 g average SAR (W/kg)	≤2
Maximum antenna size (cm^2)	≤25
Maximum antenna profile (mm)	≤1.5
Maximum antenna weight (g)	≤5

The frequency range pointed in Table 1 is chosen because most of the devices for off-body communications are designed for operation in the unregulated 2.36–2.38 GHz MBAN and 2.4–2.48 ISM frequency bands [3,5,6,10,11,22]. The minimum value of the radiation efficiency pointed in Table 1 is chosen based on a survey of radiation efficiency of internal antennas in real mobile phones [23]. The results from the measurements of the efficiency of the mobile-phone antennas show that average handset radiation efficiency varies between 4.5% and 20%. As expected, the antenna with the smallest size (antenna had a maximum length of 36 mm) has the lowest radiation efficiency (4.5%).

2.1. Initial Antenna Design

The initial design of the antenna starts with the choice of an antenna geometry and a material for the substrate. There is no single specific type of geometry for wearable antennas. The most prevalent types are dipole [5,8], monopole [1,10,12,14–17] and patch antennas [7,9,11]. The choice of geometry needs to be guided from the design requirements such as simple structure, small size, low-profile and light-weight. Consequently, the antenna needs to be physically small.

In this work, an antenna geometry based on a loop antenna was chosen. This structure is one of the simplest small radiators and provides good matching with many types of feedings, such as coaxial cables and planar transmission lines.

Another point to keep in mind during the initial design is the choice of materials for the antenna's elements and characterization of their electromagnetic properties, as shown in Figure 1. When designing antennas for potential integration into everyday clothing, the substrate material is predetermined. This is the textile from which the garment is made and in which the antenna will be embedded.

Measurements of the complex permittivity of the chosen textile can be carried out using the resonant or non-resonant methods [1]. Conductive fabrics or threads are the most widely used materials for radiating elements. The information about dc conductivity (or sheet resistance) of the conductive textiles usually is available from the datasheets provided by the manufacturers.

In this study, conventional fabrics with natural fibre (denim and cotton), as well as synthetic fibre (polyester and polyamide-elastane) have been selected as substrates to develop small, low-profile, light-weight all-textile antennas. Conductive fabric has been chosen as a material for the radiating elements and reflector. More details about the EM properties of the selected textiles can be found in our conference paper [1].

The steps to design a small all-textile antenna that meets the specifications presented in Table 1 are depicted in Figure 2. Figure 2a depicts the structure of the proposed antenna in the first step of the design process. It is composed of a rectangular loop, coplanar waveguide (CPW) transmission line, substrate, and a reflector. The choice of the CPW to feed the antenna was because it offers a single-layer manufacturing process. The substrate is denim with a real part of relative permittivity (ε_r') 1.878, and a loss tangent (tan δ) 0.0594 at 2.565 GHz. The thickness of the substrate is 1.5 mm (comprised of three-layer denim) with density from 1.54 g/cm^3 [1,2]. This substrate thickness enables us to obtain a small antenna with a low-profile. The reflector was implemented in the antenna's structure to reduce the effect of the human body over the antenna performance and SAR. It was chosen due to its simple form. Also, it is relatively easy for numerical modelling and manufacturing. The geometrical dimensions (length and width) of the loop strips and CPW were tuned to yield a resonance at around 2.47 GHz. Figure 2b shows the reflection coefficient magnitude of the proposed antenna during the design process. In the next step, an arc-shaped parasitic element was added at a close distance to the loop structure (see Figure 2a, Step 2). It is observed that the resonant frequency shifts to a lower frequency while the bandwidth and the impedance matching of the antenna are not changed. Finally, to tune the input resistance of the proposed antenna closer to 50 Ohms and broader bandwidth, four parasitic elements were inserted in the arc-shaped loop (see Figure 2a). This enables a good impedance match with a lower than −6 dB bandwidth (from 2.266 to 2.551 GHz) of approximately 285 MHz and a resonant frequency of 2.4 GHz, as shown in Figure 2b.

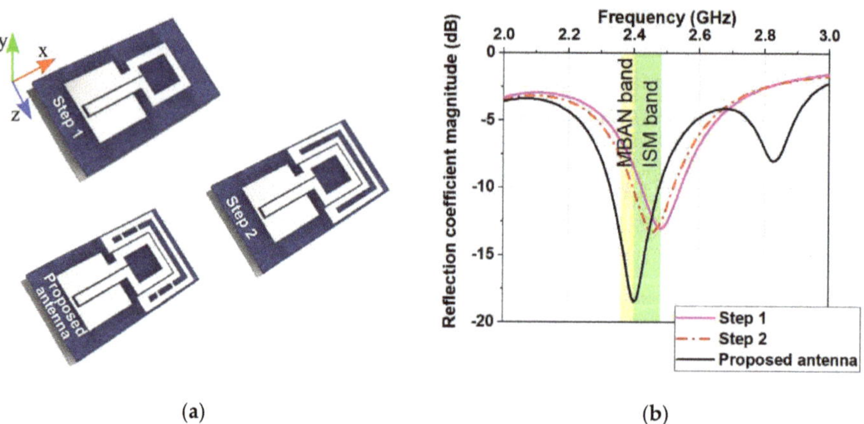

Figure 2. Antenna geometry: (**a**) Design steps; (**b**) Simulated reflection coefficient magnitudes versus frequency.

2.2. Numerical Evaluation of the Antenna Performance in Free Space and on a Model of the Human Body

Design, optimization and numerical evaluation of the antenna performance are carried out using the full-wave electromagnetic commercial software Remcom xFDTD (xFDTD, Remcom Inc.,

State College, PA, USA). This EM software packet is based on the finite difference time domain (FDTD) method which is extensively used for modelling antenna structures and human body models.

The following parameters are set in each simulation in order to obtain more accurate results. First, FDTD cell size of 0.5 mm × 0.5 mm × 0.5 mm is used for the antenna geometry. For the human body model and rest of the space, an inhomogeneous mesh having an increasing cell size from 0.5 to 1 mm is applied. All calculations continue until steady-state is reached. For the analyzed small antenna with a denim substrate, the steady-state is observed after 10,000 time-steps. A 7-layer perfect matched layer absorbing boundary condition is used.

Since the designed antenna is expected to be in close proximity or mounted directly on the different parts of the human body during the real operating conditions in a W-WSN, the design and optimization were made first in the free space. After that, geometrical dimensions of the antenna structure were optimized on a human body model. A homogeneous numerical model with dimensions 180 mm, 120 mm and 150 mm was employed to mimic the human body (ε_r' = 40.805, σ = 2.33 S/m, and ρ = 1166 kg/m^3). The selected human model dimensions are those for the flat phantom pointed in standards EN 62209-2:2010 and IEEE Std.1528-2013 for the fixed frequency of 2.45 GHz so that the effect of the power reflection at the model surface on the peak 10 g average SAR is negligible (less than 1%).

The antenna was positioned on the surface of the phantom (the distance between the antenna and the phantom is 0 mm, as shown in Figure 3d) to study the effect of the human body on the antenna performance and SAR in the worst-case scenario.

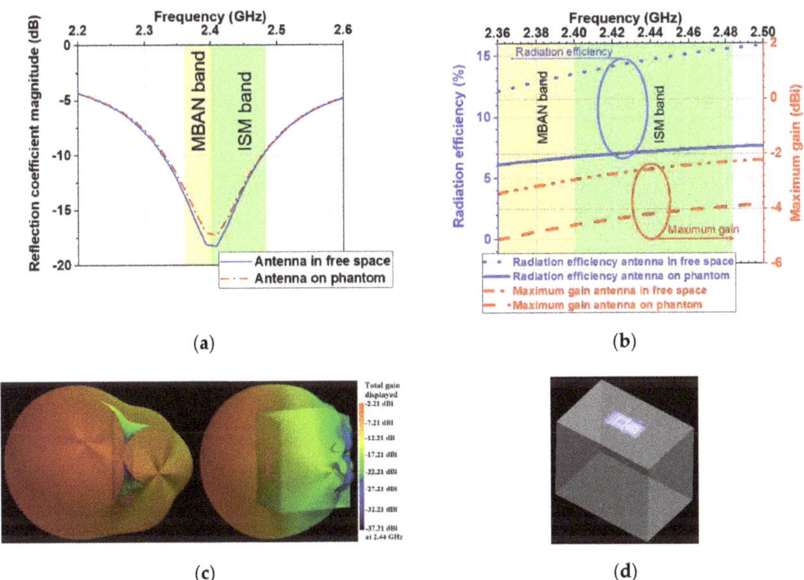

Figure 3. Simulated: (**a**) Reflection coefficient magnitudes versus frequency; (**b**) Radiation efficiency and maximum gain versus frequency; (**c**) 3D radiation patterns of the antenna at 2.44 GHz in the free space and placed on the surface of the phantom; (**d**) Model of the antenna and phantom.

The free space and on-phantom reflection coefficient magnitudes ($|S_{11}|$) of the optimized antenna are presented in Figure 3a. In the free space, the antenna bandwidth ($|S_{11}|<-6$ dB) is 285 MHz. As we can see the homogeneous phantom does not cause detuning effects on the resonant frequency. The $|S_{11}|$ curve and operating bandwidth for the case where the antenna is mounted on the phantom are the same as the case in free space.

The total radiation efficiency and maximum gain of the optimized antenna in the free space and on the phantom at the frequency bands of interest are displayed in Figure 3b. As we can see, the phantom causes a reduction in total radiation efficiency and maximum gain. The total radiation efficiency and maximum gain are decreased by a factor of about 0.5 when the antenna is placed on the phantom. Also, the total radiation efficiency and maximum gain show an increasing trend with increasing the frequency. Figure 3b shows that across the bandwidth of 2.36–2.5 GHz the simulated radiation efficiency varies from 12% to 16% when the antenna is in the free space and from 6% to 7.8% when it is placed on the phantom.

From Figure 3c we can see that the antenna exhibits unidirectional radiation pattern. The front-to-back ratio is 12.20 dB in the free space and 22.55 dB on the phantom. Another essential question to be considered in designing antennas for W-WSNs is about concerning health hazard. Hence, to address this question, a more thorough evaluation and characterization of the SAR in the human body model were carried out. According to the safety guidelines and standard [24,25], the obtained peak 1 g and 10 g average SAR values should not exceed 1.6 W/kg and 2 W/kg.

Figure 4a shows the peak 1 g and 10 g FDTD-computed average SAR generated from the antenna in the phantom, in the MBAN and ISM bands when the antenna is positioned on the surface of the phantom (the distance between the antenna and the phantom is 0 mm). The results were normalized to a net input power of 100 mW. It can be seen that the peak SAR is frequency dependent. In general, the SAR increases as the frequency increases. For the considered input power, the peak 1 g and 10 g average SAR produced in the phantom are 0.533 and 0.148 W/kg, respectively at 2.48 GHz. The peak 10 g average SAR values are much lower than the maximum allowed value of 2 W/kg as required by the ICNIRP [24]. Moreover, the peak 10 g average SAR is found to be equal to ~2 W/kg when the net input power for the proposed antenna is 1353 mW. That is, the net input power as high as 1353 mW guarantee conformance with the safety guidelines imposed by the ICNIRP [24].

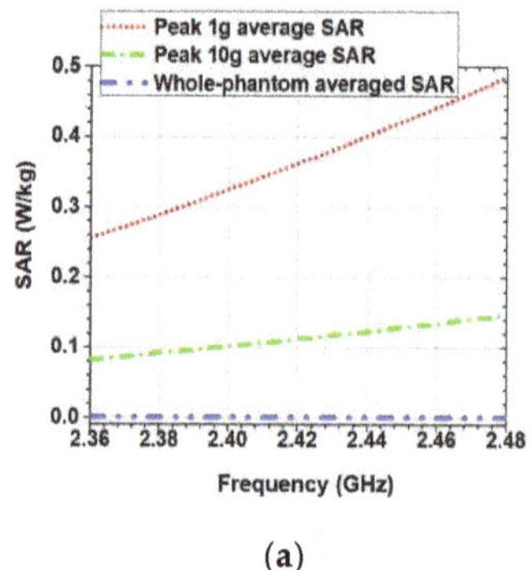

(a)

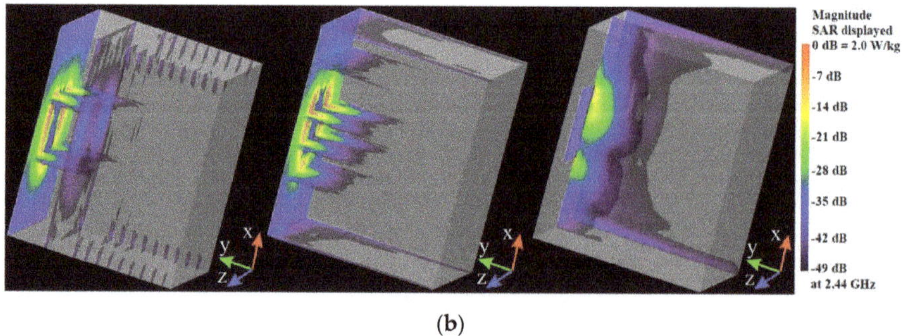

(b)

Figure 4. FDTD-computed SAR: (a) peak 1 g and 10 g average SAR, and whole-phantom averaged SAR versus frequency; (b) Distributions in xy-, yz- and zx-plane at 2.44 GHz and scale.

From the results presented in the Figure 4, it can be concluded that the proposed antenna exhibits low SAR values (peak 10 g average SAR is about ten times lower than the maximum allowed value of 2 W/kg) due to the shielding effect of the reflector and also due to the unidirectional radiation pattern of the antenna.

The SAR distributions on the surface and inside the phantom in the xy-, yz- and zx-planes are shown in Figure 4b. In the figure, we observe that the SAR values around the antenna edges are higher.

Also, under real-case scenarios, the distance between the wearable antenna and the human body may be changed, which may lead to a change in the antenna performance and SAR. Hence, numerical simulations were performed at distances 0 mm, 5 mm and 10 mm between the antenna and the human body model in order to investigate the robustness of the proposed design to effects of the human body loading.

Figure 5 presents the variation of the antenna radiation efficiency, maximum gain, peak 10 g average SAR and whole-phantom averaged SAR (at 100 mW net input power) versus the frequency for the distances 0 mm, 5 mm and 10 mm between the antenna and the phantom. As might be expected, the radiation efficiency increases with distance, while the peak 10 g average SAR and whole-phantom averaged SAR decrease. Also, the maximum gain shows an increasing trend with increasing the distance between the antenna and the phantom. From Figure 5b we see that the gain at a distance 10 mm from the phantom is larger than that in the free space. This is because the phantom, in this case, acts as a reflector, which enhances radiation in the direction opposed to the human body model.

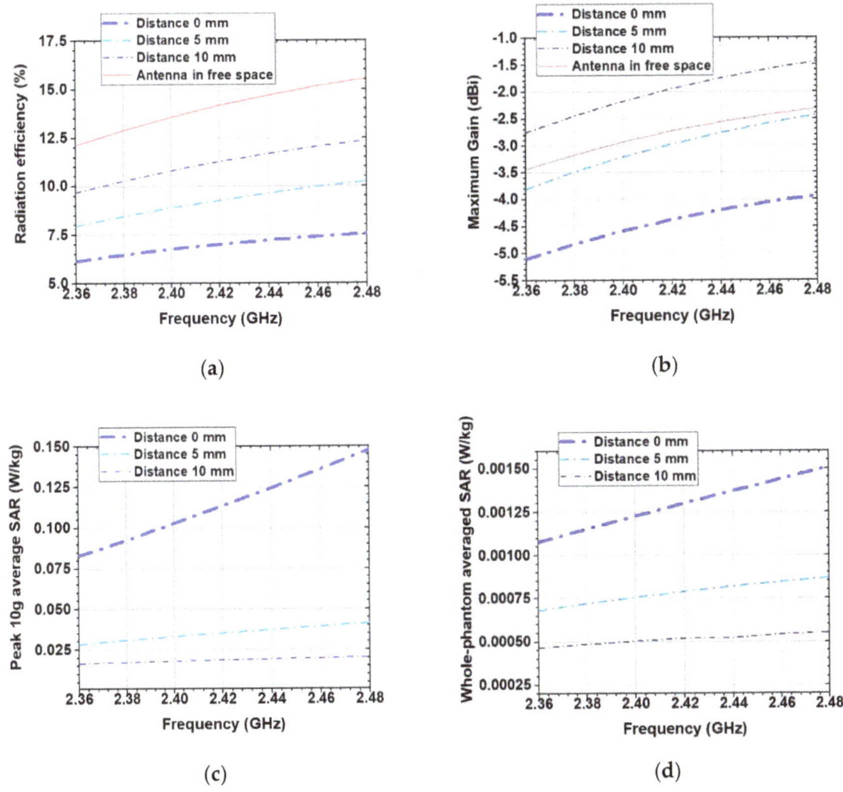

Figure 5. Simulated: (**a**) Radiation efficiency; (**b**) Maximum gain; (**c**) peak 10 g average SAR; (**d**) whole-phantom averaged SAR at distances 0 mm, 5 mm and 10 mm between the antenna and the human body model.

Moreover, when the designed antenna is a part of a wearable wireless sensor node intended for a medical purpose such as tracking, recording, and monitoring of biomedical signals (used in a medical or home healthcare environment), it is essential to create an electromagnetically compatible device to minimize interference with other devices. In this case, the wearable wireless sensor node has to be designed and validated as a medical device (or medical electrical equipment according to the definition in IEC 60601-1). The standards which specify tests and requirements for the electromagnetic compatibility of the medical electrical equipment are IEC 60601-1-2 and IEC 60601-1-11.

Finally, examining the results from numerical simulation presented in Figures 3–5 as compared to the specifications of Table 1 illustrates that, the proposed antenna satisfies the requirements for application in W-WSNs.

2.3. Fabrication of the Antenna Prototype

For the fabrication of the radiating elements of a textile antenna, it is possible to use the inkjet printing [26], embroidery [27] or cut-transfer-glueing process [2]. The choice usually is made on the base of the materials used for the radiating elements.

To fabricate the antenna's prototype, we used the cut-transfer-glueing process. By using the cutting machine, the antenna's elements were cut into the designed shapes with high accuracy (a tolerance of ±0.01 mm). A highly conductive fabric (DC conductivity 2.5×10^5 S/m and 0.08 mm of thickness) was used for the conductive elements of the antenna. Next, both radiating elements and reflector were attached to the denim substrate by using a polymer tape that is activated by ironing. A coaxial cable (diameter 1.13 mm and length of 200 mm) with a 50 Ω U.FL connector was soldered to the CPW using a low-temperature solder wire. First, the soldering points of the coaxial cable with the conductive fabric were tin-plated at 250 °C (heating time 2.5 s), as shown in Figure 6. Then, the coaxial cable inner conductor was soldered on top of the middle conductor of the CPW while the coaxial cable braid shield was soldered to the CPW ground plane.

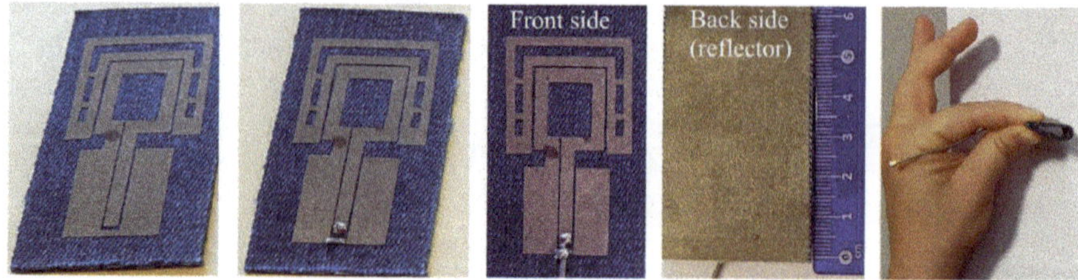

Figure 6. Photographs of the antenna's prototype during the fabrication process.

2.4. Measurements

The performance of the prototype of the designed antenna may be measured by using passive and active tests. In the passive tests, the prototype is connected to the measuring equipment (a network analyzer, signal generator, receiver, or spectrum analyzer) using a coaxial cable. In these tests, the antenna reflection coefficient magnitude ($|S_{11}|$), bandwidth, gain, radiation pattern and radiation efficiency are measured. The detailed definition of the antenna parameters and their measurement methods are presented in [2,28,29].

Moreover, a full verification of the antenna design requires more extensive tests, which represents the behaviour of the antenna under practical operating conditions, called active tests [2,29,30]. In these measurements, a network simulator (or a radio communication test module) is used to set up a connection to the antenna under test, that is connected to a sensor node to reproduce a real-world performance. In order to conduct accurate and repeatable measurements, the testing needs to perform inside a chamber (anechoic or reverberation) with a controlled environment [2,28].

For antennas used near or on the human body such as antennas intended for sensor nodes in W-WSNs experimental measurements of the antenna parameters and characteristics on a physical model of the human body (called phantom) are also required. There are three types of phantoms for experimental use: solid, semi-solid, and liquid [31]. SAR measurements can be made using a robotic system, associated test equipment and a liquid phantom or using an infrared camera, associated test equipment and a solid or semi-solid phantom.

In order to assess the performance of the proposed antenna, $|S_{11}|$ was evaluated when the antenna is placed on a semi-solid phantom and in the case of free space condition. The semi-solid phantom (dimensions 180 mm (length) × 120 mm (width) × 150 mm (depth)) was fabricated accordingly to the recipe and technique described in [31]. The measurement of $|S_{11}|$ were performed using a Tektronix TTR503A vector network analyzer. As shown in Figure 7, when the antenna is placed on the semi-solid phantom, the $|S_{11}|$ remains almost unchanged. By comparing simulated and measured results, we observe a good agreement between them with a slight shift in the resonant frequency.

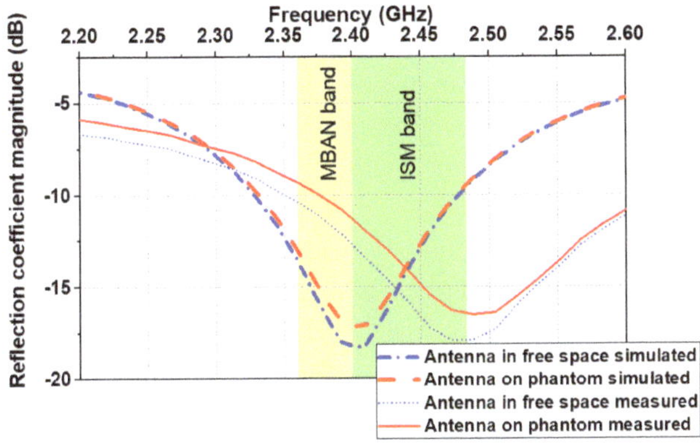

Figure 7. Simulated and measured reflection coefficient magnitudes versus frequency in the free space and on the semi-solid phantom.

In order to get a complete performance of the fabricated prototype, active tests were carried out. The first set of measurements was performed in a semi-anechoic chamber. Next, in order to take into account, the environmental variability, the measurements were repeated in a shielded room. All tests were carried out for both scenarios: (1) in the free space and (2) when the antenna is placed on the phantom.

Two XBee S1 (DiGi International, Hopkins, MN, USA) modules were used in all tests as wireless nodes. One of the nodes was connected by a coaxial cable with a dipole antenna that has a resonant frequency of 2.44 GHz. The dipole antenna was in a vertical orientation. Next, the wireless node was connected to a personal computer via a UART-to-USB controller (Figure 8a,b) and was configured to generate a continuously repeated pseudo-random signal of 100 packets (each packet contains 50 bytes). The second wireless node was connected to the fabricated antenna prototype. The wireless nodes were connected, running XCTU software. On the XCTU, the transmission power, operating frequency and data rate, were set to −1 dBm, 2.41 GHz, and 9.6 kbits/s, respectively.

The semi-anechoic chamber was divided into 15 specific positions (three columns and five rows). The dipole antenna connected to the XBee node was stationary. At the same time, the other XBee node connected to the antenna prototype was located in each of described 15 specific positions in line-of-sight to the dipole. Both antennas were placed at a high of 1.33 m.

A range test was performed in order to determine the range and link quality between the nodes representing a real-world scenario. During the range test, XCTU sends data packets from the stationary XBee node to the remote node and waits for the echo to be returned from the remote node to the stationary node. Also, during this test, the XCTU determines RSSI (Received Signal Strength Indicator) value and calculates the packet error rate.

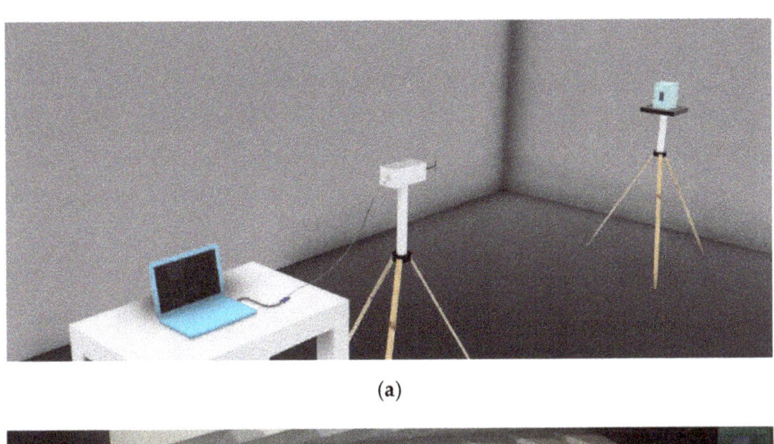

(a)

(b)

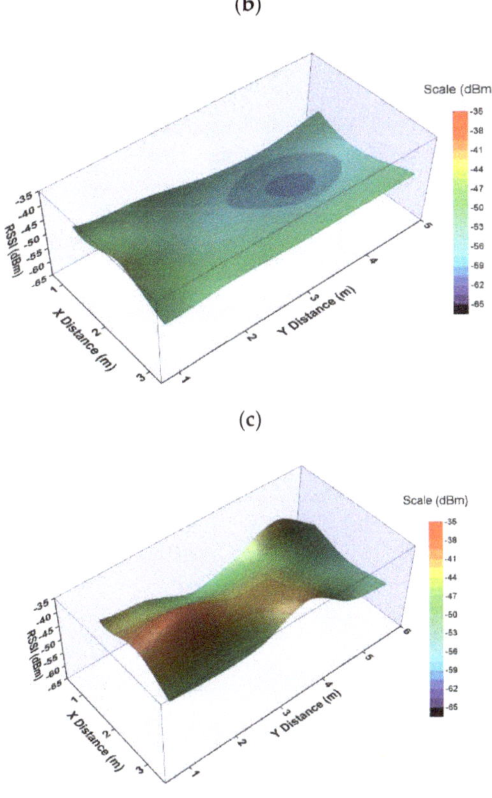

(c)

(d)

Figure 8. *Cont.*

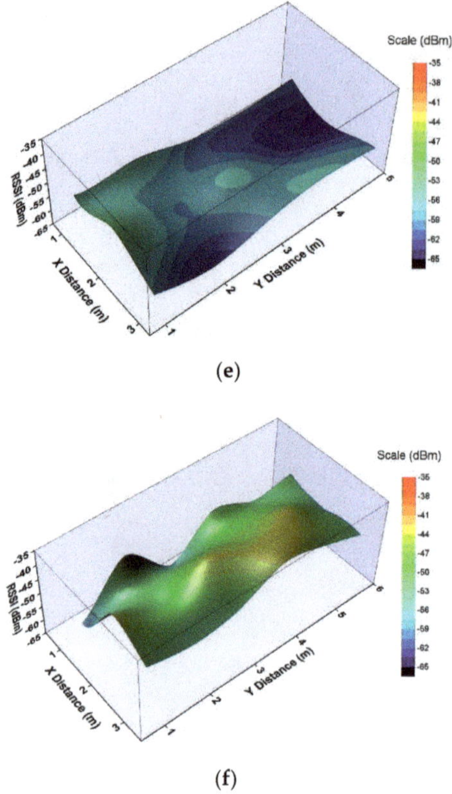

Figure 8. RSSI: (**a**) Simple drawing of the configuration of the test setup; (**b**) Photograph of the test setup in the semi-anechoic chamber; (**c**) Distribution in the semi-anechoic chamber in the free space; (**d**) Distribution in the shielded room in the free space; (**e**) Distribution in the semi-anechoic chamber when the antenna is on the semi-solid phantom; (**f**) Distribution in the shielded room when the antenna is on the semi-solid phantom.

The distributions over the xy-plane of the measured RSSI in the semi-anechoic chamber and shielded room are shown in Figure 8. We can see that in the free space the measured RSSI is between −44 and −69 dBm depending on the specific antenna position in the semi-anechoic chamber (see Figure 8c). Also, when the distance between the antennas increases, the measured RSSIs decrease. When moving from semi-anechoic to shielded room scenario, it can be seen that in the free space the measured RSSIs vary between −37 and −56 dBm. A positive interference is clearly seen in Figure 8d when the distance between the antennas is 2 m (the measured RSSI is −37 dBm). Moreover, when the distance between the antennas is 6 m, the measured RSSI decreases to −56 dBm. Fluctuations of the RSSI values in the shielded room are due to constructive and destructive interference.

When the antenna is placed on the phantom, the measured RSSIs were in the range of −50 to −64 dBm in the semi-anechoic chamber (see Figure 8e) and in the range of −44 to −60 dBm in the shielded room (Figure 8f). Comparing the results with the free space scenario, we can conclude that the measured RSSIs when the antenna is on the phantom are lower (with about 6 dB) than those in the free space. These differences in RSSIs are attributed to the reduction in the radiation efficiency and gain when the antenna is on the phantom. We also observed that at all measurements, the packet error rate was zero.

Comparing the results, we see that at the same positions, the measured RSSIs in the semi-anechoic chamber are lower than those in the shielded room. This is since the more of the reflections in the semi-anechoic chamber are eliminated while in the shielded room, the propagation occurs by multiple reflections in the environment, which results in an additional energy contribution.

From the results in Figure 8, it is possible to conclude that the proposed antenna shows very good RSSI values both in the semi-anechoic chamber and shielded room which satisfies well the requirement of the receiver sensitivity in W-WSNs (−94 dBm) [32].

3. Impact of EM Properties of the Textile Materials on the Performance of the Small Antennas for W-WSNs

3.1. Antenna Designs

This section investigates the effect of the EM properties of different textile materials on the performance of small low-profile backed antennas. Three antenna structures: (1) Antenna with a cotton substrate, (2) Antenna with a polyamide-elastane substrate and (3) Antenna with a polyester substrate were developed. Here, each antenna has a substrate thickness of 1.5 mm, with a real part of the relative permittivity of 1.6321 (cotton), 1.5493 (polyamide-elastane) and 1.6202 (polyester), respectively. The loss tangent of the textile substrates is 0.0439 (cotton), 0.0146 (polyamide-elastane) and 0.0051 (polyester). In the design of the antennas (with substrates from cotton, polyester and polyamide-elastane) the configuration of the antenna with denim substrate was used. The difference being that at each substrate, the geometrical dimensions of the loop and parasite elements were optimized for maximum radiation efficiency and optimal impedance matching in the desired bandwidth.

Figure 9 shows the structure and geometrical dimensions for each antenna. It is seen that in order to provide good impedance matching at the resonant frequency of 2.4 GHz, the perimeter of the square-loop on the denim substrate is 72 mm while the perimeter of the square-loop on the polyester substrate is 84 mm. Also, the loop strips width of the antenna with a polyester substrate are decreased to 1 mm to enhance the bandwidth.

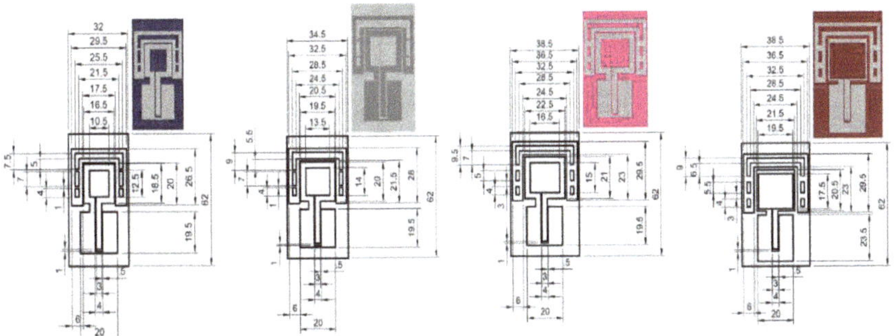

Figure 9. Configuration of the antennas with denim, cotton, polyamide-elastane and polyester substrates.

From the results presented in the Figure 9, we can conclude, that the real part of the permittivity of the antenna substrate has consequences on the overall antenna size. As expected, with the increasing of the substrate material permittivity, the overall size of the antenna decreases [1,2,28].

A comparison with previously reported designs shows that the overall size of each of the four proposed antennas is between 25% and 90% smaller than the antennas in [1,9,10,12,14,15,26]. Moreover, the proposed antennas have a lower profile than [5,8,10–12,14].

3.2. Results and Discussion

Two scenarios were numerically studied when the antennas are: (1) in the free space and (2) placed on the semisolid phantom.

As shown in Figure 10a in free space, the antennas with denim and cotton substrates have a bandwidth ($|S_{11}| < -6$ dB) of 285 and 269 MHz, respectively. In contrast, antennas with substrates from polyamide-elastane and polyester have a relatively narrow bandwidth of 178 and 130 MHz, respectively. We hypothesise that since the polyamide-elastane and polyester have a lower loss (loss tangent is between 0.005 and 0.015), the bandwidth of the antennas with substrates from these materials is more narrow versus the bandwidth of the antennas with denim and cotton substrates.

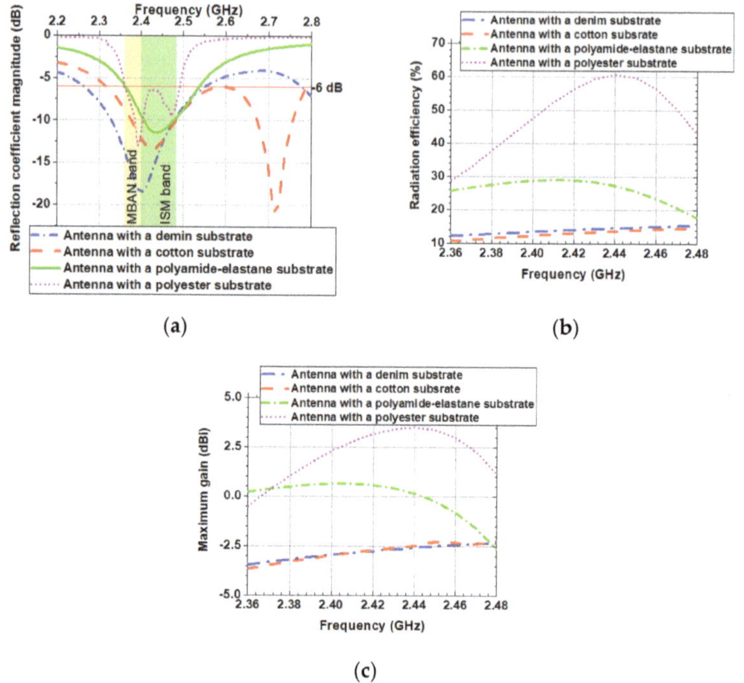

Figure 10. Simulated in the free space: (**a**) Reflection coefficient magnitudes versus frequency; (**b**) Radiation efficiency versus frequency; (**c**) Maximum gain versus frequency.

Figure 10b,c show the simulated radiation efficiency and maximum gain of the four antennas when they are in the free space. The behaviour of the radiation efficiency of the antennas, as a function of frequency, is shown in Figure 10b. From the results we can see that the radiation efficiency of the antennas with denim and cotton substrates remains almost unchanged (between 10% and 16%) across the MBAN and ISM bands. Furthermore, Figure 10b shows that the simulated radiation efficiency of the antenna with a polyamide-elastane substrate varies from 18% at 2.48 GHz to 30% at 2.42 GHz. The antenna with a substrate from polyester shows significant variations in the radiation efficiency from 28% at 2.36 GHz to 60% at 2.44 GHz.

The maximum gain of the antenna with a polyester substrate varies between −0.5 and 3.5 dBi across the MBAN and ISM bands, as seen in Figure 10c. The gain of the antenna with a polyamide-elastane substrate varies from −2.5 to 0 dBi. On the other hand, for antennas with cotton and denim substrates, small variations of only 1.5 dB in gain are observed. Also, the maximum gain of the antennas with denim and cotton substrates shows an increasing trend with increasing the frequency.

The simulated reflection coefficient magnitudes when the antennas were placed on the human body model are shown in Figure 11a. A good impedance matching is maintained for all four antennas despite the slight change in reflection coefficient magnitudes. The results show that the resonant frequency of the antennas with a substrate from cotton and denim fabric is not affected when positioned on the phantom. Similar results are observed in [1]. The resonant frequency of the antennas with a substrate from polyester and polyamide-elastane is slightly shifted up. Moreover, when the antennas are placed on the phantom, their bandwidths remain unchanged compared to the free-space scenario. It can be concluded that the resonant frequency and bandwidth are insensitive to detuning when the textile antennas backed by a reflector are positioned on the phantom.

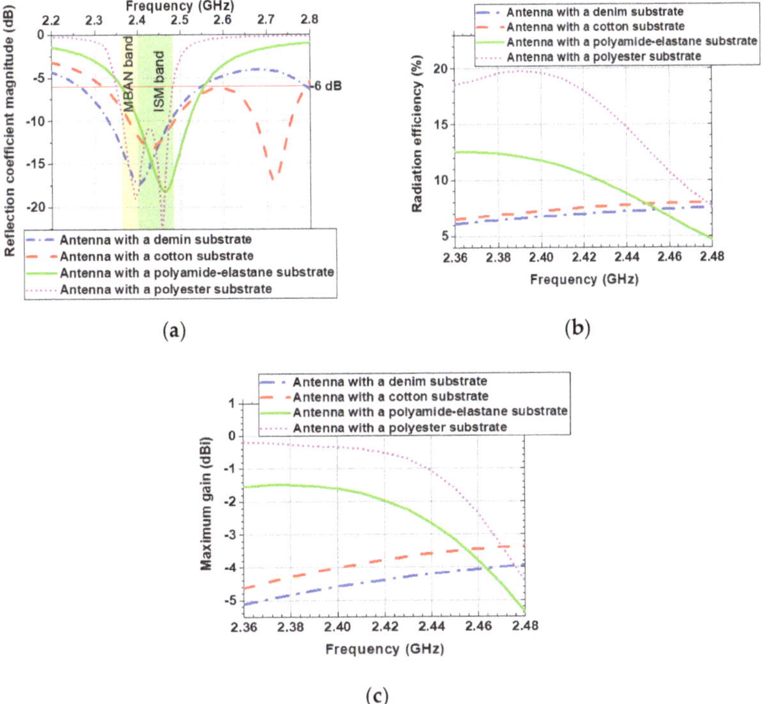

Figure 11. Simulated on the phantom: (**a**) Reflection coefficient magnitudes versus frequency; (**b**) Radiation efficiency versus frequency; (**c**) Maximum gain versus frequency.

The investigations have shown that the phantom significantly affected both the maximum gain and radiation efficiency of the antennas. For the antennas with denim and cotton substrates, the maximum gain and radiation efficiency are decreased by a factor of about 0.5 compared to the free space scenario. The simulated radiation efficiency of the antenna with a polyamide-elastane substrate falls between 5% and 13%, while simulated radiation efficiency of the antenna with a polyester substrate falls between 8% and 20%. The observed reductions in radiation efficiency and maximum gain are due to the power absorbed at the substrate and human body model.

Figure 12a,b compare the computed peak 10 g average SAR and whole-phantom averaged SAR generated from each of the four antennas in the homogeneous phantom, in the MBAN and ISM bands. The results are normalized to a net input power of 100 mW. The antennas with denim and cotton substrates exhibit very low SAR values (below 0.02 W/kg), as shown in Figure 12a. On the other hand, 10 g average SAR of the antennas with polyamide-elastane and polyester substrates varies significantly from 0.5 to 1.15 W/kg and from 0.5 to 1.95 W/kg, respectively. From the results presented in Figure 12a it can be concluded that the peak 10 g average SAR for all four antennas is lower than that in the safety limit pointed in [24].

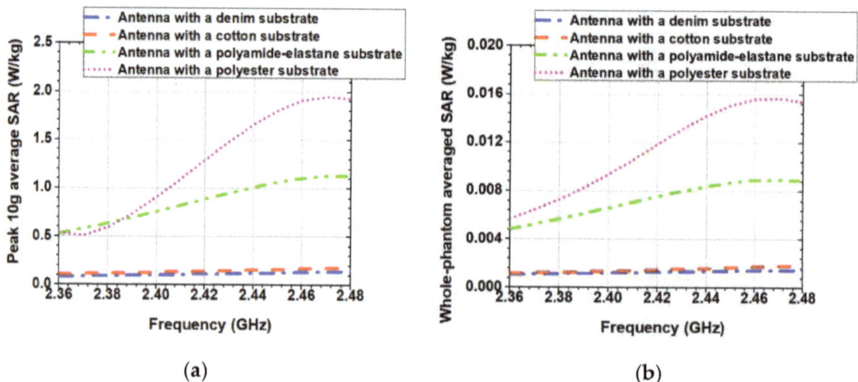

Figure 12. Simulated: (**a**) Peak 10 g average SAR (W/kg); (**b**) Whole-phantom averaged SAR (W/kg).

In MBAN and ISM bands, and for an input power of 100 mW, the whole-phantom averaged SAR for the antennas with denim and cotton substrates is below 0.002 W/kg (about 40 times lower than the maximum allowed value of 0.08 W/kg), as seen in Figure 12b. The obtained whole-phantom averaged SAR for the antennas with polyamide-elastane and polyester substrates is 10 and five times lower than the maximum allowed value of 0.08 W/kg, respectively.

Finally, the comparison between the antennas with substrates from natural fibres (cotton and denim) and substrates from synthetic fibres (polyamide-elastane and polyester) showed that the antennas with substrates from synthetic fibres give higher SAR values than the antennas with substrates from natural fibres (Figure 12).

Also, numerical simulations were performed for each antenna at distances 0 mm, 5 mm and 10 mm between the antenna and the human body model in order to investigate the effect of the EM properties of different textile materials on the robustness of the proposed designs to impact of the human body loading.

As might be expected, for each of the antennas the radiation efficiency and maximum gain increase with distance, while the peak 10 g average SAR decreases. From Figure 13b we also see that the gain at a distance 10 mm from the phantom for each antenna is larger than that in the free space. Consequently, we can conclude that the action of the phantom at a distance of 10 mm from the antenna is akin to a reflector, which enhances radiation in the direction opposed to the human body model.

Moreover, the computed maximum allowable net input powers which satisfy the ICNIRP restriction of 2 W/kg for peak 10 g average SAR for each antenna at different distances to the phantom are shown in Table 2. The maximum allowable net input power which satisfies the ICNIRP restriction at a distance of 0 mm is 1353 mW and 1062 mW for the antennas with denim and cotton substrates, respectively while for antennas with polyamide-elastane and polyester substrates is 176 mW and 104 mW, respectively.

Next, the prototypes of the antennas with cotton, polyamide-elastane and polyester substrates were fabricated using the procedure described in Section 2.3 and in [1]. Photographs of the fabricated prototypes are shown in Figure 14. The prototypes are very light (the weight of the prototype with a substrate from denim is 3.2 g, cotton 2.7 g, polyamide-elastane and polyester 2.6 g), which will allow the antennas to be easily integrated into clothing.

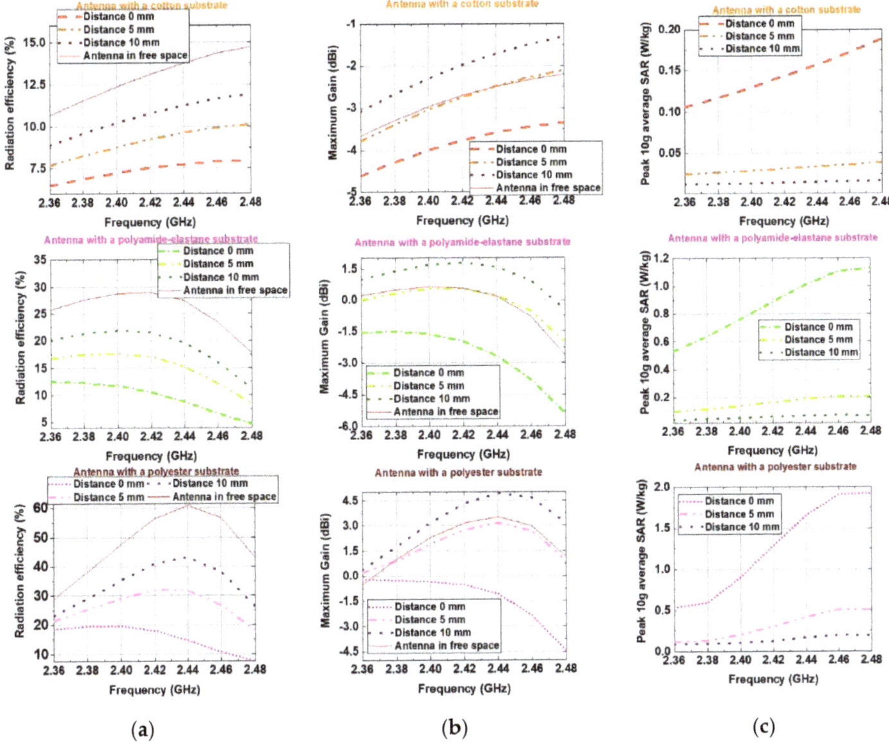

Figure 13. Simulated: (**a**) Radiation efficiency; (**b**) Maximum gain; (**c**) Peak 10 g average SAR versus frequency.

Table 2. Maximum allowable net input power which satisfies the ICNIRP restriction of 2 W/kg for peak 10 g average SAR in the homogeneous phantom at 2.48 GHz.

Distance, mm	Maximum Net Input Power, mW			
	Denim	Cotton	Polyamide-Elastane	Polyester
0	1353	1062	176	104
5	4818	5133	961	397
10	9833	12,070	2806	1065

The measured $|S_{11}|$ for all four prototypes are given in Figure 15. As can be seen, a good impedance matching is maintained in both scenarios (in free space and on the phantom). Also, a good agreement can be found between measured and simulated (see Figures 10a and 11a) reflection coefficient magnitudes.

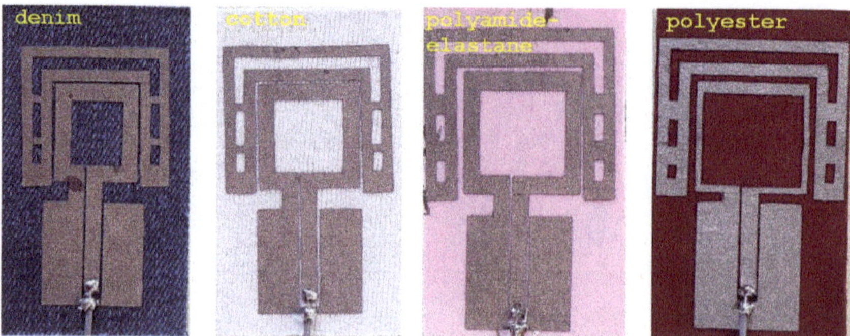

Figure 14. Photographs of the fabricated prototypes with a substrate from denim, cotton, polyamide-elastane and polyester.

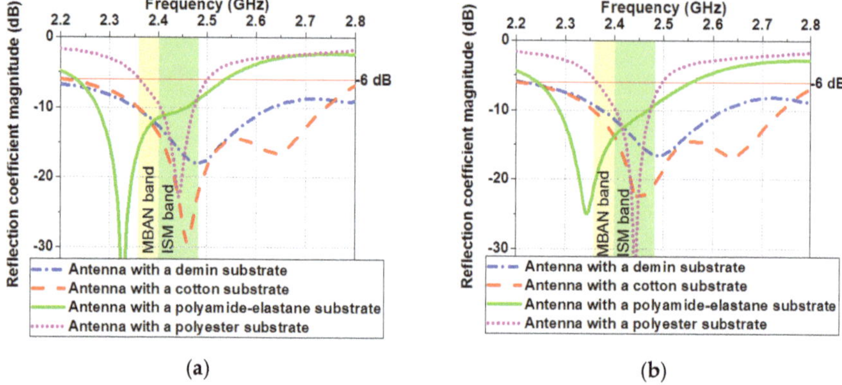

Figure 15. Measured reflection coefficient magnitudes when the antennas are: (**a**) in the free space; (**b**) on the homogeneous phantom.

4. Conclusions

In this paper, we have presented a methodology for the design, fabrication and measurement of small wearable backed antennas for application in sensor nodes. Based on the presented design methodology, a new small-sized antenna with a denim substrate was proposed. Since the designed antenna is expected to be in close proximity to the human body during the real operating conditions in a W-WSN, the design and optimization were made both in the free space and on a human body model. In the free space, the antenna exhibits bandwidth ($|S_{11}|<-6$ dB) of 285 MHz and radiation efficiency from 12% to 16% in MBAN and ISM bands. The simulated radiation efficiency varies from 6% to 7.8% when the antenna is placed on the phantom.

Concerning the health hazard, a more thorough evaluation and characterization of the SAR in the human body model were carried out. From the results, it can be concluded that the proposed antenna exhibits low SAR values (peak 10 g average SAR is about ten times lower than the maximum allowed value of 2 W/kg) due to the shielding effect of the reflector and also due to the unidirectional radiation pattern of the antenna.

The performance of the prototype of the proposed antenna was evaluated using passive and active tests. A good agreement between simulated and measured $|S_{11}|$ with a slight shift in the resonant frequency is observed. In order to get a complete performance of the fabricated prototype, range tests were performed in a semi-anechoic chamber and in a shielded room. Depending on the specific antenna position, the measured RSSIs vary between −44 and −69 dBm in the semi-anechoic chamber and between −37 and −56 dBm in the shielded room. Hence, it is possible to conclude that proposed antenna shows very good RSSIs which satisfy the requirement of the receiver sensitivity in W-WSNs (−94 dBm).

To investigate the impact of the textile materials on the antenna performance, the antenna geometry was studied on four textile substrates (denim, cotton, polyamide-elastane and polyester). The reflection coefficient magnitudes, bandwidth, maximum gain and radiation efficiency of the four antennas were numerically studied and compared to two scenarios: (1) in the free space and (2) on the semisolid phantom. The numerical investigations reveal that in MBAN and ISM bands, each of these four textile antennas achieves stable performance in both scenarios.

From the results, we can conclude, that the real part of the permittivity of the antenna substrate has consequences on the overall antenna size and resonant frequency. As expected, with increasing the substrate material permittivity, the overall size of the antenna decreases. Results show that the gain and radiation efficiency decrease with increasing dielectric losses in the textile substrate. Results for peak 10 g average SAR have revealed that the antennas with denim and cotton substrates exhibit SAR values below 0.02 W/kg. On the other hand, the peak 10 g average SAR of the antennas with polyamide-elastane and polyester substrates varies significantly from 0.5 to 1.5 W/kg and from 0.5 to 1.95 W/kg, in the frequency range of 2.36 to 2.48 GHz. Consequently, the peak 10 g average SARs are sensitive to substrate material dielectric loss.

In order to verify the results from the numerical simulations, prototypes of the antennas were fabricated, and their parameters were measured in a semi-anechoic chamber. A good agreement is found between measured and simulated reflection coefficient magnitudes of the antenna prototypes for two scenarios when the antennas are: (1) in the free space and (2) on the semisolid phantom. Therefore, the proposed small textile antennas backed by a reflector are promising candidates for integration into garments for applications in W-WSNs.

Author Contributions: Conceptualization, G.A. and N.A.; methodology, G.A. and N.A.; software, N.A.; validation, G.A. and N.A.; formal analysis, G.A. and N.A.; investigation, G.A. and N.A.; resources, G.A. and N.A.; data curation, G.A.; writing—original draft preparation, G.A.; writing—review and editing, N.A.; visualization, N.A.; supervision, N.A.; project administration, N.A; funding acquisition, N.A. and G.A. All authors have read and agreed to the published version of the manuscript.

Funding: This research was funded by BULGARIAN NATIONAL SCIENCE FUND at the Ministry of Education and Science, Bulgaria, grant number KP-06-H27/11 from 11th December 2018 "Antenna technology for wearable devices in the future communication networks".

Conflicts of Interest: The authors declare no conflict of interest.

References

1. Atnasova, G.L.; Atanasov, N.T. Impact of electromagnetic properties of textile materials on performance of a low-profile wearable antenna backed by a reflector. In Proceedings of the 2020 International Workshop on Antenna Technology (iWAT), Bucharest, Romania, 25–28 February 2020. [CrossRef]
2. Atanasov, N.T.; Atanasova, G.L.; Atanasov, B.N. Wearable Textile Antennas with High Body-Area Isolation: Design, Fabrication, and Characterization Aspects. In *Modern Printed Circuit Antennas*, 1st ed.; Al-Rizzo, H., Ed.; IntechOpen: London, UK, 2020; Volume 1, pp. 1–20. [CrossRef]
3. Antolin, D.; Medrano, N.; Calvo, B.; Pérez, F. A wearable wireless sensor network for indoor smart environment monitoring in safety applications. *Sensors* **2017**, *17*, 365. [CrossRef] [PubMed]
4. Lin, R.; Kim, H.-J.; Achavananthadith, S.; Kurt, S.A.; Tan, S.C.C.; Yao, H.; Tee, B.C.K.; Lee, J.K.W.; Ho, J.S. Wireless battery-free body sensor networks using near-field-enabled clothing. *Nat. Commun.* **2020**, *11*, 1–10. [CrossRef] [PubMed]
5. Al-Sehemi, A.G.; Al-Ghamdi, A.A.; Dishovsky, N.T.; Atanasov, N.T.; Atanasova, G.L. Design and performance analysis of dual-band wearable compact low-profile antenna for body-centric wireless communication. *Int. J. Microw. Wirel. Technol.* **2018**, *10*, 1–11. [CrossRef]

6. El Attaoui, A.; Hazmi, M.; Jilbab, A.; Bourouhou, A. Wearable wireless sensors network for ECG telemonitoring using neural network for features extraction. *Wirel. Presonal Commun.* **2019**, *111*, 1–22. [CrossRef]
7. Seimeni, M.A.; Tsolis, A.; Alexsandridis, A.A.; Pantelopoulos, S.A. The effects of ground-plane of a textile higher mode microstrip patch antenna on SAR. In Proceedings of the 2020 International Workshop on Antenna Technology (iWAT), Bucharest, Romania, 25–28 February 2020. [CrossRef]
8. Al-Sehemi, A.G.; Al-Ghamdi, A.A.; Dishovsky, N.T.; Atanasov, N.T.; Atanasova, G.L. Flexible and small wearable antenna for wireless body area network applications. *J. Electromagn. Waves Appl.* **2017**, *31*, 1–20. [CrossRef]
9. Song, Y.; Le Goff, D.; Riondet, G.; Mouthaan, K. Polymer-based 4.2 GHz patch antenna. In Proceedings of the 2020 International Workshop on Antenna Technology, Bucharest, Romania, 25–28 February 2020.
10. Ashyap, A.Y.I.; Dahlan, S.H.; Abidin, Z.Z.; Dahri, M.H.; Majid, M.K.A.R.H.A.; Kamarudin, M.R.; Yee, S.K.; Jamaluddin, M.H.; Alomainy, A.; Abbasi, Q.H.; et al. Robust and efficient integrated antenna With EBG-DGS enabled wide bandwidth for wearable medical device applications. *IEEE Access* **2020**, *8*, 56346–56358. [CrossRef]
11. Ashyap, A.Y.I.; Marzudi, W.N.N.W.; Abidin, Z.Z.; Dahlan, S.H.; Majid, H.A.; Kamaruddin, M.R. Antenna incorporated with Electromagnetic Bandgap (EBG) for wearable application at 2.4 GHz wireless bands. In Proceedings of the 2016 IEEE Asian-Pacific Conference on Applied Electromagnetics, Langkawi, Kedah, Malaysia, 11–13 December 2016. [CrossRef]
12. Ashyap, A.Y.I.; Abidin, Z.Z.; Dahlan, S.H.; Majid, H.A.; Shah, S.M.; Alomainy, A. Compact and low-profile textile EBG-based antenna for wearable medical applications. *IEEE Antennas Wirel. Propag. Lett.* **2017**, *16*, 2550–2553. [CrossRef]
13. Del-Rio-Ruiz, R.; Lopez-Garde, J.M.; Legarda, J. Planar taxtile off-body communication antennas: A survey. *Electronics* **2019**, *8*, 714. [CrossRef]
14. Kapse, S.; Gundre, S.B. Fractal shaped dual-band EBG integrated textile antenna. In Proceedings of the 2019 4th International Conference on Recent Trends on Electronics, Information, Communication & Technology (RTEICT-2019), Bengaluru, India, 17–18 May 2019. [CrossRef]
15. Atanasov, N.T.; Atanasova, G.L.; Stefanov, A.K.; Nedialkov, I.I. A wearable, low-profile, fractal monopole antenna with a reflector for enhancing antenna performance and sar reduction. In Proceedings of the 2019 IEEE MTT-S International Microwave Workshop Series on Advanced Material and Processes for RF and THz Applications (IMWS-AMP), Bochum, Germany, 16–18 July 2019; 2019. [CrossRef]
16. Alemaryeen, A.; Noghanian, S. Performance analysis of textile AMC antenna on body model. In Proceedings of the 2017 United States National Conference of URSI National Radio Science Meeting (USNC-URSI NRSM), Boulder, CO, USA, 4–7 January 2017. [CrossRef]
17. Alemaryeen, A.; Noghanian, S. On-body low-profile textile antenna with artificial magnetic conductor. *IEEE Trans. Antennas Propag.* **2019**, *67*, 3649–3656. [CrossRef]
18. Fang, R.; Song, R.; Zhao, X.; Wang, Z.; Qian, W.; He, D. Compact and loa-profile UWB antenna based on graphene-assembled films for wearable applications. *Sensors* **2020**, *20*, 2552. [CrossRef] [PubMed]
19. Androne, A.; Tamas, R.D.; Tasu, S. Influence of the substrate material on the radar cross section of square loop unit cells for frequency selective surfaces. In Proceedings of the 2020 International Workshop on Antenna Technology (iWAT), Bucharest, Romania, 25–28 February 2020.
20. Zhu, J.; Cheng, H. Recent development of flexible and stretchable antennas for bio-integrated electronics. *Sensors* **2018**, *18*, 4364. [CrossRef]
21. Nepa, P.; Rogier, H. Wearable antennas for off-body links at VHF and UHF bands: Challenges, the state of art, and future trends below 1 GHz. *IEEE Antennas Propag. Mag.* **2015**, *57*, 30–52. [CrossRef]
22. Jiang, Z.H.; Brocker, D.E.; Sieber, P.E.; Werner, D.H. A Compact, low-profile metasurface-enabled antenna for wearable medical body-area network devices. *IEEE Trans. Antennas Propag.* **2014**, *8*, 4021–4030. [CrossRef]
23. Rowell, C.; Lam, E.Y. Mobile-phone antenna design. *IEEE Antennas Propag. Mag.* **2012**, *54*, 14–34. [CrossRef]
24. ICNIRP. Guidelines for limiting exposure to time-varying electric, magnetic, and electromagnetic fields (up to 300 GHz). *Health Phys.* **1998**, *74*, 494–522.
25. IEEE Standards Coordinating Committee. *IEEE for Safety Levels with Respect to Human Exposure to Radio Frequency Electromagnetic Fields, 3 kHz to 300 GHz*; IEEE Std C95.1-2005; IEEE: New York, NY, USA, 2006.

26. Al-Naiemy, Y.; Elwi, T.A.; Khaleel, H.R.; Al-Rizzo, H. A systematic approach for the design, fabrication, and testing of microstrip antennas using inkjet printing technology. *ISRN Commun. Netw.* **2012**, 1–11. [CrossRef]
27. Zhong, J.; Kiourti, A.; Sebastian, T.; Bayram, Y.; Volakis, J.L. Conformal load-bearing spiral antenna on conductive textile threads. *IEEE Antennas Wirel. Propag. Lett.* **2016**, *16*, 230–233. [CrossRef]
28. Balanis, C.A. *Antenna Theory Analysis and Design*, 2nd ed.; John Wiley & Sons, Inc.: New York, NY, USA, 1996; pp. 839–881.
29. Arai, H. *Measurement of Mobile Antenna Systems*, 2nd ed.; Artech House: Boston, MA, USA, 2013; pp. 39–84.
30. Chen, Z.N. *Antennas for Portable Devices*, 3rd ed.; John Wiley & Sons, Ltd.: Chichester, UK, 2007; pp. 12–13.
31. Yilmaz, T.; Foster, R.; Hao, Y. Broadband tissue mimicking phantoms and a patch resonator for evaluating noninvasive monitoring of blood glucose levels. *IEEE Trans. Antennas Propag.* **2014**, *6*, 3064–6075. [CrossRef]
32. Lam, P.T.; Le, T.Q.; Le, N.N.; Nguyen, S.D. Wireless sensing modules for rural monitoring and precision agriculture applications. *J. Telecommun. Inf. Technol.* **2017**, *2*, 107–123. [CrossRef]

Article

A Fully-Printed CRLH Dual-Band Dipole Antenna Fed by a Compact CRLH Dual-Band Balun

Muhammad Kamran Khattak, Changhyeong Lee, Heejun Park and Sungtek Kahng *

Department of Information & Telecommunication Engineering, Incheon National University, Incheon 22012, Korea; Kamrankhattak01@gmail.com (M.K.K.); Antman@inu.ac.kr (C.L.); h.park.inu@gmail.com (H.P.)

* Correspondence: s-kahng@inu.ac.kr; Tel.: +82-10-3423-0817

Received: 13 July 2020; Accepted: 31 August 2020; Published: 3 September 2020

Abstract: In this paper, a new design method is proposed for a planar and compact dual-band dipole antenna. The dipole antenna has arms as a hybrid CRLH (Composite right- and left-handed) transmission-line comprising distributed and lumped elements for the dual-band function. The two arms are fed by the outputs of a compact and printed CRLH dual-band balun which consists of a CRLH hybrid coupler and an additional CRLH phase-shifter. Its operational frequencies are 2.4 and 5.2 GHz as popular mobile applications. Verifying the method, the circuit approach, EM (Electromagnetics) simulation and measurement are conducted and their results turn out to agree well with each other. Additionally, the CRLH property is shown with the dispersion diagram and the effective size-reduction is mentioned.

Keywords: dual-band dipole; CRLH antenna; dual-band balun; CRLH balun; wireless communication

---

1. Introduction

Nowadays, wireless connectivity from equipment to other equipment and technological convergence becomes more vibrant. This results from a need to push the current limits. For instance, the 2.4 GHz band in the WLAN (Wireless local area network) frequency is commonly used, but this band is not sufficient to provide proper amounts of data because of excessive use and interference with other wireless communication methods using the same frequency band (e.g., Bluetooth, DCP, and ZigBee). So, to meet the frequency requirements, a number of multiband antennas have been proposed with various structures such as IFAs (Inverted-F antennas), bow-ties, slots and monopoles [1–4]. Pushpakaran et al. proposed a dog bone shape dual-band dipole antenna for WLAN applications. They presented a method for achieving a dual-band property by using the stacking technique [5]. Deepak [6] proposed a dipole antenna along with a folded element for multi-band operation. To eliminate the return current leak of the SMA (Subminiature version A) connector, they separate the dipole arms by using the double sides. Sim et al. reported multi-band asymmetric dipole antenna for WLAN operation. For the feeding of dipole antenna, they are using the 50-ohm coaxial cable as positive and negative on the two arms [7]. J. Huang used a tapered transition the balance feed line in [8]. The trapezoid dipole arms are printed on one side of the substrate and the other single dipole is formed on the opposite side. Nair [9] presented an F-shaped slot line to feed a dual-band dipole antenna which are fed by an SMA connector. H. Azeez modified a dipole with a pair of E-shaped conductive arms to generate multiple resonance [10]. Its feeding scheme is the same as [9]. Others put their radiating elements near a large metal ground like Alekseytsev who coupled a slit with an overpass metal-line to excite an open loop and a parasitic for two resonance frequencies [11]. More complicated configurations are built by Barani, such as one PIFA (Planar Inverted-F antenna) conjoined with PIFAs, with monopoles and slits on the ground to increase the number of bands [12].

Because the ground-edge mounted antennas occupy a large footprint, multi-band antennas are made as 3D structures. Tang adopted vertical loops passing through via-holes of two parallel lines [13]. Sreelakshmy put a thick disk on a vertical feed-line, and formed two asymmetric holes splitting one angular and radial mode into two, and used them for the dual-band characteristics [14].

The previously mentioned multi-band dipole antennas adopt SMA and coaxial feeds as an unbalanced signal port. To cope with that, the balanced to unbalanced (balun) structure is needed for a dipole antenna. Besides, to feed the dual-band dipole antenna, it is necessary to combine the dual-band balun. Liu et al. proposed the CRLH (Composite right- and left-handed) transmission line balun, in which they made the virtual ground for the odd mode and shunt inductor to remove a via. So, the phase difference between the two output ports is around 180° at 1.5 and 3.6 GHz [15]. Tseng et al. developed a CRLH balun to control the phase slope. As a result, they obtained the multi-band property by using the two branches where one arm consists of the RH-TL (Right-handed transmission line) and the other part is based on the LH-TL (Left-handed transmission line) with lumped elements [16]. According to odd-mode and even-mode analysis, the dual-band baluns are proposed with open-stub and BPF (Bandpass filter) [17,18]. Huang et al. reported a microstrip-based Marchand-type balun. They could achieve the high selectivity by using the roll-off of the filter [19]. Isolation between two bands can be better by making an SIR (Step impedance resonator) feedback loop between the two output branches by Li [20]. Multiple loops causing the phase difference are formed through layers with vertical lines as bridges [21]. Kahng et al. proposed a CRLH-based balun for common-mode current indicator which is employed in the chip and PCB EMC (Printed circuit board electromagnetics compatibility) problems. The balun consists of the branch-line coupler and the compact metamaterial phase shifters that are horizontally cascaded [22]. It works for a single band.

In this paper, a compact and fully printed metamaterial dual-band balun and metamaterial dual-band dipole antenna are proposed and put into one structure. The dual-band balun has the CRLH hybrid branch line-coupler and a CRLH phase-shifter. The conventional coupler and phase shift lines are replaced by miniaturized phase-shift lines having +90° for one frequency and −90° for another frequency for dual-band and low-cost fabrication. Instead of power-divider shapes, using lumped L and C, and multiple and long stages seen in others' works, the new balun has single-stage hybrid coupler and stage- lines and phase-shifter, and complete distribute elements and its design is elaborated on in detail. The dipole antenna is different from others by having two arms in the form of hybrid metamaterial lines and have the same omni-directional pattern at the two resonance frequencies; 2.4 and 5.2 GHz are chosen as the test case and the circuit approach is done first, and followed by the EM simulation and the measurement. Additionally, the CRLH property is shown with the dispersion diagram and the size reduction effect of the proposed balun is addressed.

2. Compact CRLH Dual-Band Balun

2.1. CRLH Dual-Band Hybrid Branch-Line Coupler

In this section, the design of a dual-band balun as fully-printed distributed-elements without lumped elements is explained. The size of the entire structure will be much reduced by devising a CRLH phase shift-line to have 90° at the lower frequency band and −90° at the higher frequency band, and by forming a compact branch-line coupler.

The circuit of the CRLH phase-shift line is set up by solving the equations to get the dual-band properties as the hybrid branch line coupler. Before starting the process of the design, as a goal, the following specifications in Table 1 are given as follows.

Table 1. The specifications of the composite right- and left-handed (CRLH) dual-band balun.

Spec. Items	Feature	
Band	f_1	2.4 GHz
	f_2	5.2 GHz
Insertion Loss	Hybrid	<1.5 dB
	Balun	<1.5 dB
Return Loss	<−15 dB	
Isolation	<−15 dB	

In Figure 1, C_R, C_L, L_R and L_L are determined by generating 90° at f_1, and −90° at f_2. It is a matter of course, the zeroth-order resonance (ZOR) should also be created around $(f_1 + f_2)/2$ for the size-reduction. So, the mathematical equations for C_R, C_L, L_R and L_L are as follows [22,23]. Others end up with multiple stages, but a single-stage circuit is proposed here to reduce the size.

$$L_R = \frac{Z_c\pi[(\frac{\omega_1}{\omega_2})+1)]}{2\omega_2[1-(\frac{\omega_1}{\omega_2})^2]},\ C_L = \frac{\pi[(\frac{\omega_1}{\omega_2})+1)]}{2\omega_2 Z_c[1-(\frac{\omega_1}{\omega_2})^2]}$$

$$L_L = \frac{2Z_c[1-(\frac{\omega_1}{\omega_2}))]}{\pi\omega_1[1+(\frac{\omega_1}{\omega_2})^2]},\ C_R = \frac{2\pi[(\frac{\omega_1}{\omega_2})+1)]}{\omega_1 Z_c[1+(\frac{\omega_1}{\omega_2})^2]}$$

where $\omega_1 = 2\pi f_1$, $\omega_2 = 2\pi f_2$ and Z_c characteristic impedance in above equations. Solving the equations by setting Z_c, f_1 and f_2 at 35.5 Ω, 2.4 and 5.2 GHz, C_R, C_L, L_R and L_L are as 2.52 pF, 0.64 pF, 3.17 and 0.81 nH, respectively. Additionally, to get the 50-Ω case, tackling the equations as f_1 and f_2 at 2.4 and 5.2 GHz the values are as follows: 1.78 and 0.46 pF, 4.46 and 1.14 nH, in the same order with the 35.5-Ω case. Using these elements, the phase and dispersion diagrams are plotted and show the CRLH characteristics including the ZOR point as in Figure 2.

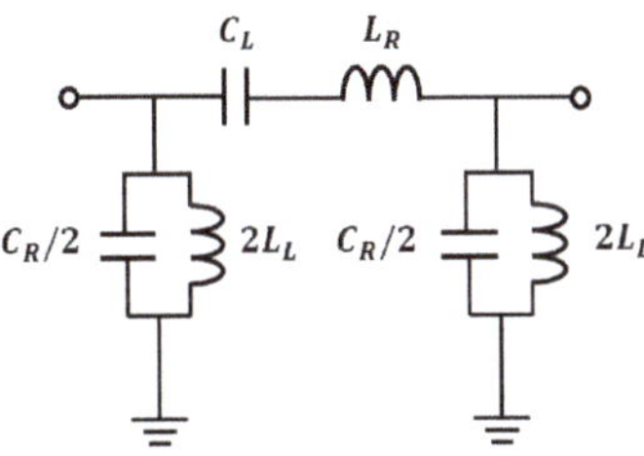

Figure 1. Equivalent circuit of the proposed dual-band CRLH phase-shifter line.

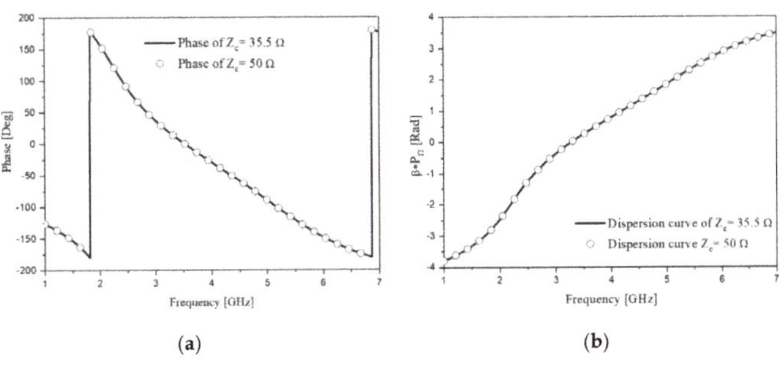

Figure 2. Circuit simulated results of the CRLH phase-shift line. (a) phase; (b) dispersion diagram.

In Figure 2a, the phases of 90° and −90° are achieved at 2.4 and 5.2 GHz. This phase-shift line complies with the specifications in Table 1 and will put into the 90°-branches of the hybrid branch-line coupler for miniaturization in terms of the physical size. Besides, there is the dispersion diagram in Figure 2b where the curve shows the LH region (β < 0) and the RH region (β > 0), along with the zeroth-order resonance (ZOR) near 3.8 GHz.

The finalized physical dimensions of the 35.5 Ω phase-shift line and geometry are given as table II and Figure 3a. Additionally, the physical dimensions of the 50 Ω phase-shift line and geometry are given as Table 2 and Figure 3b. The single-stage fully printable CRLH phase-shift line is designed using the CST-MWS as a full-wave EM simulator. Figure 3 shows the proposed CRLH phase-shift line geometry and EM simulated data of the phases. Both 35.5 and 50 Ω have an interdigital and shorting structure for C_R, C_L, L_R and L_L. The phases of 90° and −90° are realized at 2.4 and 5.2 GHz at 35.5 and 50 Ω. These will be substituted for the 90°-branches of the hybrid-branch-line coupler with a view to the effective size reduction of the dual-band balun.

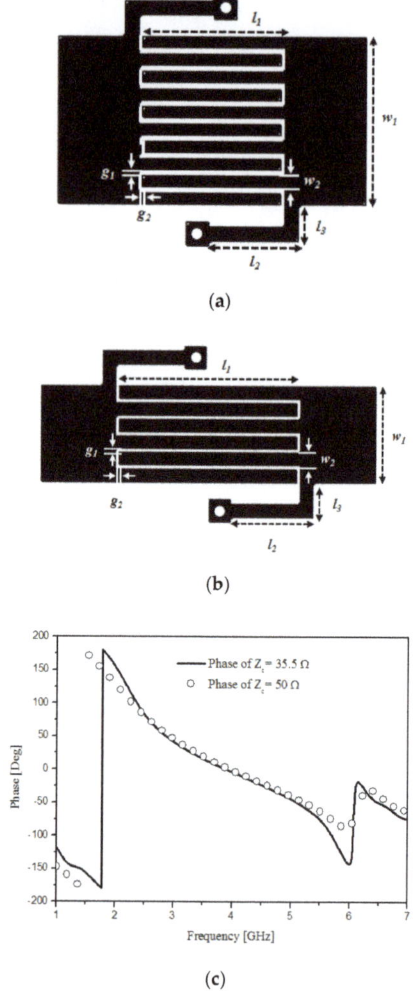

Figure 3. The physical geometries of the phase-shifter for (**a**) 35 Ω and (**b**) 50 Ω of their phases (**c**).

Table 2. Physical dimensions of the proposed CRLH phase-shift line.

Items (35.5 Ω)	l_1	l_2	l_3	w_1	w_2	g_1	g_2
Value (mm)	3.2	2	0.8	3.65	0.23	0.1	0.1
Items (50 Ω)	l_1	l_2	l_3	w_1	w_2	g_1	g_2
Value (mm)	4.4	2	0.8	2.18	0.28	0.1	0.1

Each of the segments in Figure 4a is realized with Table 2 for Figure 3a,b and becomes Figure 4b. As a crucial building block of the proposed balun, the function of the dual-band branch-line coupler is checked with the electric-field distributions at the frequencies of interest as in Figure 4c. The RF energy from port 1 is split into the output ports, and port 2 is turned on first as the 0°- field-shot, and then port 3 is turned on after 90°-lapse for 2.4 and 5.2 GHz. Nonetheless, at 4 GHz as the stopband, the electric field is not detected at the output ports in the 0°-field-shot and 90°-shot. The geometry is physically realized as in Figure 4d where the length is about 2 cm.

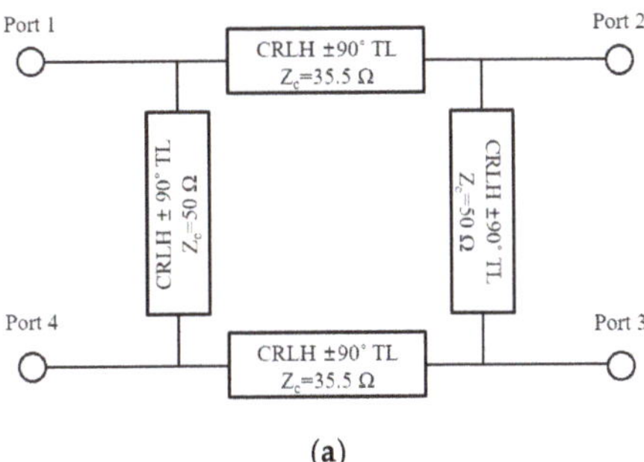

(a)

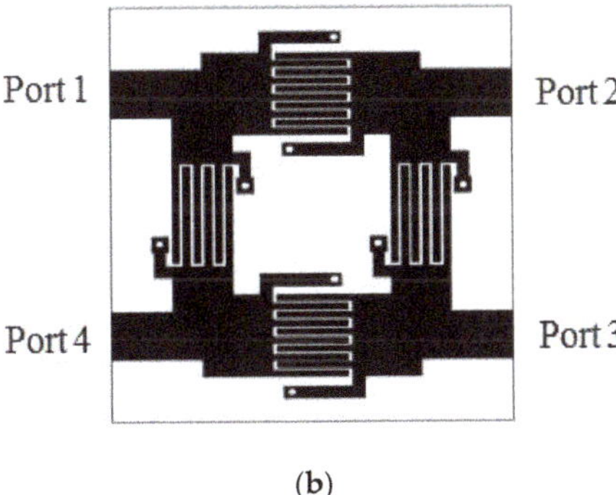

(b)

Figure 4. *Cont.*

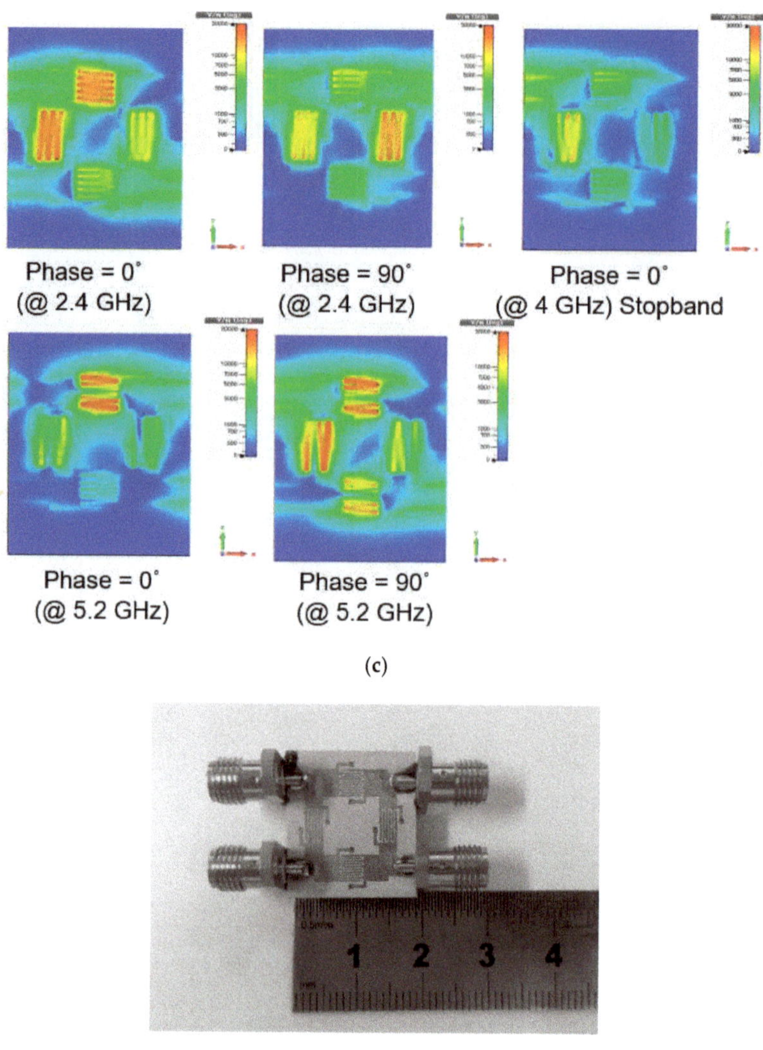

(**c**)

(**d**)

Figure 4. The physical geometry of the dual-band CRLH hybrid branch-line coupler as a single-stage geometry: (**a**) schematic; (**b**) EM (Electromagnetic) design; (**c**) E-field distribution; (**d**) fabricated prototype.

Figure 5a shows S_{11}, S_{21} and S_{31} of the EM simulation From S_{11}, the impedance is matched at the two frequencies. S_{21} and S_{31} present the equal power-division at the outputs. Figure 5b has the measured S_{11}, S_{21} and S_{31}. Similarly, to the simulated results, the power-division as well as the impedance match are obtained at 2.4 and 5.2 GHz. Figure 5c reveals the phase difference obtained as expected for the branch-line coupler. Somewhat incomplete parts of the function will be mitigated in the stage of the balun.

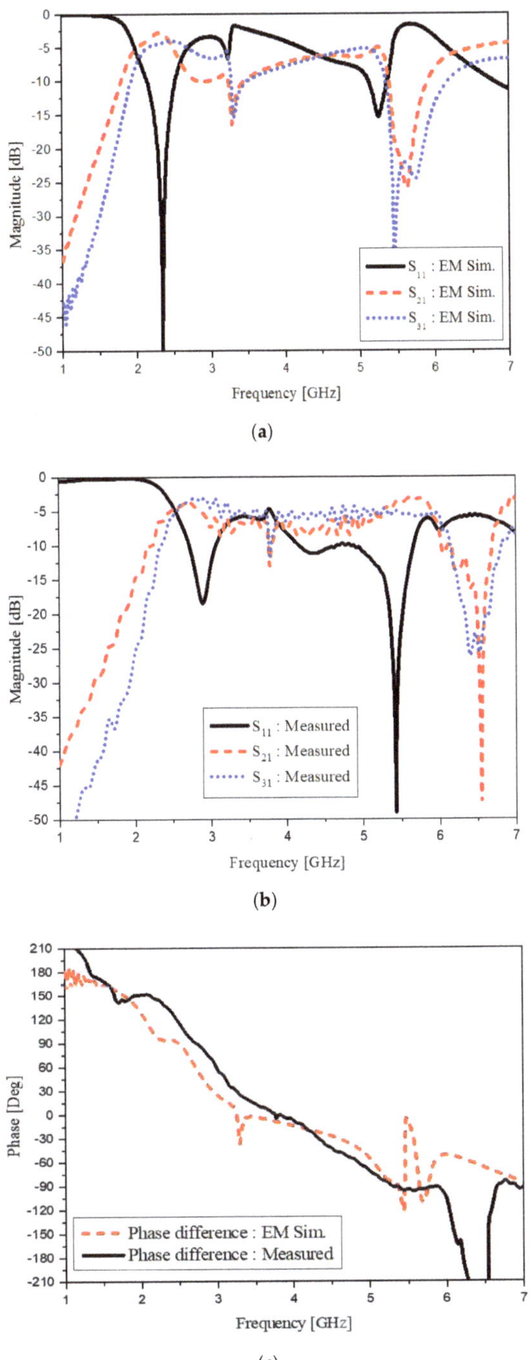

Figure 5. The frequency response of the dual-band CRLH hybrid branch-line coupler (**a**) EM simulated; (**b**) measured; (**c**) phase difference.

2.2. CRLH Dual-Band Balun

The branch-line coupler of Figure 4a is developed to the metamaterial balun of Figure 6a in the level of the schematic. The implementable geometry of the CRLH hybrid branch-line coupler extended with the CRLH phase-shift lines is shown in Figure 6b. Prior to fabrication, electromagnetic observation is conducted on the suggested device. The purpose of this observation is to see this device working as a desirable balun that shows the 180°-phase difference between ports 2 and 3 from the standpoint of the electric field. In Figure 6c, for 2.4 and 5.2 GHz, the RF energy is divided to ports 2 and 3, and port 2 and port 3 take turns by a 180°-time lapse. Meanwhile, at 4 GHz, bot output ports appear dark, which means no energy passes the circuit. As to the fabricated balun, its overall size is 14 × 20 × 1.2 mm^3 and realized on the FR4 consistent to the former design stage. Considering 27 × 27 × 1.2 mm^3 as the nominal size of other dual-band baluns with FR4, the size is reduced more than four times. The proposed balun shows similar results to circuit design as shown in Figure 7. The dual-band operating frequency is 2.4 and 5.2 GHz as satisfying the specifications in Table 1. The curves of S_{21} and S_{31} are almost the same each other as the −3.8 dB. This also results from the use of FR4 substrate. Here comes 180° as the phase difference at f_1 and f_2.

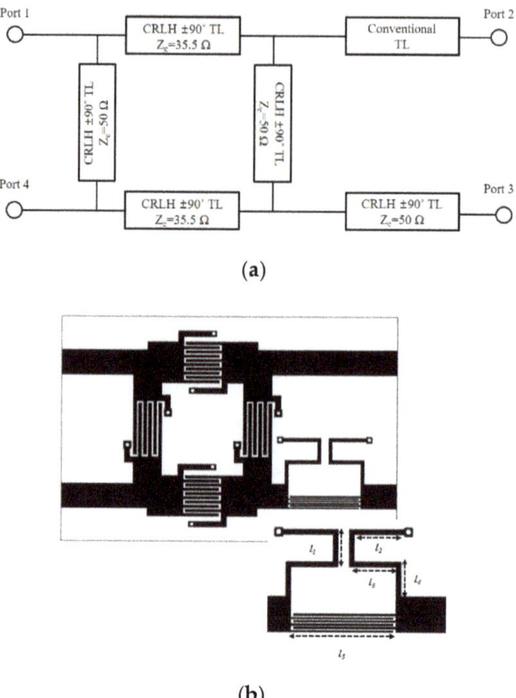

Figure 6. *Cont.*

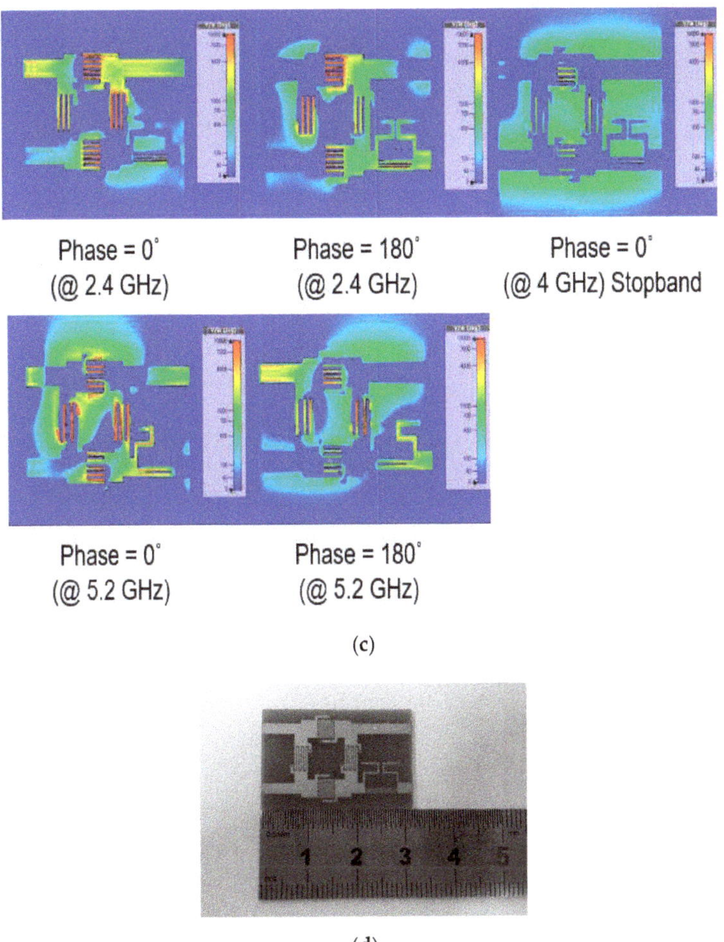

Figure 6. The physical geometry of the dual-band CRLH balun as a single-stage geometry: (**a**) schematic; (**b**) EM design; (**c**) E-field distribution; (**d**) fabricated design.

The frequency response of this compact balun works as expected. In Figure 7, both the simulated and measured s-parameters meet the design specifications. Additionally, the simulation and measurement results are in good agreement. The 180°-phase difference is seen. With regard to the size of the dual-band balun from the dual-band hybrid branch-line coupler, there is no change even though there is one new component added to the hybrid branch-line coupler. Since the new component as the dual-band phase-shift line is made as part of output port 3, it does not work as something negative in the size-reduction. Next, a dual-band dipole printable on the substrate is elaborated on.

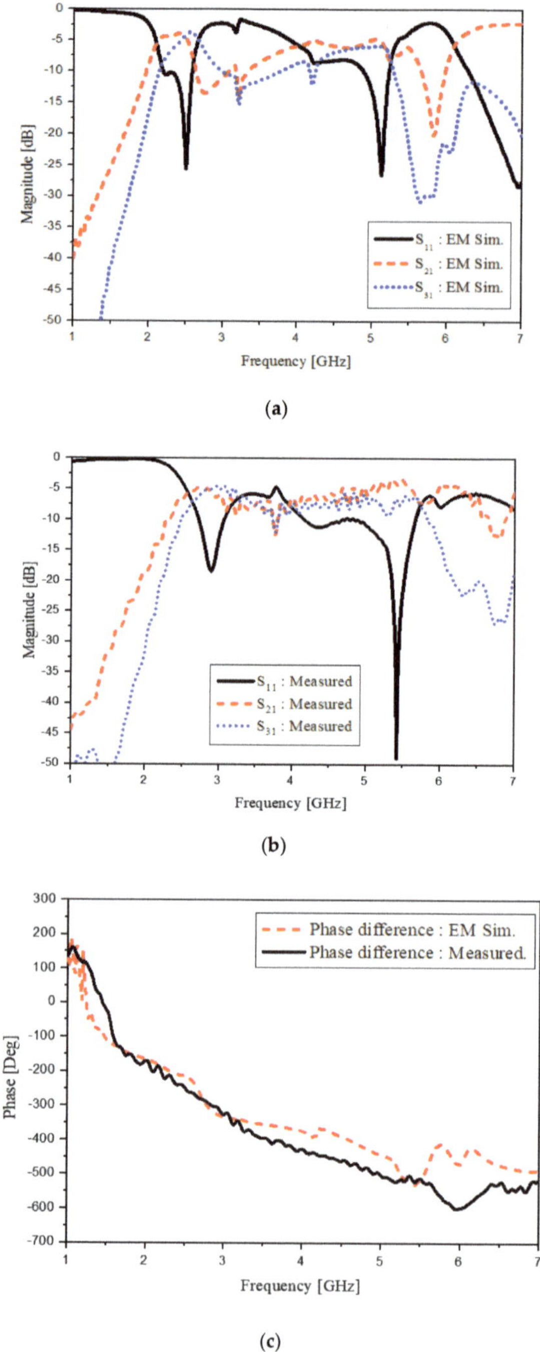

Figure 7. The frequency response of the dual-band CRLH balun: (**a**) EM simulated; (**b**) measured; (**c**) phase difference.

3. CRLH Dual-Band Dipole

The motivation of this part of the work is to make a dipole resonate at two frequencies not in the harmonic relationship and have the same omni-directional pattern at different resonance frequencies as a dipole. The proposed CRLH dipole antenna which has a dual-band property and operating at 2.4 and 5.2 GHz shown in Figure 8. In Figure 8a the highlighted cyan section indicates the upper path and yellow section express the lower path as an equivalent circuit for Figure 8b where $R_R = 67\ \Omega$, $L_{R1} = 3.9$ nH, $L_{R2} = 25.5$ nH, $L_{L1} = 12$ nH, $L_{L2} = 2$ nH, $C_R = 0.38$ pF, $C_L = 0.5$ pF. These result in the resonance at 2.4 and 5.2 GHz as in Figure 8c. In contrast to other dual-band dipoles, to maintain the omni-directional radiation pattern by suggesting a bow shape structure, a new structure is designed by adding a new path and making the entire geometry as the hybrid metamaterial arms on FR4 substrate. The upper current path is used to excite the high frequency and lower side path is matching at the low frequency. So, the proposed antenna connected from the top as the lumped inductor through the bottom path lumped capacitors as a loop shape. Geometrical elements w_s, w_f, w_1, w_2, w_g, g_1, g_2, l_s, l_f, l_1, l_2, l_3, l_4, and l_g are 100, 3, 80, 3, 150, 5, 150, 85, 20, 22, 15, 10, 85, and 3, respectively in mm. Compared to 140 mm or larger in length for commercial dipoles which are volumetric with the dielectric enclosure, the proposed antenna is smaller and planar. Using the geometrical parameters, the simulated and measured S_{11} as well as the far-field pattern is provided next.

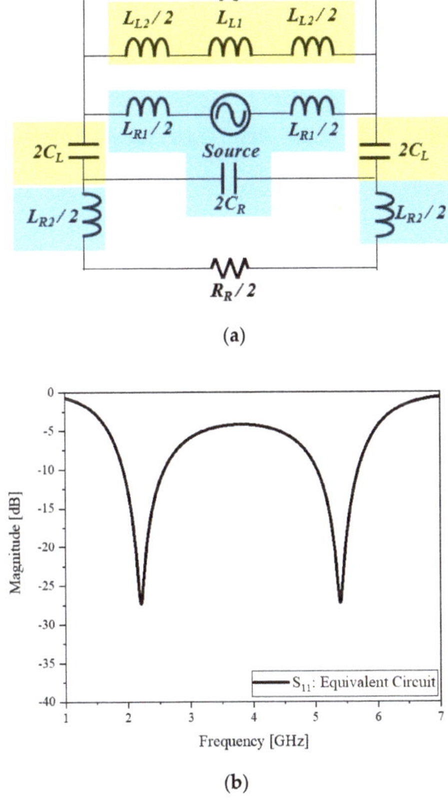

Figure 8. *Cont.*

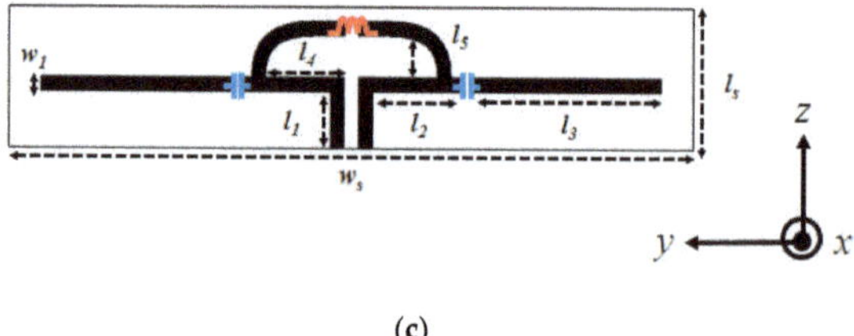

(c)

Figure 8. The CRLH dual-band dipole antenna design: (**a**) equivalent circuit; (**b**) S_{11} of the equivalent circuit; (**c**) structure in the EM CAD (Computer aided design) program.

Figure 9a,b expresses the resonant currents. The EM simulated and measured S_{11} in Figure 9c agree well and comply with the objectives. Dispersion curves are obtained from the circuit model and EM model of the new dipole as in Figure 9c. While 5.2 GHz has a positive wavenumber, 2.4 GHz has a negative wavenumber, which reveals the metamaterial property, helpful to size-reduction. Additionally, the far-field patterns with antenna gains 3.9 dBi and 2.38 dBi for the two frequencies seem very similar as given in Figure 9e.

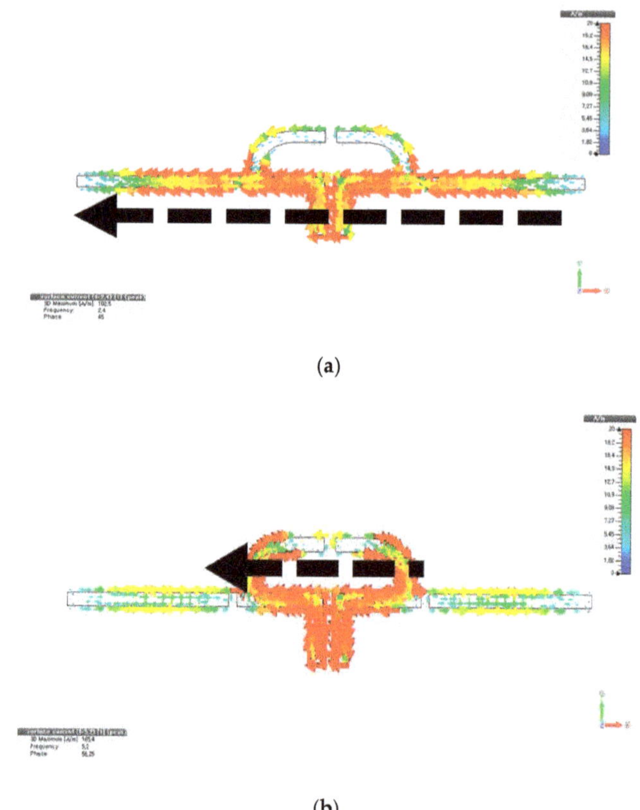

(a)

(b)

Figure 9. *Cont.*

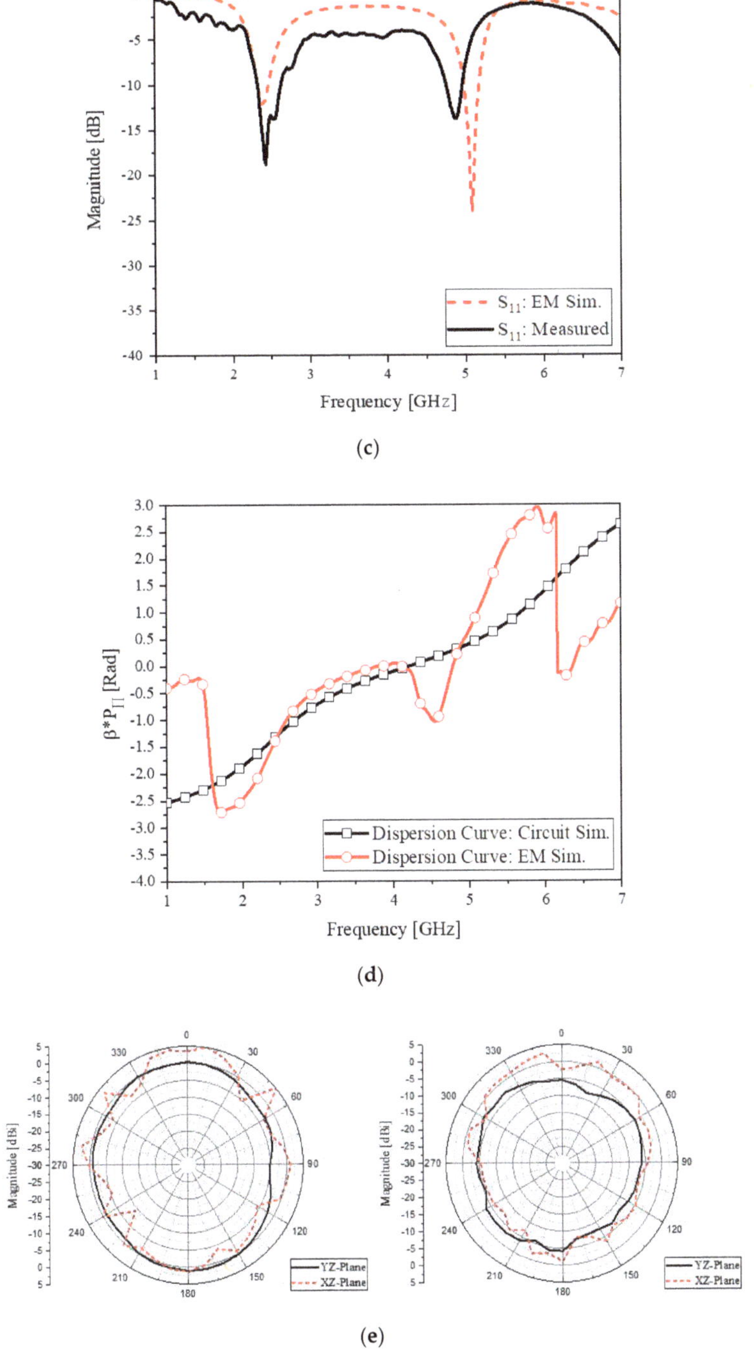

Figure 9. CRLH dual-band dipole. (**a**) current at 2.4 GHz; (**b**) current at 5.2 GHz; (**c**) S_{11}, (**d**) dispersion curves; (**e**) measured beam-patterns.

4. CRLH Dual-Band Antenna with CRLH Dual-Band Balun

Figure 10a is a sketch of the idea how the balun is used to feed the circuit model of the dipole antenna. The realized CRLH dual-band balun and antenna structure are provided in Figure 10b,c. The measurement harness at an anechoic chamber is seen in Figure 10d. The antenna gains are observed and they are above −1 dBi at the two frequencies. Table 3 presents the simulated and measured antenna gains. Figure 11a,b shows the gains and radiation characteristics of the CRLH dual-band antenna at 2.4 and 5.2 GHz are obtained similar to Figure 9. As to the antenna efficiency, it is over and similar to 50% in simulation, and becomes lower than 40% in measurement as checked in Table 3. They are changed from the simulated results in a small scale, due to the manual soldering for wiring of the semi-rigid coaxial cables and connector adapter as a major reason as well as the deviation of FR4 dielectric constant from the vendor's data. For example, it was found out the inductor was soldered in the lab which has the effect of adding an extra inductance and pushes the second resonance frequency downward by approximately 0.4 GHz. The fabricated antenna gives a not-very-high efficiency due to the aforementioned errors, but the efficiency at the two frequencies is usable for communication as guessed from practices. With regard to the approach of combining the dual-band balun with the dual-band dipole, the gain of the antenna falls by 0.7 dB at 2.4 GHz and 1.1 dB at 5.2 GHz in the simulation because of adding the balun. This is loss, but the antenna gain is nearly 1 dBi. The loss grows in the measurement as 5 dB at 2.4 GHz and 4 dB at 5.2 GHz, even though the antenna gain for the two frequencies is below 0 dBi. The loss from the combination is accounted for by the 1-dB insertion loss of the manufactured balun, and the 3.5 dB of the mismatch from cabling and connector attachment. Precision etching with a low-loss substrate and cascading all the blocks without the cables will mitigate this loss problem. As for the far-field patterns of the antenna is broad and close to the omni-directional shape. This can be an obvious advantage, as a small planar balun feed and a planar dipole enabling the same radiation function working at different frequencies, while commercial baluns and dual-band dipoles are bulky and give difficulty in being integrated to flat wireless systems for diverse use-cases such as IoT (Internet of Things) apparatus.

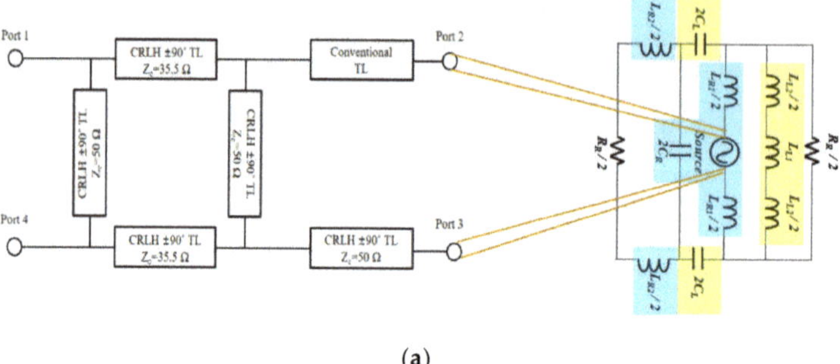

(a)

Figure 10. *Cont.*

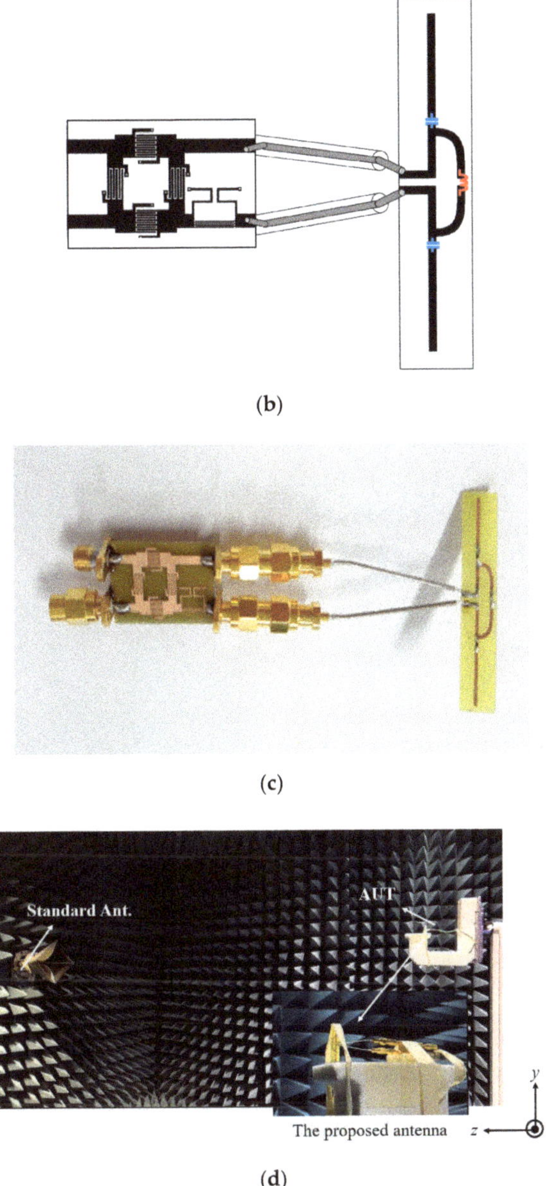

(b)

(c)

(d)

Figure 10. The CRLH dual-band dipole fed by the CRLH balun: (**a**) schematic; (**b**) hybrid schematic; (**c**) fabricated geometry; (**d**) far-field pattern measurement facility.

Table 3. Antenna gain and efficiency.

Type	Frequency	Sim./Meas.	Peak gain	Eff.
w/o Balun	2.4 GHz	Sim.	1.64 dBi	91%
		Meas.	5.13 dBi	88%
	5.2 GHz	Sim.	2.23 dBi	65%
		Meas.	2.93 dBi	47.9%
w/Balun	2.4 GHz	Sim.	0.9 dBi	35%
		Meas.	−0.48 dBi	21%
	5.2 GHz	Sim.	1.1 dBi	30%
		Meas.	−0.96 dBi	27%

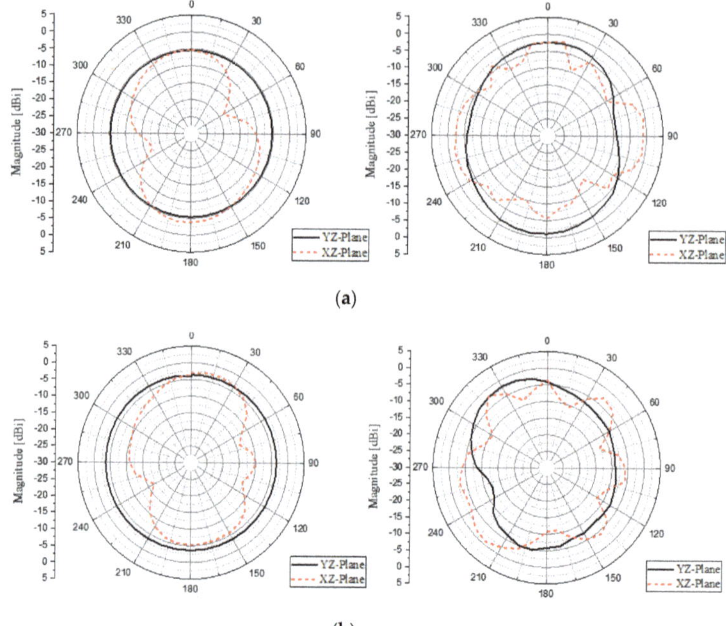

Figure 11. The radiation patterns of the CRLH dual-band dipole with CRLH balun: (**a**) simulated and measured results at 2.4 GHz; (**b**) simulated and measured results at 5.2 GHz.

5. Discussion

It is meaningful to check the technical status of the proposed idea in reflection of others' in the same subject. First, comparison is made between the new balun and other baluns aiming at multiple bands.

Through Table 4, the features of the new metamaterial balun are mentioned with those of selected papers as above. The kinds of features are type of structure, frequency bands, size, metamaterial or not, and origin of the design. While Tseng et al proposed the mixture of distributed and lumped elements, this work and the others are made from distributed elements, which provides ease of manufacture [16]. Except for this report, the circuits work for 2.4 and 5.2 GHz, and both works adopt the concept of metamaterials. They are obviously different in the following observations Tseng [16] uses multiple stages of lumped elements as a metamaterial line and a conventional delay-line to be a power-divider not a hybrid coupler. However, the proposed structure has no lumped elements and single-stage metamaterial-line segments for a hybrid coupler not a power-divider. Size-wise, the proposed balun is much smaller than others'. Huang, F et al presented a little bit smaller than the proposed geometry because their line is diagonally folded into the center of the area and the input port is connected to the common corner, but this work has not folded the branch-line coupler. The other circuits originate from the power-divider, filter, and T-junction, but the backbone of this work is the branch-line coupler similar to Yang et al's design [18]. Please note, Yang's device [18] is not a metamaterial but a coupler with stubs. This comparison is convincing that the proposed balun is novel and appropriate for the objectives. Second, it is worth checking characteristics and benefits of the proposed dual-band dipole in comparison with others in the same subject.

Table 4. Comparison of the proposed CRLH dual-band balun and the previous works.

Ref.	Type	Freq.	Size	Metamaterial	Origin
[16]	Distributed & Lumped (Hybrid)	1.24–3.58 GHz	$0.40\ \lambda_0 \times 0.53\ \lambda_0$	o	Power divider
[17]	Distributed	2.45/5.2 GHz	$0.3\ \lambda_0 \times 0.19\ \lambda_0$	x	Bandpass filter
[18]	Distributed	2.45/5.25 GHz	$0.7\ \lambda_0 \times 0.32\ \lambda_0$	x	Hybrid coupler
[19]	Distributed	2.4/5.2 GHz	$0.15\ \lambda_0 \times 0.15\ \lambda_0$	x	T-junction
[20]	Distributed	2.4/5.2 GHz	$0.65\ \lambda_0 \times 0.65\ \lambda_0$	x	T-junction
This work	Distributed	2.4/5.2 GHz	$0.24\ \lambda_0 \times 0.16\ \lambda_0$	o	Hybrid coupler

The antennas are looked into in terms of type, frequency, size, whether metamaterial or not, layering, and with or without a balun as in Table 5. Deng, Li, and Deepak's strucutres [1,2] and [6] are PIFA, bow-tie and SIR, respectively, but most of them take the form of a dipole or its modification. In contrast to all, only this work uses a metamaterial dipole. While Li, Huang, Sim, and Azeez's antennas [2,4,7] and [10] have more than three bands, 2.4 and 5.2 GHz are obtained by the other dipoles. To check the sizes as miniaturization effect, the areas of all the antennas are compared, which leads to a finding that the proposed antenna is substantially small. As given a question which antennas are metamaterials, this work (with Huang and Pushpakaran's antennas [4] and [5]) answers the question. Huang and Pushpakaran's antennas [4] and [5] are patch with a slit and slit coupled patches, respectively, and they have no evidence of metamaterials. However, this work reveals metamaterial properties with Figures 2b and 9d. With regard to layering, which is related to cost, Li, Huang, Pushpakaran, Huang, and Alekseytsev's antennas [2,4,5,8] and [11] have two or three layers. Nonetheless, this work is realized with a single layer of dielectric, which is cost effective. Most of them do not have baluns, but Huang et al's antenna [8] uses a 2-faceted balun comprising an upper tapered line and a lower line, and this work is connected with the CRLH balun of a single layer. The proposed antenna and CRLH balun have distinct features as explained.

Table 5. Comparison of the proposed CRLH dual-band dipole antenna and the previous works.

Ref.	Type	Frequency	Size	Metamaterial	Layer	Balun
[1]	PIFA	2.4/5.2 GHz	$0.15\ \lambda_0 \times 0.1\ \lambda_0$	x	Single	X
[2]	Bow-tie	2.4/3/5.2 GHz	$0.4\ \lambda_0 \times 0.36\ \lambda_0$	x	Double	X
[4]	Monopole	2.4/3.5/5.2~5.8 GHz	$0.05\ \lambda_0 \times 1\ \lambda_0$	o	Double	X
[5]	Dipole	2.4/5.2 GHz	$0.3\ \lambda_0 \times 0.32\ \lambda_0$	o	Triple	Slot line transition
[6]	SIR	2.4/5.2 GHz	$0.22\ \lambda_0 \times 0.16\ \lambda_0$	x	Single	X
[7]	Asymmetric Dipole	2.4/5.2/5.8 GHz	$0.25\ \lambda_0 \times 0.05\ \lambda_0$	x	Single	X
[8]	Dipole	2.4/5.2 GHz	$0.6\ \lambda_0 \times 0.4\ \lambda_0$	x	Double	2 faceted-balun
[9]	Dipole	2.4/5.2 GHz	$0.2\ \lambda_0 \times 0.1\ \lambda_0$	x	Single	X
[10]	F-shape Dipole	2.4/5.2/5.8 GHz	$0.65\ \lambda_0 \times 0.65\ \lambda_0$	x	Single	X
[11]	Quasi-Yagi	2.0/2.4 GHz	$1.17\ \lambda_0 \times 0.56\ \lambda_0$	x	Double	Slot line transition
This work	CRLH Dipole	2.4/5.2 GHz	$0.4\ \lambda_0 \times 0.08\ \lambda_0$	o	Single	○

6. Conclusions

A dual-band balun and a dual-band dipole antenna for the 2.4 and 5.2 GHz wireless communication are designed and combined to meet the demands on higher throughputs in IoT mobile connectivity with standing and portable electronic products. These two blocks are novel and small in that different from others, distributed type and single-stages of CRLH TX-lines are basically used to meet the requirement on the dual-band performance. In detail, the balun originates from a branch-line coupler and its segments which are 90°-lines in the conventional designs are replaced by metamaterial phase-shifters that show −90°-phase at 2.4 GHz and +90°-phase at 5.2 GHz as a small structure. The one-stage compact balun eventually makes the output ports out of phase by 180° at the target frequencies. In order to coincide with the two frequencies of the balun, an ordinary dipole is changed to a CRLH structure where a negative phase and a positive phase are generated at 2.4 and 5.2 GHz as the resonance condition of the dipole. The size does not grow, since the negative phase occurs at the lower frequency. The two bands of the dipole are fed by the balanced currents from the CRLH balun. This enables the cascaded blocks to radiate the far-field wave omni-directionally at the two frequencies with the acceptable antenna gains greater than −1 dBi, despite the error in the experiment due to the manual soldering of cabling as a major reason and the dielectric constant deviating from the vendor's data as a minor one.

Author Contributions: Conceptualization, S.K., C.L. and M.K.K.; methodology, S.K., C.L. and M.K.K.; software, C.L.; validation, C.L., and H.P.; formal analysis, H.P.; investigation, C.L.; resources, C.L., and H.P.; data curation, M.K.K.; writing—original draft preparation, M.K.K.; writing—review and editing, M.K.K.; visualization, M.K.K., and C.L.; supervision, S.K.; project administration, S.K.; funding acquisition, S.K. All authors have read and agreed to the published version of the manuscript.

Funding: The lead authors Changhyeong Lee and Sungtek Kahng on behalf of the contributors acknowledge that this work was supported by the Post-Doctoral Research Program (2017) in the Incheon National University. Besides, it should be mentioned that this work was carried out with the support of "Cooperative Research Program for Agriculture Science & Technology Development (Project No. PJ014762)" Rural Development Administration, Republic of Korea.

Conflicts of Interest: The authors declare no conflict of interest.

References

1. Deng, J.; Li, J.; Zhao, L.; Guo, L. A dual-band inverted-F MIMO antenna with enhanced isolation for WLAN applications. *IEEE Antennas Wirel. Propag. Lett.* **2017**, *16*, 2270–2273. [CrossRef]
2. Li, T.; Zhai, H.; Wang, X.; Li, L.; Liang, C. Frequency-Reconfigurable bow-tie antenna for bluetooth, WiMAX, and WLAN applications. *IEEE Antennas Wirel. Propag. Lett.* **2015**, *14*, 171–174. [CrossRef]
3. Medeiros, C.R.; Lima, E.B.; Costa, J.R.; Fernandes, C.A. Wideband slot antenna for WLAN access points. *IEEE Antennas Wirel. Propag. Lett.* **2010**, *9*, 79–82. [CrossRef]
4. Huang, H.; Liu, Y.; Zhang, S.; Gong, S. Multiband metamaterial-loaded monopole antenna for WLAN/WiMAX applications. *IEEE Antennas Wirel. Propag. Lett.* **2015**, *14*, 662–665. [CrossRef]
5. Pushpakaran, S.V.; Raj, R.K.; Vinesh, P.V.; Dinesh, R.; Mohanan, P.; Vasudevan, K. A metaresonator inspired dual band antenna for wireless applications. *IEEE Trans. Antennas Propag* **2014**, *62*, 2287–2291. [CrossRef]
6. Deepak, U.; Roshna, T.K.; Nijas, C.M.; Vasudevan, K.; Mohanan, P. A dual band SIR coupled dipole antenna for 2.4/5.2/5.8 GHz applications. *IEEE Trans. Antennas Propag.* **2015**, *63*, 1514–1520. [CrossRef]
7. Sim, C.; Chien, H.; Lee, C. Dual-/Triple-Band asymmetric dipole antenna for WLAN operation in laptop computer. *IEEE Trans. Antennas Propag.* **2013**, *61*, 3808–3813. [CrossRef]
8. Huang, J.; Lai, W. Design of a compact printed double-sided dual-band dipole antenna by FDTD for WiFi application. *Microw. Opt. Technol. Lett.* **2013**, *55*, 1845–1851. [CrossRef]
9. Nair, S.M.; Shameena, V.A.; Nijas, C.M.; Aanandan, C.K.; Vasudevan, K.; Mohanan, P. Slot line fed dual-band dipole antenna for 2.4/5.2 GHz WLAN applications. *Int. J. RF Microw. Comput. Aided Eng.* **2012**, *22*, 581–587. [CrossRef]
10. Azeez, H.I.; Yang, H.-C.; Chen, W.-S. Wearable triband E-Shaped dipole antenna with low SAR for IoT applications. *MDPI Electron.* **2019**, *8*, 665. [CrossRef]

11. Alekseytsev, S.A.; Gorbachev, A.P. The novel printed dual-band Quasi-Yagi antenna with end-fed dipole-like driver. *IEEE Trans. Antennas Propag.* **2020**, *68*, 4088–4090. [CrossRef]
12. Barani, I.; Wong, K. Integrated Inverted-F and open-slot antennas in the metal-framed smartphone for 2 × 2 LTE LB and 4 × 4 LTE M/HB MIMO operations. *IEEE Trans. Antennas Propag.* **2018**, *66*, 5004–5012. [CrossRef]
13. Tang, M.-C.; Wu, Z.; Shi, T.; Ziolkowski, R.W. Dual-Band, linearly polarized, electrically small huygens dipole antennas. *IEEE Trans. Antennas Propag.* **2019**, *67*, 37–47. [CrossRef]
14. Sreelakshmy, R.; Vairavel, G. Novel cuff button antenna for dual-band applications. *ICT Express* **2018**, *5*, 26–30. [CrossRef]
15. Liu, C.; Menzel, W. Broadband via-free microstrip balun using metamaterial transmission lines. *IEEE Microw. Wirel. Compon. Lett.* **2008**, *18*, 437–439.
16. Tseng, C.H.; Chang, C.L. Wide-Band balun using composite right/left-handed transmission line. *Electron. Lett.* **2007**, *43*, 1154–1155. [CrossRef]
17. He, Y.; Sun, L. Dual-Band balun bandpass filter using coupled lines with shunt open-ended stubs. *Microw. Opt. Technol. Lett.* **2014**, *56*, 2358–2360. [CrossRef]
18. Yang, S.Y.; Cho, C.S.; Lee, J.W.; Kim, J. A novel dual-band balun using branch-lines with open stubs. *Microw. Opt. Technol. Lett.* **2010**, *52*, 642–644. [CrossRef]
19. Huang, F.; Wang, J.; Zhu, L.; Chen, Q.; Wu, W. Dual-Band microstrip balun with flexible frequency ratio and high selectivity. *IEEE Microw. Wirel. Compon. Lett.* **2017**, *27*, 962–964. [CrossRef]
20. Li, E.S.; Lin, C.-T.; Jin, H.; Chin, K.-S. A systematic design method for a dual-band balun with impedance transformation and high isolation. *IEEE Access* **2019**, *7*, 143805–143813. [CrossRef]
21. Ali, A.; Yun, J.; Ng, H.J.; Kissinger, D.; Giannini, F.; Colantonio, P. High performance asymmetric coupled line balun at sub-thz frequency. *MDPI Appl. Sci.* **2019**, *9*, 1907. [CrossRef]
22. Kahng, S.; Lee, J.; Kim, K.-T.; Kim, H.-S. Metamaterial CRLH structure-based balun for common-mode current indicator. *J. Electr. Eng. Technol.* **2014**, *9*, 301–306. [CrossRef]
23. Caloz, C.; Itoh, T. *Electromagnetic Metamaterials: Transmission Line Theory and Microwave Applications*; Wiley: Hoboken, NJ, USA, 2006.

Letter

An Ultra-Wide Band Antenna System for Pulsed Sources Direction Finding †

Razvan D. Tamas and Stefania Bucuci *

Department of Electronics and Telecommunications, Constanta Maritime University, 900663 Constanta, Romania; tamas@ieee.org
* Correspondence: stefania.bucuci@ieee.org

† This paper is an extension of the work presented at the 2015 IEEE International Workshop on Antenna Technology, Seoul, Korea, 4–6 March 2015.

Received: 8 June 2020; Accepted: 18 August 2020; Published: 20 August 2020

Abstract: Electric discharges in high-voltage power distribution systems can be localized through their electromagnetic signature in the radio-frequency range. Since discharges produce series of short pulses, the corresponding spectrum usually covers wide frequency ranges, typically up to 1 GHz. In this paper, we propose an ultra-wide band (UWB) antenna system and a direction-finding (DF) approach based on using energy-based descriptors, instead of classical frequency-domain parameters. As an antenna system, we propose a dual-padlock configuration with a suitable pulse-matched response, featuring two unbalanced outputs. The proposed antenna system was successfully validated, both by simulations and measurements.

Keywords: ultra-wide band antennas; energy-based antenna descriptors; pulsed sources; direction finding

---

1. Introduction

Ultra-wide band applications generally include radar, communications, and direction-finding systems. Direction finding usually deals with electromagnetic source localization, including electric sparks in a power distribution system, based on their electromagnetic signature [1].

There are two types of direction-finding methods: amplitude based, and phase-based methods [2]. Specific algorithms can be applied on signals received from antenna arrays, in order to provide accurate direction finding [3].

The simplest amplitude-based method of DF consists in analyzing the received voltage after a mechanical rotation of a directional antenna, considering it as a reference of the source direction. The bearing is then found on a scale placed on the receiving antenna. In that case, the received voltage is displayed as a function of the rotation angle. By using two directional antennas and computing the sum and difference of the received signal amplitudes, one can extract the source bearing without rotating the antenna system [4].

Another amplitude-based technique consists of comparing the signal amplitudes from two orthogonal receiving antennas, in order to find the angle of arrival (AoA) [2]. The method is known as the amplitude comparison technique.

Phase-based approaches include both interferometry and Doppler direction finding [2,4]. Interferometers use antenna arrays in order to find the source direction from the phase differences between the signals received on each antenna. Conversely, a Doppler direction finder compares the phase of the received signal to that of a reference signal with the same central frequency, provided that the radiation pattern is steered either mechanically or electrically, and the frequency modulation occurs on the received signal due to the Doppler effect.

By using antenna arrays, multiple source localization is possible as well [5]. Algorithms for the direction of arrival estimation in the near-field zone with a vector sensor array are proposed in [6]. Moreover, direction and polarization can be extracted, by using improved algorithms for data processing [7].

Several types of radiating systems can be used for direction finding purposes. A direction-finding technique with generic, band limited vector sensors able to discriminate all the six field components is presented in [8]. However, most of the narrow-band designs need only three antennas [9] e.g., one dipole and two orthogonal loops [4]. Three collocated, orthogonal loops can be used instead; wideband operation is made possible by using impedance matching units [10]. A more general analysis covering all possible three-antenna configurations is conducted in [11].

Vector sensor radiating systems including thick wire loops can be used for a fractional bandwidth in the order of 1.5 [12]. Other approaches are based on using dual-band vector sensors in order to provide broadband operation [13]. Moreover, multiband antenna systems can actually be used to localize specific sources, e.g., mobile phones [14]. Planar, spiral shape antennas can provide a more compact solution with a fractional bandwidth figure of around 5 [15,16]. A four-port ultra-wide band (UWB) configuration exhibiting a fractional bandwidth figure of 2 is proposed in [17]. It should be emphasized that multimode antennas have recently been proposed [18], as an alternative to multi-element radiating systems.

As opposed to other direction-finding methods that require large antenna arrays [19], the amplitude comparison technique is based on an antenna system consisting of only two orthogonal loop-type radiators; annular shapes would grant good performance across an ultra-wide band. Since asymmetrical feed lines are generally used in practice, a monopole type antenna might be preferred. However, most of the ring-type monopoles do not preserve the radiation properties of a loop, as they are not actually fed as a loop, but as any other monopole [20,21].

Electric discharges in power distribution systems can generally be modeled as random series of UWB monocycle pulses, with a shape quite similar to the first derivative of the Gauss pulse. Classical narrow band direction finding methods would not lead to an accurate AoA estimation for such electromagnetic sources.

In this paper, we propose a novel UWB antenna system for spark detection and localization in power distribution systems. It consists of two identical, orthogonal padlock-shaped (half-ring) antennas. As opposed to other similar systems [22], each antenna can be asymmetrically driven by using a coaxial cable. This is an advantage over traditional UWB rings which would require an UWB balun. The antenna system was designed to operate with pulses, with a spectrum centered on 250 MHz. The system exhibits very good energy-based figures and an excellent agreement between simulated and measured results can be noted. In order to better quantify the mutual coupling between two UWB antennas with pulsed excitation, we introduced a new descriptor that we called energy-based coupling coefficient. We also show that such a parameter is more relevant to pulsed applications than transmission scattering parameters.

In the last section of our work, we propose a direction-finding methodology using the dual padlock antenna. The methodology combines angle averaging and time gating for a better accuracy.

2. Antenna System Design Energy Based Descriptors

The simplest design for amplitude comparison direction finding requires a system of antennas, comprising two vertically oriented loop antennas (ring-shaped), orthogonally arranged, and a "sense" antenna (omnidirectional) used to resolve the "front-back" ambiguity [4].

It has previously been shown [22] that the direction finding of UWB sources can be performed by using the amplitude comparison method, without a sense antenna. In that case, pulse polarities on the orthogonal ring type antenna system show the quadrant of the source direction. Moreover, in the case of spark localization within power plants, the half-space of the incoming wave is usually a priori known.

Thus, the radiating system will only consist of two orthogonal loop elements. Each loop should present a radiation pattern with two lobes over the most of the frequency band, while its nulls are in the antenna plane. For the application under consideration, the polarization of interest is the horizontal one.

The main disadvantage of loop antennas is the balanced input when fed through symmetrical transmission lines. Ultra-wide bandwidth baluns are generally difficult to design and manufacture. In a previous paper [23], we proposed an innovative ultra-wideband, half-ring antenna system sensitive to horizontally polarized electromagnetic field generated by sparks in power plants and energy distribution systems. The novel UWB antenna system has two unbalanced inputs. The excitation of our design was inspired from that used for a folded monopole antenna, i.e., one end of the half ring is connected to the ground plane, and the other one to the inner conductor of the coaxial line. As opposed to folded monopole antennas, our design uses a vertical ground plane since planar antennas are technologically preferred.

The system consists of two identical axially crossed "padlock" shaped antennas, as in Figure 1. Thus, the antenna system will retain the advantages of the loop antenna, but can be fed by an asymmetrical coaxial line.

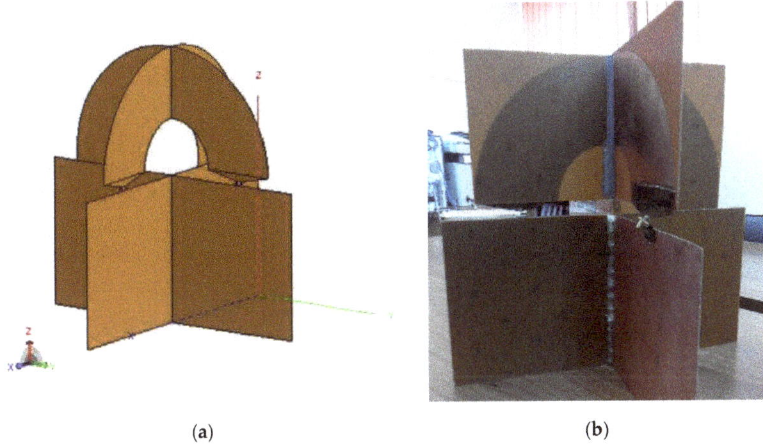

(**a**) (**b**)

Figure 1. Proposed antenna design: (**a**) Simulation design; (**b**) Experimental antenna. [23].

The size of the radiating elements depends on the spectrum of the input pulses generated by the electric discharges. In most cases, the waveform of such a pulse can be assimilated to the first derivative of the Gaussian function presented in Figure 2.

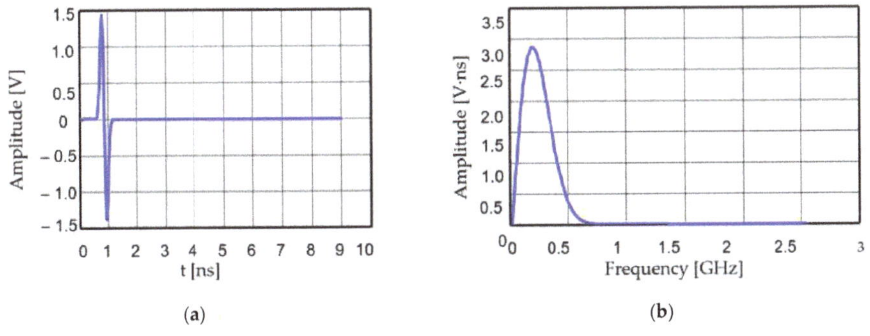

(**a**) (**b**)

Figure 2. First derivative of a Gaussian function: (**a**) waveform; (**b**) spectrum.

The reflection coefficient in the frequency domain, $\Gamma(\omega)$, is defined as the ratio between the complex amplitudes of the reflected wave and forward wave at the input of the antenna feed line.

The time-domain reflection coefficient is found by applying an inverse Fourier transform, $\gamma(t) = \mathcal{F}^{-1}\{\Gamma(\omega)\}$. The instantaneous reflected voltage at the source can be expressed as a convolution,

$$v_r(t) = \frac{\left(v_{g0} * \gamma\right)(t)}{2} \tag{1}$$

where $v_r(t)$ is the instantaneous reflected voltage at the source and $v_{g0}(t)$ is the instantaneous electromotive force of the pulse source. It should be noted that a known, time-domain reflection coefficient, $\gamma(t)$ gives the reflected output signal for any input signal.

When transmitting or receiving pulsed signals energy-based descriptors should be used instead of classical, frequency-domain antenna parameters [24]. In order to quantify the energy balance at the antenna input, two suitable energy parameters have been defined [25,26]: the pulse reflection coefficient and the pulse matching ratio, respectively.

The pulse reflection coefficient, g is defined as the square root of the ratio between the energy of the reflected signal and the energy of the forward signal:

$$g = \sqrt{\frac{\text{Reflected signal energy}}{\text{Forward signal energy}}} = \frac{\text{RMS}(v_r)}{\text{RMS}\left(\frac{v_{g0}}{2}\right)} = 2\sqrt{\frac{\int_{\text{supp } v_r(t)} {v_r}^2(t)dt}{\int_{\text{supp } v_{g0}(t)} {v_{g0}}^2(t)dt}} = 2\sqrt{\frac{\mathfrak{R}_{v,v_r}(0)}{\mathfrak{R}_{v_{g0},v_{g0}}(0)}}, \tag{2}$$

In (2), supp $v_r(t)$ and supp $v_{g0}(t)$ are the temporal supports of the instantaneous reflected voltage and instantaneous electromagnetic force, respectively; the corresponding autocorrelation functions are denoted by $\mathfrak{R}_{v_r,v_r}(0)$ and $\mathfrak{R}_{v_{g0},v_{g0}}(0)$.

As the frequency-domain reflection coefficient, g has a subunitary magnitude; a null would correspond to a perfect matching. It should be highlighted that the pulse reflection coefficient is always defined for a given waveform of the excitation. Moreover, it can be shown that $g = |\Gamma|$ for sinusoidal signals.

The pulse matching ratio, s [25] is an energy-based descriptor similar to the frequency- domain voltage standing wave ratio (VSWR),

$$s = \frac{1+g}{1-g} \tag{3}$$

The value of this parameter is greater than or equal to 1; a perfect matching is expressed by a $s = 1$. As with other energy-based figures of merit, this parameter is also defined for a given waveform of excitation.

The antenna impulse response is a function of time that characterizes the antenna as a linear system. The input parameter can be the input voltage of a transmitting antenna. The output figure is usually derived from the electric far-field by compensating the propagation effects, i.e., attenuation and delay [4]. The electric far-field can be written as:

$$\left[v_g * h_t(\hat{\mathbf{r}})\right](t) = rE(\mathbf{r}, t + r/c_0), \tag{4}$$

where $v_g(t)$ is the voltage across the antenna input, $h_t(\hat{r}, t)$ is the impulse response of the transmitting antenna, which is proportional to its effective height [27], and $E(t)$ is the far, electric field.

The energy gain is an important figure of merit in terms of ultra-wide band radiation, defined as [28]:

$$G(\hat{r}) = \frac{4\pi \text{ Energy radiated per unit solid angle } (\hat{r})}{\text{Total radiated energy}} = \frac{16\pi Z_0}{\eta} \frac{\mathfrak{R}_{e_t(\hat{r}),e_t(\hat{r})}(0)}{\mathfrak{R}_{v_{g0},v_{g0}}(0)}, \tag{5}$$

where η is the free space wave impedance, and

$$e_t(\hat{\mathbf{r}}, t) = rE(\mathbf{r}, t + r/c_0), \tag{6}$$

where $\mathbf{r}$ is the position vector of the field point, $\hat{\mathbf{r}}$ is the unit vector of the corresponding direction, and c_0 is the speed of light.

A new energy-based parameter would be necessary in order to better quantify the mutual coupling for a given activation waveform. An energy-based coupling coefficient can be computed as:

$$c = \sqrt{\frac{\text{Transmitted (output) energy}}{\text{Forward (input) energy}}} = \frac{\text{RMS}(\mathrm{v})}{\text{RMS}\left(\frac{\mathrm{v}_g}{2}\right)} = 2\sqrt{\frac{\int_{\text{supp } \mathrm{v}(t)} \mathrm{v}^2(t)dt}{\int_{\text{supp } \mathrm{v}_g(t)} v_g{}^2(t)dt}} = 2\sqrt{\frac{\mathfrak{R}_{\mathrm{v},\mathrm{v}}(0)}{\mathfrak{R}_{\mathrm{v}_g,\mathrm{v}_g}(0)}}, \tag{7}$$

where $\mathrm{v}(t)$ is the output waveform, $\mathrm{v}_g(t)$ is the input waveform, and $\mathfrak{R}_{\mathrm{v}_g,\mathrm{v}_g}$, $\mathfrak{R}_{\mathrm{v},\mathrm{v}}$ are the autocorrelation functions of the input and output signal, respectively.

3. Proposed Direction-Finding Methodology

The shape of the traditional radiation patterns changes from one frequency to another and the nulls do not always occur on the right direction. For antennas with a pulsed excitation, energy-based radiation patterns can be drawn up from the energy gain, defined as in (5).

The antenna system was designed for an excitation proportional to the first derivative of the Gauss function, as in Figure 2. Its spectrum is centered on 250 MHz. The energy radiation patterns for both antennas are shown in Figure 3 when applying the above excitation successively on each antenna.

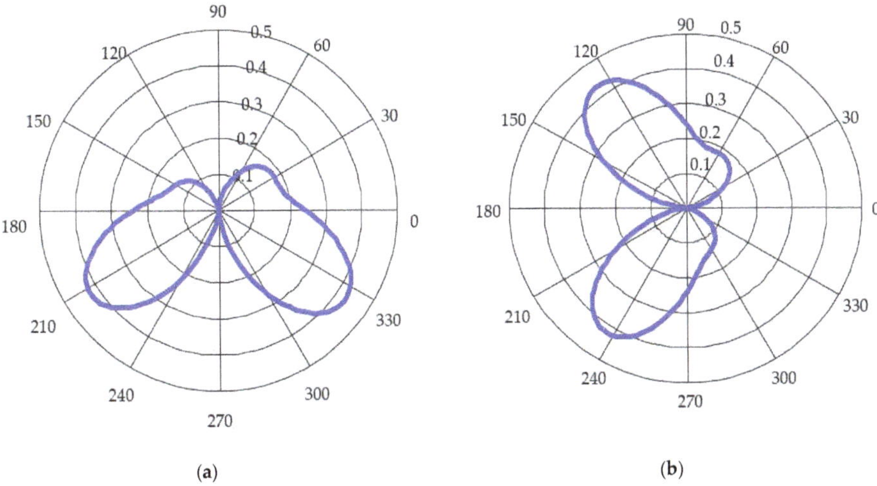

Figure 3. Energy-based pattern diagram: (**a**) First antenna; (**b**) Second antenna.

The nulls of the energy-based radiation patterns occur at $\varphi = 90°$ and $270°$ for the first antenna, and $\varphi = 0°$ and $180°$ for the second antenna. An ideal vector sensor consists of two antennas, one with a sine, and the other with a cosine shaped pattern diagram [22]. The angle of arrival can be expressed as the ratio between the signal amplitude at the first antenna output and at the second antenna output, respectively:

$$\text{AoA}^{\text{ideal}} = \tan^{-1}\frac{A_1}{A_2} = \tan^{-1}\frac{F_1^{ideal}(\varphi)}{F_2^{ideal}(\varphi)}, \tag{8}$$

where A_1, A_2 are the received signal amplitudes and F_1^{ideal}, F_2^{ideal} are the ideal radiation characteristics, i.e.,

$$F_1^{ideal}(\varphi) = |\sin\varphi|, \tag{9}$$

$$F_2^{ideal}(\varphi) = |\cos\varphi|. \tag{10}$$

As shown in Figure 3, the real energy-based radiation patterns are not proportional to the sine and cosine of the angle of incidence, respectively. By defining

$$R(\varphi) = \frac{A_1}{A_2} = \frac{F_1^{real}(\varphi)}{F_2^{real}(\varphi)}, \tag{11}$$

The AoA is then found,

$$\text{AoA}^{\text{real}} \cong R^{-1}\left(\frac{A_1}{A_2}\right). \tag{12}$$

Figure 4 presents a comparison between the real and ideal radiation characteristics.

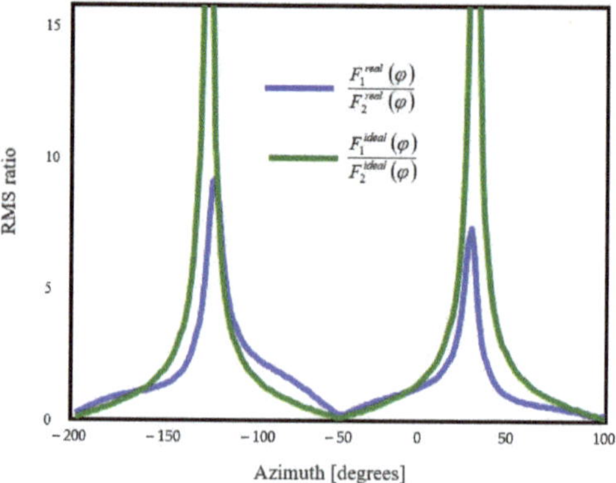

Figure 4. Ratio between RMS of the received signals.

As Figure 4 shows, computing the AoA by simply evaluating the arctangent of the signal ratio would lead to errors. Since $R(\varphi)$ may not result in an analytical form, its inverse can be computed by using a lookup table.

4. Results: Antenna System

The measuring setup (Figure 5) for the frequency-domain gain consists of the antenna system under test, a calibrated antenna, and a vector network analyzer.

Since we measured an ultra-wide band radiating system, an anechoic chamber was not necessary. An inverse Fourier transform was performed on the measured data, and the result was windowed in the time-domain, as the system response was short enough compared to the shortest indirect propagation path. Figure 6 gives a sample of time-domain response before (a) and after (b) windowing. In Figure 6a, one can easily discriminate the system response from the effect of the multipath propagation.

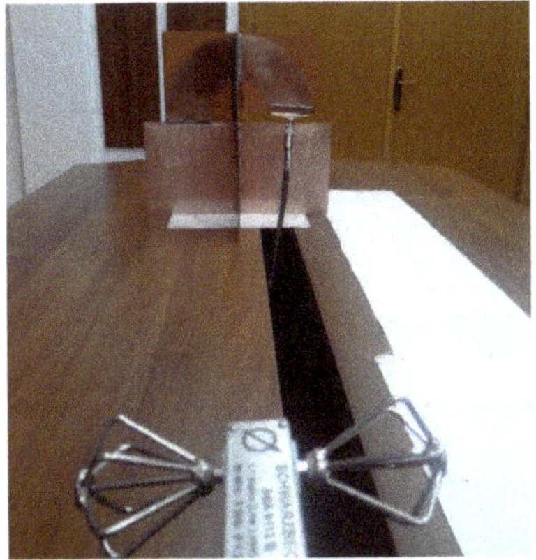

Figure 5. Measuring setup [23].

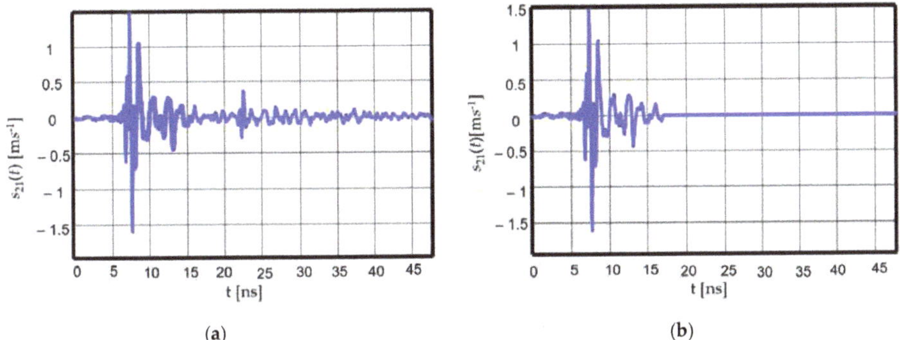

(**a**) (**b**)

Figure 6. Time-domain response: (**a**) before windowing; (**b**) after windowing [23].

In order to find the antenna gain from the measured S parameters, we evaluate the ratio between the received and transmitted power with the Friis formula:

$$\frac{P_r}{P_t} = G_t G_r \left(\frac{\lambda}{4\pi d} \right), \tag{13}$$

where d is the distance between the two antennas, G_t is the gain of the calibrated (transmitting) antenna and G_r is the gain of the measured (receiving) antenna.

By assimilating the entire setup to a two-port device terminated on $R_0 = 50\ \Omega$, as shown in Figure 7, the transfer function can be written as:

$$S_{21} = \frac{2V_2}{V_g} \tag{14}$$

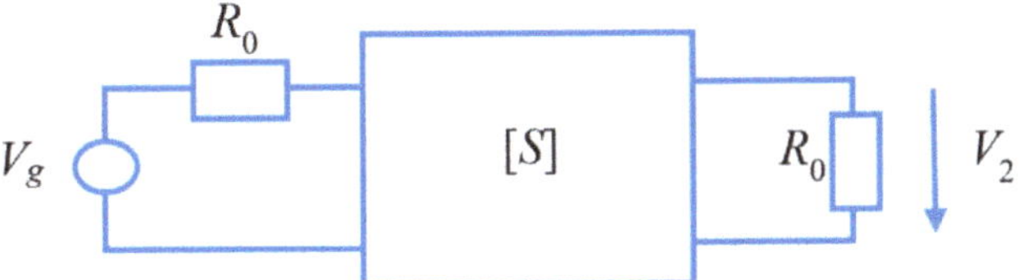

Figure 7. Equivalent two-port circuit.

By using the Friis formula and taking into account the impedance mismatch at each antenna [29]

$$\sqrt{G_t G_r} = \frac{4\pi d \cdot |S_{21}|}{\lambda\,|1 - S_{22}|} \cdot \sqrt{\frac{R_0}{R_{a2}(f) \cdot \left(1 - |S_{11}|^2\right)}} \tag{15}$$

where $R_{a2}(f)$ is the radiation resistance of the antenna under test.

Consequently, the gain of the antenna under test is

$$G_r = \frac{1}{G_t}\left(\frac{4\pi d}{\lambda}\right)^2 \frac{R_0}{R_{a2}(f)} \frac{|S_{21}|^2}{|1 - S_{22}|^2\left(1 - |S_{11}|^2\right)} \tag{16}$$

where S_{21} is the Fourier transform of the impulse response, after windowing the signal.

The S parameters of the proposed antenna system are presented in Figures 8 and 9. The simulated results (in blue) are compared to measured results (in red).

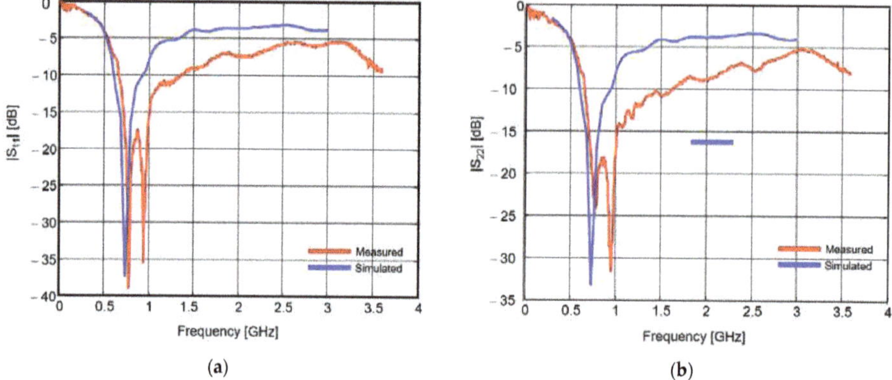

Figure 8. Reflection coefficients: (**a**) First antenna; (**b**) Second antenna [23].

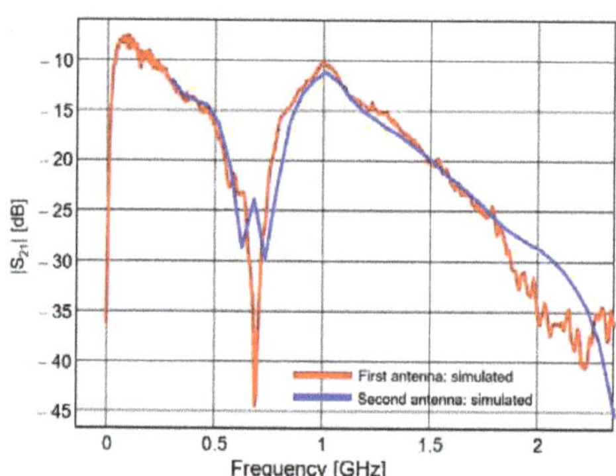

Figure 9. Transfer function between the two padlock antennas [23].

A very good agreement can be noted between the measured and simulated transmission parameters; conversely, there are some discrepancies between the simulated and measured reflection parameters. Those discrepancies are mainly due to the lack of accuracy of our simulator for modeling the excitation. The measured reflection coefficient is actually better than that resulting from simulation, as its magnitude is below −10 dB between 700 MHz and 1.4 GHz, and below −5 dB between 500 MHz and 2.2 GHz.

The magnitude of the transmission coefficient reaches a global maximum of −8 dB around 100 MHz and a local maximum of −10 dB around 1 GHz. That figure might not be satisfying for narrow band applications at such frequencies; the mutual coupling is quite high and that would impinge on the orthogonality. However, for pulsed waveforms, the mutual coupling should be assessed in straight correlation with the spectral power density of the signal.

Energy-based descriptors are actually more relevant for pulsed applications than frequency- domain parameters. For our antenna system, and for the first derivative of a Gauss function (Figure 2), as an excitation with a spectrum centered on 250 MHz, we found $g = -5.51$ dB and $s = 3.2567$, respectively. The energy-based coupling coefficient is −17.86 dB.

In order to measure the energy-based gain, the antenna was rotated horizontally with an angular pitch of 10°. Figure 10 presents the energy-gain pattern diagrams for each padlock antenna versus simulated gain. The diagrams were plotted for $\theta = 90°$ and horizontal polarization.

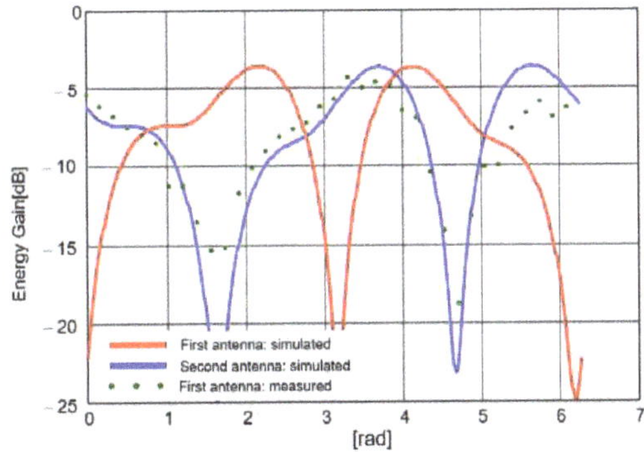

Figure 10. Energy gain pattern diagram [23].

A very good agreement between simulations and measurements can be noted. There are deep extinctions of the pattern diagrams in the plane of each antenna, which makes the proposed system appropriated to UWB direction finding purposes.

5. Results: Direction Finding Approach

In order to validate our UWB direction finding approach, we utilized a setup consisting of the padlock antenna system, a calibrated biconical antenna and a vector network analyzer (Figure 11). We measured two transfer functions, i.e., between the probe antenna and each padlock radiator, with the calibrated antenna placed successively in a matrix of 25 positions, as in Figure 12.

Figure 11. Direction finding approach: measurement setup.

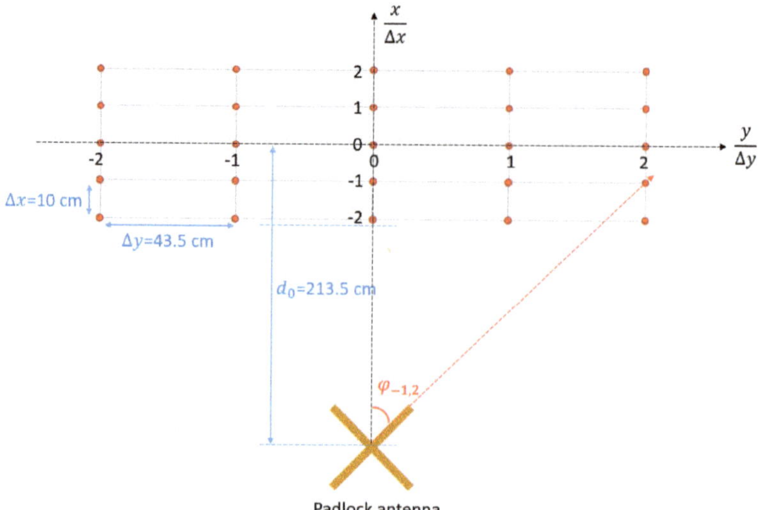

Figure 12. Matrix of measuring positions.

The measurements were performed in a multipath environment. In practice, such a matrix of measuring points would correspond to an average based methodology of direction finding with the aim to reduce the effect of the multipath propagation [30]. That is, one can take the middle row in Figure 12, as successive positions of the electromagnetic source to be found. The columns would actually be successive positions of the padlock antenna, and the measured azimuth angles in one column are subject to averaging, in order to accurately estimate the AoA. Additionally, a time domain gating is performed on each received waveform, in order to further improve the accuracy.

It should be noted that, when measuring the transfer function of one radiator, the other one was terminated on a 50 ohm load.

In Figure 13, we show the waveforms of the signals received on the two padlock radiators for the middle (reference) row in the matrix of measuring positions.

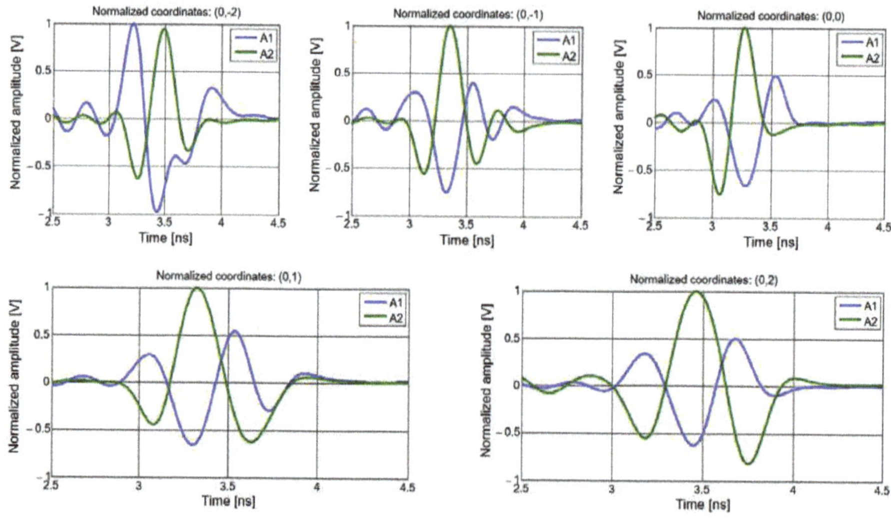

Figure 13. Waveforms of the received signals: middle (reference) row.

We computed the ratio between the RMS of the two received waveforms for each measuring position, and compared them to the simulated results in Figure 4. The comparison is shown in Figure 14.

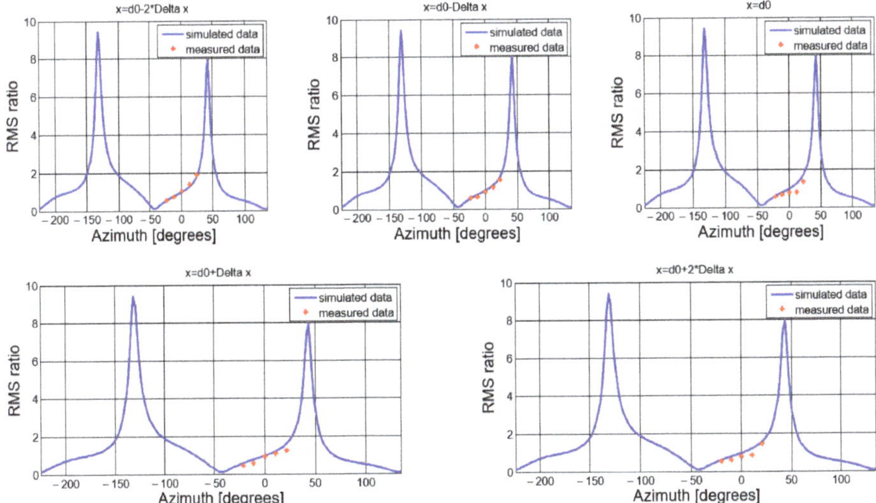

Figure 14. Ratio between RMS voltages of the received signals as a function of azimuth angle: measured and simulated results.

Real and estimated azimuth angles are given in Tables 1 and 2, respectively.

Table 1. Real azimuth angles.

φ_{ij} [°]	$j = -2$	$j = -1$	$j = 0$	$j = 1$	$j = 2$
$i = -2$	24.21	12.67	0.00	−12.67	−24.21
$i = -1$	23.15	12.07	0.00	−12.07	−23.15
$i = 0$	22.17	11.52	0.00	−11.52	−22.17
$i = 1$	21.27	11.01	0.00	−11.01	−21.27
$i = 2$	20.43	10.55	0.00	−10.55	−20.43

Table 2. Estimated azimuth angles.

$\hat{\varphi}_{ij}$ [°]	$j = -2$	$j = -1$	$j = 0$	$j = 1$	$j = 2$
$i = -2$	25.20	16.20	1.80	−14.40	−23.40
$i = -1$	19.80	9.00	−3.60	−19.80	−21.60
$i = 0$	14.40	−10.80	−10.80	−16.20	−23.40
$i = 1$	12.60	5.40	−1.80	−21.60	−27.00
$i = 2$	18.00	−7.20	−10.80	−19.80	−23.40

A mean error figure can be evaluated for each direction as

$$\overline{\varepsilon_j} = \frac{1}{2N+1} \sum_{i=-N}^{N} \left(\hat{\varphi}_{ij} - \varphi_{0j} \right) \tag{17}$$

The error vector is shown in Table 3. The resulting overall mean error is 5.32°.

Table 3. Azimuth mean error vector.

j	−2	−1	0	1	2
$\overline{\varepsilon_j}$ [°]	4.17	9.00	5.04	6.84	1.59

6. Conclusions

We introduced a new dual padlock antenna designed for UWB applications, namely for spark localization within power plants and distribution systems. The antenna system was conceived for an amplitude comparison method. The main advantage over other loop-type antennas for direction finding consists of an unbalanced input that makes it possible to use asymmetrical feed lines without inserting a balun. The antenna system was simulated and then manufactured and measured. The characterization was performed by using both frequency-domain parameters and energy-based descriptors, such as pulse reflection coefficient, pulse matching ratio, energy gain and the energy-based coupling coefficient that we have presented in a separate section.

A good agreement between measured and simulated data could be noted. For the first derivative of a Gauss function, as an excitation with a spectrum centered on 250 MHz, we found a pulse matching ratio of 3.2567, and an energy-based coupling coefficient of −17.86 dB. The energy pattern diagrams for $\theta = 90°$ and horizontal polarization have deep nulls in the planes containing each antenna, which makes our system appropriated to direction finding.

The proposed direction-finding approach was validated by locating a calibrated antenna successively placed in a matrix of 25 positions.

We found an overall, mean absolute error of 5.32°. The discrepancies between simulated and measured azimuth error are mainly due to the finite bandwidth of the measured transfer functions that results in ripples before and after the time support of the received signals that contribute to the RMS.

We should point out that monitoring electric sparks is based on random signals. Consequently, moving the padlock antenna away from the electromagnetic source may result in received signals with a spectral power density below the noise level for some measuring positions. Detection and locating may be improved by using a collaborative system [31,32], with several padlock antennas rather than one antenna scanning.

Author Contributions: Conceptualization, R.D.T.; methodology, R.D.T. and S.B.; software, R.D.T., S.B.; validation, R.D.T., S.B.; formal analysis, R.D.T.; investigation, R.D.T., S.B.; resources, R.D.T.; data curation, R.D.T.; writing—original draft preparation, S.B.; writing—review and editing, R.D.T.; visualization, R.D.T.; supervision, R.D.T.; project administration, R.D.T.; funding acquisition, R.D.T. All authors have read and agreed to the published version of the manuscript.

Funding: This research received no external funding.

Conflicts of Interest: The authors declare no conflict of interest.

References

1. Kozlowski, A.; Barski, M.; Stuchly, S. *Measurement of the Radiated Electric Fields from Electrostatic Discharge;* IEEE: Ottawa, ON, Canada, 1990; pp. 420–421.
2. Brinegar, C. *Passive Direction Finding: Combining Amplitude and Phase Based Methods;* IEEE: Dayton, FL, USA, 2000; pp. 78–84.
3. Thiele, G.A. *Electromagnetic Direction Finding Techniques;* National Technical Information Service, U.S. Deprtment of Commerce: Washington, DC, USA, 1975.
4. Radiomonitoring & Radiolocation Catalog. 2016. Available online: https://cdn.rohde-schwarz.com/ru/downloads_45/common_library_45/brochures_and_datasheets_45/Radiomonitoring_and_Radiolocation_Catalog.pdf (accessed on 4 November 2019).

5. Schmidt, R.O. Multiple emitter location and signal parameter estimation. *IEEE Trans. Antennas Propag.* **1986**, *34*, 276–280. [CrossRef]
6. Ma, H.; Tao, H.; Kang, H. Mixed far-field and near-field source localization using a linear electromagnetic-vector-sensor array with gain/phase uncertainties. *IEEE Access* **2016**, *4*. [CrossRef]
7. Li, J. Direction and polarization estimation using arrays with small loops and short dipoles. *IEEE Trans. Antennas Propag.* **1993**, *41*, 379–387. [CrossRef]
8. Nehorai, A.; Paldi, E. Vector-sensor array processing for electromagnetic source localization. *IEEE Trans. Signal Process.* **1994**, *42*, 376–398. [CrossRef]
9. Wong, K.T. Direction finding/polarization estimation–dipole and/or loop triad(s). *IEEE Trans. Aerosp. Electron. Syst.* **2001**, *37*, 679–684. [CrossRef]
10. Musicant, A.; Almog, B.; Oxenfeld, N.; Shavit, R. Vector sensor antenna design for VHF band. *IEEE Antennas Wirel. Propag. Lett.* **2015**, *14*, 1404–1407. [CrossRef]
11. Yuan, X.; Wong, K.T.; Xu, Z.; Agrawal, K. Various compositions to form a triad of collocated dipoles/loops, for direction finding and polarization estimation. *IEEE Sens. J.* **2012**, *12*, 1763–1771. [CrossRef]
12. Appadwedula, S.; Keller, C.M. *Direction-Finding Results for a Vector Sensor Antenna on a Small UAV*; IEEE: Waltham, MA, USA, 2006; pp. 74–78.
13. Lominé, J.; Morlaas, C.; Imbert, C.; Aubert, H. Dual-band vector sensor for direction of arrival estimation of incoming electromagnetic waves. *IEEE Trans. Antennas Propag.* **2015**, *63*, 3662–3671. [CrossRef]
14. Sarkis, R.; Craeye, C.; Férréol, A.; Morgand, P. *Design of Triple Band Antenna Array for GSM/DCS/UMTS Handset Localization*; IEEE: Berlin, Germany, 2009.
15. Bellion, A.; Meins, C.L.; Vergonjanne, A.J.; Monediere, T. A New Compact Dually Polarized Direction-Finding Antenna on the UHF Band. In Proceedings of the 2008 IEEE Antennas and Propagation Society International Symposium, San Diego, CA, USA, 5–11 July 2008; pp. 1–4.
16. Pack, R.N.; Lasser, G.; Filipovic, D.S. Performance characterization of four-arm MAW spiral antennas for digital direction-of-arrival sensing. *IEEE Trans. Antennas Propag.* **2018**, *66*, 2761–2769. [CrossRef]
17. Duplouy, J.; Morlaas, C.; Aubert, H.; Potier, P.; Pouliguen, P.; Djoma, C. Wideband and reconfigurable vector antenna using radiation pattern diversity for 3-D direction-of-arrival estimation. *IEEE Trans. Antennas Propag.* **2019**, *67*, 3586–3596. [CrossRef]
18. Pohlmann, R.; Zhang, S.; Jost, T.; Dammann, A. Power-based direction-of-arrival estimation using a single multi-mode antenna. In Proceedings of the 14th Workshop on Positioning, Navigation and Communications (WPNC), Bremen, Germany, 25–26 October 2017.
19. Marrocco, G.; Galletta, G. Hermite-rodriguez UWB circular arrays. *IEEE Trans. Antennas Propag.* **2010**, *58*, 381–390. [CrossRef]
20. Deodhar, K.; Baxi, P.; Naik, A.; Gupta, R. *Printed Annular Ring Monopole Antenna for UWB Applications*; IEEE: Orlando, FL, USA, 2007; pp. 1–5.
21. Guo, L.; Rehman, M.; Liang, J.; Chen, X.; Parini, C. *A Study of Cross Ring Antenna for UWB Applications*; IEEE: Cambridge, UK, 2007; pp. 424–427.
22. Schantz, H.G. Smart antennas for spatial RAKE UWB systems. In Proceedings of the IEEE Antennas and Propagation Society International Symposium, Monterey, CA, USA, 20–25 June 2004; pp. 2524–2527.
23. Topor, R.; Tamas, R.D. A Novel Dual-Padlock UWB Antenna System. In Proceedings of the 2015 iWAT, Seoul, South Korea, 4–6 March 2015; pp. 140–143.
24. Allen, O.E.; Hill, D.A.; Ondrejka, A.R. Time-domain antenna characterizations. *IEEE Trans. Electromagn. Compat.* **1993**, *35*, 339–346. [CrossRef]
25. Tamas, R.; Babour, L.; Fond, E.; Alexa, M.; Slamnoiul, G.; Cosereanu, L.; Saguet, P.; Chilo, J. Energy-based input reflection coefficient for the characterization of ultra-wide band antennas. In Proceedings of the 2008 International Workshop on Antenna Technology: Small Antennas and Novel Metamaterials, Chiba, Japan, 4–6 March 2008; pp. 330–333.
26. Topor, R.E.; Bucuci, S.; Tamas, R.D.; Danisor, A.; Dumitrascu, A.; Berescu, S. *Direction Finding Antenna System for Spark Detection and Localization*; IEEE: Constanta, Romania, 2014.
27. Shlivinski, A.; Heyman, E.; Kastner, R. Antenna characterization in the time domain. *IEEE Trans. Antennas Propag.* **1997**, *45*, 1140–1149. [CrossRef]
28. Tamas, R.; Babour, L.; Fond, E.; Slamnoiu, G.; Chilo, J.; Saguet, P. Cylindrical dipoles as ultra-wide band antennas: An energy-based analysis. *Microw. Opt. Technol. Lett.* **2008**, *50*, 917–921. [CrossRef]

29. Anchidin, L.; Ilie, C.-A.; Bucuci, S.; Tamas, R.; Caruntu, G. *A New Insight on the Distance Averaging Method: Linear Scanning Versus Matrix Scanning*; IEEE: Constanta, Romania, 2018.
30. Tamas, R.D.; Deacu, D.; Vasile, G.; Ioana, C. A method for antenna gain measurements in nonanechoic sites. *Microw. Opt. Technol. Lett.* **2014**, *56*, 1553–1557. [CrossRef]
31. Yan, J.; Pu, W.; Zhou, S.; Liu, H.; Bao, Z. Collaborative detection and power allocation framework for target tracking in multiple radar system. *Inf. Fusion* **2020**, *55*, 173–183. [CrossRef]
32. Siriwongpairat, W.P.; Han, Z.; Liu, K.J.R. Power controlled channel allocation for multiuser multiband UWB systems. *IEEE Trans. Wirel. Comm.* **2007**, *6*, 583–592. [CrossRef]

Letter

Wideband Dual-Polarized VHF Antenna for Space Observation Applications †

Alexandru Tatomirescu and Alina Badescu * 

Spl. Independentei, nr. 313, University POLITEHNICA of Bucharest, 060042 Bucharest, Romania; atatomirescu@munde.pub.ro

* Correspondence: badescuam@yahoo.com

† This paper is an extended version of the paper: Tatomirescu, A.; Badescu, A. A Wideband Cross-Polarized Antenna Array Element For Radio Detection of Cosmic Particles. In Proceedings of the 2018 IEEE Conference on Antenna Measurements & Applications (CAMA), Vasteras, Sweden, 3–6 September 2018; pp. 1–4.

Received: 4 July 2020; Accepted: 29 July 2020; Published: 4 August 2020

Abstract: This work presents the design for an antenna element that can be used in radio arrays for the monitoring and detecting of radio emissions from cosmic particles' interactions in the atmosphere. For these applications, the pattern stability over frequency is the primary design goal. The proposed antenna has a high gain over a relative bandwidth of 88%, a beamwidth of 2.13 steradians, a small group delay variation and a very stable radiation pattern across the frequency bandwidth of 110 to 190 MHz. It is dual polarized and has a simple mechanical structure which is easy and inexpensive to manufacture. The measurements show that the ground has insignificant impact on the overall radiation pattern.

Keywords: wideband antenna; VHF; cross polarized; array element; radio detection; cosmic rays

---

1. Introduction

The highest energy cosmic particles come from outside our solar system and their observation leads to a better understanding of the highest energetic processes across the Universe. Their detection is the subject of several radio experiments: CODALEMA in France [1], LOPES in Germany [2], LOFAR in Holland [3] and AERA in Argentina [4]. The work presented here is aimed to be useful for future planned experiments [5].

Depending on the cosmic particles' energy level and mass, direct and indirect detection methods are used [6] for their detection. The indirect detection principle relies on the interactions that produce a broadband electromagnetic field in the radio frequencies. The radiation is coherent from frequencies ranging from a few MHz to up to few GHz [7]. The electromagnetic field is measured with networks of antennas and based on results the characteristics of the primary particle, which are retrieved with digital beamforming techniques [6]. In most applications, the VHF band is used because of the radio interferences in the UHF bandwidth.

The radio interferometer has specific requirements for the array element design. In order to have a wide scanning beam pattern, the array element has to have a low directivity and a wide beamwidth. At frequencies lower than 100 MHz, the natural cosmic background noise (manly coming from our galaxy) is high, thus, a high antenna gain/efficiency is not necessary [8]. Nevertheless, at higher detection frequencies, the antenna gain and bandwidth limit the sensitivity of measurements thus become a crucial part of the array's design [6].

Given the number of elements in the array, the very large physical size of the network and the outdoor placement of antenna, the manufacturing and mechanical stability of each element is important and must be addressed early in the design process.

To limit the uncertainty in the reconstruction process, the array element pattern must be as stable as possible over observation angles and frequency band [6] (a stable antenna pattern facilitates the calibration procedure and the data post-processing). Each antenna should measure two orthogonal components of the electrical field [6], and this is why in this paper, a dual polarized antenna is addressed.

In previous experiments, different antenna designs were implemented and tested [6,9]. Among them, one worth mentioning is the short aperiodic loaded loop antenna (SALLA) [8] used in the Tunka–Rex experiment [10]. Although it is an inexpensive and easy-to-deploy antenna, the trade off is the narrow band pattern stability and the high dependence on the ground's conductivity, which can vary significantly [7]. The same drawbacks appear in V-shape dipoles that were used in LOPES, LOFAR and CODALEMA experiments.

A Butterfly antenna design with a wide bandwidth and high stable gain was used in CODALEMA and AERA experiments. However, it is also affected by the ground conditions, which are highly whether dependent. Finally, the logarithmic periodic dipole antenna used in AERA (LPDA) has the advantages of both a wide bandwidth and a good gain. However, it is the most expensive to manufacture and deploy, and also susceptible at winds due to its height. In addition, it has a variable group delay that distorts the wideband cosmic ray pulse, which affects the reconstruction process.

An analysis of radio frequency interference [4] pointed out that the band of 110 to 190 MHz is relatively free of interferences and could replace or complement the already existing experiments below 100 MHz [6]. The work presented here addresses this design challenge. The solution has been chosen on the considerations of the largest possible bandwidth and stability of the radiation pattern over the band of interest for all observation angles. Practical consideration was given to the manufacturing cost and, more important, the ground reflections. Compared to other available models, the proposed array element has a stable radiation pattern and a low variation of the group delay over a larger relative bandwidth, but also a higher gain.

This paper is organized as follows: the next section briefly describes the design of the antenna (proposed and detailed in [11]) and introduces the new manufactured prototype. The following section presents the free-space antenna measurement method, the experimental results obtained (all in good agreement with simulations), together with the influence of the ground on the radiation pattern and the determination of group delay variations over incoming direction and frequencies. The last part summarizes the work and highlights the improvements compared to solutions adopted in other experiments.

2. Design Considerations and Prototype

The bow-tie antenna was chosen as a basic starting point for this work because of its mechanical simplicity. Such an antenna exhibits a good bandwidth and a wide beamwidth [6,12].

Due to the large physical size of the radiating elements (necessary at such frequencies) one of the main concerns was related to the high winds at which the antenna was exposed. A practical solution was the usage of aluminum mesh for the construction of the radiating elements. However, the mesh must be mounted on a solid metallic frame to prevent gravitational deformation and this could affect the radiation properties of the antenna. Another solution was designing the radiating elements as a grid of metallic bars. Both solutions had the disadvantage of a much higher manufacturing cost, especially when considering the large number of antennas required in arrays for radio detection of cosmic particles. Considering all these limitation, we preferred to use aluminum sheets for the radiating elements and position the antenna as close as possible to the ground to make it more resistant at high winds. A cut-off was inserted in the radiating elements to lower the wind resistance, to lower weight and also to limit the higher order modes.

To avoid the effect of the ground reflection, a metallic plate was inserted between the radiating element and the actual ground, as illustrated in Figure 1. One of the dual polarized elements was oriented along the x axis (corresponding to the $\phi = 0°$) whereas the other element was oriented perpendicular, along the y axis (i.e., $\phi = 90°$). The zenith (i.e., $\theta = 0°$) is along the positive z axis. The exact dimensions of the radiating elements are shown in Figure 2 [11].

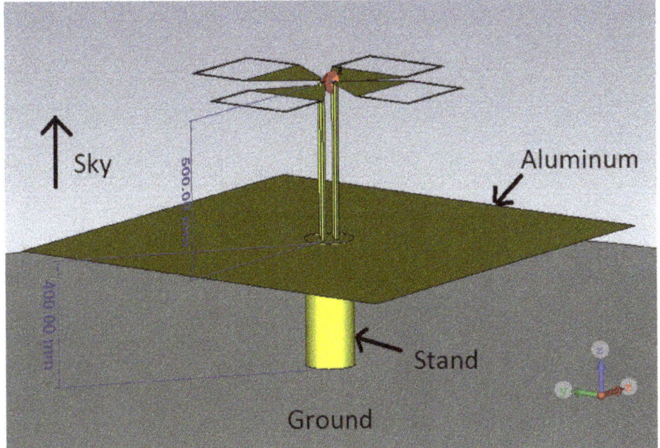

Figure 1. The simulated model of the proposed antenna, including the mounting on the ground. The feeding point is represented with red. The direction of the zenith is indicated [11] © 2018 IEEE.

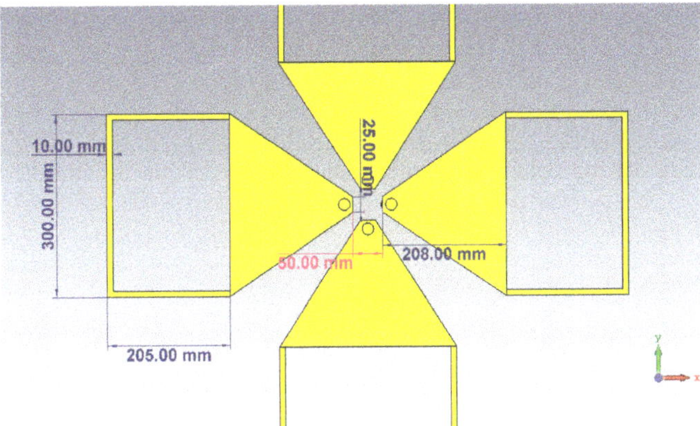

Figure 2. The dimensions of the radiating elements [11] © 2018 IEEE.

The prototype is presented in Figure 3 and has an the overall height of less than 1 m. For the prototype, common construction materials were used: the radiating elements were plasma cut from a 2 mm thick aluminum sheet. The antenna was attached to the ground plate (a 3 mm thick aluminum square plate with a side of 1.4 m) using 15 mm diameter copper pipes and inserts. Holes in the ground plane and radiating elements were cut and threaded for easy assembly of the antenna. Usually, the dimensions of the metallic plate should exceed the dimensions of the antenna with $\lambda/4$ [13], however, large conductive plates also increase the edge effects. Decreasing the size of the metallic plate will degrade the peak gain at lowest frequencies and increase the back lobes towards the ground making it more susceptible to the conductivity of the ground. The final design represented the best compromise in terms of directivity pattern of the antenna and input reflection coefficient.

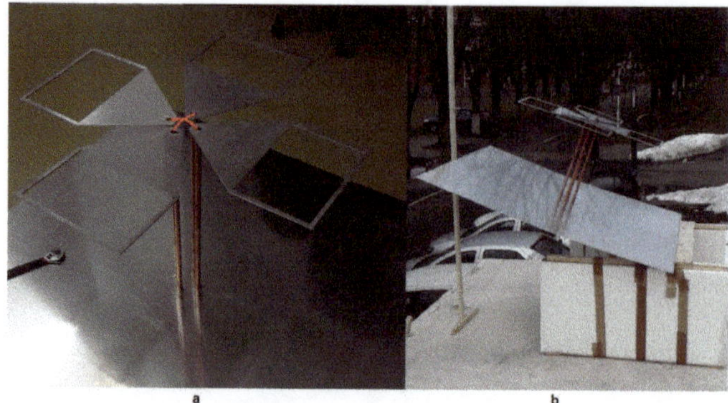

Figure 3. The antenna prototype: feeder (**a**) and a caption during measurements (**b**). The position of the feeding inner conductor of the coaxial cable is illustrated with red (**a**). The inner conductor is soldered on one blade and the outer conductor on the other, similar to [12] © 2012 IEEE.

3. Results

3.1. Simulation Results

The numerical model of the antenna (with the metallic ground plane) was simulated with a commercial finite difference time domain electromagnetic solver, CST 2016. The first point of interest was the influence of the soil, so we performed simulations over an infinite ground plane with a finite conductivity in the 0.001 S/m–0.2 S/m interval [14] to cover different possible humidities of the soil. Figure 4 presents comparative results at 110 MHz for $\sigma = 0.001$ S/m and $\sigma = 0.2$ S/m. At higher frequencies, the effect of the ground was even smaller due to the larger electrical distance between the antenna and ground. It can be concluded that the influence of the soil's conductivity on the radiation pattern is not higher than 0.3 dBm, thus, practically negligible.

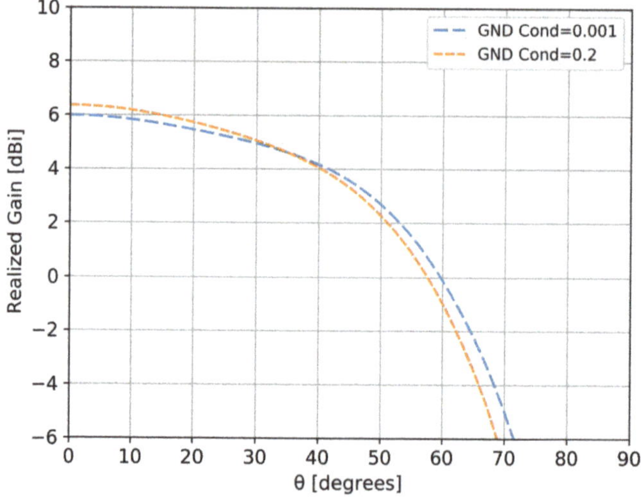

Figure 4. Variations in the elevation plane of the antenna situated on an infinite ground plane with different conductivities ($\sigma = 0.001$ S/m is represented with blue line and $\sigma = 0.2$ S/m with orange). Results are obtained at a frequency of 110 MHz, for an azimuth angle of $\phi = 0^\circ$.

As mentioned in the first section, a stable pattern over the frequency range of interest is mandatory for an accurate detection and estimation of the wideband pulse generated by the incoming cosmic particles. Figures 5 and 6 show the variation of the gain with frequency in two azimuth perpendicular planes for the radiating element positioned along the x axis. Due to symmetry reasons, both elements (in x and y directions) had identical behavior.

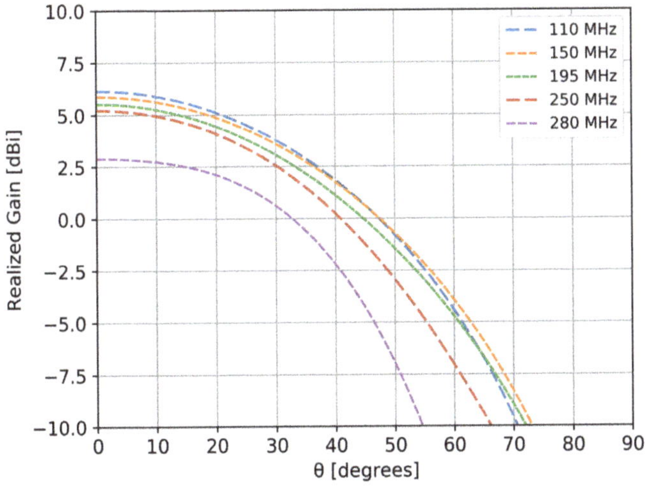

Figure 5. Radiation pattern in the elevation plane at different frequencies for the element along the x axis (cut in the $\phi = 0°$ plane).

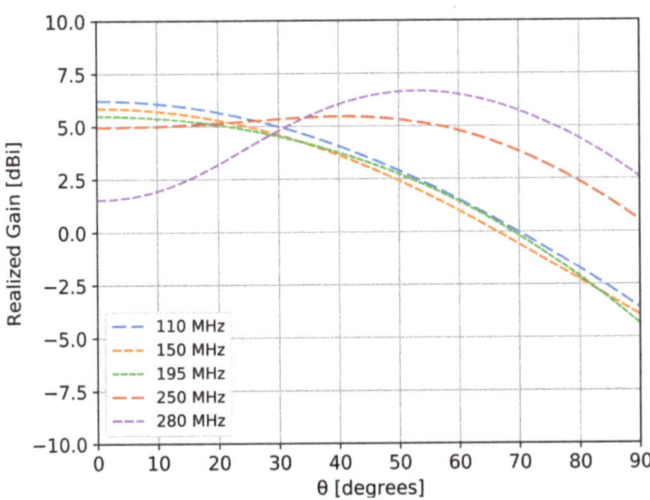

Figure 6. Radiation pattern in the elevation plane at different frequencies for the element along the x axis (cut in the $\phi = 90°$ plane).

Depending on the frequency of interest, the half power beamwidth in one plane was around 70° (Figure 5) and around 100° (Figure 6) in the other perpendicular one. An anti-resonance effect can be observed in Figure 6, around 300 MHz. The gain towards zenith is decreased due to the in-phase summation of the direct and reflected component by the metallic ground. However, in the 110–190 MHz band of interest the pattern is fairly stable, with variations of less than 1 dB.

In order to measure the incoming wideband field without distortions, the antenna should have a small variation of the group delay across the entire band of interest. The group delay itself is very important parameter as wavefront measurements totally relay on it.

The simulated group delay characteristic of the proposed antenna is presented in Figure 7 for various incoming zenith angles. The variation of the group delay is less than 2.5 ns at each frequency of simulation. This is a very important improvement compared to other experiments that shown a 80 ns variation [6]. Similar results are observed for other azimuth angles.

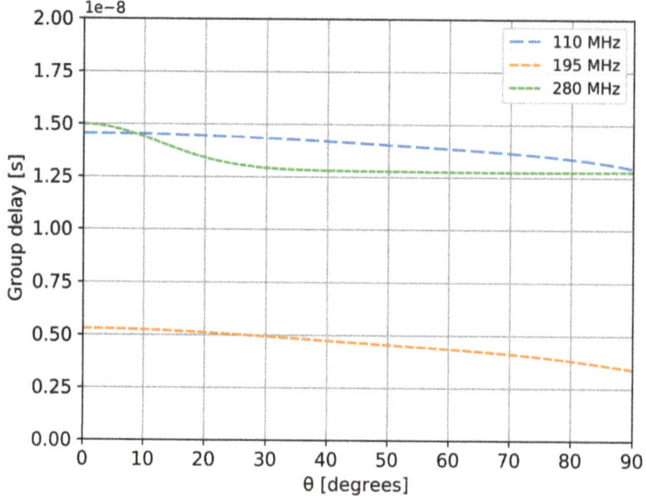

Figure 7. Simulated group delay variation in the elevation plane ($\phi = 0°$).

The variation of the group delay with frequency is shown in Figure 8 (for the zenith incoming direction). Similar results are observed for other angles. Results from other experiments shown a variation of about 100 ns for Small Black Spider antennas, 25 ns for Butterfly antennas and about 10 ns for Salla antennas [9].

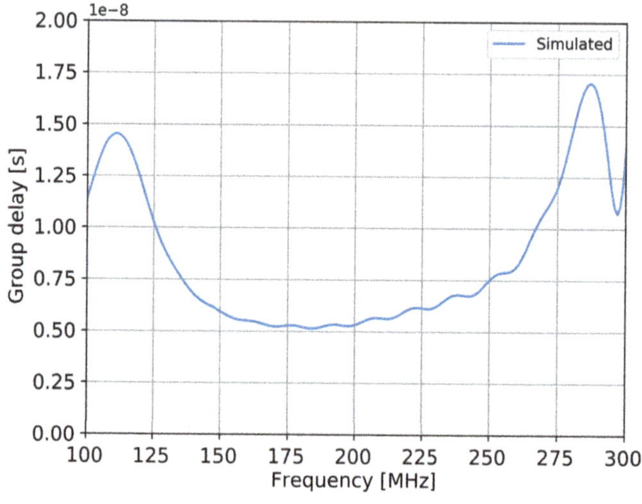

Figure 8. Simulated group delay variation with frequency for the zenith direction ($\phi = 0°$).

3.2. Experimental Results

Measurements of the radiation pattern were performed in an outdoor environment to meet the far-field conditions. This requirement arise due to the large dimensions of the antenna that would otherwise demand usage of a very large anechoic chamber.

The far-field slant elevated range method was adopted [15] to characterize the antenna. This method allows for transmission-loss free-space measurements of the gain product of two antennas, when the directive source is at the proper height. The accuracy is limited at ±0.35 dB [15]. The frequency range over which the method was proven to work is from 100 MHz to GHz, for various antenna types (simple dipoles, horns, etc.) [15].

The measurement setup is presented in Figure 9. A calibrated emitting probe antenna (model BicoLOG 5070 produced by Aaronia) was fixed onto a 4 m pole outside the window of a building, approximately 15 m above ground. The pole was inclined at a fixed angle of 45°. The emitting antenna is represented as point P in Figure 9). The prototype (device under test, DUT, in Figure 9) was placed on an elevated platform (the elevated platform is also shown in Figure 2, right side). The DUT can be inclined at various θ angles values using a wooden support which allows for the determination of the elevation characteristics of the antenna. Both antennas were connected to a vector network analyzer (VNA) (from Deviser, model 7300 NA) that allows for the elimination of cable effects through calibration.

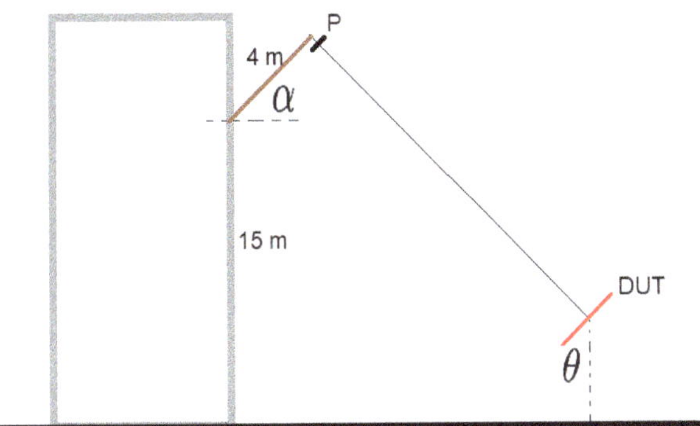

Figure 9. Slant elevated range principle (not at scale). The angle α is fixed at 45°. The inclination θ of the platform was varied where the device under test (DUT) was placed.

The realized gain $Gr_{DUT}(f)$ of the antenna under test can be calculated using Equation (1). It includes the mismatch losses, radiation losses and the directivity of the antenna compared to an isotropic radiator:

$$Gr_{DUT}(f) = \frac{Gr_{Probe}(f) * H_{est}(f)}{S_{21}(f)} \tag{1}$$

where $Gr_{Probe}(f)$ is the realized gain of the antenna and is given by its manufacturer, $H_{est}(f)$ is the estimate of the channel using a reference antenna and $S_{21}(f)$ is the measured transfer parameter between the DUT and the probe antenna. The channel estimator was determined for a band of 400 MHz using calibrated antennas (for both emitter and receiver). A time domain gate was also applied [16] on the measured transfer parameter, and the method and results are carefully presented in [17]. The gating window introduces a 0.5 dB uncertainty [16].

The prototype antenna was aligned to match polarization of the probe antenna and the DUT platform was inclined at three elevation angles using a wooden support. The results presented in Figure 10 show a good agreement with simulation (included for comparison). Even though low elevation directions show a lower gain, the gain is still higher than the usual requirements for such applications (i.e., −10 dB [6]). To eliminate the statistical errors (i.e., positioning errors and polarization mismatch loss), the measurements were repeated three times and the results in Figure 10 represent the mean values obtained.

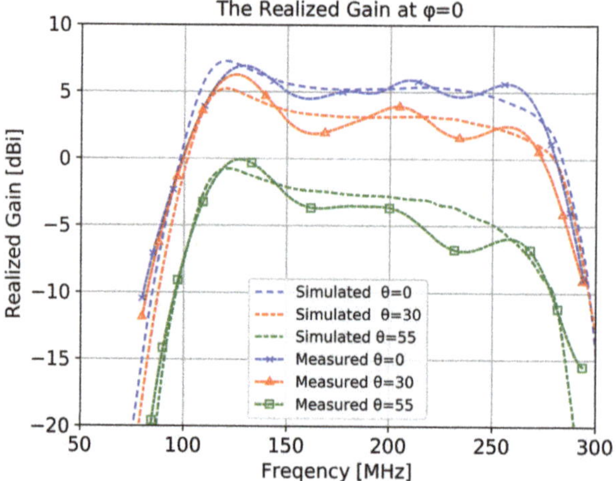

Figure 10. Simulated and measured radiation pattern versus frequency.

The measured input reflection coefficients for the two perpendicular elements of the antenna are presented in Figure 11. Simulation results are also included in the plot for comparison. In the simulation, due to the symmetry of the structure, the two ports had a similar performance. The input reflection coefficient was smaller than −5 dB for the interval of 110 MHz to 190 MHz with a fractional impedance bandwidth of 88%. The elements also exhibit a good isolation between each other (i.e., higher than −27 dB).

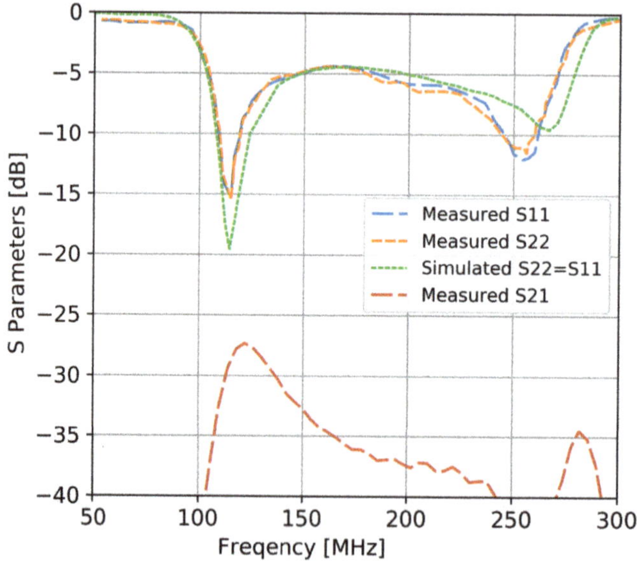

Figure 11. The simulated and measured S parameters of the prototype antenna.

To increase the stability of the gain in the band of interest (necessary in the reconstruction process of particles' characteristics) a simple L matching network was designed, consisting of a series lumped capacitor of 25 pF and a shunt wire inductor of 90 nH. An almost constant value of −7 dB is observed for the input reflection coefficient in Figure 12. The slight decrease at the frequencies around 300 MHz (i.e., outside the band of interest) is compensated by the decrease in gain thus a stable gain is acquired, without performance loss.

Around 300 MHz the field radiated by the bow-tie element and its reflection from the metal plate cancel each other forming an anti-resonance. The effect can be shifted to a higher frequency by lowering the element's height.

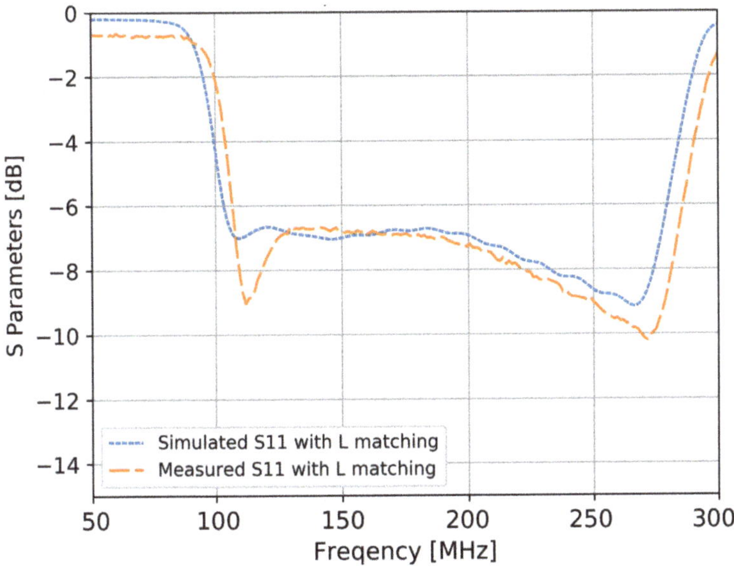

Figure 12. The simulated and measured S parameters of the antenna with a simple L matching network.

4. Conclusions

In this work, we present the design of an antenna that can be used in an array configuration for detection and estimation of the electromagnetic radiation produced by cosmic particles that interact in the atmosphere. The main focus was on the 110–190 MHz frequency range, however, performances of the antenna were analyzed outside this regime for future possible extensions.

Low loss, lightweight, stable and cheap materials were used in the design, with low manufacturing complexity. The antenna has a small profile of less than a meter, so wind impact should be minimal. The influence from the ground was also minimal compared to other available solutions for similar applications (e.g., CODALEMA, AERA, etc.) [6].

The prototype has a very good bandwidth of 88% with a stable radiation pattern, with a good agreement between the simulations and measurement results. The performances in terms of the group delay are excellent, as the group delay is crucial for wavefront measurements and for the beamforming processes. Since the proposed prototype has a small Q factor, its group delay is small compared with the LPDA antennas used in previous experiments [6]. The variation of the group delay with frequency are much less than for Small Black Spider antennas and Butterfly antennas, and comparable with Salla antennas [9]. The disadvantage of the latter is a lower gain that increases the detection threshold in terms of cosmic-ray energy by about 30% [6].

Author Contributions: Conceptualization, A.T. and A.B.; methodology, A.T. and A.B.; software, A.T.; formal analysis, A.B.; All authors have read and agreed to the published version of the manuscript.

Funding: This work was supported by a grant of Ministry of Research and Innovation, CNCS-UEFISCDI, project number PN-III-P1-1.1-TE-2016-0530, within PNCDI III.

Conflicts of Interest: The authors declare no conflict of interest.

References

1. Ardouin, D.; Belletoile, A.; Charrier, D.; Dallier, R.; Denis, L.; Eschstruth, P.; Gousset, T.; Haddad, F.; Lamblin, J.; Lautridou, P.; et al. Radio-detection signature of high-energy cosmic rays by the CODALEMA experiment. *Nucl. Instrum. Methods Phys. Res. Sect. A Accel. Spectrom. Detect. Assoc. Equip.* **2005**, *555*, 148–163. [CrossRef]
2. Falcke, H.; Apel, W.D.; Badea, A.F.; Bähren, L.; Bekk, K.; Bercuci, A.; Bertaina, M.; Biermann, P.L.; Blümer, J.; Bozdog, H.; et al. Detection and imaging of atmospheric radio flashes from cosmic ray air showers. *Nature* **2005**, *435*, 313–316. [CrossRef] [PubMed]
3. Van Haarlem, M.P.; Wise, M.W.; Gunst, A.W.; Heald, G.; McKean, J.P.; Hessels, J.W.T.; de Bruyn, A.G.; Nijboer, R.; Swinbank, J.; Fallows, R.; et al. LOFAR: The low-frequency array. *Astron. Astrophys.* **2013**, *556*, A2.
4. The Pierre Auger Collaboration. The Pierre Auger Cosmic Ray Observatory. *Nucl. Instrum. Methods Phys. Res. Sect. A Accel. Spectrom. Detect. Assoc. Equip.* **2015**, *798*, 172–213. [CrossRef]
5. Badescu, A. A radio detection system for cosmic observations. *Astron. Geophys.* **2015**, *56*, 1.30–1.33. [CrossRef]
6. Schröder, F.G. Radio detection of cosmic-ray air showers and high-energy neutrinos. *Prog. Part. Nucl. Phys.* **2017**, *93*, 1–68. [CrossRef]
7. Ohta, I.S.; Akimune, H.; Fukushima, M.; Ikeda, D.; Inome, Y.; Matthews, J.N.; Ogio, S.; Sagawa, H.; Sako, T.; Shibata, T.; et. al. Measurement of microwave radiation from electron beam in the atmosphere. *Nucl. Instrum. Methods Phys. Res. Sect. A Accel. Spectrom. Detect. Assoc. Equip.* **2016**, *810*, 44–50. [CrossRef]
8. LOPES Collaboration. New antenna for radio detection of UHECR. In Proceedings of the 31st International Cosmic Ray Conference, Łódz, Poland, 7–15 July 2009.
9. Abreu, P.; Aglietta, M.; Ahlers, M.; Ahn, E.J.; Albuquerque, I.F.M.; Allard, D.; Allekotte, I.; Allen, J.; Allison, P.; Almela, A.; et al. Antennas for the detection of radio emission pulses from cosmic-ray induced air showers at the Pierre Auger Observatory. *J. Instrum.* **2012**, *7*, P10011.
10. Schröder, F.G.; Besson, D.; Budnev, N.M.; Gress, O.A.; Haungs, A.; Hiller, R.; Kazarina, Y.; Kleifges, M.; Konstantinov, A.; Korosteleva, E.E.; et al. Tunka-Rex: A radio antenna array for the Tunka experiment. *AIP Conf. Proc.* **2013**, *111*, 1535.
11. Tatomirescu, A.; Badescu, A. A Wideband Cross-Polarized Antenna Array Element For Radio Detection of Cosmic Particles. In Proceedings of the 2018 IEEE Conference on Antenna Measurements & Applications (CAMA), Vasteras, Sweden, 3–6 September 2018; pp. 1–4.
12. Li, B.; Yin, Y.; Hu, W.; Ding, Y.; Zhao, Y. Wideband Dual-Polarized Patch Antenna With Low Cross Polarization and High Isolation. *IEEE Antennas Wirel. Propag. Lett.* **2012**, *11*, 427–430.
13. Kraus, J.K.; Marhefka, R.J. *Antennas for All Applications*; McGraw-Hill: Singapore, 2003.
14. Sternberg, B.K.; Levitskaya, T.M. Electrical parameters of soils in the frequency range from 1 kHz to 1 GHz, using lumped-circuit methods. *Radio Sci.* **2001**, *36*, 709–719 [CrossRef]
15. Arnold, P. The "Slant" antenna range. *IEEE Trans. Antennas Propag.* **1966**, *14*, 658–659. [CrossRef]
16. Piasecki, P.; Strycharz, J. Measurement of an omnidirectional antenna pattern in an anechoic chamber and an office room with and without time domain signal processing. In Proceedings of the 2015 Signal Processing Symposium (SPSympo), Debe, Poland, 10–12 June 2015; pp. 1–4.
17. Tatomirescu, A. UHF Band Antenna Radiation Pattern Measurements in Multipath Channel using Time Domain Gating. In Proceedings of the 2020 International workshop on antenna and propagation, Bucharest, Romania, 25–28 February 2020; pp. 1–4.

Letter

Evaluation and Impact Reduction of Common Mode Currents on Antenna Feeders in Radiation Measurements †

Andreea Constantin [1,2,*] and Razvan D. Tamas [1,2]

1 Department of Electronics and Telecommunications, Constanta Maritime University, 900663 Constanta, Romania; tamas@ieee.org

2 Doctoral School of Electronics, Telecommunications and Information Technology, University Politehnica of Bucharest, 061071 Bucharest, Romania

* Correspondence: andreea.constantin@ieee.org

† This paper is an extended version of Constantin, A.; Anchidin, L.; Tamas, R.D.; Caruntu, G. A New Method to Reduce the Impact of the Common Mode Currents for Field Measurements on Symmetrical Antennas. In Proceedings of the 2019 International Workshop on Antenna Technology (iWAT), Miami, FL, USA, 3–6 March 2019; pp. 87–90.

Received: 15 June 2020; Accepted: 11 July 2020; Published: 13 July 2020

Abstract: Common mode currents on antenna feeders usually occur when feeding a symmetric radiator through an asymmetric line, or when the ground plane is electrically small. Such currents may have magnitudes comparable to the feed currents and therefore have a major impact on the total radiated field. For antenna radiation measurements, both assessment and reduction of the common mode currents on antenna feeders are crucial. Techniques to discriminate antenna and feeder radiation are mainly needed for design and optimization purposes. Antenna gain measurements in a multipath site can be performed by using the distance averaging method. In this paper, we show that the distance averaging technique can be applied to reduce the effect of common mode currents for measuring the field radiated by symmetrical antennas. Two measuring configurations are proposed depending on the number of symmetry degrees of the antenna under test, and a differential approach for extracting the field created by the common mode currents was also developed. The experimental validation was performed by measuring a simple wire dipole and a log-periodic dipole array (LPDA) with a small square loop as a probe, both on the feeder side and on the feeder free side.

Keywords: antenna radiation measurements; common mode current; distance averaging; multipath site; small antenna; loop probe; log-periodic dipole array

---

1. Introduction

The radiation originating from common mode currents has been thoroughly studied, mostly on cables attached to printed circuits boards [1–5]. The purpose of such studies is mainly related to the electromagnetic compatibility. When feeding symmetrical antennas or electrically small antennas through asymmetrical transmission lines (e.g., coaxial cables), common mode currents may occur on the outer conductor of the feeder. Common mode currents should normally be kept at least ten times smaller than the feed currents, in order to avoid undesirable effects. However, it has been shown [6] that common mode currents may have magnitudes comparable to the feed currents when the antenna size is comparable to the ground size or to the feed line length, and therefore have a major impact on the total radiated field.

For antenna radiation measurements, both the assessment and reduction of the common mode currents are crucial. However, there are relatively few studies focusing on the contribution to the radiation [7–10] or suppression of common mode currents [11,12]. Some authors have proposed the replacement of the coaxial cable with optical fiber for eliminating the large distortion associated with the unwanted radiation from the feed line [13].

In order to obtain accurate measurements of antenna radiation, anechoic chambers with proper shielding and absorbing material are generally employed to reduce both the interference from the external environment and the effect of the multiple propagation paths [14].

The electromagnetic field generated by common mode currents is generally measured in a single-path site such as an anechoic chamber or an open area test site (OATS).

Antenna measurements within a reverberation chamber, semi-anechoic chamber, or even a non-ideal environment will include the effect of the multipath propagation, which will result in a distortion of the measured radiation pattern [15]. In such cases, an improvement of the measuring accuracy can be achieved through removal [16,17] or compensation [18,19] of the undesired contributions.

Most of the work treating the radiation from common mode currents at a symmetrical antenna input has focused on characterizing or properly designing a balun [6,20], rather than reducing that effect by post-processing. However, using a balun in a measuring setup not only increases the cost, but would also impinge on the global frequency response.

When measuring the radiation of a symmetrical antenna such as a linear dipole, the field measured over a direction orthogonal to the antenna symmetry axis has a different magnitude on the feeder side compared to the feeder free side. The difference results from the radiation originating from the common mode currents. For a symmetrical, directive antenna (e.g., log-periodic dipole array, LPDA), common mode currents on the feeding line are commonly associated with unwanted phenomena such as the asymmetry resonance [21–23]. Moreover, the effect of the common mode currents on feed lines might also be of interest when designing compact LPDAs [24].

We have previously presented [25] a distance averaging method for measuring the antenna gain in a multipath site. The method basically consists of reducing the effect of the indirect paths based on the variability of their contributions to the total field compared to the direct path.

In a previous conference paper [26], we introduced a method to reduce the effect of the common mode currents for measuring the field radiated by symmetrical antennas. Our approach is based on the distance averaging technique. In this paper, we extend on our previous work as follows: (1) We propose two different approaches for this technique, depending on the number of the symmetry degrees of the antenna under test; (2) we present a distance averaging method to extract the effective area of the loop antenna that we used as a probe; and (3) we develop a differential approach for evaluating the magnetic field generated by the common mode currents on an antenna feeder.

An experimental validation was performed by measuring a simple wire dipole and a LPDA with a small square loop as a probe, both on the feeder side and on the opposite side.

2. Impact Reduction of the Common Mode Currents in Antenna Measurements

We considered a typical two-antenna measuring system consisting of a probe antenna (PA) and an antenna under test (AUT), respectively. As an AUT, we successively used two types of symmetrical radiators fed through coaxial cables: a two-symmetry degrees antenna (i.e., a dipole) and a one-symmetry degree antenna (i.e., a log-periodic dipole array). As a PA, we took a small, square loop antenna.

We designated as the "cable side" the field points in a direction orthogonal to the antenna, along the feed line. The "antenna side" will include field points in the same direction, but on the cable free side.

The field on the "antenna side" is entirely due to the radiation of the AUT, conversely, on the "cable side", the field is due both to the radiation of the AUT and the cable.

We propose two different measuring methodologies depending on the number of symmetry degrees of the antenna under test. When using a simple wire dipole, measurements will be performed by placing the probe on each side of the antenna (Figure 1).

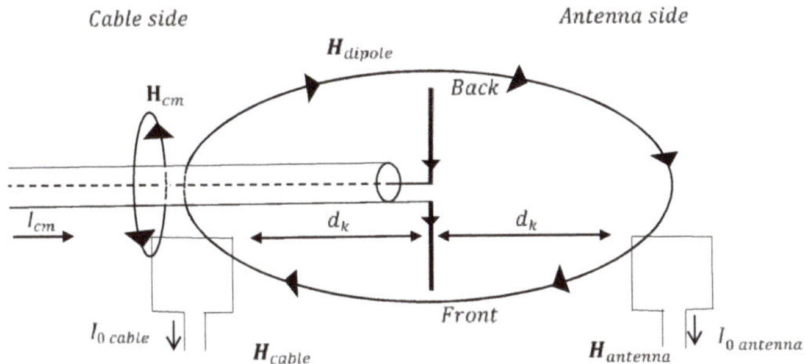

Figure 1. Measuring the methodology for a dipole antenna.

When using the log-periodic dipole array as an antenna under test, measurements are performed by successively placing the coaxial cable on both sides of the feed point (Figure 2).

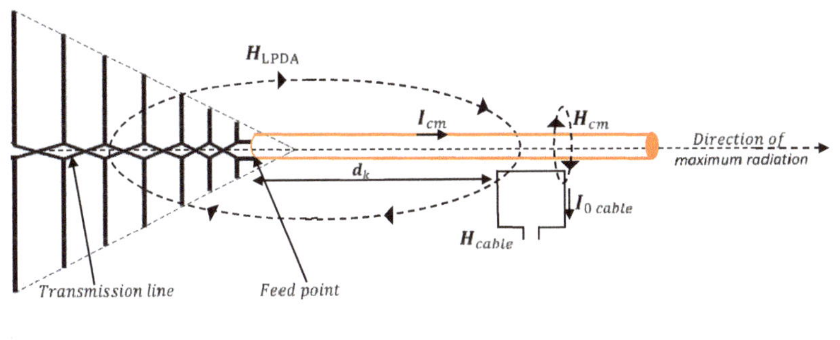

(a)

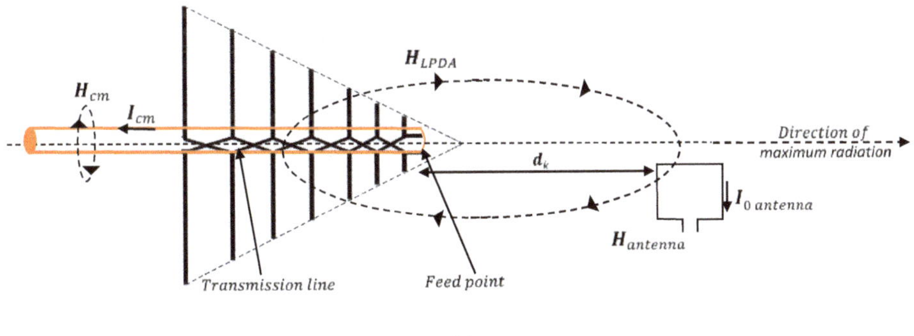

(b)

Figure 2. Measuring methodology for a log-periodic dipole array (LPDA): "cable side" (**a**) and "antenna side" (**b**).

On the "cable side", the PA will measure the field created by the antenna and the common mode currents on the feeder. By placing the coaxial cable in the opposite direction, the loop antenna will only measure the magnetic field generated by the log-periodic antenna.

For each type of AUT, the effect of the common mode current on the coaxial line can be assessed by subtracting the field measured at the same distance with the probe placed on the "cable side" and on the "antenna side", respectively.

Such measurements are performed at several distances between the antenna under test and the probe, in order to apply the distance averaging approach [25].

Referring to Figures 1 and 2, the magnetic field measured by the loop on the "cable side" and "antenna side" can be expressed as

$$H_{cable} = H_{cm} + H_{dipole/LPDA}, \tag{1}$$

where H_{cm} is the magnetic field component generated by the common mode currents and $H_{dipole/LPDA}$ is the field component generated by the antenna, activated by the feed currents.

The contribution of the common mode current to the magnetic field can be found as

$$H_{cm} = H_{cable} - H_{antenna}. \tag{2}$$

where

$$H_{antenna} = H_{dipole/LPDA}. \tag{3}$$

where $I_{0\ antenna/cable}$ is the probe output current depending on the position of the loop with respect to the AUT; A_e is the effective area of the loop; η is the free space wave impedance; and R_0 is the normalizing impedance considered as a load at the probe output.

The received power can then be expressed either by integrating the incident wave power density, S_p over the loop (Figure 3)

$$P_r = S_p A_e = \frac{1}{2}\frac{E^2}{\eta}A_e = \frac{1}{2}\eta H^2_{cable/antenna}A_e, \tag{4}$$

or as the power dissipated into the load at the antenna output,

$$P_r = \frac{1}{2}R_0 I^2_{0\ cable/antenna}. \tag{5}$$

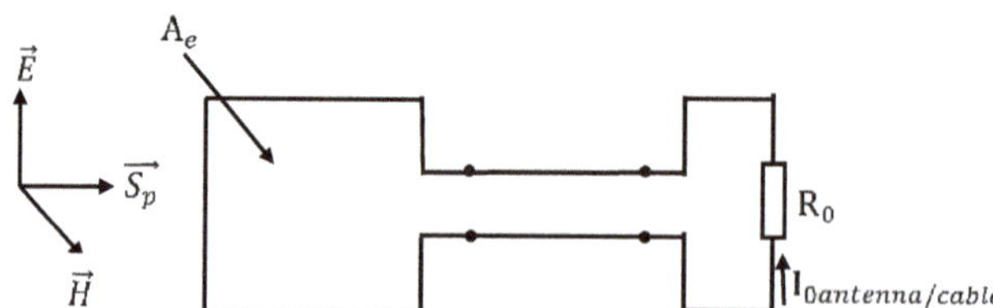

Figure 3. Power balance for the loop probe.

Consequently, the magnetic field can be written as follows:

$$H_{antenna/cable} = \sqrt{\frac{R_0 I^2_{0\ antenna/cable}}{\eta A_e}}. \tag{6}$$

Since the circuit consisting of the probe and the AUT is terminated on the normalizing impedance at both ports, the output current can be computed as

$$I_{0antenna/cable} = \frac{V_g\left|S_{21antenna/cable}\right|}{2R_0}, \tag{7}$$

where V_g is the electromotive force of the excitation at the AUT input. The contribution S_{21cm} of the common mode currents to the transfer function $S_{21cable}$ can be derived from Equation (2),

$$S_{21\ cm} = S_{21\ cable} - S_{21\ antenna}. \tag{8}$$

The contributions of the common mode current to the output current and to the magnetic field are given in Equations (9) and (10), respectively:

$$I_{0\ cm} = \frac{V_g|S_{21\ cm}|}{2R_0}, \tag{9}$$

$$H_{cm} = \sqrt{\frac{R_0 I_{0\ cm}^2}{\eta A_e}}. \tag{10}$$

The effective area of the loop antenna in Equation (10) should include the impedance mismatch effect both at the transmitting and receiving antennas.

When transfer functions are measured at N different distances, an average can be computed over that dataset by compensating the effects of the propagation in terms of attenuation and delay; we have previously used such a distance averaging technique [25] with the aim to reduce the effects of the multipath propagation for antenna gain measurements. As the field corresponding to indirect propagation paths, common mode currents also have a distance variant distribution. The application of the distance averaging (Figure 4 for dipole antenna and Figure 5 for log-periodic dipole array) might therefore significantly reduce the impact of the common mode current on the antenna radiation measurements.

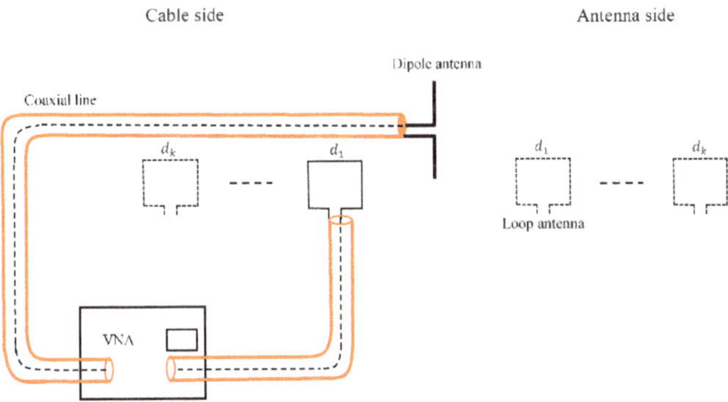

Figure 4. Distance averaging technique applied for reducing the effect of the common mode current on the dipole antenna radiation measurements.

The average transfer function can be computed from the transfer functions $S_{21\ cm}^{d_k}$ measured at each distance d_k between antennas,

$$\overline{S}_{21\ cm} = \sum_{k=1}^{N} \frac{d_k}{d_0} \exp(jk_0 d_k) S_{21\ cm}^{d_k}, \tag{11}$$

where d_0 is the reference distance (set at 1 m) and k_0 is the free space wavenumber. Average figures can then be derived both for output currents and magnetic field components; such figures can be defined for the "antenna" and "cable side", and for the common mode contribution, respectively:

$$\overline{I}_{0\ cm/cable/antenna} = \frac{V_g \left| \overline{S}_{21\ cm/cable/antenna} \right|}{2R_0}, \tag{12}$$

$$\overline{H}_{cm/cable/antenna} = \sqrt{\frac{R_0 \overline{I}^2_{0\ cm/cable/antenna}}{\eta A_{e\_cm/cable/antenna}}}. \tag{13}$$

The effect of the common mode currents should be reduced for the "cable side" measurements and therefore, corrected figures should be calculated,

$$H^{d_k}_{cable\ corr} = \frac{d_k}{d_0} \exp(-jk_0 d_k) \overline{H}_{cable}. \tag{14}$$

Relation (14) gives the field value at a given distance by simply multiplying the result by that distance, provided that the average figure corresponds to a distance of 1 m between antennas.

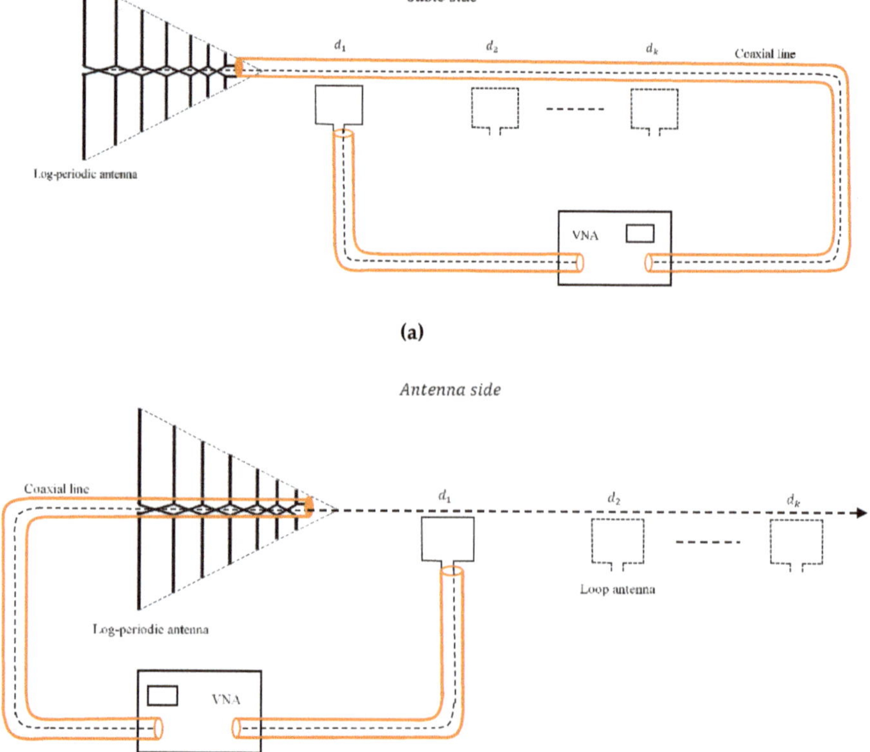

Figure 5. Distance averaging technique applied for reducing the effect of the common mode current on LPDA radiation measurements: "cable side" (**a**) and "antenna side" (**b**).

3. Probe Antenna Calibration

The gain of the loop antenna that we used as a probe can be found by characterizing the transmission between the probe and a calibrated antenna (Figure 6).

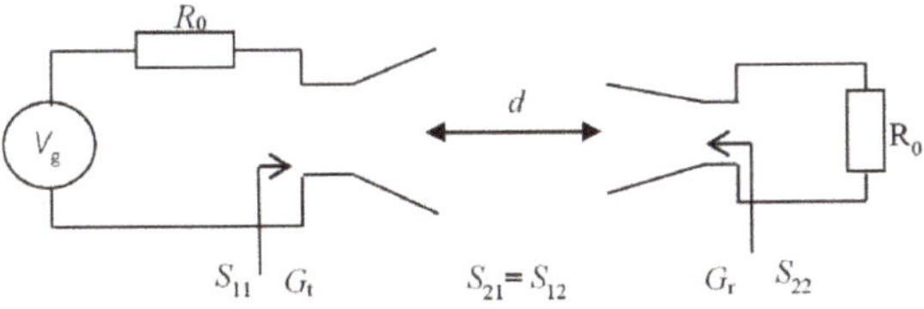

Figure 6. Probe calibration setup.

One of the AUTs (i.e., the LPDA) has previously been calibrated inside a professional, compact range in an "antenna side" setup (Figure 7). The measuring system consists of a circular array of probe antennas placed inside an anechoic chamber, a calibrated RF generator, and a calibrated receiver, respectively. The AUT (i.e., the LPDA) was placed on a turntable. As a result, the near-field was measured on a closed surface, and the realized gain of the AUT could be accurately extracted through near-field to far-field transformations.

Figure 7. LPDA calibration.

As a result, that AUT could itself be used as a probe for calibrating the loop when the LPDA is in an "antenna side" configuration. Furthermore, the loop calibrated as described previously will be able to measure the radiation of any other configuration (e.g., with the LPDA) in a "cable side" setup, or the dipole in an "antenna side" and "cable side" setup.

With the notations in Figure 6 and taking into account the impedance mismatch at both antennas, the gain of the receiving antenna (i.e., the AUT) can be found from the Friis formula; since part of the measurements are usually performed at the Fresnel zone ranges, a field-zone correction factor, $F(f,d)$ should be applied on the measured results, as we have proposed in a previous paper [27]. Since the loop is used as a probe (receiving) antenna, one should characterize it through its effective area, rather than its gain, that is,

$$A_e = \frac{4\pi r^2}{G_t}\frac{R_0}{R_{a2}}\frac{\left|F(f,d)S_{21}\right|^2}{|1-S_{22}|^2\left(1-|S_{11}|^2\right)}. \quad (15)$$

where R_{a2} is the radiation resistance of the AUT and λ is the wavelength.

When both antenna aperture sizes ($2h_1$ and $2h_2$ in Figure 8) are comparable to the measuring range, the lower limit of the Fraunhofer zone is found as

$$d \geq \frac{8(h_1 + h_2)^2}{\lambda}. \quad (16)$$

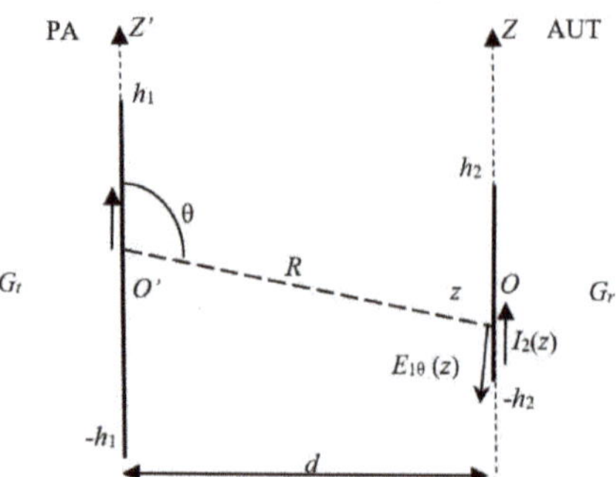

Figure 8. Transmission between two linear antennas.

In order to accurately evaluate the effective area of the probe antenna in a multipath environment, we used the distance averaging method [25,28]. The setup in shown in Figure 9.

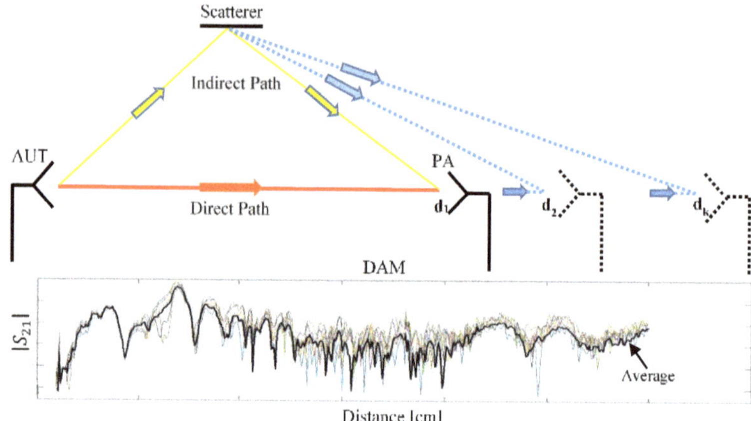

Figure 9. Distance averaging method.

The average transfer function can be expressed as in Equation (11) and the loop effective area can be found from Equation (15).

4. Results

In order to validate our approach, we measured a dipole and a LPDA, respectively, by using a square loop probe (Figure 10). The dipole was resonating around 1.2 GHz and had a total length of 9 cm. The LPDA was designed for the frequency range 800 MHz–3 GHz and was 13 × 13 cm in size. As a probe, we used a square loop with a side length of 2 cm. Both probe and AUT were connected to a VNA for measuring the scattering parameters. The measurements were performed in a non-anechoic environment (a regular room inside a building). Measured data were then processed with a MATLAB code implementing relations (9) to (14), in order to apply our distance averaging approach, and to further extract the contribution of the common mode currents.

Since the LPDA was calibrated inside a compact range in an "antenna side" type configuration, we first extracted the effective area of the loop when placed in the same configuration. We used the distance averaging approach as the measurements were performed in an environment with multiple propagation paths.

Figure 10. Antennas under test and probe antenna.

Figure 11 shows the normalized transfer functions of the antenna system measured at eight different distances ranging between 25 and 60 cm, and the average transfer function computed as in (11).

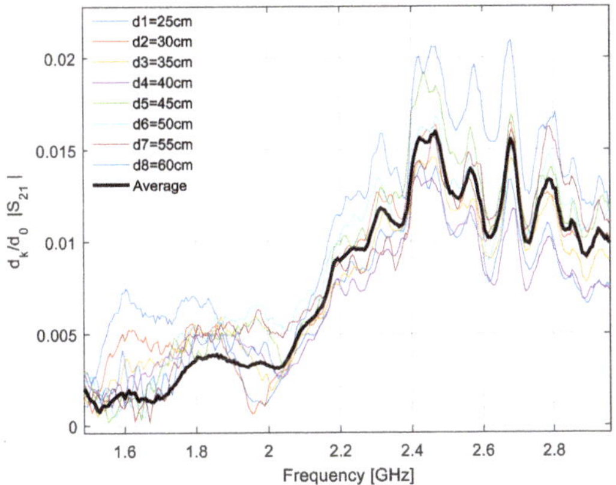

Figure 11. Normalized transfer functions and the average figure.

The effective area of the loop as a function of frequency is given in Figure 12.

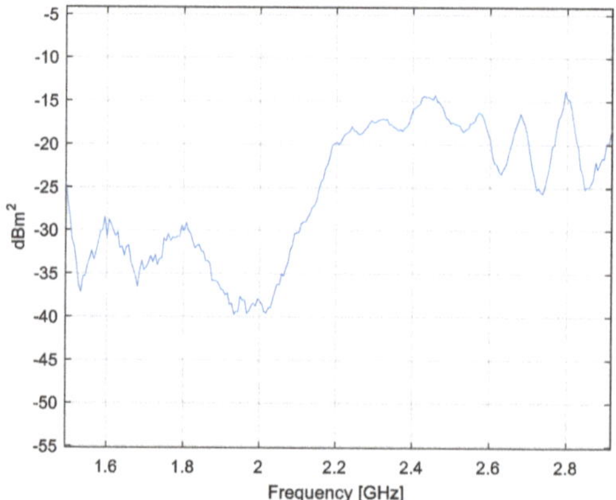

Figure 12. Effective area of the loop probe.

Once the loop probe was calibrated, we assessed the effect of the common mode currents by measuring the transfer functions on both the "antenna side" and "cable side". The setup for each AUT is presented in Figures 13 and 14, respectively.

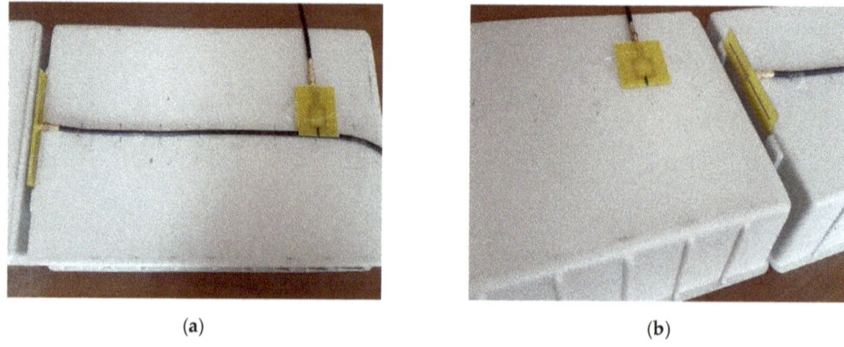

(**a**) (**b**)

Figure 13. Measuring setup for a dipole antenna: "cable side" (**a**) and "antenna side" (**b**).

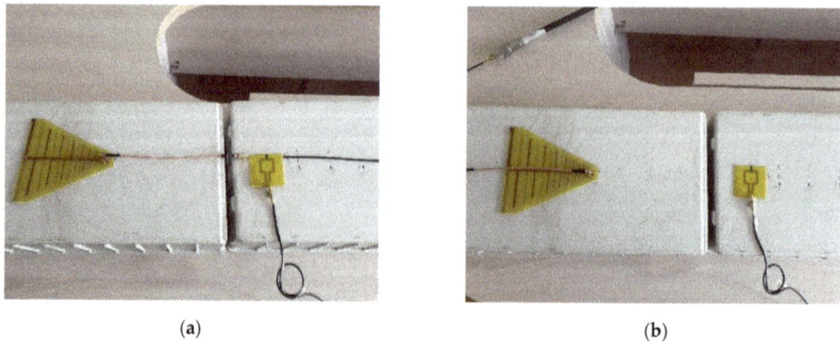

(**a**) (**b**)

Figure 14. Measuring setup for a LPDA: "cable side" (**a**) and "antenna side" (**b**).

For the dipole antenna, the measurements on the "cable" and "antenna side" were performed at distances between antennas ranging from 5 to 40 cm with a distance increment of 5 cm. For the LPDA, the distance to the probe ranged between 25 and 60 cm with the same increment. All the distances corresponded at least to the Fresnel zone [27].

The measurements on both antennas were performed between 1.5 and 3 GHz, a frequency range where the loop has a good radiation efficiency. The effect of the impedance mismatch at the probe output was corrected on the measured data. The magnetic field generated by the common mode currents can be evaluated as in (3) by subtracting the results measured on the "antenna side" from those measured on the "cable side" (Figure 15).

In Figure 16, we show the contribution of the common mode current to the output current, as given by (9), with S_{21} measured at each of the eight distances between the loop and AUT. On the same diagram, we give the average figure resulting from (12) after evaluating the normalized, distance averaged transfer function as defined in (11). As Figure 16 shows, the common mode contribution to the output current can be dramatically diminished by applying the distance averaging technique.

The magnetic field on the "cable side" can be expressed by using (13), and the figure corrected with the common mode current effect by using (14). In Figure 17, we give the variation of the corrected, magnetic field strength as a function of distance and frequency.

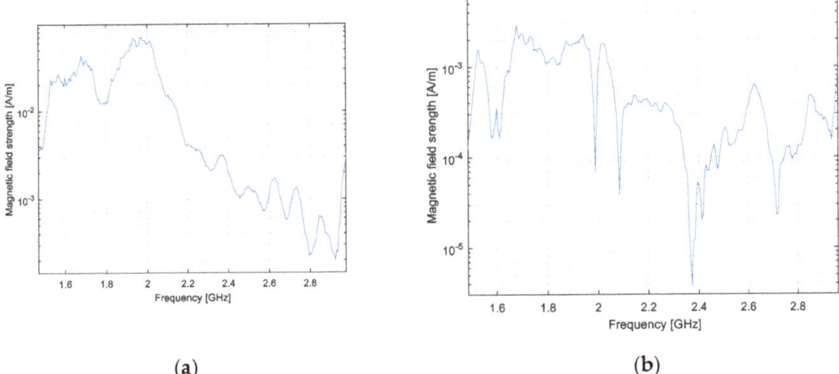

Figure 15. Magnetic field generated by common mode currents: dipole (**a**) and LPDA (**b**).

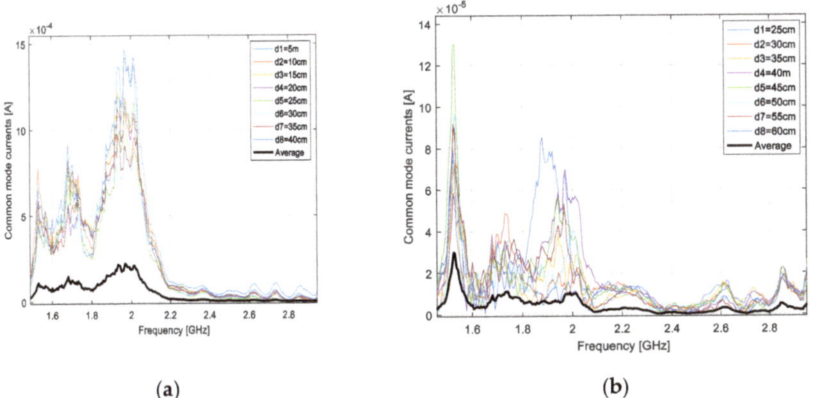

Figure 16. Contribution of the common mode current to the output current versus distance averaged figure: dipole (**a**) and LPDA (**b**).

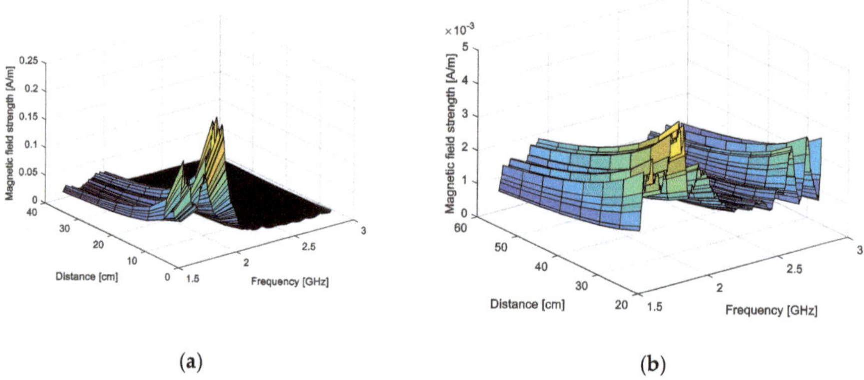

Figure 17. Magnetic field measured on the "cable side" after correction, as a function of distance and frequency: dipole (**a**) and LPDA (**b**).

Figure 18 shows a comparison between the magnetic field measured on the "cable side" at 40 cm, with and without correction of the common mode current effect, and the magnetic field on the "antenna side" at the same distance.

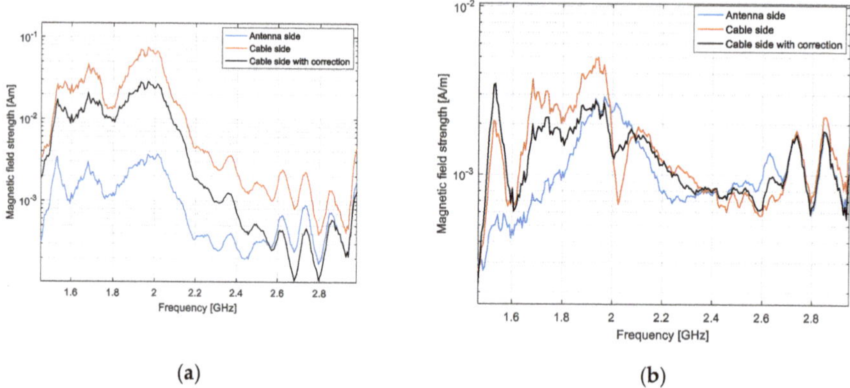

Figure 18. Magnetic field measured on "antenna side" and "cable side", with and without correction: dipole antenna (**a**) and LPDA (**b**); the distance between the probe and the AUT was set at 40 cm.

It appears that by applying our distance averaging technique, the corrected magnetic field magnitude on the "cable side" got closer to the magnetic field strength measured on the "antenna side". We defined a root mean square error by taking the field strength on the cable free side as a reference. The error decreased from 71% down to 29% for the dipole and from 6.2% down to 3.1% for the LPDA.

5. Conclusions

In this paper, we proposed a differential approach for evaluating the magnetic field generated by the common mode currents on an antenna feeder by subtracting the magnetic field magnitude on the "antenna side" from the same figure measured on the "cable side".

In order to extract the effective area of the loop probe, we applied a distance averaging technique derived from an approach originally developed for antenna gain measurements in a multipath site.

We also developed a distance averaging approach for correcting the field radiated by a symmetrical antenna fed through a coaxial line, with the effect of the common mode current. The common mode current has a distance variant distribution and therefore, its effect on the field measured aside the feeder can be diminished by averaging the results acquired at different distances between the probe and the antenna under test. By applying the proposed technique, a magnetic field value corresponding to a reference distance of 1 m was first derived; the actual corrected field value at the "cable side" distance was then found by multiplying the result by the distance and the corresponding phase factor.

The method was successfully validated on a symmetric wire dipole as an antenna under test and a square loop as a probe. Our distance averaging approach could also be applied to reduce the radiating effect of the common mode currents on the feed line of a LPDA. The root mean square error on the measured field magnitude was reduced by a factor of two for both antennas under test. However, common mode currents had a stronger impact for the dipole antenna since for the LPDA, we used a more balanced feeding circuit.

Several factors may impact on the accuracy of our approach, and will be investigated in a future work. First, our field zone extrapolation method might not be accurate enough for distances between the probe and AUT falling in the near-field zone. Second, a low ratio between the field generated by the common mode currents and that radiated by the antenna would make the former less discernable. Finally, one should bear in mind that the loop probe was calibrated on the "antenna side" configuration, but the probe radiation properties may change on the "cable side" due to the feeder proximity.

Author Contributions: Conceptualization, A.C. and R.D.T.; methodology, A.C. and R.D.T.; software, A.C. and R.D.T.; validation, A.C. and R.D.T.; formal analysis, A.C. and R.D.T.; investigation, A.C. and R.D.T.; resources, R.D.T.; data curation, A.C. and R.D.T.; writing—original draft preparation, A.C. and R.D.T.; writing—review and editing, R.D.T.; visualization, A.C. and R.D.T.; supervision, R.D.T.; project administration, R.D.T.; funding acquisition, R.D.T. All authors have read and agreed to the published version of the manuscript.

Funding: This research received no external funding.

Conflicts of Interest: The authors declare no conflict of interest.

References

1. Paul, C.R.; Bush, D.R. Radiated Emissions from Common-Mode Currents. In Proceedings of the 1987 IEEE International Symposium on Electromagnetic Compatibility, Atlanta, GA, USA, 25–27 August 1987; pp. 1–7.
2. Sørensen, M.; Hubing, T.H.; Jensen, K. Study of the impact of board orientation on radiated emissions due to common-mode currents on attached cables. In Proceedings of the 2016 IEEE International Symposium on Electromagnetic Compatibility (EMC), Ottawa, ON, Canada, 25–29 July 2016; pp. 36–40.
3. Drewniak, J.L.; Hubing, T.H.; Van Doren, T.P. Investigation of fundamental mechanisms of common-mode radiation from printed circuit boards with attached cables. In Proceedings of the IEEE Symposium on Electromagnetic Compatibility, Chicago, IL, USA, USA, 22–26 August 1994; pp. 110–115.
4. Toyota, Y.; Sakai, Y.; Torigoe, M.; Koga, R.; Watanabe, T.; Wada, O. Fast and Accurate Estimation of Radiated Emission from Printed Circuit Board Using Common-mode Antenna Model Based on Common-Mode Potential Distribution. In Proceedings of the 2007 IEEE International Symposium on Electromagnetic Compatibility, Honolulu, HI, USA, 9–13 July 2007; pp. 1–6.
5. Sakai, Y.; Watanabe, T.; Wada, O.; Matsushima, T.; Iokibe, K.; Toyota, Y.; Koga, R. EMI antenna model based on common-mode potential distribution for fast prediction of radiated emission. In Proceedings of the 2006 IEEE International Symposium on Electromagnetic Compatibility, Portland, OR, USA, 14–18 August 2006; Volume 2, pp. 280–284.
6. Vorobyov, A.V.; Zijderfeld, J.H.; Yarovoy, A.G.; Ligthart, L.P. Impact common mode currents on miniaturized UWB antenna performance. In Proceedings of the European Conference on Wireless Technology, Paris, France, 4–6 October 2005; pp. 285–288.
7. Jerse, T.A.; Paul, C.R. A hybrid method for efficiently estimating common-mode radiation from transmission-line structures. In Proceedings of the International Symposium on Electromagnetic Compatibility, Atlanta, GA, USA, 14–18 August 1995; pp. 145–149.
8. Paul, C.R. A comparison of the contributions of common mode and differential-mode currents in radiated emissions. *IEEE Trans. Electromagn. Compat.* **1989**, *31*, 189–193. [CrossRef]

9. Daijavad, S.; Rubin, B.J. Modeling common-mode radiation of 3D structures. *IEEE Trans. Electromagn. Compat.* **1992**, *34*, 57–61. [CrossRef]
10. Hardin, K.B.; Paul, C.R. Decomposition of radiating structures using the ideal structure extraction methods (ISEM). *IEEE Trans. Electromagn. Compat.* **1993**, *35*, 264–273. [CrossRef]
11. Tang, Z.; Liu, J.; Lian, R.; Li, Y.; Yin, Y. Wideband Differentially Fed Dual-Polarized Planar Antenna and Its Array With High Common-Mode Suppression. *IEEE Trans. Antennas Propag.* **2019**, *67*, 131–139. [CrossRef]
12. Liu, Y.-Y.; Tu, Z.-H. Compact Differential Band-Notched Stepped-Slot UWB-MIMO Antenna with Common-Mode Suppression. *IEEE Antennas Wirel. Propag. Lett.* **2017**, *16*, 593–596. [CrossRef]
13. Alexander, M.; Loh, T.H.; Betancort, A.L. Measurement of electrically small antennas via optical fibre. In Proceedings of the 2009 Loughborough Antennas Propagation Conference, Loughborough, UK, 16–17 November 2009; pp. 653–656.
14. Hsiao, Y.T.; Lin, Y.Y.; Lu, Y.C.; Chou, H.T. Applications of time-gating method to improve the measurement accuracy of antenna radiation inside an anechoic chamber. In Proceedings of the IEEE Antennas and Propagation Society International Symposium. Digest. Held in Conjunction with: USNC/CNC/URSI North American Radio Sci. Meeting (Cat. No.03CH37450), Columbus, OH, USA, 22–27 June 2003; pp. 794–797.
15. Loredo, S.; Leon, G.; Zapatero, S.; Las-Heras, F. Correction of multipath effects in measured patterns by using FFT and time-gating. In Proceedings of the 2008 IEEE Antennas and Propagation Society International Symposium, San Diego, CA, USA, 5–11 July 2008; pp. 1–4.
16. Fourestie, B.; Altman, Z.; Wiart, J.; Azoulay, A. Correlate measurements at different test sites. *IEEE Trans. Antennas Propag.* **1999**, *47*, 1569–1573. [CrossRef]
17. Loredo, S.; Pino, M.R.; Las-Heras, F.; Sarkar, T.K. Echo identification and cancellation techniques for antenna measurement in non-anechoic test sites. *IEEE Antennas Propag. Mag.* **2004**, *46*, 100–107. [CrossRef]
18. Black, D.N.; Joy, E.B. Test zone field compensation. *IEEE Trans. Antennas Propag.* **1995**, *43*, 362–368. [CrossRef]
19. Leather, P.S.H.; Parsons, D. Equalization for antenna-pattern measurements: Established technique—New application. *IEEE Antennas Propag. Mag.* **2003**, *45*, 154–161. [CrossRef]
20. Suto, K.; Matsui, A. Effects of the common mode on radiation patterns of the tapered slot antenna. In Proceedings of the 2017 International Symposium on Antennas and Propagation (ISAP), Phuket, Thailand, 30 October–2 November 2017; pp. 1–2.
21. McLean, J.; Sutton, R. Asymmetry anomalies and resonances in hybrid LPDA-broadband dipole antennas. In Proceedings of the 2003 IEEE Symposium on Electromagnetic Compatibility. Symposium Record (Cat. No.03CH37446), Boston, MA, USA, 18–22 August 2003; Volume 1, pp. 64–68.
22. Hilbert, M.; Tilston, M.A.; Balmain, K.G. Resonance phenomena of log-periodic antennas: Characteristic-mode analysis. *IEEE Trans. Antennas Propag.* **1989**, *37*, 1224–1234. [CrossRef]
23. Balmain, K.; Nkeng, J. Asymmetry phenomenon of log-periodic dipole antennas. *IEEE Trans. Antennas Propag.* **1976**, *24*, 402–410. [CrossRef]
24. Gooran, P.R.; Lalbakhsh, A.; Moradi, H.; Jamshidi, M. (Behdad) Compact and wideband printed log-periodic dipole array antenna using multi-sigma and multi-Tau techniques. *J. Electromagn. Waves Appl.* **2019**, *33*, 697–706. [CrossRef]
25. Tamas, R.D.; Deacu, D.; Vasile, G.; Ioana, C. A method for antenna gain measurements in nonanechoic sites. *Microw. Opt. Technol. Lett.* **2014**, *56*, 1553–1557. [CrossRef]
26. Constantin, A.; Anchidin, L.; Tamas, R.D.; Caruntu, G. A New Method to Reduce the Impact of the Common Mode Currents for Field Measurements on Symmetrical Antennas. In Proceedings of the 2019 International Workshop on Antenna Technology (iWAT), Miami, FL, USA, 3–6 March 2019; pp. 87–90.
27. Anchidin, L.; Tamas, R.D.; Androne, A.; Caruntu, G. Antenna gain evaluation based on weighting near-field measurements. In Proceedings of the 2017 International Workshop on Antenna Technology: Small Antennas, Innovative Structures, and Applications (iWAT), Athens, Greece, 1–3 March 2017; pp. 78–81.
28. Anchidin, L.; Tamas, R.D.; Caruntu, G.; Ilie, C.-A. Near-Field Gain Measurements Using the Distance Averaging Method: Linear Scanning Versus Matrix Scanning. In Proceedings of the 2018 USNC-URSI Radio Science Meeting (Joint with AP-S Symposium), Boston, MA, USA, 8–13 July 2018; pp. 107–108.

Article

RF Channel-Selectivity Sensing by a Small Antenna of Metamaterial Channel Filters for 5G Sub-6-GHz Bands

Muhammad Kamran Khattak, Changhyeong Lee, Heejun Park and Sungtek Kahng *

Department of Information and Telecommunication Engineering, Incheon National University, Incheon 22012, Korea; k.khattak@huawei.com (M.K.K.); Antman@inu.ac.kr (C.L.); h.park.inu@gmail.com (H.P.)

* Correspondence: s-kahng@inu.ac.kr

Received: 27 February 2020; Accepted: 31 March 2020; Published: 2 April 2020

Abstract: In this paper, a new small antenna is suggested for 5G Sub-6-GHz band mobile communication. It can change the channel among the three given bands (called the 3.5-GHz area), as a wide-band antenna is connected to a small multiplexer comprising three metamaterial channel filters. The function of channel selection of this antenna system is experimentally demonstrated to prove the validity of the presented scheme. The channel selection for 5G mobile communication is conducted from f_1 (channel 1) through f_2 (channel 2) to f_3 (channel 3), when TX and RX antennas with gains over 0 dBi and S_{11} less than −10 dB are located far-field apart ($R_{Far} \gg 2.1$ cm), and result in the transmission coefficient (S_{21}) being the greatest at the selected channel, which is detected by a vector network analyzer.

Keywords: 5G mobile communication; Sub-6-GHz; compact antenna; channel selection; channel filters; metamaterials

1. Introduction

Over half of the last decade, 5G has been a catchphrase in IT industry with another that is the 4th industrial revolution. It has been driving the world's first-class IT institutes to translate its hidden potential use-cases into practices in the shape of IoT devices, driverless cars, faster cell-phone-featuring networking services, and so on. It is a matter of course that RF components and antennas corresponding to the given bands are necessary to these wireless devices [1,2].

As a way to increase the speed of data transfer, the use of millimeter-wave areas was suggested [3,4] and connectivity had to be added. Hence, beamforming becomes a basic virtue to millimeter-wave antennas which are made as part of a chip for controlling the beam from an array. Besides beam-control, a wide-band is expected from the antenna to take care of multiple wireless network service providers as one device. The 5G spectrum is grouped into millimeter-waves and sub-6-GHz bands, and each group is divided into three channels. As to the sub-6GHz-group or simply speaking, the frequency range, f = 3.5 GHz area which consists of three channels from frequencies f = 3.46 to 3.65 GHz, when the antenna is purchased with a wide-band, it should be able to distinguish one channel from another for being adaptive to the change in the mobile environment.

A channel is distinguished from others with a filter in the wireless communication system. In spite of there being many RF filters of the conventional design methods for wireless systems, intrinsic limitations are given to size-reduction and extension to an upper level circuit combining or dividing channels. One cause of the limitations is the half-wavelength resonance condition. This limitation is eased by the metamaterial filter design method which can minimize the size of the structure and improve the passband and stopband characteristics [5,6]. G. Jang et al. showed a sub-wavelength-sized filter of the zeroth order-resonance (ZOR) effect of the CRLH TX-line, which has a lower loss and a wide-stopband [5]. C. Liao made a filter of a new shape which obtains a stronger coupling in a limited area [6]. On the contrary to the metamaterial configuration for smaller geometries, the size of a filter is reduced by curling a straight stepped transmission-line resonator (SIR) and adding stubs [7–15]. Hou et al. folded SIRs and attached stubs to have multiple bands such as 2.13 and 3.47 GHz [7]. Similarly, Sun et al. used SIRs to generate multi-mode resonance (MMR) at harmonic frequencies of 1.5 GHz [8]. Like them, Cho and Chen utilized SIRs flanked by TX-lines for the generation of triple bands [9,10]. Xu, Zhang, and Chu short-ended one of the MMRs to reduce the area of triple-band filters [11–13]. MMRs are used only as ports and enclose smaller resonators to make passbands at 2.3, 3.7, and 5.3 GHz [14]. Li et al. coupled two MMRs with gaps to resonate at 1.8, 2.4, and 5.8 GHz [15]. The passbands of each are spaced apart with large frequency gaps. In addition, these kinds of dual- or triple-band filters alone cannot distinguish channels effectively if the channels are closely located side by side. Selectivity in the contiguous channel case can be realized as the following. Multiplexers like a duplexer and more channel elements are made up with filters and make the system effectively choose a channel [16–18]. One example of this kind is introduced in M. Chen et al.'s work where the conventional filters are assembled [16]. Tantiviwat et al. showed a multiplexer comprising stub-loaded MMRs as non-metamaterial filters and coupled port, one with two filters as parallel-edge coupling unlike the junction-type multiplexer [17]. Open-gapped SIR ring filters for effective size-reduction gathered to the common TX-line to implement a triplexer [18]. Meanwhile, new metamaterial filters are adopted as channel elements and put together along a line merged to the common port [19].

In this paper, a 3.5-GHz-area wide-band antenna is grafted onto a new metamaterial multiplexer to distinguish received signals to different sub-6-GHz channels. Channel selectivity can be given by a stacked structure like an antenna that has multiple bands like [20,21], but they have radiating elements modified with slits which may cause the tilted far-field patterns at a frequency, and their bands are not closely located and cannot cover the 5G service. Different from them, separately designed filters and multiplexers are combined with the wide-band antenna without slits which will distort the beam. The channels of the multiplexer are secured by three metamaterial bandpass filters consistent with the three mobile service providers' frequencies. The ability of channel selection is checked by a test setup where all the ports of the filter, multiplexer, and antenna are set at the 50 Ω impedance and a wide-band metamaterial array antenna as the TX from [22] sends the signals of the three channels over the 3.5 GHz area and the aforementioned triplexer-fed wide-band antenna as the RX receives the signals of frequencies f_1, f_2, and f_3 at the common port and pushes them to their own channel filter ports. The TX and RX antennas with gains greater than 0 dBi and S_{11} below −10 dB in the operational bands can make a wireless link for a full-band of the 5G sub-6-GHz communication.

2. TX and RX Antennas with Novel Components

The wireless communication link for the 5G sub-6-GHz band is investigated between the TX and RX antennas in the proposed manner depicted as below in Figure 1.

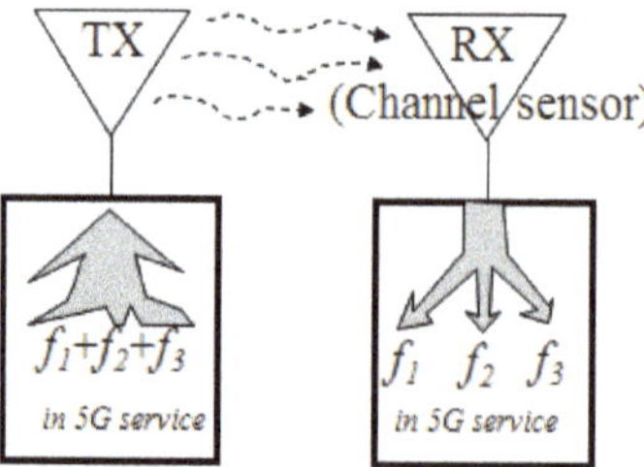

Figure 1. Three channel signals combined in the TX antenna will be split and selected to their corresponding bands in the RX antenna system.

The signals from the three bands of the 3.5 GHz region for the 5G service, say, f_1, f_2, and f_3 are transmitted from the TX side to the RX side. The received frequencies f_1, f_2, and f_3 are split into channels 1, 2, and 3 each, which is channel selection.

The left schematic of Figure 2 is the antenna system loaded with switches, where S_{11} is the reflection coefficient in the figure. The wide-band radiating element has multiple slits where PIN diodes are attached. The PIN diodes are turned on by the controller and DC bias circuit to catch one signal among frequencies f_1, f_2, and f_3. Synchronization with the TX is required and probabilistic. The center frequency, not the bandwidth, of each band is decided, and one frequency is usable at a time. To overcome such problems in realization and operational cost, the right schematic is suggested. The radiating element is not loaded with active components. It should be a wide-band and compact structure. The feed circuitry should have a multiplexer that divides the collection frequencies into different bands through filters. Then, the bandwidth can be controlled with the center frequency. Synchronization is not needed, which enables all three bands to be used, regardless of moments in time. Also, small filters are required to make the multiplexer small. The channel filter is realized as a metamaterial structure as follows.

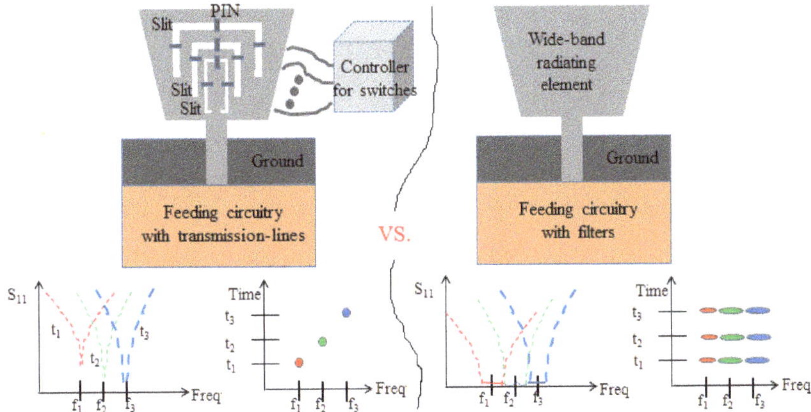

Figure 2. Two schemes of channel selection: switch-loaded RX antenna with the ordinary feed (**Left**), and non-active device loaded RX antenna with the multiplexer feed circuit (**Right**).

In Figure 3, the shape of the proposed channel filter and its performances are given. First of all, the channel filter has a gate-shaped resonator comprising a short strip on FR4 substrate and vias. The strip as the series inductance, its gap as the series capacitance, the via as the shunt inductance along with the shunt capacitance between the ground and strip can make a CRLH metamaterial resonator. The equivalent-circuit of this CRLH resonator is handled to create the ZOR at the center frequencies of the channels, as in Figure 3a. At this moment that a relatively large series capacitance is needed, the gap looks very small compared to the other parts of the 3D structure. Second of all, the resonators are coupled from the input port through the middle part to the output port to work as a filter. Other metamaterial filters use horizontal SRRs or align longitudinal resonators, but the new filter uses vertical structures and cascade them transversely between the ports. The physical dimensions of the filter are $l_s, l_1, l_2, l_3, l_{4,1}, l_{4,2}, l_{4,3}, l_{5,1}, l_{5,2}, l_{5,3}, w_s, w_1, w_2$, and w_3 set as 19.3, 3.5, 6, 14, 7.75, 7.4, 6.9, 7.5, 7.2, 6.7, 23.7, 1.7, 1.3, and 2.6 in mm for Figure 3a. These values are put in Table 1 and used for one channel and tuned to form the three channels. Figure 3b shows the passbands centered at 3.46, 3.55, and 3.65 GHz, compliant with the industrial requirement. Each channel's S_{11} is less than −10 dB and S_{21} is close to −0.5 dB in operating frequency bands. The ZOR occurs coinciding with frequencies f_1, f_2, and f_3, along with the LH- and RH-regions in the dispersion diagram presenting the metamaterial characteristics of the geometry. These filters are assembled by the following multiplexer.

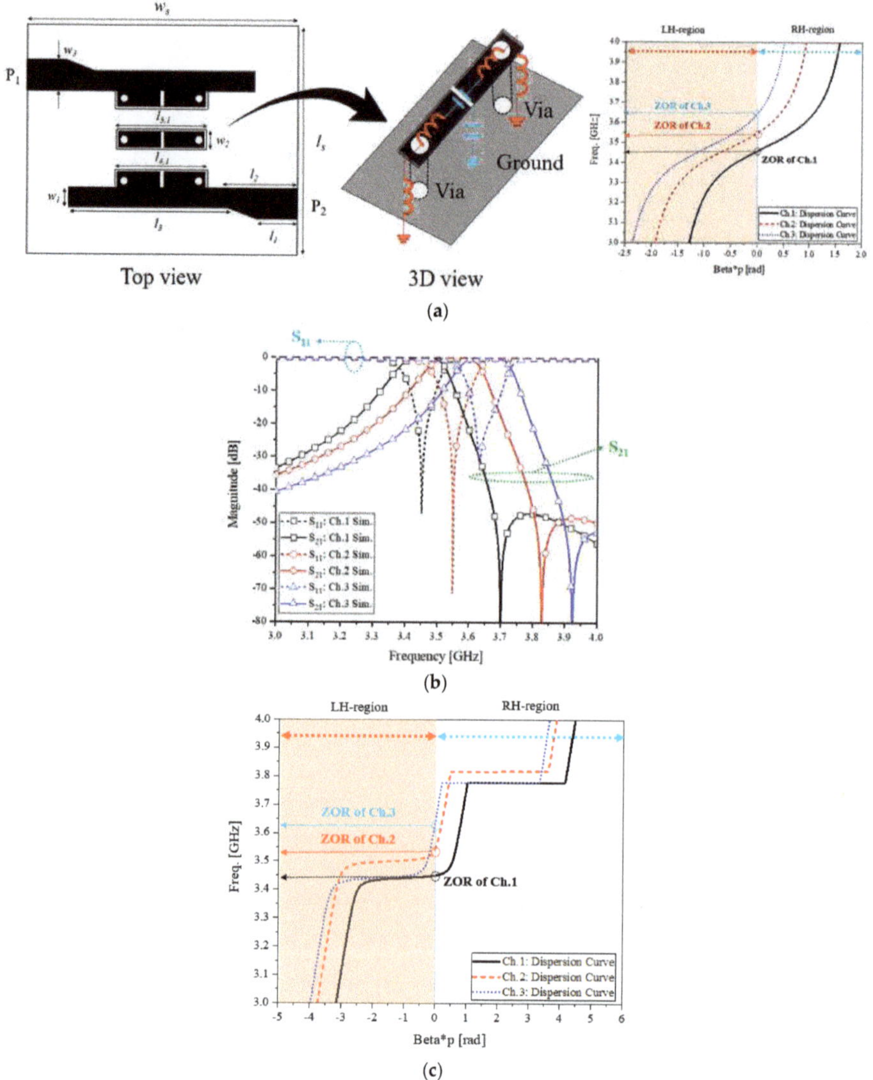

Figure 3. The metamaterial channel filter (**a**) Geometry and equivalent circuit (**b**) Frequency responses of the three bands (**c**) Dispersion diagram of the channel filters as the metamaterial property.

As mentioned with Figure 2, a multiplexer is needed to combine the channels. Figure 4a shows the schematic and geometry of the multiplexer where P_1 is the port connected to the RX antenna to receive all the signals having an electrically small footprint. The schematic describes the signal flow from P_1 to the impedance-matched branches. Its geometrical parameters l_1, l_2, l_3 and l_4 are 18, 23.5, 22.9, and 21.5 in mm with FR4 substrate, respectively. In order to keep the area from being larger, a multi-branch junction shape is chosen and the positions of the channel branches are determined to retrieve all the passbands the same as individual channel filters. The P_2, P_3, and P_4 are output ports of f_1, f_2, and f_3, in that order. Figure 4b presents the passbands obtained at the ports P_2 for Ch. 1, P_3 for Ch. 2, and P_4 for Ch. 3. Also, the electric field distributions as the insets prove that only the right frequency signal enters the channel. The results are consistent with those in Figure 3b. It is time to mention wide-band antennas for transmitting and receiving covering the three bands.

Table 1. Geometrical parameters of the filters, multiplexers and antennas.

Filters									
Items	l_s	l_1	l_2	l_3	$l_{4,1}$	$l_{4,2}$	$l_{4,3}$		
Val. (mm)	19.3	3.5	6	14	7.75	7.4	6.9		
Items	$l_{5,1}$	$l_{5,2}$	$l_{5,3}$	w_s	w_1	w_2	w_3		
Val. (mm)	7.5	7.2	6.7	23.7	1.7	1.3	2.6		
Multiplexers and Antennas									
Items	l_s	l_g	l_1	l_2	l_3	w_s	w_1	w_2	
Val. (mm)	30	10	5.5	3	7	20	1	10	
Items	g_{rad12}	l_{rad2}	g_{notch1}	g_{notch2_w1}	g_{notch2_l1}	w_s	l_s	l_{cmch1}	l_{cmch2}
Val. (mm)	11.7	10	3	1.75	10	53.7	67.5	34.2	56
Items	l_{cmch3}	l_{ch1_reso1}	l_{ch1_reso2}	$g_{ch1\text{-}12}$	$g_{ch1\text{-}23}$	$w_{ch1\text{-}reso1}$	$w_{ch1\text{-}reso2}$	w_s	l_s
Val. (mm)	29.6	7.1	6.9	2	2	1	1	70	82

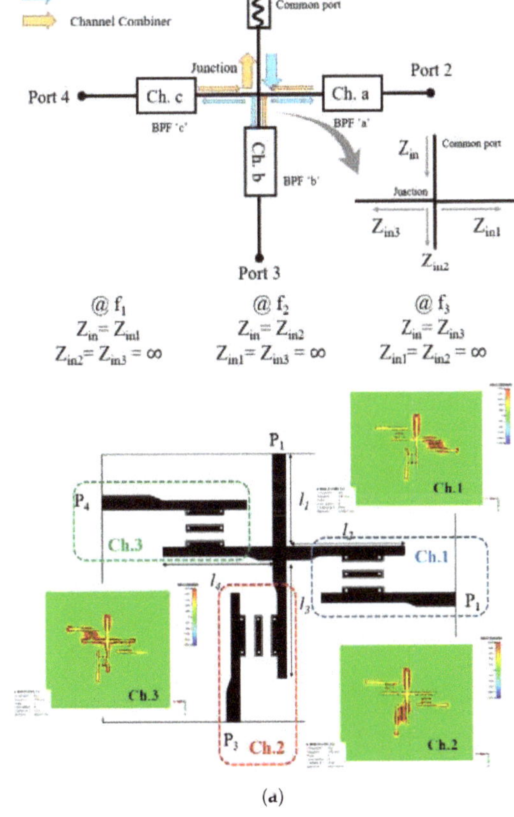

(a)

Figure 4. *Cont.*

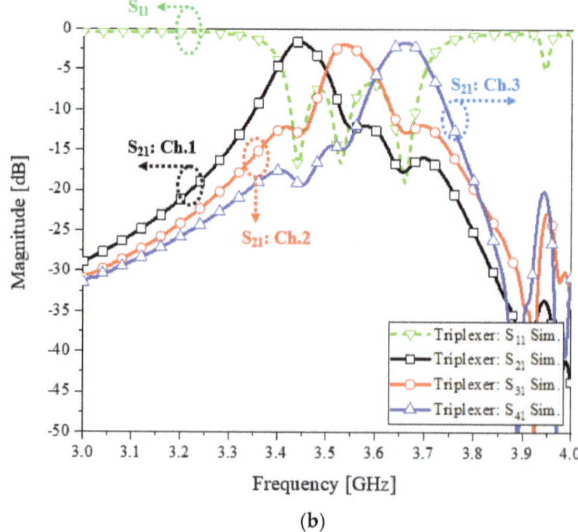

(b)

Figure 4. The compact multiplexer comprising the three channel filters. (**a**) Geometry. (**b**) The overall frequency response of the three bands.

Figure 5a,b are a wide-band RX antenna and an ultra-wide-band array TX antenna. The TX antenna is adopted by the authors' former work to have an increased directivity [22]. The TX antenna elements which consist of a pentagonal radiator and a triangular ground with a dent for impedance match and show a wide-band impedance match, as shown in S_{11} of Figure 5b, stem from a metamaterial power-divider. This arraying is intended to increase the directivity of the azimuth-plane radiation-pattern as appearing in Figure 5b. Meanwhile, the RX antenna should have an omni-directional radiation to respond to an arbitrary angle of incidence and have a wide-band function from 3.2 to 5.5 GHz as S_{11} of Figure 5a. Explaining the geometry, Figure 5a as a stub-loaded stepped patch for RX has 30, 10, 5.5, 3, 7, 20, 1 and 10 mm as l_s, l_g, l_1, l_2, l_3, w_s, w_1, and w_2 with FR4 substrate, respectively, and is fed by a multiplexer of three metamaterial channel filters. g_{rad12}, l_{rad2}, g_{notch1}, g_{notch2_w1}, g_{notch2_l1}, w_s, l_s, l_{cmch1}, l_{cmch2}, l_{cmch3}, $l_{ch1\text{-}reso1}$, $l_{ch1\text{-}reso2}$, $g_{ch1\text{-}12}$, $g_{ch1\text{-}23}$, $w_{ch1\text{-}reso1}$, $w_{ch1\text{-}reso2}$, w_s, and l_s are 11.7, 10, 3, 1.75, 10, 53.7, 67.5, 34.2, 56, 29.6, 7.1, 6.9, 2, 2, 1, 1, 70, and 82 in mm, respectively, with Figure 5c. The values are shown in Table 1. As another parameter, α_r, which is the angle of two tapering sides of the TX antenna element, is set at 160.6°, which needed to match the impedance over the wide-band. Figure 5d shows how the two blocks are placed a far-field distance apart.

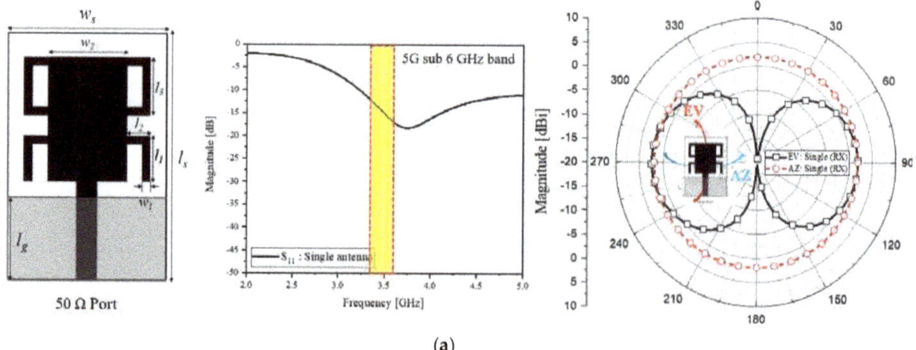

(a)

Figure 5. *Cont.*

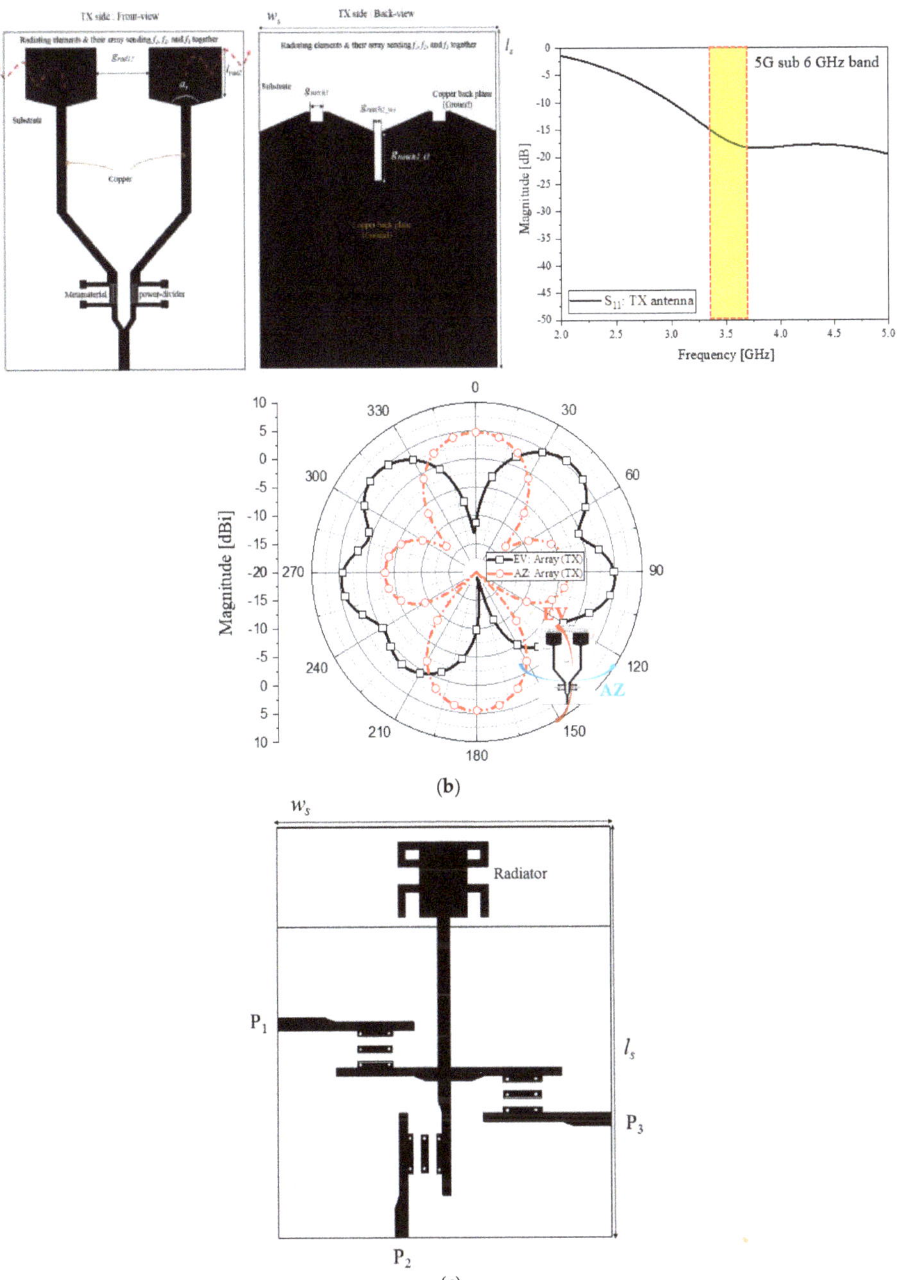

(**b**)

(**c**)

Figure 5. *Cont.*

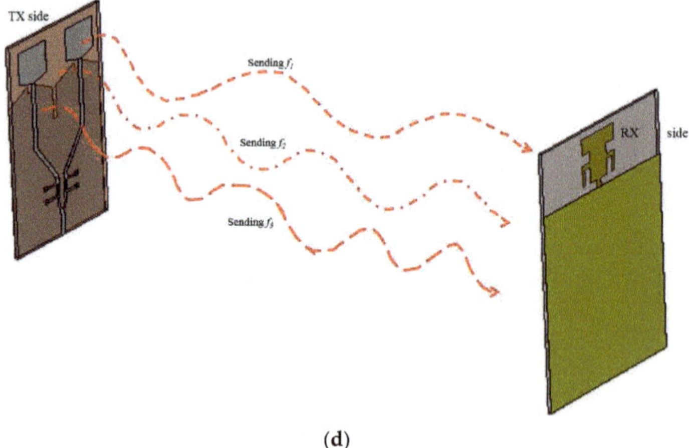

(d)

Figure 5. Shapes and interactive configuration of the TX and RX antennas. (**a**) A compact wide-band RX antenna and far-field patter for signal. (**b**) Front and bottom views of the wide-band TX antenna. (**c**) Wide-band RX antenna connected to multiplexer of three channel filters. (**d**) Scheme of linking the TX and RX antennas.

3. Experiment of Channel Selectivity Sensing

The components and circuits above are manufactured. From the filters of the multiplexer to antennas, power transfer coefficients are investigated.

Throughout the test, there are no amplifiers for both the TX and RX blocks. Before sensing each channel of the wireless link, the measured functions of the filtering blocks of the multiplexer are looked into, Figure 6 shows the three passbands of the 5G sub-6-GHz area. Frequencies f_1, f_2, and f_3 are shifted from 3.46, 3.55, 3.65 GHz on account of manufacturing errors and connector assembly errors. However, it is revealed that the errors pose insignificant problems in observing the channel selectivity of the system. The measured far-field pattern of the RX antenna fed by the multiplexer, when each of the three channels is turned on, is plotted in comparison with its simulated version as in Figure 6b. There is a matrix of the beam-patterns of the RX channels with respect to the TX channels in Figure 6c. Each column of the matrix means a TX signal and each row implies an RX channel. For example, as for the first column as TX f_1, the first row as RX channel 1 has the greatest antenna gain and efficiency over the entire elements of the first column. Hence, the diagonal elements outperform the rest of the matrix. The co-channel link works much better than cross-channel links. For the final experiment, the filters and multiplexer are adopted to the RX block which realistically faces the TX block as seen in Figure 6d. Figure 6e–g present a channel chosen in turn from channel 1 to channel 3, the impedance of the RX antenna is matched to the chosen frequency (S_{22}) and only accepts the desired frequency signal out of many incident frequencies emanated from the ultra-wide-band (S_{11}) of the TX. The peak of the red curve as S_{21} (power transfer coefficient) appears at frequencies f_1, f_2, and f_3 by taking turns, as a proof of the RF link and proposed channel-selection scheme. Besides, the contributions of this work are mentioned in Table 2 where the differences of this work and the referenced works are compared.

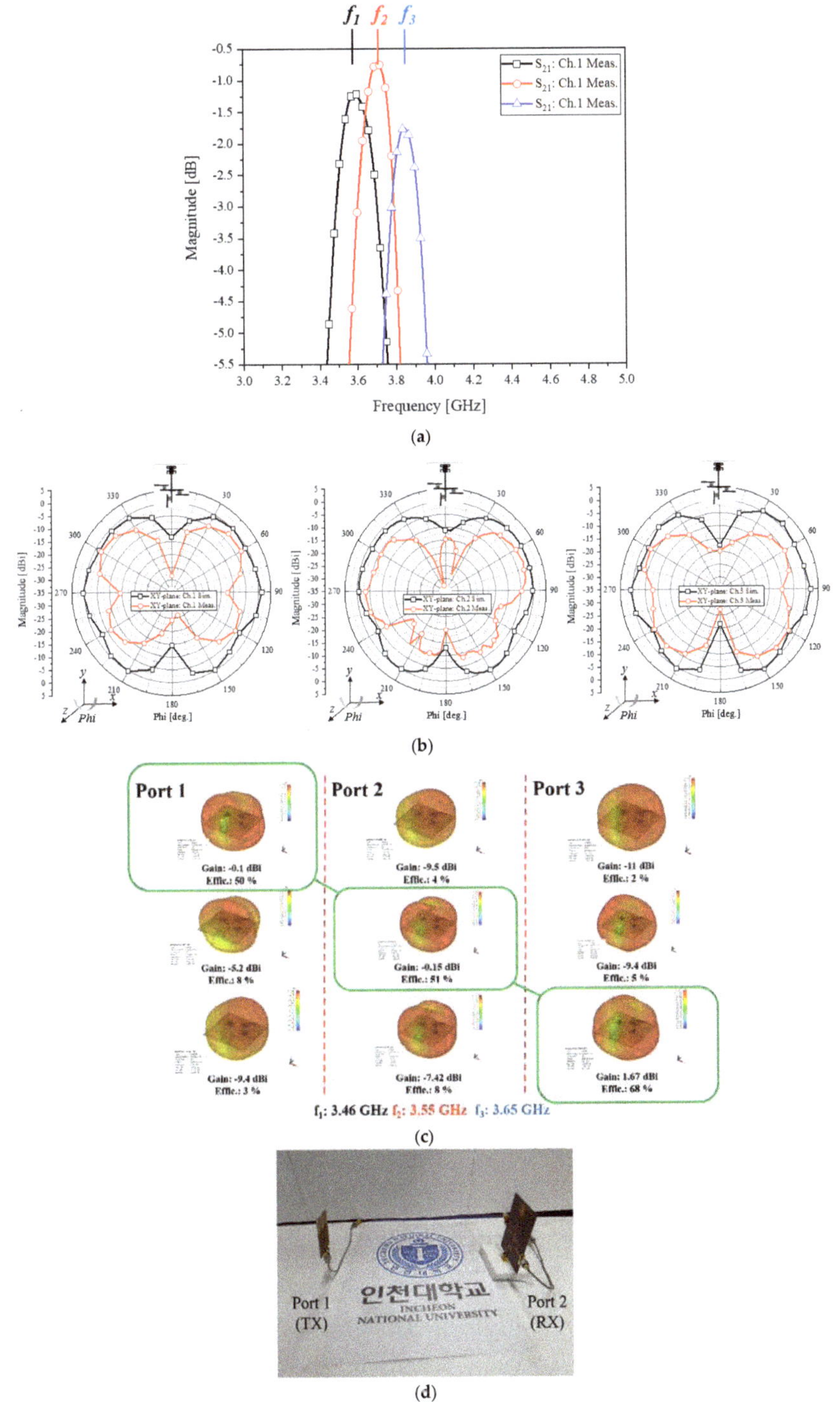

Figure 6. *Cont.*

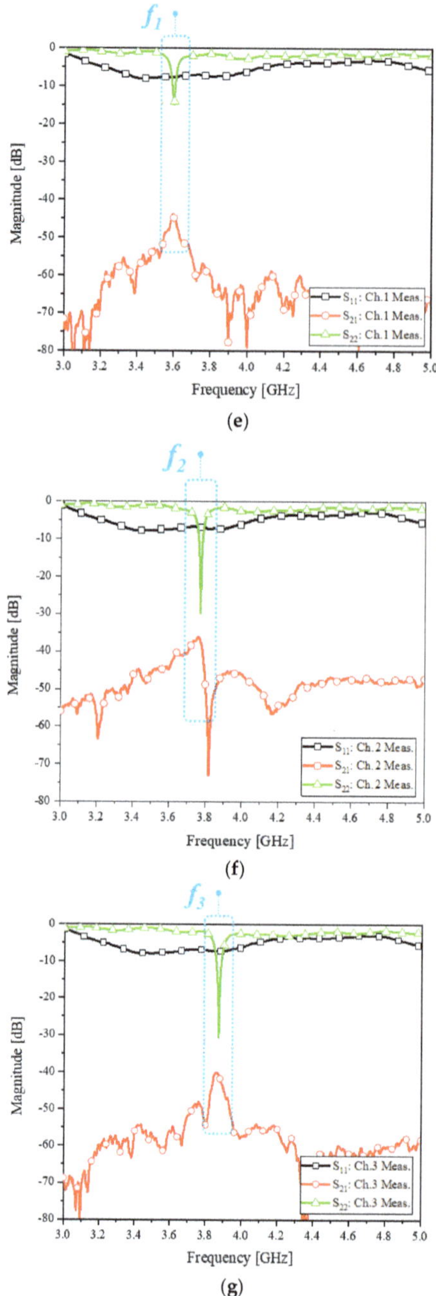

Figure 6. Test results of channel selection (**a**) Measured passbands of the filters (**b**) Simulated and measured far-field patterns of the RX antenna at f_1, f_2, and f_3 (**c**) Antenna gains of the RX antenna with columns as TX signals and rows as RX channels (**d**) Photograph of the realistic the TX to RX antenna wireless link test-setup (**e**) Antennas' S_{ii} and received f_1 signal (**f**) Antennas' S_{ii} and received f_2 signal (**g**) Antennas' S_{ii} and received f_3 signal.

Table 2. Comparing the differences of the proposed work and others'.

Ref. No.	Single-Band Selectivity	Contiguous/Ch. Selection	Metamaterial Concept	5G Service Coverage	Multiplexing	Size (λ_g @ Center Freq.)
[7]	×	×	×	×	×	$0.225\lambda_g \times 0.226\lambda_g$
[8]	×	×	×	×	×	$0.16\lambda_g \times 0.16\lambda_g$
[9]	×	×	×	×	×	$0.26\lambda_g \times 0.36\lambda_g$
[10]	×	×	×	×	×	$0.23\lambda_g \times 0.12\lambda_g$
[11]	×	×	×	×	×	$0.18\lambda_g \times 0.27\lambda_g$
[12]	×	×	×	×	×	$0.108\lambda_g \times 0.521\lambda_g$
[13]	×	×	×	×	×	$0.08\lambda_g \times 0.15\lambda_g$
[14]	×	×	×	×	×	$0.44\lambda_g \times 0.39\lambda_g$
[15]	×	×	×	×	×	$0.16\lambda_g \times 0.17\lambda_g$
[17]	×	×	×	×	×	$0.18\lambda_g \times 0.20\lambda_g$
[21]	×	×	×	×	×	$0.75\lambda_0 \times 0.75\lambda_0$
This work	○	○	○	○	○	* $0.32\lambda_g \times 0.41\lambda_g$ ** $1.19\lambda_g \times 0.89\lambda_g$

* Filter: Ch.1; ** Triplexer; ○ means Yes; X means No.

4. Conclusions

For the 5G sub-6-GHz communication, a small antenna system is introduced to have the ability of channel selection to avoid the interference from other neighboring channels. The channel is distinguished by devising a compact multiplexer of three metamaterial channel filters. A TX antenna as an ultra-wide-band metamaterial array and the proposed antenna as the RX are put along to check the wireless links between them and the channel-selection function. According to the experiment, change from frequencies f_1 through f_2 to f_3 is detected clearly by the RX side. This will lead to a way to check the channel selection ability of 5G mobile services.

Author Contributions: Conceptualization, S.K., C.L. and M.K.K.; methodology, S.K., and C.L.; software, C.L.; validation, C.L., and H.P.; formal analysis, C.L.; investigation, M.K.K.; resources, C.L., and H.P.; data curation, M.K.K.; writing—original draft preparation, M.K.K.; writing—review and editing, M.K.K.; visualization, M.K.K., and C.L.; supervision, S.K.; project administration, S.K.; funding acquisition, S.K. All authors have read and agreed to the published version of the manuscript.

Funding: The lead authors Changhyeong Lee and Sungtek Kahng on behalf of the contributors acknowledge that this work was supported by the Post-Doctoral Research Program (2017) in the Incheon National University.

Conflicts of Interest: The authors declare no conflict of interest.

References

1. Shafi, M.; Molisch, A.F.; Smith, P.J.; Haustein, T. 5G: A Tutorial Overview of Standards, Trials, Challenges, Deployment, and Practice. *IEEE J. Sel. Areas Commun.* **2017**, *35*, 1201–1221. [CrossRef]
2. Kim, Y.; Roh, W. Feasibility of Mobile Cellular Communications at Millimeter Wave Frequency. *IEEE J. Sel. Top. Signal Process.* **2016**, *10*, 589–599. [CrossRef]
3. Lee, C.; Khattak, M.K.; Kahng, S. Wideband 5G Beamforming Printed Array Clutched By LTE-A 4 × 4-Multiple-Input–Multiple-Output Antennas With High Isolation. *IET Microw. Antennas Propag.* **2018**, *12*, 1407–1413. [CrossRef]
4. Stutzman, W.L.; Thiele, G.A. *Antenna Theory and Design*, 3rd ed.; John Wiley & Son: Hoboken, NJ, USA, 2012.
5. Jang, G.; Kahng, S. Compact Metamaterial Zeroth-Order Resonator Bandpass Filter for a UHF Band and Its Stopband Improvement by Transmission Zeros. *IET Microw. Antennas Propag.* **2011**, *5*, 1175–1181. [CrossRef]
6. Liao, C.; Chi, P.; Chang, C. Microstrip Realization of Generalized Chebyshev Filters with Box-Like Coupling Schemes. *IEEE Trans. Microw. Theory Tech.* **2007**, *55*, 147–153. [CrossRef]
7. Hou, Z.; Liu, C.; Zhang, B.; Song, R.; Wu, Z.; Zhang, J.; He, D. Dual-/Tri-Wideband Bandpass Filter with High Selectivity and Adjustable Passband for 5G Mid-Band Mobile Communications. *Electronics* **2020**, *9*, 205. [CrossRef]

8. Sun, S.J.; Su, T.; Deng, K.; Wu, B.; Liang, C.H. Shorted-Ended Stepped-Impedance Dual-Resonance Resonator and Its Application to Bandpass Filters. *IEEE Trans. Microw. Theory Tech.* **2013**, *61*, 3209–3215. [CrossRef]
9. Cho, Y.H.; Yun, S.W. A Tri-Band Bandpass Filter Using Stub-Loaded SIRs with Controllable Bandwidths. *Microw. Opt. Technol. Lett.* **2014**, *56*, 2907–2910. [CrossRef]
10. Chen, F.; Qiu, J.; Chu, Q. Design of Compact Tri-Band Bandpass Filter Using Centrally Loaded Resonators. *Microw. Opt. Technol. Lett.* **2013**, *55*, 2695–2699. [CrossRef]
11. Xu, K.; Zhang, Y.; Li, D.; Fan, Y.; Li, J.L.W.; Joines, W.T.; Liu, Q.H. Novel Design of a Compact Triple-Band Bandpass Filter Using Short Stub-Loaded SIRs and Embedded SIRs Structure. *Prog. Electromagn. Res.* **2013**, *142*, 309–320. [CrossRef]
12. Zhang, S.; Zhu, L. Compact Tri-Band Bandpass Filter Based on $\lambda/4$ Resonators with U-folded Coupled-line. *IEEE Microw. Wirel. Compon. Lett.* **2013**, *23*, 258–260. [CrossRef]
13. Chu, Q.X.; Wu, X.H.; Chen, F.C. Novel Compact Tri-Band Bandpass Filter with Controllable Bandwidths. *IEEE Microw. Wirel. Compon. Lett.* **2011**, *21*, 655–657. [CrossRef]
14. Chen, C.F.; Huang, T.Y.; Wu, R.B. Design of Dual- and Triple-Passband Filters Using Alternately Cascaded Multiband Resonators. *IEEE Trans. Microw. Theory Tech.* **2006**, *54*, 3550–3558. [CrossRef]
15. Li, Q.; Zhang, Y.H.; Feng, X.; Fan, Y. Tri-Band Filter with Multiple Transmission Zeros and Controllable Bandwidths. *Int. J. Microw. Wirel. Technol.* **2016**, *8*, 9–13. [CrossRef]
16. Chen, M.H.; Assal, F.; Mahle, C. A Contiguous Band Multiplexer. *COMSAT Tech. Rev.* **1976**, *6*, 285–307.
17. Tantiviwat, S.; Ibrahim, S.Z.; Razalli, M.S. Design of Quad-Channel Diplexer and Tri-Band Bandpass Filter Based on Multiple-Mode Stub-Loaded Resonators. *Radio Eng.* **2019**, *27*, 129–135. [CrossRef]
18. Lee, J.N.; Kahng, S.; Jang, G.; Park, J.K. Three-Channel Output Multiplexer Design Using Band-Pass Filter and Ultra-Wideband Antenna. *J. Electromagn. Eng. Sci.* **2017**, *17*, 111–112. [CrossRef]
19. Lee, B.; Jang, G.; Kahng, S.; Eom, D.; Yang, I.; Kahng, K.; Kim, H. Compact Duplexer for the UWB System Using Novel CRLH Bandpass Filters. In Proceedings of the 2012 Asia Pacific Microwave Conference Proceedings, Kaohsiung, Taiwan, 4–7 December 2012; pp. 247–249.
20. Li, Y.; Zhao, Z.; Tang, Z.; Yin, Y. A Low-Profile, Dual-Band Filtering Antenna with High Selectivity for 5G Sub-6 GHz Applications. *Microw Opt. Technol Lett.* **2019**, *61*, 2282–2287. [CrossRef]
21. Li, Y.; Zhao, Z.; Tang, Z.; Yin, Y. Differentially-Fed, Dual-Band Dual-Polarized Filtering Antenna with High Selectivity for 5G Sub-6 GHz Base Station Applications. *IEEE Trans. Antennas Propag.* **2019**. [CrossRef]
22. Kahng, S.; Eom, D.; Lee, B.; Yang, I.; Kahng, K. A Compact Metamaterial UWB Power-Divider Fed Wide-Band Array Antenna. In Proceedings of the 2012 International Symposium on Antennas and Propagation, Nagoys, Japan, 29 October–2 November 2012; pp. 1506–1509.

MDPI
St. Alban-Anlage 66
4052 Basel
Switzerland
Tel. +41 61 683 77 34
Fax +41 61 302 89 18
www.mdpi.com

Sensors Editorial Office
E-mail: sensors@mdpi.com
www.mdpi.com/journal/sensors

www.ingramcontent.com/pod-product-compliance
Lightning Source LLC
LaVergne TN
LVHW071545170726
843515LV00008B/2221
* 9 7 8 3 0 3 6 5 1 8 2 8 2 *